Epizootic Ulcerative Fish Disease Syndrome

Epizootic Ulcerative Fish Disease Syndrome

Devashish Kar
Department of Life Science
Assam (Central) University
Silchar, India

AMSTERDAM • BOSTON • HEIDELBERG • LONDON • NEW YORK • OXFORD • PARIS
SAN DIEGO • SAN FRANCISCO • SINGAPORE • SYDNEY • TOKYO

Academic Press is an imprint of Elsevier

Academic Press is an imprint of Elsevier
125 London Wall, London EC2Y 5AS, UK
525 B Street, Suite 1800, San Diego, CA 92101-4495, USA
225 Wyman Street, Waltham, MA 02451, USA
The Boulevard, Langford Lane, Kidlington, Oxford OX5 1GB, UK

Notices

Knowledge and best practice in this field are constantly changing. As new research and experience broaden our understanding, changes in research methods, professional practices, or medical treatment may become necessary.

Practitioners and researchers must always rely on their own experience and knowledge in evaluating and using any information, methods, compounds, or experiments described herein. In using such information or methods they should be mindful of their own safety and the safety of others, including parties for whom they have a professional responsibility.

To the fullest extent of the law, neither the Publisher nor the authors, contributors, or editors, assume any liability for any injury and/or damage to persons or property as a matter of products liability, negligence or otherwise, or from any use or operation of any methods, products, instructions, or ideas contained in the material herein.

ISBN: 978-0-12-802504-8

British Library Cataloguing-in-Publication Data
A catalogue record for this book is available from the British Library

Library of Congress Cataloging-in-Publication Data
A catalog record for this book is available from the Library of Congress

For information on all Academic Press publications
visit our website at http://store.elsevier.com/

Working together
to grow libraries in
developing countries

www.elsevier.com • www.bookaid.org

Publisher: Janice Audet
Acquisition Editor: Patricia Osborn
Editorial Project Manager: Jaclyn Truesdell
Production Project Manager: Caroline Johnson
Designer: Ines Cruz

Typeset by TNQ Books and Journals
www.tnq.co.in

Printed and bound in the United States of America

Dedication

Dedicated to the memory of a person who was a highly talented, generous soul, but had left this mundane world prematurely without seeing the fruit of my humble efforts. He is the late Devapriya Kar, FCA, LLB, my respected elder brother.

Also dedicated to the loving memory of my respected parents, teachers, and to my beloved students and fishermen.

Contents

About the Author

Dr Devashish Kar is a pioneer and preeminent researcher in India in the fields of wetlands, rivers, fisheries, and aquaculture. He completed his master's program at the University of Gauhati with specializations in Fishery Science and Aquaculture. He was awarded a PhD by the University of Gauhati for his outstanding work in the "beel" (wetlands) fisheries of Assam. On a prestigious British Council Study Fellowship Award, Dr Kar was in King's College London for its nine-month Advanced Training in Science Education program. Dr Kar was awarded the prestigious Biotechnology National Associateship by the Indian government's Department of Biotechnology (DBT) for his pioneering research in the field of fish disease, particularly in tackling the dreadful epizootic ulcerative syndrome (EUS) fish disease in India and for conducting further research in defining EUS in collaboration with the National Institute of Virology, Pune. As convener, Dr Kar has organized a number of national and international symposia and workshops in the fields of wetlands, fisheries, and aquaculture, including ornamental fishes, in collaboration with DBT, DST, CSIR, ICAR, MOEF, UGC, and

MPEDA. These events have been attended by preeminent scientific personalities, notably Prof. Asis Datta, Prof. Samir Bhattachryya, Dr K.C. Jayaram, and others. Dr Kar has presented papers and chaired scientific sessions at a large number of national and international symposia both in India and abroad, notably the Gordon Research Conference in the United States; 2nd International Symposium on GIS/Spatial Analysis in Fisheries and Aquatic Sciences in England during 2002; Lake Symposium at IISc, Bangalore (2000, 2002, 2010, 2012, and 2014, including chairing sessions); and Indian Science Congress (2012 and 2013, including chairing sessions), to name a few. He also has published in more than 190 national and international journals. Of particular note, 16 research scholars have been awarded MPhil and PhD degrees under his supervision. Professor Kar has authored 34 books (14 books single-authored by him), including one published by Springer (London) and one in press with Elsevier (USA). Also, the book *Community Based Fisheries Management* is in press with Apple Academic Press (USA).

As President of Conservation Forum, Dr Kar has made a profound contribution to the Society in Environment-related works in collaboration with Prof. Madhav Gadgil of the Indian Institute of Science (Retd.), Bangalore. In addition to being Editor of the *Conservation Forum Journal*, Dr Kar is a Scientific Fellow of the Zoological Society of London and a Fellow of the Linnean Society of London; Fellow of the Zoological Society, Calcutta; Fellow of the Applied Zoologists Research Association; Fellow of the Society of Environmental Biologists; Fellow of the Inland Fisheries Society of India; and others.

At the moment, Dr Devashish Kar is seniormost Professor and the Dean of the School of Life Sciences in Assam (Central) University at Silchar, India.

Foreword

There is now increasing emphasis on nutrition security, which covers the qualitative aspects of food. It is in this context that fishes occupy an important position in the food chain. Like all living organisms, fishes are also subject to infection by diseases. Epizootic ulcerative syndrome, or EUS, is causing severe mortality among our fishes. This is an exotic disease that entered our country in July 1988. Since then, many fishery enterprises have been closed. It is in this context that the study of Professor Devashish Kar of the Assam (Central) University at Silchar assumes great importance. The present book on epizootic ulcerative fish disease syndrome, which is being published by Elsevier Inc., USA, is a timely contribution. It can help to revive many fishery enterprises.

I therefore hope that the book will be widely read, and used to control EUS on the one hand and revive capture and culture fisheries on the other. If this is done, it will be a very important contribution to strengthening nutrition security among children, women, and men.

Prof. M.S. Swaminathan
Founder Chairman, M.S. Swaminathan
Research Foundation
Chennai, India

Preface

'Water' is 'life' and must be protected and conserved. Having its origin in water, life has evolved itself into an enthralling world of coveted and bewilderingly diverse flora and fauna. The dependence of the living world, notably humankind, on the biological wealth of rivers, lakes, seas and oceans, cannot be overemphasized and, possibly, does not need any elucidation. Therefore, there is, perhaps, a requirement to broaden and deepen our comprehension about the aquatic ecosystem with regard to its physical, chemical and biological features and interactions. The water, as an ecosystem, performs a number of significant environmental functions, notably, re-cycling of nutrients, re-charging of ground water, augmentation and maintenance of stream flow and recreation of people, to exemplify a few.

Notwithstanding the above, fresh water is one of the most important natural resources crucial for the survival of all living beings. It is even more important for humans as they depend on it also for food production, industrial growth, hydropower generation and waste disposal as well as for cultural requirements. Limnology is the science which deals with the freshwater environments, their physico-chemical characteristics, their biota and the ecosystem processes therein. Limnology is, therefore, universal in its significance.

Fishes are significant living components of both lotic and lentic systems. They constitute almost half the total number of vertebrates in the world. Of the c 39,900 vertebrate species known to exist so far in the world, 21,723 are living species of fishes; of which, c 8411 are of freshwater and the rest, 11,650 are marine species. In the Indian region alone, of the c 2500 species, 930 are freshwater inhabitants and 1570 are marine.

A living body is prone to suffer from a disease being attacked by pathogen(s) or parasite(s). This is true also with a fish body. Often, a disease may be so virulent, that it could sweep unabated in an epidemic dimension. In this connection, it may be said that, patterns and long-term trends are important when deciding whether or not an epidemic exists in the present period and in predicting future epidemics.

Snieszko (1974) had stated that, an overt infectious disease could occur when a susceptible host is exposed to a virulent pathogen under stress. The influence of each sub-set of the environment could be variable; and, disease outbreak (*e.g.*, EUS, in this case) may occur, only if there is sufficient relationship among them. Although a filterable biological infectious agent is thought to be the primary cause of EUS outbreak, it is generally accepted that, certain abiotic factors which result in sub-lethal stress of the fish, are also important in initiation of disease outbreaks. Snieszko cited temperature, eutrophication, sewage, metabolic products of fishes, industrial pollution and pesticides as potential sources of stressful environmental conditions.

Life and disease processes of fishes are similar in many ways to those of other vertebrates, in that, most animals have muscles, skeletons, skins and internal organs which function in approximately similar ways. However, there is one major difference: 'fish live in water'. As such, all of their physiological structures and functions are influenced by this fact. Thus, to understand what water means to fish is to lay a foundation for a more complete understanding of what fish need for health and how disease processes are related to physiological and environmental requirements.

In order to understand the influence of environmental factors on fish disease, one has to realize that, in aquatic environment, there is a much greater chemical and physical variability than exists in the terrestrial environment. In very large bodies of water, such as oceans, conditions may be relatively stable; but, in the coastal and estuarine areas, there is a greater magnitude of environmental change(s). In small bodies of water, this variability is even greater; while, in fish hatchery operations, many man-made stresses are added. Therefore, in the aquatic environment, life goes on under dynamic and unstable circumstances and fishes must continually adapt to changes in population density pressure, temperature, dissolved gases, light, pH, etc. However, it may be noted here that, the effects of these parameters could be far more severe than those ever faced by the terrestrial animals. Such environmental changes could impose stresses of considerable magnitude on the somewhat limited homeostatic mechanisms of fishes.

Epizootic Ulcerative Syndrome or EUS is a *hitherto* unknown, dreadful, virulent and enigmatic disease among the freshwater fishes which had been sweeping the water bodies in an epidemic dimension almost semi-globally causing large-scale mortality among the freshwater fishes; thus, rendering many of them endangered and throwing the life of the fishermen out of gear through loss of avocation and causing difficulties to the fish eaters through scarcity of fish flesh as a protein-rich source of nutrition. This dreaded fish disease, thus, has been a major concern in several countries of the world, particularly, the Asia-Pacific region.

As a historical resume, in Queensland, Australia, an epizootic of marine and estuarine fishes, characterized by shallow hemorrhagic ulcers, had occurred in 1972 with recurrence(s) in subsequent years. The disease had been named as 'red spot disease'. A similar type of disease, characterized by dermal ulcers, had been reported from Papua New Guinea: (a) from the rivers of the south during 1975–76; and, (b) from the north during 1982–88. Concomitantly, Indonesia had also reported similar type of disease in Bogor during 1980, which, subsequently, had spread to West, Central and East Java. This disease was named as infectious dropsy or 'hemorrhagic septicemia'. Later, the disease was reported from Malaysia during 1981–83. The affected fishes had red or necrotic areas of ulceration all over their bodies and were called 'Webak Kudes'. The disease was, subsequently, reported from fishing areas of Kampuchea, during early 1984, along with a significant decrease in the natural fish stock. Also, a similar type of disease had been reported from the southern and central parts of Lao PDR during 1984. Myanmar had experienced the outbreak of EUS during 1984–85 affecting both wild and cultured fish stocks. In Thailand, the disease epizootic was first reported in 1980 in the natural water systems and the disease had recurred somewhat regularly almost every year from 1980 to 1985 in different water bodies. In Sri Lanka, the disease was first reported in 1988 in the Kelani river, Dandugan Oya, and in the nearby streams causing extensive fish mortality. In Bangladesh, the first outbreak of EUS had occurred during February-March 1988 in the rivers Meghna, Padma and Jamuna; as well as, in the adjoining water bodies with colossal loss of the commercial fish stock. After sweeping semi-globally in epidemic dimension, this *hitherto unknown* enigmatic and virulent fish disease entered India, during July 1988, through Barak valley region of Assam and started causing large-scale mortality among the FW fishes. It had caused panic among the fish consumers and had thrown the life of the poor fishermen out of gear and rendering them starving for days together in view of no sale of fish. And, the author of the present Treatise had decided to start researching on this challenging fish disease problem (sacrificing his remuneration leave, and with the help of many generous souls), as part of discharging his social responsibility, at this crucial juncture of the Society and the People. In India, the outbreak of the virulent EUS fish disease had been first encountered during July, 1988 among the fishes in the Freshwater bodies of North East (NE) India. In 1989, Nepal had also been affected by EUS (Das and Das, 1993; Kar, 2007, 2013).

EUS had been a *hitherto* unknown enigmatic virulent, epidemic among the fishes since about 4 decades. It had been designated by various names, some of which are often colloquial. Egusa and Masuda (1971) had, perhaps, at first, described it in Japan as *Aphnomyces* infection. The infection had also, possibly, occurred among other fishes, being named as Mycotic Granulomatesis (MG). Such a nomenclature was mainly based on histopathological (HP) results (Miyazaki and Eugusa, 1972). Concomitantly, a *hitherto* unknown EUS condition had been encountered, since 1972, mainly on the skin of estuarine fishes in Australia. Such an element had been named as Red Spot Disease (RSD) (McKenzie and Hall, 1976). Subsequently, almost similar conditions had occurred among the freshwater fishes throughout South-East (SE) and South Asia since the late 70's. The main symptoms included dermal ulcerations and there had been large scale mortalities among the fishes. The condition was designated as 'Epizootic Ulcerative Syndrome' (EUS) at the consultation of Experts on Ulcerative Fish Diseases held at Bangkok during 1986 (FAO, 1986). In addition to the above, almost similar ulcerative condition had also been reported among the estuarine fishes in the east coast of USA since around 1978. Such a condition was termed as Ulcerative Mycosis (UM) (Noga and Dykstra, 1986).

Lack of a comprehensive treatise on 'Epizootic Ulcerative Fish Disease Syndrome' on a global scenario, has prompted this humble piece of work and the author will consider himself amply rewarded if this humble piece of work proves to be useful to those for whom it is meant.

Devashish Kar
Professor, Department of Life Science
School of Life Sciences
Assam (Central) University
Silchar, Assam, India

Acknowledgments

"Today's pansy is tomorrow's hawthorn." With the fervent prayer that this humble piece of work will benefit the people of this land, the undersigned offers his heartfelt gratitude to the numerous benevolent souls who never grudged him any help in his strenuous and hazardous fieldwork spread over more than 40 years. The village folk who kept a vigil with him on many a stormy night can neither be named nor their help fully acknowledged.

May the undersigned utilize the privilege of this opportunity to express his profound regards and dutiful reverence to his respected Late Parents, Dr Himangshu Jyoti Kar (Father) and Mrs Sunity Kar (Mother), without whose blessings this humble piece of work would not possibly have seen the light of day.

The undersigned utilizes the privilege of this opportunity to pay his very humble profound regards to Professor M.S. Swaminathan, FRS, Ex-Director General of the Indian Council of Agricultural Research, for his very kind Foreword.

The undersigned also records his deep sense of gratitude and profound regards to his respected teachers, notably Dr Subhas Chandra Dey, Retired Professor of Gauhati University, Dr Naresh Chandra Datta, Retired Professor of Calcutta University, and Dr Martin Monk and Dr R.G. Bailey of King's College London for their constant encouragement, guidance, and blessings.

The undersigned also expresses his deep sense of appreciation and gratitude to his family members, notably Professor Mrs Radha Rani Dev, Late Mr Devapriya Kar, Mr Devajyoti Kar, and Dr Deva Prasad Kar for their help and encouragement. Last but not least, this humble piece of work would perhaps have been nipped in the bud if my wife, Doctor and Professor Mrs Swarupa Kar, and my children, Miss Devarati Kar (daughter) and Master Dyutiman Kar (son) had not been with me like shadows. I must say here that I am fortunate to have the intellectual stimulation, companionship, and association with a band of excellent, intelligent, and obedient students who have been with me at different stages of preparation of this humble piece of work. In particular, Ratnabir, Romen Singh, and Binku helped in some typing work. Sulata, Papia, and Uma also helped me much in typing the arrangement of references. But it is Biplab Das who has been with me like a shadow and helped me at almost every step since I started writing the manuscript. Biplab helped me substantially in arranging the text and numbering the illustrations, and also helped me much in computational work. My very dear student, Mr Satyajit Das, enormously helped in arranging the chapters and references. The undersigned also wishes to acknowledge the help and cooperation received from his esteemed colleagues, friends, well-wishers, and all those benevolent souls connected with this humble piece of work. Particularly, I would like to mention Mr Govinda Datta, Proprietor of Jayanti Press, Silchar, who not only gave me encouragements and blessings as my brother-in-law, but also guided me in many aspects of preparation of this piece of work. Further, contemporary literature has been consulted with gratitude.

The author utilizes this opportunity to express his deep sense of gratitude to Elsevier Inc. for acting as a pillar of logistic support, for its excellent manuscript delivery system, and for its kind efforts in the publication of this humble piece of work.

Devashish Kar
Silchar
June 1, 2015

Chapter 1

Introduction

1.1 ABOUT FISH

"Fishes" are primarily adapted cold-blooded aquatic vertebrates with paired fins for locomotion and gills for respiration.

1.2 ABOUT FISH DIVERSITY

Fish constitutes almost half of the total number of vertebrates in the world. They live in almost all conceivable aquatic habitats. Approximately 21,723 living species of fishes have been recorded out of 39,900 species of vertebrates (Jayaram, 1999). Of these, 8411 are freshwater (FW) and 11,650 are marine (Jayaram, 1981). India is one of the megabiodiverse countries in the world and occupies ninth position in terms of FW megabiodiversity (Mittermeier and Mittermeier, 1997). In India, there are about 2500 species of fishes (Jayaram, 2003), of which *c* 930 live in FW and *c* 1570 are marine (Kar, 2003). The bewildering ichthyodiversity of this region has been attracting many ichthyologists from both India and abroad. Concomitantly, the Northeastern region of India has been identified as a "Hotspot" of biodiversity by the World Conservation Monitoring Centre (WCMC, 1998). The rich diversity of this region could be assigned to certain causes, most notably the geomorphology and tectonics of this zone (Jayaram, 2010). The hills and the undulating valleys of this area give rise to a large number of torrential hill streams, which lead to big rivers and finally become part of the Ganga−Brahmaputra−Barak−Chindwin−Kolodyne− Gomati−Meghna system (Kar, 2000, 2013).

Based on IUCN categories, the CAMP Workshop (CAMP, 1998) for FW fishes has identified certain fish species that have attained threatened/endangered status. At the same time, there has been little study with regard to details of endemism and species richness in Northeast India. As such, a detailed study related to Germplasm inventory, evaluation, and gene banking, along with the health and disease of FW fishes, would not only help us to land at a concrete decision on the above-mentioned aspects, but also contribute to fulfilling India's obligation to the Convention on Biological Diversity, with special emphasis on Articles 6 and 8 (UNEP, 1992). Further, development of the database of biological parameters is a prerequisite for preparation of a detailed fish inventory. Genetic characterization and gene banking is a step forward toward further confirmation of the species at the molecular level.

1.2.1 An Account of the Ichthyospecies

Zoogeographically, FW fishes have been classified differently by different workers. Although the classification made by Myers (1949) has proved to be the most useful and widely accepted one, FW fishes of marine origin had been further classified as "peripheral FW forms" by Nichols (1928) and Darlington (1957), which has also been accepted by many recent fish geographers. Incidentally, the ichthyofaunas of this region have by and large been found to belong to the following categories (Kar, 1990):

1.2.1.1 Primary Freshwater Fishes

Genera-wise, the breakup of species under this group includes, among others, *Notopterus, Chitala, Labeo, Cirrhinus, Catla, Cyprinus, Puntius, Rasbora, Aspidoparia, Amblypharyngodon, Barilius, Danio, Esomus, Salmophasia, Botia, Lepidocephalichthys, Acanthocobotis, Rita, Mystus, Wallago, Ompok, Ailia, Eutropiichthys, Clupisoma, Pangasius, Gagata, Glyptothorax, Clarias, Heteropneustes, Badis, Nandus, Anabas, Trichogaster, Mastacembelus, Macrognathus,* and *Tetraodon.*

1.2.1.2 Peripheral Freshwater Fishes

Genera-wise, the breakup under this group includes, among others, *Gudusia, Tenualosa, Pisodonophis, Chanda, Xenentodon, Aplocheilus, Monopterus, Sicamugil, Rhinomugil,* and *Glossogobius.*

In addition to the above, on the basis of Indian and extra-Indian fish distribution (Motwani et al., 1962), the ichthyospecies of this region could be significantly incorporated under the following two groups:

Widely Distributed Species

Genera-wise, the breakup under this group includes, among others, *Esomus, Puntius, Rasbora, Ompok, Wallago, Clarias, Xenentodon, Channa, Glossogobius, Anabas, Macrognathus,* and *Mastacembelus.* Incidentally, these ichthyospecies, in addition to being from this region, also

Epizootic Ulcerative Fish Disease Syndrome. http://dx.doi.org/10.1016/B978-0-12-802504-8.00001-8

occur in many other parts of India, Pakistan, Bangladesh, Sri Lanka, and Malaya.

Species of Northern India

Species under this group include, among others, *Megarasbora elanga, Botia dario, Lepidocephalichthys guntea, Glyptothorax telchitta, Parambassis baculis, Rhinomugil corsula, Sicamugil cascasia*, and *Tetraodon cutcutia.*

In addition to the foregoing analyses, ecomorphologically (Dey, 1973) the fishes of this region could further be categorized into four distinct groups as follows:

True Hill-Stream or Rheophilic Forms Fishes with strong body musculature and adapted to torrential abodes: *Garra, Psilorhynchus, Balitora, Glyptothorax*, etc.

Semitorrential Forms Fishes with less body modifications compared with rheophilic forms: *Botia, Lepidocephalichthys, Acanthocobitis, Schistura, Canthophrys, Gagata*, etc.

Migratory Forms Well-built fish having the power of overcoming adverse ecological conditions: *Tenualosa, Barilius, Channa, Badis*, etc.

Plain Water Forms Fishes having minimum body modifications and insignificant migratory habits: *Pisodonophis, Gudusia, Notopterus, Chitala, Amblypharyngodon, Aspidoparia, Catla, Cirrhinus, Cyprinus, Danio, Esomus, Labeo, Puntius, Rasbora, Salmophasia, Mystus, Sperata, Ompok, Wallago, Rita, Clupisoma, Eutropiichthys, Silonia, Pangasius, Clarias, Heteropneustes, Chaca, Xenentodon, Aplocheilus, Monopterus, Chanda, Nandus, Sicamugil, Rhinomugil, Anabas, Trichogaster, Glossogobius, Macrognathus, Mastacembelus*, and *Tetraodon.*

1.3 ABOUT FISH DISEASE

The teleosts are said to inhabit an alien world, and their utilization by humans for food or sport has until recently been virtually dependent on the hunting of wild species. It is only in the last 100 years that intensive cultural methods for a few species have evolved. As such, the diseases that result from the intensive cultural techniques have been a major constraint on the economic development of aquaculture (Hatai,1994).

The aquatic environment encompasses a wide variety of parameters, virtually all of which influence the stability of the homeostasis that is essential for the growth and reproduction of fishes. If altered beyond acceptable limits, these may predispose to or actually cause disease.

All surface waters may contain species of wild fish that could act as reservoirs of infectious disease. Further, toxin-producing algae are found in freshwaters in addition to marine and brackish waters. Under suitable environmental conditions, they grow to considerable cell densities, called "blooms" or "tides"; during or after blooms, toxins may be produced that are lethal to fish.

The anatomy and physiology of fish are modified principally toward the two major ecological factors that control their existence, viz., the aquatic environment and the poikilotherm's inability to control its temperature. These factors are of overriding significance in dictating the chain of events following any pathological attack such as microbial infection, traumatic damage, or nutritional deficiency.

In fish disease, more readily than in disease of any other farmed or wild species, it is possible to recognize the significance of "stress" factors. The word "stress" has different meanings for different groups of workers. As originally defined by Selye (1950), it was the sum of all the physiological responses by which an animal tries to maintain or reestablish a normal metabolism in the face of a physical or chemical force. Brett (1958) gave a definition that is correlated more readily with the fish disease situation when he suggested that "stress" is a stage produced by environmental or other factors that extends the adaptive responses of an animal beyond the normal range, or that disturbs normal functioning to such an extent that chances of survival are significantly reduced. The changes that occur in response to environmental stress are termed general adaptation syndrome (GAS). They are said to be nonspecific physiological and biochemical changes that take place in three phases:

1. The alarm reaction
2. The stage of resistance, during which adaptation to achieve homeostasis under the changed circumstances takes place
3. The stage of exhaustion, when adaptation has ceased to be adequate and homeostasis is not achieved

In addition to the above, the specific immune system is one component of the protective system of all vertebrates that enables the individual to survive and maintain its homeostasis in an environment that is innately hostile. The property that distinguishes the immune defense system from other defense systems is its specificity and ability to "remember" a particular infectious agent and thus confer subsequent resistance to an individual who has recovered from an infection. Thus, a salmon that has recovered from furunculosis may be refractory to a subsequent infection by *Aeromonas salmonicida*, but this immunity is irrelevant to subsequent infection by *Mycobacterium piscium*. In fish, generally the thymic tissue develops embryologically from primordial cells associated with the epithelium of the pharyngeal pouches. Lele made a comparative study of the morphology of thymus in fishes and found the organ to have a variable origin and variable duration in different species of fishes.

Diseases in FW fishes are generally studied under the following heads:

1. Parasitology of teleosts
2. Bacteriology of teleosts
3. Mycology of teleosts
4. Virology of teleosts

1.3.1 Parasitology of Teleosts

Many phyla of the animal kingdom have representatives that are parasitic on fish. The number of species of fish parasites reported to date are in the thousands, and many more remain to be discovered. Very few are seriously harmful to fish. Most individual fish in wild or cultivated populations are infested with parasites, but in the great majority of cases, no significant harm appears to be caused to the host fish. Although there are surprisingly few reports of parasites causing mortality or serious damage to wild fish populations, perhaps such effects often go unnoticed. Parasites in wild fish are frequently only remarked upon when they are so obvious as to lead to rejection of fish by fishers or consumers.

No attempt could be made to present a detailed classification because of the great diversity of taxonomic groupings of the parasites found in fish and the dispute that surrounds their taxonomy.

1.3.1.1 Protozoan Parasites

Ichthyophthiriasis or white spot disease is caused in Indian major carps (IMC) by the protozoan ciliate *Ichthyophthirius multifilis*, which infects different regions of the body externally and causes simple hyperplasia of the epidermal cells around the site of infection that causes the formation of pustules. Tripathi (1955) experimentally introduced *Ichthyophthirius* in the fingerlings of *Labeo bata* and *Cirrhinus mrigala* in order to study the effect.

The symptoms are the presence of small whitish cysts (c 1.0 mm in diameter) on the skin, gills, and fins.

Affected fish can be dipped in 1:5000 formalin solution for an hour for 7–10 days, or in 2% NaCl solution for more than 7 days (Gopalakrishnan, 1963, 1964). Affected ponds can be disinfected with salt or quicklime (Hora and Pillay, 1962). Examples of other ciliate infections observed in India are as follows:

1. *Trichodina indica* infect different species of fishes
2. *Scyphidia pyriformes* affect rohu, catla, mrigal, etc. (Tripathi, 1954)
3. *Cyclochaeta* spp. infect pond fishes (Hora and Pillay, 1962)

Both *Trichodina indica* and *Scyphidia pyriformes* are killed in 5–10 min by 2%–3% common salt solution.

4. Costiasis caused by *Costia necatrix* is the most common mastigophoran infection observed in IMC. Symptoms include the presence of a bluish coat on the skin of the host fish and the presence of a large amount of mucosa. The parasite causes irritation and disturbs respiration. A bath in 3% NaCl solution for 10 min appears to be quite effective (Gopalakrishnan, 1964).

Other flagellate parasites recorded are as follows: *Bodomonas rebae* in rohu, catla, and mrigal (Tripathi, 1954); *Trypanosoma clariae* in *Clarias batrachus*; and *Trypanosomae punctati* in *Channa punctatus* (Hasan and Qasim, 1962)

5. Trichodinosis is caused by *Trichodina, Trichodinella, Chilodonella*, etc. These organisms affect carp skin and gills. Multiple infections are commonly recognizable as spherical organisms, ovoid or long with rows of cilia. Chelated copper compounds like "Argent," "AquaVet," etc. have recently been found effective against protozoan parasites.
6. Myxosporidians constitute typical fish parasites known to produce cysts on different regions of the body and viscera. Some common myxosporidian genera are *Leptotheca, Myxobolus, Myxidium*, etc.

General symptoms include weakness, emaciation, falling of scales, perforation of scales, raising of scales along the posterior margins, loss of chromatophores, etc. (Tripathi, 1952). Cysts on the scales show an inner and an outer fibrous layer of epidermal origin in fishes affected with *Myxobolus mrigalae* (Chakravarty, 1939). However, myxosporidian infections have been commonly observed among IMCs in waters having <400 ppm chloride content (Basu, 1950). Gill myxoboliasis among IMCs (rohu, catla, and mrigal) caused large-scale mortality in Bangladesh.

Myxosporidian cysts are not easily eliminated by chemical treatment. Disinfection with Candy's Fluid or a weak solution of NaCl, to some extent, controls *Myxobolus mrigalae* affecting mrigal fingerlings (Sarkar, 1946). Concomitantly, myxosporidian infections encountered in overstocked ponds are controlled by thinning the population and using yeast pellets at 1 g/kg of feed (Pal, 1975).

1.3.1.2 Helminth Parasites

Worm diseases are caused by trematodes (monogenetic and digenetic), cestodes, nematodes, acanthocephalans, and hirudineans. A brief account of helminth parasites in fish is given below:

Monogenetic Trematodes

Among the various monogenetic trematodes reported in fishes, *Gyrodactylus* and *Dactylogyrus* (Chauhan, 1953; Tripathi, 1957; Gupta, 1961; Gopalakrishnan, 1968) cause serious infections. The former infects skin and gills, while the latter affects only gills and is further reported to feed on the blood of the thin capillaries of the branchial region (Jain, 1959).

A *Gyrodactylus* attack causes fading of colors, falling of scales, accumulation of mucous on the caudal peduncle region and fins, peeling of skin, etc. When the gills are attacked, the fish dies due to the hypersecretion of mucus on the gill surface brought about by irritation. Infected fishes may be seen rubbing their bodies against hard surfaces in order to get rid of the parasites.

Baths alternately in 1:2000 acetic acid and NaCl solutions have been reported to be effective in IMCs. Dip treatment in 5% NaCl for 5 min has also been found to be effective (Gopalakrishnan, 1964).

Digenetic Trematodes

Diplostomiasis or black spot disease is caused by *Diplostomum* spp. This disease is generally prevalent in some rearing and stocking ponds. Symptoms generally include the presence of small black nodules over almost the entire body of the fish, having an average diameter of *c* 1.3 mm.

Dip treatment of isolated infected specimens for about an hour in 3:100,000 picric acid has been found to be effective in controlling the parasite (Gopalakrishnan, 1963).

Cestodes

Fish mortality due to cestodes was once reported in a tank at Nagpur (Chauhan and Ramakrishna, 1958). Affected fish generally appear dull and sick, and part of their gut becomes generally swollen or completely choked by cestode cysts. In some instances, the gall bladder is also affected.

Bothriocephalus infection affects intestinal epithelium. *Ligula* and *Schistocephalus* sometimes take shelter in the peritoneal cavity of carps.

Nematodes

A widely occurring nematode that affects the scale pockets of common carps in Europe is the *Philometra lusiana*. However, some nonpiscian nematode parasites include *Camallanus, Cucullanus, Heliconema,* and *Spiroptera*. Not much information is available on the details of nematode infections and their remedies.

Acanthocephalans

Among acanthocephalans, *Acanthocephalus, Centrorhynhus,* and others have been reported to affect fishes (Tripathi, 1957).

Hirudinean Parasites

IMCs are sometimes attacked by leeches (Khan, 1944). Carps have been reported to be affected by *Hemiclepsis marginata* (Saha and Sen, 1955). The leeches feed on the blood of the host fish, causing irritation and abnormal movements of the host.

Dip treatment in a 1:1000 solution of glacial acetic acid and disinfection with 1:10,000 $KMnO_4$ has been reported to be quite satisfactory (Khan, 1944). There have also been suggestions of dip treatment in 2.5% common salt solution for about 30 min.

1.3.1.3 Crustacean Parasites

Argulus (carp lice), *Lernaea*, and *Ergasilus*, belonging to the class Crustacea, are considered important ectoparasites on fishes. The parasitic copepods *Argulus* and *Lernaea* attach themselves to the body of the fish, with their body buried into the scale pockets and with paired egg sacs protruding free. *Argulus* in particular attaches itself to the body of the fish by means of suckers and hooks, but can also swim freely in water. There have been many instances of fish mortality in Indian ponds due to argulosis. The affected fish become very weak and emaciated.

Prominent symptoms include stunted growth, peeling off of the scales, and the appearance of red spots at the sites of infection.

Treatment measures could include the following:

Ponds showing severe *Argulus* infection be drained and dried. Sometimes, vertical wooden bamboo poles are fixed inside the water so that the affected fish may rub their bodies against them to get rid of their ectoparasites. Lime is applied in the affected pond at 0.1−0.2 g/L after all the fish are removed and the pond bottom has been exposed to sun for at least 24 h before the application of lime (Hora and Pillay, 1962).

Argulus sp. reproduce rapidly at a temperature range of 20−28 °C. Positive results in *Argulus* treatment could be obtained by the application of Lindane (benzene hexachloride) at 0.02 mg/L.

In addition to *Argulus*, *Lernaea* (also known as anchor worm) is a minute rodlike ectoparasite that attaches to the host fish anywhere on the body by means of the anchor-like appendages present on the parasite's head. Many carps are attacked by this parasite, particularly during the summer.

When only a few fishes are affected, mechanical removal of the parasites could be possible by pulling them out of their anchorage in the host body with the help of a pair of forceps. This could be followed by a bath in weak permanganate solution for 2−3 min (Alikunhi, 1957). Dip treatment for a short period in a 5 ppm solution of $KMnO_4$ is suggested in the event of a large number of fishes being affected (Gopalakrishnan, 1963).

Tripathi and Chaturvedi (1974) recently recorded an isopod parasite, *Ichthyoxenus jellinghausii* from *L. bata* and *L. gonius* at Pariat Lake in Jabalpur. This parasite does not generally cause much harm to the host. Both dip treatment and group treatment are generally recommended to remove this parasite.

1.3.2 Bacteriology of Teleosts

Bacterial diseases are responsible for heavy mortality in both wild and cultured fish. The actual role of these microbes may vary from that of a primary pathogen to an opportunist invader of a host. Water, especially where organic loads are high, is an environment in which many genera of bacteria could thrive. It has been shown by many workers that the normal bacterial flora of fish is a direct reflection of the normal bacterial flora of the water in which they swim. However, few bacterial species appear to be obligatory parasites and are unable to survive for long outside the fish host.

The "red spot" of eels is said to be the first bacterial disease to be described in fish. According to Hofer (1904), it was first described by Bonaveri as early as 1718 from the Commacchio lagoons of the Adriatic coast of Italy.

During the last century, two factors played significant roles in the development of knowledge of bacterial diseases of fish. The first was the establishment of a Parliamentary Committee of Enquiry into furunculosis in Scotland in 1929—the disease had caused severe losses in Scottish salmon rivers. This committee produced three detailed reports in 1930, 1933, and 1935 that even today provide the basis for our knowledge of this disease; these reports also established the criteria for subsequent critical studies.

In many fish diseases, bacteria have been found to be associated with the host, generally as secondary invaders and sometimes as primary causative agents. Sometimes, disease diagnosis becomes complicated. For example, pathogenic *Aeromonas* spp. have been associated with infectious dropsy in IMCs. However, the disease could be reproduced by inoculation of pure bacterial cultures. Further, the possibility of viral involvement cannot be easily ruled out. Diagnoses of ulcerative skin conditions and fin or gill rot are particularly difficult because a wide range of fungi, bacteria (*Aeromonas, Pseudomonas,* and *Vibrio* spp.), and protozoans may be present. Such ulcerative conditions and fin rots have recently been described in common carps and rohu in the Punjab. The conditions have been termed hemorrhagic septicemias and are generally complicated by severe infections with the fungus *Saprolegnia*. Further, the pathogen *Aeromonas punctata* had been isolated from fish blood and dried ulcers and held to be the primary cause. Rohu in particular were severely affected with lesions in the head region. The condition had occurred during the coldest time of the year (about 17 °C). It may be noted here that most bacterial pathogens in fishes are gram-negative.

A bath treatment for 30 min in 20 ppm proflavine hemisulfate is generally quite satisfactory for treatment of external lesions. However, if vibrios are suspected to be involved, a nifurpirinol (nitrofurazone-furnace) bath for 1 h at 10 ppm (active ingredient) is recommended.

Mycobacterial gill diseases are best treated with baths; usually with Quaternary ammonium compounds for 1 h in the range 1—4 ppm. If other compounds are not available, a dip in 500 ppm of copper sulfate for 30 s to 1 min is worth trying. Further, kanamycin sulfate is said to be effective against certain gram-positive and gram-negative fish bacterial pathogens including *Aeromonas, Vibrio,* and *Flexibacter,* as well as mycobacterial infections.

A brief account of some of the bacterial diseases found to occur among FW fishes of India is given below:

1.3.2.1 Fin and Tail Rot Disease

This disease affects both adults and young fish. In the early stage, the infection appears like a white line on the margin of the fin, spreading and imparting a frayed appearance to the appendage, which eventually putrifies and disintegrates. The disease may spread through contact and cause heavy damage.

Treatment of the disease may include a bath in 1:2000 solution of copper sulfate for 1—2 min until the fish shows signs of distress (Hora and Pillay, 1962). Some amount of cure could be possible by painting the site of infection with a concentrated solution of copper sulfate (Pal and Ghose, 1975).

1.3.2.2 Ulcer Disease

Hemorrhagic ulcers may be caused by *Flexibacter columnaris*, which is evidenced by red and white plaques, often with a reddish peripheral zone.

As a first treatment, badly affected fishes should be destroyed and the pond water disinfected with 0.5 ppm of $KMnO_4$. In the case of fishes showing early stages of infection, dip treatment in 1:2000 copper sulfate solution for 3—4 days could be useful.

1.3.2.3 Carp Erythrodermatitis (CE)

Carp erythrodermatitis (CE) was once upon a time considered a part of infectious carp dropsy syndrome. A severe skin ulceration in common carp stock was reported from West Java during 1980. Although not very much treatment has been recommended, terramycin mixed with feeds at 5—7 g/100 kg fish daily for 7—10 days could be useful to some extent (Jhingran and Pullin, 1988).

1.3.2.4 Dropsy

Protrusion of scales, hemorrhagic ulcers on the skin and fins, inflammation of the intestine, exophthalmic condition, and accumulation of fluid within the body are some of the symptoms of dropsy.

As a treatment measure, disinfection with 1 ppm potassium permanganate solution or dip treatment in 5 ppm of the same solution for 2 min have been found to be useful (Gopalakrishnan, 1963). No food should be given during the treatment process.

1.3.2.5 Eye Disease

A variant of the bacterium *Aeromonas liquefaciens* causes an eye disease, which often assumes epidemic dimension, affecting medium- and large-sized catla. The affected sites are generally the eyes, optic nerves, and brain of the fish (Gopalakrishnan and Gupta, 1960). Further, *Channa marulius* showing an eye disease, also in epidemic form and caused by *Staphylococcus aureus,* has been recently encountered—this disease causes cataract-like symptoms in fish. Initially, the cornea becomes vascularized and later opaque. Subsequently, the eyeball gets putrefied and results in the death of the fish.

Treatment with KMnO₄ (dose: 0.1 ppm) for disinfecting the environment, followed by application of lime (300 ppm), gives useful results (Kumaraiah et al., 1977).

1.3.3 Mycology of Teleosts

A wide variety of Phycomycetes and Fungi imperfecti have been known to be associated with diseases in fishes. However, in many cases there is no direct evidence of a primary etiological role. The integumentary mycoses associated with members of the order Saprolegniales are by far the most significant fungal diseases in teleostean fishes.

1.3.3.1 Integumentary Mycoses

Saprolegniasis

This term is often used to describe fungal infections of skin and gills. Although strictly speaking, the term should be used only after definitive identification of the fungus as a member of the order Saprolegniales, it could involve a variety of fungi. Clinically, similar lesions could occasionally be formed by the genus *Pythium.* Willoughby (1970) had also isolated *Leptomitus lacteus* from such lesions.

Considerable interest developed toward the end of nineteenth century in an epizootic in Atlantic salmon caused by fungal invasion (Huxley, 1882; Murray, 1885). *Saprolegnia ferax* was thought to be responsible for the condition at that time. Later, it was felt that these workers were studying the final stages of a condition now known as "ulcerative dermal necrosis" (UDN). In 1932, Kanouse had described the sexual form of *Saprolegnia parasitica,* which she had obtained on a hemp seed substrate after original growth on fish eggs. Seymour (1970) had presented a comprehensive account of the genus *Saprolegnia.*

Saprolegniaceae are "water molds" of the class Oomycetes, which possess profusely branched nonseptate mycelium appearing as cotton- and woollike tufts in water. The hyphae vary considerably in diameter between species, but all contain cellulose.

A number of predisposing factors are involved in the development of fungal infection in fish—the factors affect both the fish and fungus, and a combination of factors rather than any single condition ultimately leads to infection. It has long since been believed that the fungi responsible for saprolegniasis are secondary pathogens, and lesions are commonly seen after handling and after traumatic damage to the skin that could be caused by overcrowded conditions or in conjunction with pollution or bacterial or viral infections.

Primary saprolegniasis had been reported by Hoshima et al. (1960) in cultured eels without any visible prior injury to the fish. Tiffney (1939) found that macroscopic injury significantly increased the likelihood of fungal infection. Richards and Pickering (1978) had shown that in outbreaks of saprolegniasis in spawning brown trout, a form of *Saprolegnia* with a low degree of homothallic fungus appears incapable of producing sexual structures despite prolonged incubation on a variety of media, except to a limited degree at low temperature. This supports the findings of Willoughby (1968), who had consistently isolated a similar sterile form of *Saprolegnia* from lesions of UDN.

Temperature plays a significant role in the development of *Saprolegnia* infections. While infection following trauma may occur at any temperature compatible with fish life, most epizootics occur when the temperature is low. Hoshima et al. (1960) pointed out that saprolegniasis of eels ceased when the water temperature rose above 18 °C.

Clinical features of saprolegniasis are as follows:
Saprolegnia lesions are focal gray-white patches on the skin of the fish that, when examined under water, have a cotton- and woollike appearance where the hyphal filaments extend out into the water. The early lesions are often almost circular and grow by radial extension around the periphery until lesions merge. At this later stage, the fungal patches are often dark gray or brown as the mycelium traps muds or silt. Skin and gill lesions are by far the most frequently observed but there have been reports of infection of internal organs. Agersborg (1933) reported intestinal infection in fingerlings of brook trout with *S. ferax,* and a similar infection with *Aphanomyces* spp. was reported by Shanor and Saslow (1944).

Some of the HP features of saprolegniasis are given below:

The fungus usually establishes itself focally, invading the stratum spongiosum of the dermis and then extending laterally over the epidermis, eroding it as it spreads. Relatively superficial invasion of the dermis rapidly leads to fluid imbalance and peripheral circulatory failure due to the inability to maintain circulating blood volume.

In hematoxylin—eosin stained sections of the skin infected with *Saprolegnia,* numerous hyphae are seen on the skin surface enmeshing cellular debris and material trapped from the water by the hyphal strands.

With regard to isolation of the fungus, a variety of methods may be used to obtain bacteria-free colonies, all of which involve the use of agar media. A brief account of prophylaxis and treatment is given below.

Maintenance of fishes in good husbandry conditions is a prerequisite. Correct feeding, avoidance of crowded conditions, and maintenance of good water quality are also essential. Should the fish develop saprolegniasis, a variety of external disinfectants may be used. These include use of malachite green, copper sulfate, potassium permanganate, and formalin.

1.3.3.2 Other Fungal Diseases

Systemic Mycoses

Fijan (1969) described a condition in channel catfish in which skin ulcers were seen that were 2—15 mm in diameter and up to 5 mm deep, with no gross evidence of inflammation. Nodules up to 25 mm in diameter were found in association with adhesions and peritonitis, suggesting both hematogenous spread and local extension. The disease had been reproduced in the channel catfish. In experimental fish, abscesses containing hyphae and mixed purulent and gaseous materials were found.

Cerebral Mycetoma

Carmichael (1966) described a *Phialophora*-like fungus that had caused epizootics of so-called cerebral mycetoma in cutthroat trout. The organism was named *Exophiala salmonis,* the lesion being a chronic nonsuppurative granuloma with the presence of numerous giant cells in the brain and cranial area.

Scolecobasidium Humicola Infection

Ross and Yasutake (1973) had described a systemic mycotic infection in coho salmon held for experimental purposes. In affected fish, an enlarged abdomen was usually seen, often together with skin lesions. Ascites, adhesions, and gray areas in internal organs, especially the kidney, were also often seen. The disease could not be transmitted experimentally by incorporating the fungus into normal diets, but transmission was achieved when ground glass was added to the diet. Doty and Slater (1946) described a species of *Heterosporium* that was pathogenic to Chinook salmon.

Sphaeropsidales Infection

Phoma herbarum, a fungal plant saprophyte, had been isolated from three species of diseased salmonids in Washington state and Oregon (Ross et al., 1975). An earlier outbreak in Chinook salmon was briefly described by Wood (1968). Detailed morphology of the fungus was described by Boerema (1970). It has branched septate hyphae, and young cultures on Sabouraud dextrose agar are light buff in color, changing to light pink and then to black with aging and the formation of pycnidia, which produce hyaline unicellular conidia.

In outbreaks, morbidity is rarely >5%, and the disease usually affects fry and fingerlings. When the intensity of the disease is high, fishes are seen to swim abnormally and are unable to maintain equilibrium. They often develop swollen vents with hemorrhagic fin and skin lesions. Early internal lesions are confined to the swim bladder and are small (1—2 mm) white areas in the anterior ends of the organs; the pneumatic duct area is probably the first affected. In more advanced stages of infection, the lumen becomes filled with mycelium and the epithelium of the swim bladder becomes hyperplastic. The wall is rapidly destroyed and adjacent internal organs are affected. There is an extensive acute inflammatory response or a chronic granulomatous reaction. PAS- and Giemsa-positive hyphae are usually evident and are approximately 50—100 μm in length and 2—3 μm in width.

Pure cultures of the organism could be obtained by aseptically removing material from the abdominal cavity and plating out onto Sabouraud dextrose agar. The disease was reproduced by Ross et al. (1975), but the organism appeared to be only weakly contagious and the condition was found to be related to diet.

Ichthyosporidium (*Ichthyophonus hoferi*)

This organism was assigned to Haplosporidia and named *Ichthyosporidium gasterophilium* when it was described by Caullery and Mesnil (1905). Later, Laveran and Pettit (1910) recognized a similar organism as a fungus and named it *Ichthyophonus hoferi* after Hofer, who had first described it as a haplosporidian affecting flounder, sea trout, etc. Sprague and Vernick (1974) described the appearance of the organism *Ichthyosporidium* under electron microscope and classified it among Microsporidia.

Its clinical features, briefly, are as follows:

The disease is a systemic granulomatosis and is found in both freshwater and marine fishes of many species. It has been the cause of death in a large number of Atlantic herring in frequent epizootics along the East Coast of North America. Outbreaks in herrings usually occur in winter and spring, and infection is evident as a roughened skin texture described as the "sandpaper effect" that occurs principally on the lateroventral tail region. This effect is caused by the loss of epithelium over proliferating dermal fungal granulomata. Further growth of the fungus causes local necrosis and results in the formation of either abscesses or ulcers.

In certain other species of fishes, the internal organs are more frequently affected than the skin, and such internal infection is evident as raised white nodules very similar to granulomata or tuberculosis.

Its histopathology, briefly, is as follows:

Most frequently observed is the "resting stage." The germinating spore is frequently seen in section and consists of a cytoplasmic elongation bounded by the inner spore wall that herniates through the thicker outer wall. Heart, liver, muscle, kidney, spleen, and even the brain may be affected, and signs will obviously vary according to the extent of the damage and the organ or organs implicated. Host response to the parasite is variable, but a severe granulomatous response is the usual finding with a large number of epitheloid cells, macrophages, and occasional giant cells. In the early stages, cells of the inflammatory series are seen in large numbers. The granulomata usually have a well-developed capsule of connective tissue, and occasionally spores are found surrounded only by a capsule of fibrous tissue.

The isolation of the fungus is briefly described below:

The organism may be cultured in Sabouraud dextrose agar slants with 1% bovine serum added (Sindermann and Scattergood, 1954). Growth is abundant in 7−10 days at an optimum temperature of 10 °C.

The prophylaxis and treatment, briefly, could be as follows:

A method of treatment of aquarium fishes using Phenoxethol is quoted in Reichenbach-Klinke and Elkan (1965), but this is only thought to be effective in the early stages of the disease.

Branchiomycosis

This disease, otherwise known as gill rot, is characterized by areas of infarctive necrosis in the gill due to intravascular growth of *Branchiomyces* spp. of fungi. Two species of *Branchiomyces* spp. are said to be involved in the disease.

The first record of branchiomycosis was reported by Plehn in 1912. Carps are the most frequently affected fishes. *B. sanguinis* is usually localized in the blood vessels of the gills, and nonseptate branched hyphae have been recorded.

Histologically, the results are hyperplasia, fusion of gill lamellae, and areas of massive necrosis resulting from thrombosis of vessels by fungal hyphae. Affected fish may succumb as rapidly as two days after infection, and morbidity of up to 50% may occur.

Schäperclaus (1954) suggested that the disease was encouraged by waters rich in organic fertilizers, algal blooms, and temperatures exceeding 20 °C.

There is no suggested treatment of the disease. Prevention could only be achieved by strict hygiene, removal of dead fish, and avoidance of overfeeding, especially at high water temperatures. An increase in the water supply could prove beneficial during attacks. Eradication of the disease may be attempted by draining out the affected water followed by liming of the affected ponds.

Aphanomyces

The genus *Aphanomyces* belongs to the family Saprolegniaceae and class Oomycetes. Today, Oomycetes are not always considered to be true fungi, but fungi-like protists. They are now often classed alongside diatoms, brown algae, and xanthophytes within the phylum Heterokonta as part of the third botanical kingdom, the Chromista. They are sometimes called Pseudofungi either as a general term or as a formal taxon (Cavalier-Smith, 1987). Nevertheless, some authors still regard them as "fungi."

The pathogenic *Aphanomyces piscicida* had been isolated from mycotic granulomatosis (MG)−affected fish in Japan. An *Aphanomyces* fungus was subsequently obtained from red spot disease (RSD) outbreaks in Australia in 1989. *Aphanomyces* sp. had also been independently isolated from epizootic ulcerative syndrome (EUS)−affected fishes in Southeast (SE) Asia.

1.3.4 Virology of Teleosts

The study of virus infections in fish has been, and in many instances still is, a poorly developed subject. Only when viral infections cause disease conditions in fishes that hamper the economy, health, and nutrition of humans does information comparable to that available for viral infections in human beings and domesticated animals become available.

1.3.4.1 Nature of Viruses

Viruses are very small infectious agents that multiply only within the living cells of a host. The distinctive characteristics of viruses are their simple structure and their mechanism of replication. These are briefly summarized below:

Small Size

Viruses usually range from 20 to 300 nm, while bacteria generally range from 200 to 2000 nm. The particle size of any one virus type is effectively constant.

Genetic Material (Genome)

This may be DNA or RNA but never both together.

Replication

Viruses cannot multiply on simple media but only within and with the assistance of the components (e.g., ribosomes) of living host cells. Inside the host cell, the virus genome directs the production of new virus. Virus particles do not grow in size, nor do they undergo binary fission.

1.3.4.2 Viral Morphology and Structure

There is a limited range of morphological types, and these are used as a basis for classification. However, as the number of distinctive morphological features is small, viruses with very different biological properties may appear to be morphologically indistinguishable.

Structurally, a "virus particle" or a "virion" consists of a core containing the nucleic acid surrounded by a coat largely or wholly composed of protein. Survival under adverse environmental conditions depends on the toughness of this coat. The coat protein is antigenic and specific for each virus type. The components of a virion, briefly, are as follows:

Capsid: Protein coat
Capsomeres: The protein structural units of which the capsid is composed
Core: Inner structure containing nucleic acid
Envelope: Some viruses are surrounded by a lipoprotein envelope containing viral antigens but also partially derived from the membranes of the host cells.

1. Nature and number of strands of nucleic acid
2. Symmetry of capsid
3. Nature of capsid—naked or enveloped
4. Number of capsomeres, diameters of capsid, and period of helix or helical symmetry

Further, for more diagnostic characterization, the following characters could be useful:

5. The molecular weight of the nucleic acid
6. Proportions of nucleotides in the nucleic acid
7. Number, molecular weight, and antigenicity of viral proteins
8. Liability to chemical and physical agents—e.g., acids, alkalis, solvents, and heat

To date, however, only a few viruses isolated from fish have been characterized in sufficient detail to allow classification according to the scheme used for other animal viruses. Among fish viruses, only rhabdoviruses and channel catfish viruses are sufficiently characterized. Some of the viruses found in fishes are briefly tabulated below:

Classification of Viruses of Teleostean Origin

Virus name and classification	Morphology	Size (nm)	Chemistry	Replication site
IPN unknown	Icosahedral	55–65	RNA, protein	Cytoplasm
VHS rhabdovirus	Bullet-shaped	180 × 60–70	RNA, protein, lipid	Cytoplasm
IHN	—	160 × 90	—	—
CCV complex Herpes virus	Icosahedral	175–200	DNA, protein, lipid	Nucleus
Herpes virus *Herpesvirus salmonis*	-Do-	175	DNA, protein	Nucleus
Lymphocystis unknown	-Do-	130–260	DNA, protein, lipid	Nucleus

Virions show three basic types of shape or symmetry, viz.,

1. Cubic: Here, virions resemble small crystals. More precisely, the shape of the virion is icosahedral. The capsid is constructed by assembling capsomeres according to a rigorous geometrical pattern as in adenoviruses.
2. Helical: Here, virions are long rods, the capsomeres being arranged around a spiral of nucleic acid as in myxoviruses. In addition, a lipoprotein envelope of one or more layers may surround a nucleocapsid of either icosahedral or helical symmetry.
3. Complex: In this case, the basic geometry is not immediately apparent, as in pox viruses.

Classification

As yet, no single scheme of virus classification is entirely satisfactory. Almost all of the following characteristics are generally used for virus classification:

Virus Replication

Interactions between viruses and host cells are not well understood for many viral infections, even today. Exact details vary, particularly according to the nature of the nucleic acid of which the viral genome is composed.

The normal "productive" virus cycle forms the basis of acute infections. It is demonstrated by inoculating a suitably chosen tissue culture monolayer with sufficient virus to ensure that cells are infected; adequate time is allowed for the virus particles to attach to the cells; excess inoculum is removed; nutrient medium is added; and the cultures are incubated at an appropriate temperature.

Initially, there is a period called the "latent period" during which infectivity remains at a low level, representing "residual inoculum." This is followed by a period of exponential increase in infectivity (the "rise period"), and finally a point is reached thereafter when there is no further increase in the yield of virus (the "plateau"). The length of these phases and the size of the final yield depend on the virus, the cells, and the conditions of incubation.

A unique feature of viral growth is that during the early period (the "eclipse"), the virus becomes undetectable, or at least barely detectable, either microscopically or in terms of infectivity. The nature of this phenomenon becomes clear when one remembers that the essence of the virus is contained in its nucleic acid, with the capsid serving merely as protection for this genetic material between generations. New virus is produced not by growth and division of parent organisms, but by assembly from pools of subunits whose synthesis is directed by the infecting genome. This time between the end of the eclipse and the start of the rise period is known as the "maturation phase" and indicates the time necessary for the assembly and release of new virus.

The initial attachment of virus to host cell is not dependent on temperature and is probably based on electrostatic interactions. The subsequent engulfing of virus is often called "viropexis," a process similar to pinocytosis. Infecting virus may often be seen in the cytoplasmic vacuoles, though the entry of membrane-bound viruses (e.g., myxoviruses) may involve fusion of the outer layer of the virus with the cell membrane, with only the viral core (nucleocapsid) being released into the cell.

From this point on, events are geared to the fulfillment of two major requirements:

1. The replication of the viral genome to give daughter copies exactly like the original
2. The provision of viral structural proteins that will form capsids around these daughter genomes

Viruses have in common an inherent ability to synthesize proteins; ribosomes and much of the other apparatus for this process, as well as the relevant amino acids and energy resources, must be supplied by the host. The degree of host dependence, however, is virus-specific, with some agents having a greater capacity than others to provide their own enzymes and to direct autonomous metabolic activities. Those with limited capacity appear to mitigate their deficiency by modifying host cell enzymes.

An account of the different diseases caused by viruses in fish is briefly discussed below:

1.3.4.3 RNA Viral Diseases

Infectious Pancreatic Necrosis (IPN)

M'gonigle (1940) described a disease of salmonid fingerlings in Canada that bears a striking resemblance to the disease that is now called infectious pancreatic necrosis (IPN). There was a controversy about its cause (viral or nutritional) that was resolved when Wolf et al. (1960) isolated a virus and demonstrated it to be the etiological agent of the disease. IPN was first recognized in Europe by Bellet in French hatcheries.

Morphology and Chemical Composition The IPN virion is icosahedral, nonenveloped, and c 70 nm in diameter. The single protein shell encloses an RNA genome consisting of two segments of large-size class double-stranded RNA. The virion contains three size classes of polypeptide (large, medium, and small). Classification of the IPN virus (IPNV) is controversial. It cannot be included in the family of reoviruses because it contains only two RNA segments as against 10 to 15 in other animal viruses with double-stranded RNA genomes.

Antigenic Properties Several serotypes of the virus exist as revealed from the neutralization tests. Of these, four have been identified in North America and two in Europe. Strong cross-reaction between the two American strains and the European strains indicates considerable similarity. However, the sensitivity of the European strains to freezing and thawing distinguishes them from the American strains. Further, the virulence of the strains, as well as their ability to produce cytopathogenic effect (CPE) in tissue cultures, varies. The reference strain held in the American Type Culture Collection (ATCC) is called VR299.

Viral Multiplication and Development Multiplication of the virus occurs in the host cell cytoplasm, where the virus can be observed under electron microscope (EM) as individual particles and sometimes in crystalline arrays.

Laboratory Cultivation and Diagnosis of the Virus The virus grows in a variety of permanent cell lines (rainbow trout gonad (RTG-2), fat head minnow (FHM), and bluegill fry-2 (BF-2)) and has also been reported to grow in several other poikilothermic primary cell cultures. At temperatures of 15—20 °C, CPE is evident from 1 to 20 days postinoculation depending on the size of the inoculum and strain of the virus. The CPE starts as focal areas of necrosis evidenced by the formation of "blisters" of detached "stringy" foci of cells in the monolayer. Cells lose their fusiform shape and develop a granular cytoplasm, and complete destruction of the monolayer follows. Identification of the virus is usually done by the neutralization technique using pooled antiserum, but complement fixation and fluorescent antibody tests have also been described.

Infection and Disease Infection can apparently occur at any age in most salmonid species of fish that have not been previously exposed to virus. In young farmed fish, this results in the majority suffering from a persistent chronic infection throughout their lives. The virus is demonstrable in many organs, most notably the kidney. Clinical disease may not necessarily follow infection. However, where it does, it is largely restricted to first feeding fry up to 20 weeks after first feeding. Field observations suggest that stress such as that resulting from overcrowding, poor environment, or bad husbandry is an important disease-precipitating factor.

Clinical Signs and Gross Pathology Clinically acute or chronic disease condition occurs in very young cultured salmonids. Older fishes are affected only exceptionally. Variations are revealed in the clinical signs of the outbreak. Usually, the largest fry succumb first. Accumulation of an abnormal number of such fry against exit grids is usually the first sign observed. The affected fish are generally darker in color and display a characteristic spiral type of swimming behavior associated with nervous signs, exophthalmos, and abdominal distension.

Histopathology The only consistent histopathological (HP) feature of this disease is the focal or generalized destruction of exocrine pancreatic acinar tissue. The necrotic changes involve rupture of infected cells with release of zymogen granules. The EM shows membrane-bound areas of assembly of virus. Kidney and liver have been reported as occasionally developing necrotic foci. Microscopic lesions have been consistently reported to be found in the intestine.

Mortality Pattern and Antibody The mortality pattern depends partly on husbandry, age, and environmental factors, and partly on the serotype of the virus strain involved. A significant proportion of the fish surviving an acute outbreak could slowly develop degeneration-fibrosis changes in the pancreas that result in poor growth performance.

A neutralizing antibody is elicited in surviving fish, but virus could still be detected, indicating multiplication in sites protected from humoral antibody and presumably cell-mediated immune mechanism. IPNV has been reported to induce interferon production in FHM cells, but no reports of in vitro interferon production in fish are recorded.

Geographical Distribution and Host Range The virus has been isolated from North America, Europe, and Japan. It has been demonstrated from many species of salmonid fish—e.g., brook trout, rainbow trout, brown trout, cutthroat trout, Arctic char, Atlantic salmon, and coho salmon.

Transmission Transmission of IPN could occur vertically, with the virus inside the egg, could occur horizontally via the water, or via fish consuming other moribund fish. The resistance of the virus to the low pH of the stomach probably aids infection via oral route. Nevertheless, the most important reservoir of infection is the infected fish. Piscivorous birds may act as mechanical vectors, although their significance in the epizootiology of the disease is doubtful.

Control There is no means of treating the infected fish to eliminate virus or alleviate the severity of the disease. Avoidance is the only effective control measure. This requires that water supplies are free of infected fish and that all eggs and fish introduced to such waters are similarly free from IPNV. Eradication of virus from infected farms has been successful when all infected or presumed infected fish are slaughtered and holding facilities adequately disinfected, provided that a wild fish source was not present to cause reinfection.

Viral Hemorrhagic Septicemia (VHS) (Egtved Disease)

This disease is said to have been first detected in rainbow trout in 1949 in a village called Egtved in Denmark. However, conclusive evidence regarding its viral etiology was first given by Jensen (1963) by isolating the virus for the first time. The name "Viral Hemorrhagic Septicemia" was recommended by the Office d'Epizootie (OIE) in 1966 in view of confusion over its nomenclature in various countries.

Morphology and Chemical Composition This virus displays typical rhabdovirus morphology, the bullet-shaped particles measuring $180 \times 60-70$ nm with one end rounded and the other flat. The virion has a central hollow axis surrounded by a spiral core. In turn, it is surrounded by a lipoprotein envelope. The viral genome is composed of a single-stranded RNA.

Antigenic Properties Neutralization tests distinguish only two serotypes of VHS virus. The reference strain is called F1. There is said to be no difference in pathogenicity between the two serotypes (Vestergaard-Jørgensen, 1974).

Viral Multiplication and Development Virus matures at the cell surface, budding off from the cell membrane. Cytoplasmic inclusion bodies can be observed, consisting of helices that may represent viral nucleoprotein. Viral antigens could only be detected in the cytoplasm, indicating this to be the site of virus replication.

Laboratory Culture and Diagnosis The virus grows in both the RTG-2 and FHM cell lines. In RTG-2 infected cultures, the cells swell and become spherical and granular before detaching from the glass. Virus infection of tissue culture cells and subsequent CPE occurs only within a narrow pH range of 7.6–7.8. The virus develops at a temperature range of 6–18 °C, the optimum being 14 °C, where CPE may occur within 48–72 h. The virus is usually identified by the neutralization technique. However, the fluorescent antibody technique has also been described.

Infection and Disease Infection of previously noninfected rainbow trout results in acute disease within 7–15 days under experimental conditions. However, infection may occur up to one year later in the case of a whole population exposed to virus. "Temperature effect" is generally attributed as the cause of delay in the clinical appearance of the disease. The outbreaks usually occur below 15 °C. This indicates that in trouts, this disease is quite common during winter and ceases to occur as the temperature rises toward summer. The virus could be detected in trouts that are about to suffer, in those that reveal clinical symptoms, or in those just recovering.

Clinical Signs and Gross Pathology The clinical development of an outbreak in a previously disease-free population of rainbow trout usually follows a triphasic pattern (Ghittino, 1965). The first phase is characterized by high mortality, with affected fish becoming dark and lethargic, and showing hemorrhages at the base of the fins and in the gills. The second phase is one of chronic disease, where the predominant features are very black coloration of the fish body and anemia. The gills become very pale with exophthalmos as a common feature. The third phase is associated with the cessation of mortality, but affected fish display nervous signs manifested by a looping swimming behavior. There is also swelling and discoloration of the kidney. Virus is not generally isolated at this phase.

Histopathology Microscopic features are those of an endotheliotropic virus disease. Hemorrhages may be found in any tissue but could obviously be more frequent in highly vascular tissues and are associated with hemopoietic necrosis. The anemia of the chronic disease results in heavy deposition of hemosiderin in the melanomacrophage areas.

The VHS virus is apparently only weakly antigenic in trouts, and recovered fish often show no detectable humoral antibody response. In trouts, the virus is said to invoke an interferon response that protects passively immunized fish from VHS infection and also protects tissue cultures from infection with IPNV.

Geographical Distribution and Host Range The virus has been isolated from disease outbreaks in various countries like Denmark, Germany, France, Italy, Switzerland, and Scandinavia, but it has perhaps never been isolated from the British Isles. It is also said to be unknown in the Americans continent and the Far East.

Transmission Carriers are said to be significant reservoir of infection as extensive circumstantial evidence suggests a persistent latent infection. Eggs from infected fish are said to initially carry the virus on their surface, but the virus is lost after several days of incubation in running water. Water thus could act as a "mechanical vector" because of significant maintenance of infectivity by virus in water.

Control There is said to be no means as yet of treating infected fish to remove the virus or alleviate the severity of the disease. The only effective control measure is perhaps to ensure that water supplies are free of virus and that all fish and eggs introduced are free from VHS virus. Eradication of virus from infected trout farms could be successful only when all the infected fishes are slaughtered, the water bodies are disinfected, and other quality control measures are adopted.

Infectious Hemopoietic Necrosis (IHN)

A disease causing serious losses in sockeye salmon was shown by Rucker et al. (1953) to have a viral etiology. Similarly, a disease of Chinook salmon had recurred from when hatcheries on the Sacramento River were first established in 1941. This was shown by Ross et al. (1960) to also have a viral etiology. Although clinically and pathologically similar, the cross-infectivity between the two viruses was low, and the two were considered to be different until 1967. Then another virus, similar in many respects but low in cross-species infectivity, was isolated from rainbow trout (Amend et al., 1969). Subsequent investigations (Amend and Chambers, 1970) have shown that the three viruses are morphologically and biochemically similar. However, each has a specificity for the species from which it was isolated. All three are now considered strains of one virus type. However, in deference to its pathology, it is called infectious hematopoietic necrosis (IHN) virus.

Morphology and Chemical Composition The virus shows rhabdovirus morphology, appearing as a bullet-shaped particle measuring c 160×95 nm with one end rounded and the other end flat. The virion has an axial pore 20 nm in diameter, a core 60 nm in diameter, and an outer lipoprotein coat 15 nm thick with fine spicules showing on the surface. The viral genome is composed of single-stranded RNA.

Antigenic Properties No antigenic difference has been detected by serological methods between any of the viruses isolated from the three species of fish.

Viral Multiplication and Development The virus apparently matures at the cell surface budding from the plasma membrane. Assembly occurs in the cytoplasm.

Laboratory Culture and Diagnosis The virus replicates and causes CPE in Chinook and sockeye salmon, steelhead trout, and FHM cell cultures. Most but not all isolates grow and produce a characteristic CPE in RTG-2 cells, which round up and form clusters. Growth occurs over the range 4—18 °C, the optimum temperature being 12—15 °C. The IHN virus is usually identified in tissue culture by serum neutralization test.

Infection and Disease Infection occurs in previously noninfected fish and results in a persistent infection when the virus is normally latent and undetectable, but is isolable from sexually mature fish, particularly postspawning. Disease could occur whenever water temperatures are suitable and the virus is present.

Clinical Signs and Gross Pathology The clinical features include lethargy and sporadic hyperactivity, dark coloration, exophthalmos, anemia, abdominal distension, and hemorrhages at the base of fins.

Histopathology (HP) The main HP feature of this disease is focal hemorrhage and degeneration of all cell types in the hemopoietic tissue. The early cytological changes, characterized by swelling and nuclear pyknosis, are soon overtaken by necrosis and dissolution of wide areas of hemopoietic tissue. In addition, the eosinophilic granule cells (EGC) of the submucosa of the intestine are destroyed.

Geographical Distribution and Host Range The virus is endemic in wild salmonid fishes of the Pacific Coast of North America. Disease in hatcheries is common and Williams and Amend (1976) showed that it could cause epizootics in wild sockeye salmon. The disease has been found to occur in Chinook and sockeye salmon and in rainbow trout. Coho salmon has been reported to be resistant to this disease. They are said to act as natural reservoirs of infection.

Transmission This virus could be transmitted by direct contact via water from infected fish and by feeding infected carcasses.

Control There is perhaps no method of eliminating virus from infected or diseased fish. Acute disease in infected stock could be prevented by elevating the water temperature above 15 °C, but the infection persists and may occur as soon as the water temperature drops. The only effective measure could be to keep the stocks free of virus and to ensure that water supplies are not contaminated.

Spring Viremia of Carp (SVC)

The etiology of the disease complex of carp known as infectious dropsy has been the subject of considerable dispute for many years. The first rationalization of the confused clinical picture was achieved when Fijan et al. (1971) had

isolated a virus. Fijan subsequently separated the dropsical syndrome, as he (Fijan, 1972) recognized it, into (1) spring viremia of carp (SVC), a condition caused by the virus *Rhabdovirus carpio*, and (2) CE, a condition said to be of putative bacterial origin.

Pike Fry Virus Disease

A new rhabdovirus was isolated by Kinkelin et al. (1973) from an outbreak of disease in juvenile northern pike in the Netherlands. Infection experiments (Bootsma et al., 1975) have shown that this virus is the agent of the condition known as hydrocephalus or red disease (Bootsma, 1971).

1.3.4.4 DNA Viral Diseases
Channel Catfish Virus (CCV) Disease

This virus was isolated by Fijan (1968) and found to be a herpes virus (Wolf and Darlington, 1971).

Morphology, Chemical Composition, and Development of the Virus This virus is c 175—200 nm in diameter overall, but inside an outer lipoprotein envelope there is a symmetrical icosahedral structure of c 100 nm made up of 162 capsomeres. The genome is composed of DNA. The virus is produced in the cell nucleus and enveloped as it passes through the nuclear and cytoplasmic membranes.

Laboratory Culture and Diagnosis This virus grows well in the brown bullhead (BB) cell line at 29 °C, giving rise to syncytia. Plaques are formed when CCV-infected BB monolayers are overlaid with agar. The virus is characterized by serum neutralization.

Infection and Disease The disease occurs in fry and fingerlings during the summer months when water temperatures rise to 25—30 °C. Fishes who survive the disease are suspected of harboring a persistent latent infection.

Clinical Signs and Gross Pathology Clinical signs include loss of equilibrium, spiral swimming movements, and hanging vertically in the water. The main gross pathological features are hemorrhages of gills, skin, and internal organs, as well as abdominal distension.

Histopathology The lesions appear to begin in the posterior kidney with an increase in the number of lymphoid cells, and proximal renal tubular necrosis. Extensive necrosis of the renal hemopoietic tissue also occurs, and EM study has shown that there are a large number of viruses within the renal cells (Wolf et al., 1972). Subsequently, focal necrotic lesions develop in the liver, spleen, and digestive tract. The pancreatic acinar cells generally remain unaffected, quite in contrast to almost all other generalized fish viral diseases (Plumb and Gaines, 1975).

Both artificially and naturally infected adult catfish produce neutralizing antibodies, the former within 10–15 days of infection (Plumb, 1973). However, virus may not be detected subsequently.

Geographical Distribution and Affected Species This virus has been reported only in the southern United States in the cultivated channel catfish. Many ictalurids may be reservoirs of infection in view of the ready growth of the virus in a continuous cell line (BB) of another ictalurid fish.

Transmission The virus is shed by diseased fish. Persistent infection, though suspected, is yet to be proven. Diseased fish, those surviving disease, and those suspected to have been in contact with the virus are the major vectors of spread, but the portal of entry is unknown.

Control There is yet no means of treating infected fish. As such, control by avoidance is perhaps the only possible control measure. Eggs should be derived from brood fish free of virus, as the disease principally occurs in the young fish. Further, virus-free water should be used for growing beyond the fingerling stage.

Lymphocystis Disease

This viral disease results in chronic verrucose skin lesions and occurs in many species of wild and cultured marine and FW fishes. Lowe (1874) made the earliest record of the condition in flounders.

Wessenberg (1914), on the basis of transmission studies, was the first to postulate that the cause was a virus. However, it was Walker (1962) who had described the virus under EM, followed by its isolation by Wolf in 1966 when its viral etiology was confirmed beyond any dispute.

Morphology and Chemical Structure The virion is hexagonal in outline with no recognizable capsomeres. However, its form had been described as icosahedral. Depending on the host, the size of the virus differs, varying from 130 to 260 nm. The nucleoid core is a tangled mass of DNA threads, and there is a recognizable capsid. It is likely that there could be a lipid component, as lipid solvents rapidly destroy infectivity. No information is available on antigenic variation between lymphocystis disease virus isolates.

Diseases of Putative Viral Etiology

Papillomatosis (Cauliflower Disease of Eels) This disease was first described in 1910 among eels from the Baltic coast. Subsequently, it had been reported elsewhere in the Baltic, around Denmark, and in the coastal waters of Holland and Belgium where the incidence had increased dramatically over the last few decades. A virus of polyhedral symmetry with 52–56 nm diameter had been isolated from the blood of affected eels by Pfitzner (1969). Later, Koops et al. (1970) had isolated it in RTG-2 and FHM cells. It had been reported to be thermolabile and is generally known to lose infectivity rapidly in water. It is believed to replicate in the nucleus, and is therefore regarded as a DNA virus.

Carp Pox It was described as a disease among the cultivated carps about 400 years ago, but its probable viral etiology was demonstrated around 1952 (Schubert, 1966). Further, the suspected virus had not been isolated. However, electron micrographs show areas of chromatin within the nucleus, but differ from the nucleolus in which a viral-like genome is apparently assembled and bound in a single-layered protein-like coat. The virion becomes membrane-bound as it passes out of the nucleus into the cytoplasm. This process of replication and the general morphology and size of the particle suggests that carp pox virus belongs to the herpes virus group.

Piscine Erythrocytic Necrosis (PEN) Erythrocytic inclusions in poikilothermic vertebrates observed under light microscope had originally been believed to be protozoan parasitic infections. Subsequent EM studies had displayed that some of these inclusions are viral (Johnston and Davies, 1973).

Cytoplasmic inclusions of infected cod erythrocytes are made up of icosahedral particles adjacent to a viroplasm (a DNA center where viral components are synthesized and assembled). The viroplasm is a granular-fibrillar spheroid that may contain other structures. The virion (average diameter 330 nm) has a well-defined outer envelope and a wrinkled inner envelope enclosing an electron-dense core 230 nm in diameter. At present, PEN is classified loosely as an icosahedral cytoplasmic deoxyribonucleic acid virus.

There is no direct evidence that cod PEN virus causes disease. Heavily affected fish appear quite normal. Only mature erythrocytes are affected, and as the infection progresses, the proportion of infected erythrocytes rises as high as 99%. Herring PEN virus infects both immature and mature erythrocytes, causing degeneration of the cell, the largest inclusions being found in the youngest erythrocytes. There is little information available on the transmission of PEN virus.

Pandey et al. (2012) (Sharma, M., Shrivastav, A.B. Sahni, Y.P., Pandey, G., 2012. Overviews of the treatment and control of common fish diseases. International Research Journal of Pharmacy, 3 (7), 123–127) gave a vivid overview of the treatment and control of common fish diseases. They had stated that a fish suffers from different kinds of diseases like other animals. Diseases kill fishes including the juveniles. Fish pathogens could include bacteria, fungi, viruses, protozoans, water molds, and so on.

Fishes are also exposed to different kinds of pollutants in water. Some of the common diseases among FW fishes are columnaries, gill disease, ich, dropsy, tail and fin rot, popeye, cloudy eye, swim bladder disease, lice and nematode infestations, water quality-induced diseases, leech-induced ailments, etc. Antibiotics are frequently used to control bacterial fish diseases. This could result, however, in the formation of antibiotic-resistant strains among the bacteria. Concomitantly, nonspecific immune functions such as bacteriolytic activity and leucocytes' functions in fish could be improved by some herbs. In fact, medicinal plants have been in use since time immemorial to regulate different kinds of diseases such as bacterial and fungal among fishes.

1.4 EPIZOOTIC ULCERATIVE SYNDROME (EUS)

1.4.1 Background of EUS

1.4.1.1 Introduction

> *Disease, per se, is not an entity of an end in itself. Disease is the end result of an interaction between a noxious stimulus and biological system. In fact, it is necessary to understand all aspects of the biology of the species in order to understand a disease. The interdisciplinary approach, which has been currently characterizing so many branches of scientific research, is also in evidence in the field of diseases of fish.*
> *Mawdesley-Thomas (1972).*

Fishes are poikilothermic aquatic animals. Their existence and performance are regulated by the quality of the environment in which they live. As such, the success of fish rearing in both culture and natural water bodies depends on their quality and management as well as an understanding of the biology of fishes. The conditions in large water bodies such as seas and oceans are usually said to be quite stable and uniform. Conversely, the conditions in smaller bodies of water—for example, the hatchery ponds and aquaria in which the fishes are generally raised artificially—are generally influenced by a variety of conditions. However, all species of fishes perform the best under certain optimal conditions. The amplitude of such conditions is usually quite narrow. Understanding the relationships among the fishes, their environments, the pathogens, and the parasites is a prerequisite to achieving success in pisciculture, and is also the basis for rational management.

Severe outbreaks of disease may result from the introduction of parasites, pathogens, malnutrition, chemical and physical alterations of the water, the genetic makeup of the fish, and the interrelationships of any or all these factors. This could be presented graphically as a set of circles representing the fishes, pathogens, or parasites, and the environment. If the circles intersect, the conditions are favorable for an outbreak of disease. This relationship could also be expressed in the form of an equation as follows:

$$H + P + S^2 = D$$

where H represents the "Host"; P the "Pathogen"; S the "Stress" caused by the environment; and D the resulting "Disease."

1.4.1.2 Interaction between Disease and Environment

It may be noted here that the stress caused by changes in the environment may increase very rapidly after the acceptable range for a particular species of fish is exceeded. As such, the effect of such a stress could be expressed as a quantity that increases exponentially, while the concomitant changes in the environment could be represented arithmetically—for example, temperature by degrees; dissolved oxygen (DO) by mg/L, and so on. The stress created by such changes and the resulting mortalities are likely to increase geometrically or logarithmically.

Notwithstanding the above, intensive culture of aquatic organisms often encounters disease problems that pose a major threat to the aquaculture industry.

Interest in fish farming continues to grow throughout SE Asia, where extensive areas potentially suitable for aquaculture remain unexplored. On the other hand, intensive fish culture calls for high stocking densities, nutritious supplementary feeds, control of water quality and systematic management. Though there has been a growing interest in polyculture (simultaneous culture of several noncompetitive ichthyospecies), monoculture is also a common practice. As such, a high concentration of a single species encourages the spread of infection and infestation. Performance among nations for relatively few attractive species has stimulated international trade in fish seed with the concomitant hazard of spreading infectious diseases.

Fish forms a dependable source of animal protein for people of developing countries such as India, Bangladesh, Sri Lanka, and other countries in South and SE Asia. Large-scale fish mortality often occurs due to environmental stress followed by parasitic afflictions. Fungi, bacteria, and certain viruses are often found to be associated with such diseases, which often become epizootic and epidemic, resulting in cutaneous ulcers followed by hemorrhagic inflamed areas. Under such conditions, large-scale mortality occurs among the fishes, causing a tremendous amount of loss to the nation.

Cutaneous ulcerative diseases, often involving a number of pathogens, are a common problem in wild and cultured fish. For the last two decades, a group of epizootic syndromes, all involving a severe ulcerative mycosis, have been reported from Australia (Rodgers and Burke, 1981), the southeastern United States (Noga and Dykstra, 1986), and Asia, stretching from the Philippines (Reantaso, 1990) in the East to India (Kar and Dey, 1988a,b,c,d, 1990a,b,c,1989; Kar, 2013; Das et al., 1990) in the West.

In addition to the above, it could be said that for over 25 years, outbreaks of an ulcerative disease characterized histologically by mycotic granulomas have affected freshwater and estuarine fishes over much of Asia and Australia. The disease has been given various names, but is most commonly known as mycotic granulomatosis (MG) in Japan, red spot disease (RSD) in Australia and epizootic ulcerative syndrome (EUS) in SE and South Asia. MG, RSD, and EUS have in the past been described separately as distinct conditions. However, recent studies of some workers have shown that the same pathogenic *Aphanomyces* fungus is involved in each case, although there are other workers who subscribe to its primary viral etiology. However, it is now apparent that an account of the history of EUS would remain incomplete without consideration of outbreaks in Japan and Australia.

An account of each of the three epidemic conditions is given below:

Mycotic Granulomatosis (MG)

The first report of an EUS-like condition came in the summer of 1971 in farmed ayu (*Plecoglossus altivelis*) in Oita Prefecture, Japan (Egusa and Masuda, 1971). The characteristic lesion, a granulomatous response to invasive hyphae, was described and the disease was named mycotic granulomatosis (MG) (Miyazaki and Egusa, 1972). It had rapidly spread to several other prefectures and had affected various fish species, predominantly the cultured ayu and the goldfish (*Carassius carassius auratus*), the Formosan snakehead (*Channa maculate*), and the gray mullet (*Mugil cephalus)* (Miyazaki and Egusa, 1972). However, the common carp (*Cyprinus carpio*) was not affected. Hatai et al. (1977) had isolated the invasive oomycete fungus from the affected fish and subsequently called it *Aphanomyces piscicida* (Hatai, 1980). *Aphanomyces piscicida* is now known to be conspecific with the EUS pathogen *Aphanomyces invadens* (Lilley et al., 1997a). Serious MG epidemics had not been reported in Japan since 1973. However, outbreaks of various dimensions (not serious though) have been occurring time and again. Recently, Hatai et al. (1994) have reported a similar disease in imports of ornamental dwarf gourami (*Colisa lalia*) from Singapore. Fungus belonging to genus *Aphanomyces* also have been isolated from such consignments (Lilley et al., 1997b).

Red Spot Disease (RSD)

In 1972, outbreaks of a cutaneous ulcerative condition called red spot disease (RSD) affected estuarine fishes,

particularly the gray mullet (*M. cephalus*), in Queensland, Australia (McKenzie and Hall, 1976). The disease later progressed to affect the FW and estuarine fishes in coastal rivers in New South Wales (NSW) (Callinan et al., 1989), Northern Territory (Pearce, 1990), and Western Australia (Callinan, 1994a).

A fungus of the genus *Aphanomyces* was isolated from diseased fish by Fraser et al. (1992) and was shown to reproduce the disease in fish using bath challenges, but only when the skin of the experimental fish was artificially abraded (Callinan, 1994b). Therefore, certain other factors were considered to be involved in the disease outbreak process. Later, Virgona (1992) showed that RSD outbreaks in the estuarine fish in the Clarence River, NSW were associated with lower catchment rainfall. Subsequently, Callinan et al. (1995a) related this to runoff from acid sulfate soils. Ultrastructural examination of fish gills and skin showed that the low pH and increased concentrations of monomeric aluminum (which are indications of estuarine acidifications) might induce significant lesion formation in fishes (Sammut et al., 1996). However, in aquarium trials, RSD was subsequently induced in fishes exposed to *Aphanomyces* spores sublethally to artificially acidified water at a pH value of both 3 and 5, as well as at low concentrations of monomeric aluminum (Callinan et al., 1996; Callinan, 1997). As with *Aphanomyces piscicida,* the pathogenic RSD-*Aphanomyces* has been shown to be the same species as the EUS pathogen *Aphanomyces invadans* (Callinan et al., 1995a; Lilley et al., 1997a,b).

Epizootic Ulcerative Syndrome (EUS)

The origin of EUS could be said to be a matter of controversy and speculation.

Following reports of outbreak of MG in Japan during the 1970s, EUS was reported in Australia and in SE and South Asia during subsequent decades (OIE, 1997).

Having considered the mycotic genus *Aphanomyces* as the root cause of the initiation of EUS, the question arises, where did the fungus originate? Probably in a tropical or subtropical region in South America. Africa could be the other main possibility. But the fact that the African catfish, which is widely farmed in Asia, is highly susceptible to EUS, suggests that EUS most probably did not originate from Africa.

Nevertheless, *Aphanomyces* fungus probably had entered Australia through the import of ornamental aquarium fishes. There is bulk movement of ornamental and cultured fish around the Australian region in connection with the ornamental fish trade.

Notwithstanding the above, pathogens belonging to all three groups, viz. bacteria, fungus, and virus, have been found to be involved in the pathogenesis of EUS. However, the exact etiology primarily responsible for the initiation of EUS has not yet been correctly determined.

REFERENCES

Agersborg, H.P.K., 1933. Salient problems in the artificial rearing of salmonid fishes, with special reference to intestinal fungisitosis and the cause of white spot disease. Trans. Am. Fish. Soc. 63, 240–250.

Alikunhi, K.H., 1957. Fish culture in India. F_m Bull. Indian Coun. Agric. Res. 20, 144.

Amend, D.F., Yasuteake, W.T., Mead, R.W., 1969. A haemopoietic virus disease of rainbow trout and sockeye salmon. Trans. Am. Fish. Soc. 98, 796–804.

Amend, D.F., Chambers, V.C., 1970. Morphology of certain viruses of salmonid fishes, I. *In vitro* studies of some viruses causing haemato-poietic necrosis. J. Fish. Res. Bd. Can. 27, 1285–1295.

Basu, S.P., 1950. Myxosporidian parasites of Indian carps, *Catla catla (Ham.), Labeo rohita (Ham.), Cirrhina mrigal (Ham.)* and their seasonal variations. Proc. Indian Sci. Congr. 37, 237.

Boerma, G.H., 1970. Additional notes on *Phoma herbarum*. Persoonia 5, 15–48.

Bootsma, R., 1971. Hydrocephalus and red disease in pike-fry (*Esox lucius* L.). J. Fish. Biol. 3, 417–419.

Bootsma, R., dekinkelin, P., Le Perre, M., 1975. Transmission experiments with pike fry (*Esox lucius* L.) rhabdovirus. J. Fish. Biol. 7, 269–276.

Brett, J.R., 1958. Implications and assessment of environmental stress. In: Larkin, P.A. (Ed.), The investigations of fish power problems, HH MacMillian Lecture Series in Fisheries. University of British Columbia, Vancouver, pp. 69–83.

Callinan, R.B., Fraser, G.C., Virgona, J.L., 1989. Pathology of red spot disease in sea mullet, *Mugil cephalus* L. from Eastern Australia. J. Fish Dis. 12, 467–479.

Callinan, R.B., 1994a. A comparative review of *Aphanomyces* sp. associated with epizootic ulcerative syndrome, red spot disease and mycotic granulomatosis. In: Roberts, R.J., Campbell, B., MacRae, I.H. (Eds.), Proc. ODA Regional Seminar on Epizootic Ulcerative Syndrome, 25–27 January, 1994. Aquatic Animal Health Research Institute, Bangkok, pp. 248–252.

Callinan, R.B., 1994b. Red spot disease-epizootic ulcerative syndrome in Australia. In: Roberts, R.J., Campbell, B., MacRae, I.H. (Eds.), Proc. ODA Regional Seminar on Epizootic Ulcerative Syndrome, 25–27 January, 1994. Aquatic Animal Health Research Institute, Bangkok, pp. 82–88.

Callinan, R.B., Paclibare, J.O., Bondad-Reantaso, M.G., Chin, J.C., Gogolewski, R.P., 1995. *Aphanomyces* sp. associated with epizootic ulcerative syndrome (EUS) in the Philippines and red spot disease (RSD) in Australia: preliminary comparative studies. In: Symposium on Diseases in Asian Aquaculture, 26–29 November 1990, Bali, Indonesia. Asian Fisheries Soc., Manila (The Philippines), pp. 233–238.

Callinan, R.B., Sammut, J., Fraser, G.C., 1996. Epizootic ulcerative syndrome (Red Spot Disease) in estuarine fish: confirmation that exposure to acid sulphate soil runoff and an invasive fungus, *Aphanomyces* sp., are causative factors. In: Proceedings of the National Conference on Acid Sulphate Soils. Robert J. Smith and Associates and ASSMAC, Australia.

Callinan, R.B., 1997. Pathogenesis of Red Spot Disease (Epizootic Ulcerative Syndrome) in Estuarine Fish in Eastern Australia and the Philippines. University of Queensland, Australia, pp. 232.

CAMP, 1998. In: Report of the Workshop on 'Conservation, Assessment and Management Plan (CAMP) for Freshwater Fishes of India'. Zoo Outreach Organization and NBFGR, pp. 156, 22–26 September, 1997 (Lucknow).

Carmichael, J.W., 1966. Cerebral mycetoma of trout due to a philophora-like fungus. Sabouraudia 6, 120–123.

Caullery, M., Mesnil, F., 1905. Sur les Haplosporidies parasites de poisons marins. C.r. Se'anc. Soc. Biol 58, 610–612.

Chakraborty, M.M., 1939. Studies on myxosporidia from the fishes of Bengal with a note on the myxosporidian infection in aquaria fishes. Arch. Protistenk 92, 169–178.

Chauhan, B.S., 1953. Studies on the trematode fauna of India: Parts 1–4. Rec. Indian Mus. 51 (2), 113–393.

Chauhan, B.S., Ramakrishna, G., 1958. On the occurrence of fish mortality due to helminthic infestation by cestode cysts in a stocking tank at Nagpur (India). Indian J. Helminth. 10 (1), 53–55.

Cavalier-Smith, T., 1987. The origin of fungi and pseudo-fungi. In: Rayner, A.D.M., Brasier, C.M., Moore, D. (Eds.), Evolutionary Biology of the Fungi. Cambridge University Press, UK, pp. 339–353.

Darlington Jr., P.J., 1957. Zoogeography: The Geographical Distribution of Animals. John Wiley and Sons, New York, pp. 675.

Dey, S.C., 1973. Studies on the Distribution and Taxonomy of the Ichthyofauna of the Hill Streams of Kamrup-Khasi-Garo Regions of Assam with Special Reference to the Functional Morphology of Some Rheophillic Fishes (D.Sc. Thesis, xi + 299). University of Calcutta, India.

Doty, M.S., Slater, D.W., 1946. A new species of *Heterosporium* patho-genic on young Chinook salmon. Am. Midl. Nat. 36, 663–665.

Das, M.K., Pal, R.N., Ghosh, A.K., Das, R.K., Joshi, H.C., Mukhopadhyaya, M.K., Hajra, A., 1990. Epizootic ulcerative syn-drome: a comparative account. In: Proc. National Workshop on Ulcerative Disease Syndrome in Fish, 6–7 March, 1990 (Calcutta).

Egusa, S., Masuda, N., 1971. A new fungal disease of *Plecoglossus altivelis*. Fish. Pathol. 6, 41–46.

Fijan, N.N., 1969. Systematic mycosis in channel cat fish. Bull. Wildl. Dis. Ass. 5, 109–110.

Fijan, N.N., 1972. Infectious dropsy in carps: a disease complex. In: Mawdesley-Thomas, L.E. (Ed.), Diseases of Fish. Proc. Symp. No. 30, Zoological Society, London, May, 1971. Academic Press and The Zoological Society, London, pp. 39–51.

Fijan, N.N., Petrinec, Z., Sulimsnovic, D., Zwillenberg, L.O., 1971. Isolation of viral causative agent from the acute form of infectious dropsy in carp. Vet. Arch. 41, 125–135.

Fijan, N.N., 1968. The survival of *Chondrococcus columnaris* in the waters of different quality. Bull. Off. Int. Epizoot. 69, 1159–1166.

Fraser, G.C., Callinan, R.B., Calder, L.M., 1992. *Aphanomyces* species associated with red spot disease: an ulcerative disease of estuarine fish of Eastern Australia. J. Fish Dis. 15, 173–181.

Ghittino, P., 1965. Viral haemorrhagic septicemia (VHS) in rainbow trout in Italy. Am. N.Y. Acad. Sci. U.S.A. 126, 468–478.

Gopalakrishnan, V., 1963. Controlling pests and diseases of cultured fishes. Indian Livestk 1 (1), 51–54.

Gopalakrishnan, V., 1964. Recent developments in the prevention and control of parasites of fishes cultured in Indian waters. Proc. Zool. Soc. Calcutta 17, 95–100.

Gopalakrishnan, V., 1968. Diseases and parasites of fishes in warm-water ponds of Asia and the Far East. FAO Fish. Rep. 5 (44), 319–344.

Gopalakrishnan, V., Gupta, P.D., 1960. An eye disease which causes mortality of the *Catla catla* (Ham-Buch). Curr. Sci. 29 (6), 240.

Gupta, S.P., 1961. A reference list of trematode parasites of freshwater fishes of India with a discussion on their systematic position. Indian J. Helminth. 13 (1), 35–60.

Hasan, R., Qasim, S.Z., 1962. *Trypanososma punctati* n.sp. from the *Ophicephalus punctatus* Bloch, common freshwater murrel of India. Z. Parasitenk 22 (2), 118–122.

Hatai, K., 1980. Studies on the Pathogenic Agents of Saprolegniansis in Freshwater Fishes. Special Report of Nagasaki Prefectural Inst. of Fisheries, No. 8, Matsugae-Cho, Nagasaki, Japan.

Hatai, K., 1994. Mycotic granulomatosis in ayu (*Plecoglossus altivelis*) due to *Aphanomyces piscicida*. In: Roberts, R.J., Campbell, B., MacRae, I.H. (Eds.), Proc. ODA Regional Sem. on Epizootic Ulcerative Syndrome, 25–27 January 1994. Aquatic Animal Health Research Institute, Bangkok.

Hatai, K., Egusa, S., Takashashi, S., Ooe, K., 1977. Study on the pathogenic fungus of mycotic granulomatosis: I: Isolation and pathogenicity of the fungus from cultured ayu infected with the disease. Fish. Pathol. 11 (2), 129–133.

Hatai, K., Nakamura, K.,, Rha, S.A., Yusa, K., Wada, S., 1994. *Aphanmyces* infection in dwarf gourami (*Colisa lalia*). Fish. Pathol. 29 (2), 95–99.

Hofer, B., 1904. Handbuch der Fischrankhereiten. Munich: Allg. Fischerie-Zeitung.

Hora, S.L., Pillay, T.V.R., 1962. Handbook on fish culture in the Indo-Pacific region. FAO Fish. Biol. Tech. Pap. 14, 204.

Hoshima, T., Sano, T., Sunamayo, M., 1960. Studies on the saprolegniasis of eel. J. Tokyo Univ. Fish. 47, 59–79.

Huxley, T.H., 1882. *Saprolegnia* in relation to salmon disease. Q. Fl. Microsc. Sci. 22, 311.

Jain, S.L., 1959. Some observations on the monogenetic trematodes from the gill filaments of some Indian freshwater Fishes. Curr. Sci. 28 (8), 332–333.

Jayaram, K.C., 1981. The Freshwater Fishes of India, Pakistan, Bangladesh, Burma, Sri Lanka: A Handbook. Zoological Survey of India, Calcutta pp. xxii + 475.

Jayaram, K.C., 1999. The Freshwater Fishes of the Indian Region, xvii + 551. Narendra Publishing House, Delhi.

Jayaram, K.C., 2003. Ecostatus and conservation strategies for Mahseer fishes of India with special reference to Deccan species, pp. 3–12. In: Kar, D., Dey, S.C., Datta, N.C. (Eds.), Welfare Biology in the New Millennium. Allied Publishers Pvt. Ltd, Bangalore, pp. xx + 97.

Jayaram, K.C., 2010. The Freshwater Fishes of the Indian Region. Narendra Publishing House, New Delhi pp. xxiv + 614.

Jensen, M.H., 1963. Preparation of fish tissue culture for virus research. Bull. Int. Epizoot. 59, 131–134.

Jhingran, V.G., Pullin, R.S.V., 1988. A Hatchery Manual for Common, Chinese and Indian Major Carps. ICLARM Studies and Reviews II, Manila, Philippines.

Johnston, M.R.L., Davies, A.J., 1973. A pirhemocyton-like parasite of the blenny, *Blennius pholis* L. (Teleostei: Blenniidae) and its relationship to immuno-plasma Neumann, 1909. Int. J. Parasit. 3, 235–241.

Kar, D., 1990. Limnology and fisheries of Lake Sone in the Cachar district of Assam (India). Matsya 15–16, 209–213.

Kar, D., 2000a. Present status of fish biodiversity in south Assam and Tripura, pp. 80–82. In: Ponniah, A.G., Sarkar, U.K. (Eds.), Fish Biodiversity of North-East India. NBFGR-NATP Publication No. 2, Lucknow, pp. 228.

Kar, D., 2003. Fishes of barak drainage, Mizoram and Tripura, pp. 203–211. In: Kumar, A., Bohra, C., Singh, L.K. (Eds.), Environment, Pollution and Management. APH Publishing Corporation, New Delhi. pp. xii + 604.

Kar, D., 2013. Wetlands and Lakes of the World. pp. xxx + 687. Springer, London, ISBN 978-81-322-1022-1. e-Book ISBN: 978-81-322-1923-8.

Kar, D., Dey, S., 1988a. Preliminary Electron Microscopic studies on Diseased fish tissues from Barak valley of Assam. In: Proc. Annual Conference of Electron Microscopic Society of India, vol. 18, p. 88.

Kar, D., Dey, S.C., 1988b. A critical account of the Recent Fish Disease in the Barak valley of Assam. In: Proc. Regional Symp. on Recent Outbreak of Fish Diseases in North-East India, vol. 1, p. 8.

Kar, D., Dey, S.C., 1988c. Impact of recent fish epidemics on the fishing communities of Cachar district of Assam. In: Proc. Regional Symp. on Recent Outbreak of Fish Diseases in North-East India, vol. 1, p. 8.

Kar, D., Dey, S.C., 1988d. Epizootic ulcerative syndrome in fishes of Assam. J. Assam Sci. Soc. 32 (2), 29–31.

Kar, D., Dey, S.C., 1989. Fish disease syndrome: a preliminary study from Assam. Bangladesh J. Zoology 18, 115–118.

Kar, D., Dey, S.C., 1990a. Fish disease syndrome: a preliminary study from Assam. Bangladesh J. Zool. 18, 115–118.

Kar, D., Dey, S.C., 1990b. A preliminary study of diseased fishes from Cachar district of Assam. Matsya 15–16, 155–161.

Kar, D., Dey, S.C., 1990c. Epizootic ulcerative syndrome in fishes of Assam. J. Assam. Sci. Soc. 32 (2), 29–31.

Khan, H., 1944. Studies in diseases of fish, infestation of fish leeches and fish lice. Proc. Indian Acad. Sci. 19 B (5), 171–175.

Kinkelin, P., Le Berre, M., Lenoir, G., 1973. Rhabdovirus des poisons proprieties in tro du virus ge la maladie ronge de l'alevin de brochet. Ann. Inst. Pasteur/Microbiol 125A, 93–111.

Koops, H., Mann, H., Pfitzner, I., Schmid, O.J., Schubert, G., 1970. The cauliflower disease of eels. In: Snieszko, S.F. (Ed.), Symposium on Diseases of Fishes and Shellfishes. American Fisheries Society, Washington D.C., pp. 291–295. Spl. Publication No. 5.

Kumaraiah, P., Murgesan, V.K., Dehadrai, P.V., 1977. A new record of bacterium causing eye disease and mortality in Channa Mauratius (Ham.). J. Inland Fish Soc. India 9, 214–215.

Laveran, A., Pettit, A., 1910. Sur une epizootic des truites. C.r. hebd. Se'anc. Acad. Sci. Paris 151, 421–423.

Lilley, J.H., Hart, D., Richards, R.H., Roberts, R.J., Cerenius, L., Söderhäll, K., 1997a. Pan-Asian spread of single fungal clone results in large-scale fish-kills. Vet. Rec. 140, 653–654.

Lilley, J.H., Thompson, K.D., Adams, A., 1997b. Characterisation of *Aphanomyces invadans* by electrophoretic and Western blot analysis. Dis. Aquatic Org. 30, 187–197.

Lowe, J., 1874. Fauna and flora of Norfolk, Part 4: fishes. Trans. Norfolk Norwich Nat. Soc. 21, 56.

Mawdesley-Thomas, L.E. (Ed.), 1972. Some tumours of fish. In Diseases of Fish. Symposium of the Zoological Society of London, No. 30. Academic Press, London.

Mckenzie, R.A., Hall, W.T.K., 1976. Dermal ulceration of mullet (*Mugil cephalus*). Aust. Vet. J. 52, 230–231.

M'Gonigle, R.H., 1940. Acute catarrhal enteritis of salmonid fingerlings. Trans. Am. Fisheries Soc 70, 297–302.

Mittermeier, R.A., Mittemeier, C.G., 1997. Megadiversity: Earth's biologically wealthiest nation. In: McAllister, D.E., Hamilton, A.L., Harvery, B. (Eds.), Global Freshwater Biodiversity, Sea Wind, vol. 11. Cemex, Mexico City, pp. 1–140.

Miyazaki, T., Egusa, S., 1972. Studies on mycotic granulomatosis in fresh water fishes—I. Mycotic granulomatosis in goldfish. Fish Pathology 7, 15–25.

Motwani, M.P., Jayaram, K.C., Sehgal, K.L., 1962. Fish and fisheries of Brahmaputra river system, Assam, I. Fish fauna with observation on their zoogeographical significance. Trop. Ecol. 3, 17–43.

Murray, G., 1885. Notes on the inoculation of fishes with *Saprolegnia ferax*. J. Bot. 23, 302–308.

Myers, G.S., 1949. Salt tolerance of freshwater fish groups in relation to geographical problems. Bijdr tot de Dierk 28, 315–322.

Nichols, J.T., 1928. Fishes of the White Nile (With Table of World's Freshwater Fish Faunae). American Mus. Novitates No. 319.

Noga, E.J., Dykstra, M.J., 1986. Oomycete fungi associated with ulcerative mycosis in Menhaden, *Brevoortia tyrannus* (Latrobe). J. Fish Dis. 9, 47—53.

OIE, 1997. Epizootic Ulcerative Syndrome: Diagnostic Manual for Aquatic Animal Diseases. Office International Des Epizootics (Paris), pp. 127—129.

Pal, R.N., 1975. Role of fish pathology in aquaculture. J. Inland Fish. Soc. India 7, 131—136.

Pal, R.N., Ghosh, A.K., 1975. An effective method of controlling tail and fin rot in Indian major carps. J. Inland Fish. Soc. India 7, 98—99.

Pearce, M., 1990. Epizootic ulcerative syndrome technical report December 1987—September 1989. Northern Territory Department of Primary Industry and Fisheries. Fisheries Report no. 22. Northern Territory, Australia, p 82.

Pfitzner, I., 1969. Zür Atiologie der Bhemenkohler krankheit deer. Aale. Arch. Fisch. 20, 24—35.

Plumb, J.A., 1973. Neutralisation of Channel cat fish virus by serum of Channel cat fish. J. Wildl. Dis. 9, 324—330.

Plumb, J.A., Gaines, J.L., June 1975. Channel cat fish virus disease. In: Ribelin, W.E., Migaki, G. (Eds.), The Pathology of Fishes. University of Wisconsin Press, Madison (USA), pp. 287—302.

Reantaso, M.B., 1990. Philippines report. In: Regional Research Programme on Relationship between Epizootic Ulcerative Syndrome in Fish and the Environment, 13—26 August 1990. NACA, Bangkok, pp. 48—53.

Reichenbach-Klinke, H., Elkan, E., 1965. The Principal Diseases of Lower Vertebrates: Book I: Diseases of Fishes. Academic Press, New York.

Richards, R.H., Pickering, A., 1978. Frequency and distribution patterns of *Saprolegnia* infection in wild and hatchery-reared brown trout, *Salmo trutta* L. and Char, *Salvelinus alpinus* (L). J. Fish Dis. 1, 69—82.

Rodgers, L.J., Burke, J.B., 1981. Seasonal variation in the prevalence of red spot disease in Estuarine fish with particular reference to the sea mullet, *Mugil cephalus* L. J. Fish Dis. 4, 297—307.

Ross, A.J., Yasutake, W.T., 1973. *Scolecobasidium humicola*, a fungal pathogen of fish. J. Fish. Res. Bd. Can. 30, 994—995.

Ross, A.J., Pelnar, J., Rucker, R.R., 1960. A virus-like disease of Chinook salmon. Trans. Am. Fish. Soc. 89, 160—165.

Ross, A.J., Yasutake, W.T., Leek, S., 1975. *Phonna herbaum*, a fungal plant saprophyte as a fish pathogen. J. Fish. Res. Bd. Can. 32, 1648—1652.

Rucker, R.R., Whipple, W.J., Parvin, S.R., Evans, C.A., 1953. A contagious disease of salmon possibly of virus origin. U.S. Fish Wildl. Serv. Spc. Sci. Rep. Fish. 138, 36.

Saha, K.C., Sen, D.P., 1955. Gammexane in the treatment of *Argulus* and fish leech infection in fish. Biochem 15, 71—72.

Sammut, J., White, I., Melville, M.D., 1996. Acidification of an estuarine tributary in Eastern Australia due to drainage of acid sulphate soils. Mar. Freshw. Res. 47 (5), 669—684.

Sarkar, H.L., 1946. On a protozoan parasite *Myxobolus mrigalae* Chakravarty, infecting the fry of *Cirrhinus mrigala* (Ham.). Curr. Sci. 15 (4), 111—112.

Schä perclaus, W., 1954. Fischkrankheiten, third ed. Akademie Verlag, Berlin (Germany).

Schubert, G.H., 1966. The infective agent of carp pox. Bull. Off. Inst. Epizoot. 65, 1011—1022.

Selye, H., 1950. Stress and general adaptation syndrome. Br. Med. J 1, 1383—1392.

Seymour, R.L., 1970. The genus *saprolegnia*. Nova Hedwig. 19, 1—124.

Shanor, Q.L., Saslow, H.B., 1944. *Aphanomyces* as a fish parasite. Mycologia 36, 413—415.

Sindermann, C.J., Scattergood, L.W., 1954. Diseases of fishes of the western North Atlantic. II. Ichthyosporidium disease of the sea herring (Clupea harengus). Res. Bull. Sea Shore Fish. 19, 1—40.

Sprague, V., Vermick, S.H., 1974. Fine structure of the cyst of some sporulation stages of *Ichthyosporidium* (Microsporidia). J. Protozool. 21, 667—677.

Tiffney, W.N., 1939. The host range of *Saprolegnia parasitica*. Mycologia 31, 310—321.

Tripathi, Y.R., 1952. Studies on parasites of Indian fishes: Protozoa: Myxosporidia together with a checklist of parasitic Protozoa described from Indian fishes. Rec. Indian Mus. 50 (1), 63—88.

Tripathi, Y.R., 1954. Studies on parasites of Indian fishes, 3. Protozoa, 2. (Mastigophora and Ciliophora). Rec. Indian Mus. 52 (2/4), 221—230.

Tripathi, Y.R., 1955. Experimental infection of Indian major carps with *Ichthyophthirius multifilis* Fouquet. Curr. Sci. 24 (7), 236—237.

Tripathi, Y.R., 1957. Studies on paasites of Indian Fishes, 5. Acanthocephala. Rec. Indian Mus. 54 (1/2), 61—92.

Tripathy, S.D., Chaturvedi, G.K., 1974. A new record of an isopod, *Ichthyoxenus jellinighansii* Hocklots, from Pariat Lake, Jabalpur, India. J. Inland Fish. Soc. 6, 85—87.

UNEP, 1992. The Convention of Biological Diversity (Nairobi).

Vestaergaard-Jφrgensen, P.E., 1974. A Study of Viral Diseases in Danish Rainbow Trout: Their Diagnosis and Control. Carl Fr. Mortensen, Copenhagen.

Virgona, J.L., 1992. Environmental factors and the prevalence of a cutaneous ulcerative disease (Red Spot) in the sea mullet (*Mugil cephalus* L) in the Clarence River, New South Wales, Australia. J. Fish Dis. 15, 363—378.

Walker, R., 1962. Fine structure of lymphocystis virus in fish. Virology 18, 503—508.

WCMC, 1998. Freshwater biodiversity: a preliminary global assessment. In: A Document Prepared for the 4th Meeting of the Conference of the Practices to the Convention of Biological Diversity, World Conservation Monitoring Centre.

Weissenberg, R., 1914. Uber infectiose zellhypertrophic bei Fischen (Lymphocystiserkrankung). Sber. Preuss. Akad. Wiss. 30, 792—804.

Williams, I.V., Amend, D.F., 1976. A natural epizootic of infectious haematopoietic necrosis in fry of sockeye salmon (*Oncorhynchus nerka*) at Chilko Lake, British Columbia. J. Fish. Res. Bd. Can. 33, 1564—1567.

Willoughby, L.G., 1970. Mycological aspects of a disease of young perch in Windermere. J. Fish. Biol. 2, 113—116.

Willoughby, L.G., 1968. Atlantic salmon disease fungus. Nat. Lond. 217, 872—873.

Wood, J.W., 1968. Diseases of Pacific Salmon: Their Prevention and Treatment. Department of Fisheries, Hatchery Division, State of Washington (USA), Olympia,Washington.

Wolf, K., Quimby, M.C., Pyle, E.A., Dexter, R.P., 1960. Preparations of monolayer cell cultures from tissues of some lower vertebrates. Sci. N.Y. 132, 1890—1891.

Wolf, K., Herman, R.L., Carlson, C.P., 1972. Fish Viruses: histopathologic changes associated with experimental channel cat fish virus disease. J. Fish. Res. Bd. Can. 29, 149—150.

Wolf, K., Darlington, R.W., 1971. Channel cat fish virus: a new herpes virus of ictalurid fish. J. Virol. 8, 525—533.

Chapter 2

Origin of the Epizootic Ulcerative Syndrome Problem

There have been widespread outbreaks of a hitherto unknown enigmatic and virulent fish disease of epidemic dimensions, leading to large-scale mortality among freshwater fishes in a big portion of the globe since 1972 and rendering many species endangered. Along with various other names, this disease has been termed "epizootic ulcerative syndrome" or "EUS."

2.1 DISTRIBUTION

Following the report of mycotic granulomatosis (MG) in Japan sometime around 1972, the earliest reports of EUS outbreak date to Australia during 1972 and Papua New Guinea during 1974; from there, EUS swept unabated in an almost chronological manner through most SE and South Asian countries. The progression by country was Indonesia (1980), Malaysia (1979−83), Thailand (1985), Kampuchea and Lao PDR (1984), Myanmar (1984−85), Sri Lanka (1987), and Bangladesh (March 1988), until EUS reached India through the Barak Valley region of Assam during July 1988. It has been sweeping the region even today, causing large-scale mortality among freshwater fishes. A similar disease has been reported in the eastern United States.

2.2 ORIGIN OF THE PROBLEM

The origin of EUS could be said to be a matter of controversy and speculation.

Following the reported outbreak of MG in Japan during the 1970s, EUS was reported in Australia as well as in SE and South Asia during subsequent decades.

Having quite widely considered the mycotic genus *Aphanomyces* the root cause of EUS initiation, the question arises, where did the fungus originate? Probably in a tropical or subtropical region in South America. Africa could be the other main possibility. The African catfish living in Africa was not found to be affected by EUS, but the same fish, when farmed in Asia on a large scale, was found to be affected by EUS. This suggests that EUS most probably did not originate from Africa. Nevertheless, the *Aphanomyces* fungus probably entered Australia through the import of ornamental aquarium fishes. There is bulk movement of ornamental and cultured fishes around the Australian region in connection with the ornamental fish trade.

Notwithstanding the above, the story of EUS has an interesting parallel in Europe. It seemed that apparently healthy American crayfish were introduced to some waterways in Europe during the second half of the nineteenth century. A fungus, probably a different species of *Aphanomyces* than *A. invadens*, was brought in along with the crayfish consignment and exerted devastating impact, progressively wiping out the freshwater crayfish from Continental Europe. Then it spread to Great Britain and Ireland, followed by Turkey. The destruction had continued (Lehane, 1996).

2.3 CHRONOLOGY, STATUS, AND MAJOR OUTBREAKS IN THE WORLD

An **outbreak** can be defined as a short-term epidemic or a series of disease events clustered in space and time. Such disease events are usually new cases of a disease occurring at a higher frequency than normally expected. Throughout this section, the terms outbreak and epidemic are used more or less interchangeably.

An outbreak investigation is a systematic procedure to help identify causes and sources of epidemics, with a view to controlling the existing epidemic and preventing future ones. Usually, the primary objective of an epidemic or disease outbreak investigation is to identify ways of preventing further transmission of the disease-causing agent.

"Outbreaks" are often referred to as being either "common source" (cases resulting from exposure to a common source, as in intoxications) or "propagated source" (animal-to-animal transmission, as in most infectious diseases). In some EUS outbreaks, it is conceivable that both types of sources could be involved—the initial cases resulting from exposure to a common source (such as contaminated water or equipment) and the secondary cases resulting from fish-to-fish spread.

Notwithstanding the above, the duration of an outbreak is influenced by:

1. The number of susceptible animals exposed to a source of infection that become infected
2. The period of time over which susceptible animals are exposed to the source of infection
3. The minimum and maximum incubation periods of the disease

Epizootic Ulcerative Fish Disease Syndrome. http://dx.doi.org/10.1016/B978-0-12-802504-8.00002-X

2.3.1 History of the Spread: International and National Status and Major Outbreaks in the World

Outbreaks of an ulcerative disease, characterized histologically by mycotic granulomas, have affected freshwater and estuarine fishes over much of Asia and Australia since >40 years. The disease had been given various names, but is said to be most commonly known as **mycotic granulomatosis (MG) in Japan, red spot disease (RSD) in Australia, and epizootic ulcerative syndrome (EUS) in SE and South Asia**. MG, RSD, and EUS had in the past been described separately as distinct conditions. However, recent studies had revealed that the same pathogenic *Aphanomyces* fungus is generally involved in each case. It is therefore now apparent that an account of the history of EUS would be incomplete without taking into account the outbreaks in Japan and in Australia.

2.3.1.1 Mycotic Granulomatosis

The first report of an EUS-like condition had appeared during the summer of 1971 among farmed ayu (*Plecoglossus altivelis*) in Oita Prefecture, Japan (Egusa and Masuda, 1971). The characteristic lesion with a granulomatous response to invasive hyphae had been described, and the disease was named mycotic granulomatosis (Miyazaki and Egusa, 1972). It had fast spread to several other prefectures and affected various species of fishes, predominantly cultured ayu and goldfish (*Carassius carassius auratus*), wild Formosan snakehead (*Channa maculata*), crucian carp (*Carassius auratus*), bluegill (*Lepomis macrochirus*), and gray mullet (*Mugil cephalus*) (Miyazaki and Egusa, 1972, 1973a,b,c). Importantly, common carp (*Cyprinus carpio*) were not affected. Hatai et al. (1977a,b) isolated the invasive oomycete fungus from affected fish and subsequently called it *Aphanomyces piscicida* (Hatai, 1980a,b). *A. piscicida* is now known to be conspecific with the EUS pathogen *Aphanomyces invadans* (Lilley et al., 1997a,b). Although serious MG epizootics had not been reported in Japan since 1973, outbreaks had continued to occur periodically. Subsequently, Hatai et al. (1994) had reported a similar disease in imports of ornamental dwarf gourami (*Colisa lalia*) from Singapore. The involvement of the same *Aphanomyces* pathogen was again found (Lilley et al., 1997a).

2.3.1.2 Red Spot Disease

Outbreaks of the cutaneous ulcerative condition RSD, affecting estuarine fish and particularly the gray mullet (*M. cephalus*), in Queensland, Australia (McKenzie and Hall, 1976) was recorded in 1972. The disease had subsequently progressed to affect freshwater and estuarine fishes in coastal rivers in New South Wales (Callinan et al., 1989),

the Northern Territory (Pearce, 1990), and Western Australia (Callinan, 1994a). An *Aphanomyces* fungus was recovered from diseased fish by Fraser et al. (1992) and had been shown to reproduce the disease in fish using bath challenges, but only when the skin of experimental fish was artificially abraded (Callinan, 1994b). Therefore, some other factors were believed to be involved in the disease process. Virgona (1992) had shown that RSD outbreaks in estuarine fish in the Clarence River, New South Wales were associated with lower catchment rainfall. Callinan et al. (1995b) related this to runoff from acid sulfate soils. In fact, ultrastructural examination of fish gills and skin showed that the low pH and elevated concentrations of monomeric aluminum, which is said to be representative of estuarine acidification, had induced significant lesions in the fish (Sammut et al., 1996). In aquarium trials, RSD was subsequently induced in fish exposed sublethally to artificially acidified water (at both pH 3 and 5) and pathogenic *Aphanomyces* spores, even at low concentrations of monomeric aluminum (Callinan et al., 1996; Callinan, 1997). As with *A. piscicida,* the pathogenic RSD-*Aphanomyces* had been believed to be the same organism as the EUS pathogen *A. invadans* (Callinan et al., 1995b; Lilley et al., 1997a,b).

2.3.1.3 Epizootic Ulcerative Syndrome

Concomitant to the above, following the outbreaks of MG and RSD there had been a progressive spread westward across Asia of a syndrome associated with dermal ulceration that was causing large-scale mortalities among a large number of freshwater and estuarine fish species. The syndrome was given its present name, epizootic ulcerative syndrome (EUS), in 1986 at the Consultation of Experts on Ulcerative Fish Diseases in Bangkok (FAO, 1986). Outbreaks of EUS had been reported from 18 countries of the Asia—Pacific region, although not all may be positively confirmed as EUS according to various national and international guidelines.

The chronology of EUS outlined briefly above is discussed below in more detail:

Papua New Guinea

Concomitant to the above, an ulcerative disease outbreak believed to be EUS had occurred in the rivers of southern **Papua New Guinea** (Haines, 1983) during 1975—76. In 1982—83, there were high mortalities among the gudgeons (*Ophieleotris aporos* and *Oxyeleotris heterodon*) from inland areas and among the mullets from estuaries in northern Papua New Guinea (Coates et al., 1989). Introduced tilapia (*Oreochromis mossambicus*) had proved to be resistant, although they are said to be quite common in these areas. The preserved affected fish were later examined by Roberts et al. (1986) and the disease had been confirmed as pathologically identical to EUS.

Indonesia

Following the above episodes of fish disease, outbreaks of an epizootic hemorrhagic condition had occurred in **Java, Indonesia** during 1980 that affected primarily cultured cyprinid and clariid fish, although it was not well confirmed whether this was EUS or not (Roberts et al., 1986). Following this, typically ulcerated snakeheads and catfish had subsequently been reported in the **Indonesian** states of Sumatra, Sulawesi, and Kalimantan (Widagdo, 1990). Invasive hyphae had also been identified from the sand gobies (*Oxyeleotris marmoratus*) from eastern Kalimantan (Rukyani, 1994; D. Bastiawan, possibly unpublished).

Singapore

Roberts et al. (1986) had discussed unconfirmed accounts of ulcerated walking catfish (*Clarias batrachus*) in **Singapore** in 1977 and of subsequent occurrences thereafter. It may be noted here that there had been no records of colossal losses of fish in the aquaculture industry due to EUS, despite the fact that Singapore is said to enjoy the status as a center of trade in EUS-susceptible ornamental fishes.

Malaysia

The first reported typical EUS outbreaks in **Malaysia** had been recorded in December 1980 in rice-field fishes in northern Malaysia (Jothy, 1981), although there were reports of high mortality rates in fish in southern peninsular Malaysia in 1979 (Shariff and Law, 1980; described by Roberts et al., 1986). These had recurred almost regularly and annually ever since, albeit to a lesser extent (Shariff and Saidin, 1994). Major species affected were snakeskin gourami (*Trichogaster pectoralis*), striped snakehead (*Channa striata*), climbing perch (*Anabas testudineus*), and walking catfish (Shariff and Saidin, 1994).

Thailand

Thailand is a country that had faced frequent outbreaks of EUS, resulting in large-scale mortality of fishes. As such, quite a host of works had been done on EUS by different workers in Thailand. Significant, well-documented epizootics had occurred annually in **Thailand** since 1981 (Ulcerative Fish Disease Committee, 1983; Chulalongkorn University, 1983, 1985, 1987). The second (1982–83) and third (1983–84) outbreaks were particularly devastating, as they had affected the intensive fish culture systems of central Thailand as well as wild fish in natural waterways. It may be noted here that some of the most severe mortalities were among the farmed snakeheads and the rice-field fishes. The original outbreaks had started toward the end of the rainy season (September) and had persisted throughout the cool season to March. Outbreaks of late, in general, tend to be restricted to the coolest months of December and January. However, during December 1996, EUS was experienced in northeast (NE), central, and southern provinces (S. Kanchanakhan, unpublished). The isolation of the pathogenic fungus A. *invadans*, from EUS-affected snakeheads in Suphanburi province was described by Roberts et al. (1993).

Concomitant to the above, according to Kamonporn Tonguthai, a major epizootic had spread from the south of Thailand to the north and northeast of the country in 1981. However, the most significant and best documented instances of EUS had occurred in Thailand between September and March of each year from 1981 to 1984. Most severely affected were the intensively cultured snakeheads. Delineating the clinical and pathological features of EUS-affected fish, Supranee Chinabut of the National Inland Fisheries Institute, Bangkok, had pointed out that the clinical signs of the affected fish were the same in the case of all species. In the earliest stages of the disease, lesions had shown some damage to the epidermal layer of the skin, resulting in an area of epithelial necrosis with some inflammatory cell infiltration. On the other hand, severe cases had shown large ulcerative lesions with degenerative changes in the muscle. Marked diffuse systemic necrotizing granulomatous mycosis had spread in the necrotic muscle bundles. The fungus was extremely invasive throughout the necrotic muscular lesions. In the latest stages, advanced lesions were observed, with fungal hyphae having infiltrated into some internal organs and produced a tremendous number of mycotic granulomata in those organs.

Further, according to Chinabut, many species of parasites had been identified, including species of *Palisentis, Triancloratus, Dactylogyrus, Gyrodactylus, Henneguya, Epistylis,* and *Trichodina.* The predominant bacterial isolations from the later stages of the diseased fish were *Aeromonas hydrophila, Aeromonas sobria, Pseudomonas* spp., *Micrococcus* spp., *Flavobacterium* spp., *Vibrio* spp., etc. They all had been occasionally isolated from affected fish. Ulcerative disease rhabdovirus was also found in the diseased fish. In addition, mycotic forms like *Aphanomyces, Achlya,* and *Sapralegnia* were identified from the surfaces of ulcerated fish.

While the causes of EUS had possibly not been clearly identified, Nontawith Areechon of the Faculty of Fisheries, Kasetsart University, Bangkok, pointed out that they may be related to stressful conditions of aquatic animals. Stress could be caused by, among other factors, environmental alterations and toxic substances. These could weaken the fish and make them more vulnerable to pathogens.

Treatment for therapeutic purposes, and sometimes in practice, includes antibiotics and chemicals. Though these might bring some level of so-called satisfactory results, they have a number of undesirable side effects, too. These usually include the residues, cost increases, development of bacterial resistance, the negative impact on the environment, and so on.

Based on the Thai experience, reported Areechon, the keys to a successful crop in the case of cultured fish are healthy fry, proper pond preparation, and good management. However, these measures would not be applicable to wild fish populations. Prevention of any condition stressful to aquatic animals could be very important as well as difficult. This could hinge around good water body management, optimum stocking rates, and the proper combination and proportion of fish species. In support of this advice, Tonguthai pointed out that EUS-affected fish had improved their health condition when removed and put into a normal freshwater pond.

Myanmar, Lao PDR, and Cambodia

Following the above, **Myanmar, Lao PDR, and Cambodia** first reported major outbreaks of EUS during 1983 or 1984 (Lilley et al., 1992). Subsequent epizootics were said to be less extensive (e.g., EUS is said to have affected 35 townships in Myanmar during 1984–85 and 11 townships during 1989–90: Soe, 1990). However, given the importance of susceptible fish to rural communities in these countries, the impact continued to be significant. It may be mentioned here that diseased snakeheads from Lao PDR were confirmed at AAHRI, Bangkok as suffering from EUS, in 1996.

Vietnam, China, and Hong Kong

Several episodes of EUS outbreaks and occurrences of EUS-affected fish had also been known from **Vietnam, China, and Hong Kong**, although many of these still might not be confirmed. The first report of ulcerated snakeheads in Vietnam came from the Mekong Delta in 1983, which could most likely be the first occurrence of EUS in that country (Xuan, 1990). Ulcerated *Labeo rohita* were first observed at the Pearl River Fisheries Institute in Guangzhou, South China in 1982 (Lian, 1990). Clariid catfish were affected in the same area in 1987–88 (Lian, 1990), and *C. auratus* were reportedly affected over much of Eastern China in 1989 (Guizhen, 1990). Wilson and Lo (1992) had reported seasonal mortalities of up to *c* 70% of snakeheads (*C. maculata*) in late summer in Hong Kong since 1988.

The Philippines

An estimated 5%–40% of snakeheads, gobies, gouramies, catfish, crucian carp, *Arius* sp., and *Therapon* sp. were ulcerated. On the other hand, milkfish, bighead carp, and tilapia were unaffected (Llobrera and Gacutan, 1987). The disease continued to spread to at least 11 other provinces, affecting wild fish in lakes, rice fields, and swamps, as well as pond-cultured fish (Bondad-Reantaso et al., 1994). Mullet, goatfish (*Upeneus bensai*), croaker (*Johnius* sp.), *Psettodes* sp., and *Scanthophagus argus* in a lagoon in Cagayan Province were reported to have suffered an outbreak in 1990 that was confirmed as EUS by histological examination (Reantaso, 1991; S. Chinabut, unpublished). The occurrence of EUS in these brackish water and marine species provided an explanation as to how the condition might have spread between the islands. The severity of outbreaks is believed to have decreased since 1993. Several *A. invadans* isolates were recovered from EUS-affected fish in the Philippines (Paclibare et al., 1994).

Sri Lanka

A major outbreak of EUS in freshwater and estuarine fish in western **Sri Lanka** occurred during December 1987. It was perhaps before any outbreak had occurred in the mainland of the Indian subcontinent (Costa and Wijeyaratne, 1989). It was believed that the disease was imported from SE Asia in shipments of infected fish, possibly ornamental angelfish (*Pterophyllum scalare*), some of which were ulcerated and had suffered high mortalities (Balasuriya, 1994). Snakeheads with large necrotic ulcers were the most visible signs of the disease, but tilapia, the main commercial species, was not affected. EUS was reportedly still active in Batticaloa Lagoon in 1996 (P. Vinobaba and M. Vinobaba, unpublished).

Mainland of South Asia

Over the past 10 years, EUS had a serious effect on fisheries throughout the **mainland of South Asia**, causing enormous losses to important capture fisheries areas and damaging confidence among people in the aquaculture industry, which had many operations that were still in the early stages of development.

Bangladesh

The disease was first reported in Chandpur district of **Bangladesh** during February 1988. This first outbreak had lasted for 13 months, during which it had spread rapidly throughout the country, and was said to be seemingly aided by the flood of September 1988 (Barua, 1994). Ulceration was observed in many wild species, predominantly the snakeheads, *Puntius, Clarias, Mystus,* and *Mastacembelus.* Cultured IMCs were also affected, although large-scale mortalities due to the disease were probably restricted to fingerlings (Roberts et al., 1989). EUS prevalences are believed to have subsequently declined. However, there were reports that from 1995, the severity of outbreaks had been on the increase in Bangladesh (G.U. Ahmed, unpublished report). In January 1993, *A. invadans* cultures were isolated from farmed IMC (*L. rohita*) in NW Bangladesh and wild fish in the productive flood plain areas of NE Bangladesh.

India

The first outbreak of EUS occurred in India in the northeastern province of Assam during July 1988, followed by its occurrence in the province of Tripura during September 1988, and was considered to be linked with the floods from Bangladesh. Since then, EUS had spread to almost all other provinces with the possible exceptions of Himachal Pradesh and Punjab. The outbreaks had usually occurred between June and December, predominantly during the post-monsoon months. The incidence was more in confined waters (10%—55%) than in rivers (4%—15%). The fish genera most susceptible to EUS had been *Channa* (20%—100%), *Puntius* (5%—100%), *Clarias* (10%—30%), *Heteropneustes* (5%—12%), *Mystus* (5%—75%), *Nandus, Cyprinus,* and *Glossogobius* (10%—60%), *Anabas* (10%—55%), and *Mastacembelus* (15%—20%).

Outbreaks of EUS in **India** have been comprehensively reviewed (Zoological Society of Assam, 1988; Jhingran and Das, 1990; National Workshop on Ulcerative Disease Syndrome in Fish, 1990; Kumar et al., 1991; ICSF, 1992; Das and Das, 1993; Mohan and Shankar, 1994; Kar, 1999, 2005, 2007, 2010, 2012, 2013, 2014). The **northeast (NE) Indian** states were the first to report losses in May 1988. The disease appeared to have spread through reservoirs and paddy fields to most of the provinces, affecting some IMC farms as well. Notably, EUS had a serious impact on fish in low-salinity areas of the rich brackish-water fisheries of Chilika Lake, Odisha in November 1990 (Raman, 1992), and the reservoirs and backwaters of Kerala in June 1991 (Sanjeevaghosh, 1991). *Aphanomyces* isolates, consistent with *A. invadans*, had been recovered from EUS-affected fish in southern India.

It had been determined that the intensity of incidence was higher (40%—65%) in areas of low alkalinity (13—30 mg/L) and hardness (6—45 mg/L) through environmental monitoring of the affected waters. Heavy metal analyses had revealed significant values for zinc and copper, but not perceptibly high enough to create stress for fish. The possibly more important stress factors had been chemicals like BHC, DDT, and their metabolites, which had been detected in water, plankton, and fish samples collected from the affected areas.

Etiological investigations revealed spherical viruslike particles. A wide variety of bacterial forms had also been recorded from afflicted fishes, predominantly the *Aeromonas* spp., *Pseudomonas* spp., *Micrococcus* spp., *Vibrio* spp., *Klebsiella* spp., *Citrobacter* spp., *Staphylococcus* spp., *Arthobacter* spp., and *Corynebacterium* spp. The ulcers had been found to be invariably associated with the pathogenic fungus *Saprolegnia.*

Scientific investigations on remedial measures had been recommending quicklime at 200—600 kg/ha as providing encouraging results in containing the disease. However, limited success had possibly come from antibiotic therapy. CIFAX, a drug formulated by the Central Institute of Freshwater Aquaculture (CIFA), had occasionally been found to be helpful in containing EUS.

Similarly, bleaching powder was found to be useful in some cases.

More details about the works done on EUS in India are given later in this chapter.

Bhutan and the Eastern Terai of Nepal

The fishes of **Bhutan and the eastern Terai of Nepal** were said to be first affected by EUS during 1989, and by 1993, EUS had spread to the Himalayan valley regions including Pokhara and Kathmandu, where cold-water species including *Tor* spp. were affected (Phillips, 1989; Shresta, 1994). It was estimated that *c* 20%—30% of Nepalese pond fish production (*c* 3000 mt) was lost every year through EUS (Pantha, unpublished report).

Pakistan

Notwithstanding the above, **Pakistan** was said to have been affected by EUS since about the mid-1990s. In Pakistan, EUS was confirmed in snakeheads from Punjab Province during April 1996, and in *Cirrhinus mrigala* from Sindh Province during January 1998 (DFID, 1998). The blotched snakehead or mud murrel (*Channa punctata*) was the most commonly affected species, with *Puntius* spp., *L. rohita*, and *Cirrhinus reba* also reportedly affected. An estimated 20% of farms were affected in Sialkot Division, Punjab, with the incidence being higher in ponds that were inundated by flooding in 1996 (AAHRI, ACIAR, IoA, and NACA, 1997). Reported losses were said to be not very high in the Punjab, possibly due to the extensive use of tube well water for fish farms, as well as elevated salinities (AAHRI, ACIAR, IoA, and NACA, 1997). However, EUS had been said to be well established in parts of the Indus River. Further, given its apparent rapid spread across the country (DFID, 1998), there had been fears of the potentially serious future impacts of EUS to fisheries and aquaculture development.

Other Similar Diseases

Mention may be made here of other similar ulcerative fish diseases, although their relationship with EUS is presently not very well known. Incidentally, The two genera *Channa* and *Ophicephalus* were said to be united as *Channa* by Myers and Shapovalov (1931, cited by Clark, 1991).

2.3.2 Ulcerative Mycosis

Noga (1994) postulated that ulcerative mycosis (UM) of coastal fish populations of the western Atlantic might be the part of the same syndrome as EUS, given the similarity in the clinicopathological features of both diseases, and also considering that predominantly *Aphanomyces* fungi were recovered from UM-affected diseased fish (Dykstra et al., 1986). However, fish challenged with these *Aphanomyces* isolates had failed to develop lesions consistent with UM (Noga, 1993; Lilley and Roberts, 1997). Fish had developed UM, however, when lesion material was used as an inoculum, suggesting that some other unidentified agent, possibly another fungus, was required for infection (Noga, 1993).

2.3.2.1 United States

Notwithstanding the above, UM was first observed in April 1984 among the menhaden (*Brevoortia tyrannus*) in the Pamlico River, **North Carolina.** Further, during November of the same year, a massive kill was reported (Noga and Dykstra, 1986). Epidemics of similar diseases were later recognised in the estuaries along the eastern seaboard of the United States from Connecticut (Noga, 1993) to Florida (McGarey et al., 1990). However, it was not certain whether these were first occurrences and had represented a spread of the same disease. Several fish species were reported to contract UM-like diseases in Pamlico River (Noga et al., 1991), but the prevalence in these species was markedly lower than it was in menhaden (Levine et al., 1990a). In menhaden, a larger proportion of age-zero fish were shown to be affected than age-one fish (Levine et al., 1990b). Levine et al. (1990b) had also provided evidence that specific regions of low salinity within the Tar-Pamlico estuary were said to have harbored higher levels of diseased fish. In addition, Noga (1993) had observed that the most damaging outbreaks in the Pamlico River coincided with years of unusually high rainfall and reduced salinity (1984 and 1989). Outbreaks had continued to occur, with infection rates of menhaden of up to 100% (Levine et al., 1990b). Noga et al. (1996) had further shown that sublethal exposure to toxins, produced by a subsequently identified "phantom" dinoflagellate, *Pfiesteria piscicida*, might also have been responsible for high mortalities in the Pamlico River, **North Carolina** (Burkholder et al., 1992). These probably had resulted in dermatitis and subsequent development of UM.

2.3.2.2 Tasmania

Cod Ulcer Disease

Munday (1985) had reported the presence of severely ulcerated red cod (*Pseudophycis bachus*) in the River Tamar near Launceston, **Tasmania** in November 1980 and 1981. Although a variety of bacteria and parasites were identified from the fish, pollution was considered to be the main cause of the disease. Munday perhaps had believed that the ulcer disease was the same syndrome as EUS, although it had occurred in higher-salinity water. However, he had added that ever since Launceston's sewerage system had been improved, the disease had no longer been reported.

Works done on EUS in other countries of the world, along with further works done on EUS in the countries already mentioned above, have been indicated later in this chapter.

Notwithstanding the above, the following workers significantly contributed to the advancement of EUS research, including reports of major EUS outbreaks:

Callinan et al. (1995a), Callinan (unpublished); Chinabut et al. (1995); Chinabut (unpublished); DFID (1998); Fraser et al. (1992); Hanjavanit et al. (1997); Hatai (1994); Kanchanakhan (1996a); Lilley and Roberts (1997); Miyazaki (1994); Mohan and Shankar (1995); Pearce (1990); Reantaso (1991); S. Chinabut (unpublished); Fraser et al. (1992); Hanjavanit et al. (1997); Hatai (1994); Kanchanakhan (1996a,b); Lilley and Roberts (1997); Miyazaki (1994); Mohan and Shankar (1995); Pearce (1990); Reantaso (1991); Roberts et al. (1989); Vishwanath et al. (1997, 1998).

The foregoing account indicates that there have been widespread outbreaks of a hitherto unknown enigmatic and virulent fish disease of epidemic dimensions, leading to large-scale mortality among freshwater fishes in a big portion of the globe since 1972 and rendering many species endangered. As indicated, this disease had been termed "epizootic ulcerative syndrome" or "EUS." The historical resume of EUS episodes (as detailed above) is briefly summarized below:

Following the report of **MG in Japan** sometime around 1972, the earliest report of EUS outbreaks date to Australia during 1972 and Papua New Guinea during 1974; from there, EUS swept in an almost chronological manner through most of SE and South Asia, including Indonesia (1980); Malaysia (possibly 1977, and then 1979−83); Vietnam (possibly 1983); China (possibly 1982); Hong Kong SAR (possibly 1988); the Philippines (1985); Thailand (1981 and 1985); Kampuchea and Lao PDR (1983 and 1984); Myanmar (1983 and 1984−85); Sri Lanka (1987); and Bangladesh (March 1988); until it reached India through the Barak Valley region of Assam during July 1988, and has been sweeping the region ever since, causing large-scale mortality among freshwater fishes. Thereafter, EUS spread further to Bhutan and Nepal (1989); Pakistan (1996); Botswana and Namibia (possibly 2006, and then 2007); and Zambia (possibly 2007, and then 2008). Further, it is said to have spread to North Carolina, Florida, and Connecticut in the United States during 1984. At the moment, there are said to be at least 24 countries affected by the disease (FAO, 2009). More than 50 species of finfish are susceptible to EUS.

2.3.3 Details of Works Done on Outbreaks of Epizootic Ulcerative Syndrome in the Indian Scenario

A host of various studies have been done on EUS by workers in India, as referred to the above; India is a country that has been severely affected by EUS.

Anon (1992) held a "Consultation on Epizootic Ulcerative Syndrome vis-à-vis the Environment and the

People" under the aegis of the International Collective in Support of Fishworkers.

2.3.3.1 West Bengal

As reported by Central Inland Fisheries Research Institute (CIFRI, ICAR), West Bengal, the socioeconomic impact of the disease was staggering. A sample survey of 500 affected fish farmers in an EUS-afflicted area of West Bengal revealed that the maximum respondents (30%–40%) had suffered loss of fish. It may be noted here that fish consumption had decreased from *c* 44.4% before the outbreak of the disease to *c* 15% after the outbreak. Most of the respondents had used lime as a remedy, with 68% success. In fact, the provincial government had distributed Rs 8.0 million worth of lime to the farmers, who were believed to have used it prophylactically in the postmonsoon period. In West Bengal, All-India Radio of the government of India played a significant role in disseminating information on EUS.

2.3.3.2 Manipur

Manipur has the biggest freshwater wetland in the North-eastern region of India, called Loktak Lake, which has a water spread area of about 28,000 ha. Fish is usually an essential diet item of the Manipuris, *c* 90% of whom eat fish in either dried or fresh form.

The first information on the spread of EUS in the northeast reached the province of Manipur probably during June 1988. Accordingly, the state government immediately banned the import and sale of both dried and fresh fish as a precautionary measure.

However, the first true outbreak of EUS was reported from the capital, Imphal, during May 1989. Subsequently, reports of EUS outbreaks from other affected parts of the province started pouring in. Technical officers of the Manipur's Fisheries Department visited the affected areas for investigations. It was found that the species mainly infected by EUS were the local indigenous fishes, viz., *Channa, Clarias, Anabas, Puntius,* etc. At that time, however, EUS possibly had not affected the culturable IMCs such as Rohu, Catla, Mrigal, common carp and Chinese carps. Further, unusually enough, EUS had affected the wild waters at that time, but not human-made ponds.

The Fisheries Department initiated EUS control cells in all of the district headquarters to combat further spread of the disease. Training camps had also been organised for public awareness.

Among the measures recommended were:

1. Disinfection of nets by sun drying, boiling in water, etc.
2. Treatment of ulcerated fish in 3% solution of common salt for 5–10 min or in 500–1000 ppm of potassium permanganate for 1 min

3. Disinfection of affected ponds with unslaked lime (quicklime) at 150–200 kg/ha, depending upon soil pH

It may be noted here that the control measures adopted by the fisheries department were found to be effective and there were no further reports of the spread of EUS in the province. This seemed to be exceptional, since almost all other provinces of India had reported a second outbreak or even multiple outbreaks of EUS.

2.3.3.3 Odisha

In the province of Odisha, the first report of EUS outbreak had occurred perhaps during February 1989 from ponds in Balasore and Mayurbhanj (adjoining West Bengal), and then from Cuttack. The disease was said to have affected about 80 development blocks at that time. But surprisingly, river systems and human-made reservoirs appeared not to be affected. Such incidents probably elide the pious assertion that EUS usually spreads from wild waters to ponds, or at least usually occurs simultaneously in both types of ecosystems, as the conventional wisdom supposes. It was therefore unclear how EUS in Odisha had appeared to be confined to ponds only.

The outbreak was severe, however, and led to panic. The province was estimated to have lost about 186 metric tons (tonnes or t) of fish worth more than Rs 3.0 million. In view of this, the provincial government targeted 5500 farmers in four districts for free supplies of lime. In this exercise, 11 of the 13 districts were covered between 1989 and 1991. Further, the government also tried to disseminate information on EUS at the Panchayat level.

Also in Odisha, Khatri et al. (2004) made an assessment on the prevalence of fish diseases in Sambalpur, Western Orissa.

2.3.3.4 Tamil Nadu

Four to seven districts of the province of Tamil Nadu (TN) were said to have been affected by EUS. However, EUS was first reported from the Chengai MGR, Thanjavur, and South Arcot districts during January, February, and March 1991, respectively. The species mainly affected were indigenous fish varieties belonging to the families Channidae, Mastacembelidae, Cyprinidae, and Bagridae. However, the intensity of infection was said to be mild among the major carps and exotic carps, as well as *Tilapia* species, which are said to be predominant in inland aquaculture.

Notwithstanding the above, the Animal Sciences Department of the University of Madras conducted a microbial investigation to identify the causative agent, because the province's fisheries department lacked investigating facilities. The Tamil Nadu Veterinary and Animal Sciences University had constituted a "disease investigating team" comprising officials of the Madras Veterinary College, the

Fisheries College, the city of Tuticorin, and the province's fisheries department. During March 1991, this team toured the most affected district, viz., the Thanjavur district, collecting samples of the viscera, notably the kidneys and the livers of the diseased fish. There had been a good number of works to identify the primary causative agent.

Concomitantly, scientists from ICAR (CIFA and CIFRI) also visited the affected sites to collect samples. Further, collaborative efforts with the School of Tropical Medicine, Calcutta, had identified the causative agents as *Citrobacter intermedius* and *Klebsiella aerogens* from samples collected during February and April 1991.

It is important to mention here that the affected fish were treated with the poultry drug Bifuran (Nitrodurozone 100 mg; Furazolidone 14.5 mg/100 mg) at the concentration of 25 ppm in water, based on the drug sensitivity pattern of the isolate during investigation. It should also be mentioned that the ulcerative wounds had probably healed after five to seven days of therapy. However, cost considerations prevented this from being tried in larger areas.

Subsequently, EUS erupted extensively in many districts of TN in 1992. It was found that the outbreaks had coincided with the low temperatures prevailing during the post-monsoon rains. Analytical reports of water quality indicated low alkalinity and softness as predisposing factors.

A collaborative effort was initiated with the Department of Animal Disease Investigation and Control, and Madras Veterinary College, to identify the primary etiological agent and secondary invaders. Further, a "disease monitoring cell" controlled by the Assistant Director of Fisheries (Hydrology) was formed to monitor the disease and liaise with the nodal agency and the central government at Delhi.

Further, Muthu Ramakrishnan (2009) had worked on microbial flora from EUS-infected murrel (*Channa striatus* (Bloch, 1785–1795)) in the Tirunelveli region.

2.3.3.5 Andhra Pradesh

In Andhra Pradesh, EUS was probably reported for the first time in October 1990, from the Kolleru Lake of West Godavari and the Krishna districts. Subsequently, outbreaks occurred in almost all types of water such as irrigation canals, drains, swamps, ponds, and wetlands. EUS in Andhra Pradesh, had affected the following species widely—viz., *Channa, Clarias, Heteropneustes, Puntius,* and IMCs such as Catla, Rohu, and Mrigal. The IMCs were said to be the first to succumb to EUS attack in wild waters. The disease subsequently invaded cultured ponds.

The typical symptoms were discoloration of the skin, red spots, hemorrhagic lesions, deep ulcers, and slow and unbalanced movement of fish swimming with their heads out of the water. High mortality was found in wild waters, especially among air breathers, catfish, and *Puntius.* However, EUS was said to be not very prevalent among the IMCs in culture ponds, possibly due to preventive measures and the prophylactic use of antibiotics by area farmers.

The following preventive measures had been adopted against EUS in Andhra Pradesh:

1. Application of lime at 50 kg/acre
2. Application of salt at 10 kg/acre through gunny bags hanging from feed poles in the fish tank, and 2 kg of salt/100 kg feed
3. Dip treatment with 0.5%–2% potassium permanganate
4. Application of antibiotics like oxytetracycline, doxycycline, and terramycin at 5 gm/100 kg of fish for 10 days
5. Preventing the entry of diseased fish into the tanks by the use of a mesh
6. Addition of mineral and vitamin mixtures to the feed
7. Avoiding the exchange of water when neighboring tanks and canals had been affected
8. Periodic monitoring of the health of the fish

The following treatment was advised:

1. Application of malachite green
2. Use of antibiotics like oxytetracycline, doxycycline, and terramycin at 10–20 gm/100 kg of fish for 10 days, or erythromycin at 60–100 mg/1 kg of fish feed
3. Addition to the feed of a mineral mixture at 2%, and vitamin mixture at 100 g/tonne of feed
4. Stopping the use of manure in the tank during the disease period

2.3.3.6 Kerala

The province of Kerala is said to have been affected by EUS during the mid-1990s. Kerala has rich fishery resources, and fisheries are an important sector of the province's economy. According to the province's fishery department, the sector employs about 3% of the population of the province and contributes around 2.5% to its net domestic product.

Concomitantly, the total annual inland fish production of the province had been estimated to be 36,000 t from 355,000 ha of inland waters. Around 200,000 people belonging to 33,000 fisher families were said to be dependent almost entirely on inland fishery resources for their livelihood. They were usually subsistence fisherfolk, who are perhaps not much better off than agricultural workers.

In addition to the above, EUS was first reported from Pookote Lake in the Banasurasagar Reservoir area in the northern district of Wynad in June 1991. The disease had caused surprise to the department of fisheries, since it affected even fingerlings cultured in the lake by the department itself. In three weeks' time, the disease had also spread to wells and ponds in Wynad.

EUS subsequently spread, by the end of 1991, to the fresh and the brackish waters of Kuttanad and Vembanad

Lakes, as well as the rivulets in Kottayam, Alappuzha, and Pathanamthitta in the south. The Central Marine Fisheries Research Institute, Kochi, estimated that EUS had probably affected *c* 25% of fish in the Vembanad Lake; while cultured fish perhaps had been affected in only a few ponds.

Notwithstanding the above, the disease did not subside with the monsoon, contrary to claims by the provincial government and experts. By October, EUS had moved to Thrissur and was reported from ponds and canals inside "kole" fields. In November 1991, EUS had inflicted Kuttanad again, appearing in Kumarakom and affecting *c* 30% of the fish there. Subsequently, EUS spread to the southernmost district, Trivandrum, by January 1992, where *c* 15%–30% of the catch in Veli Lake was affected. EUS was subsequently reported from the Achenkoil River two months later.

It was estimated that as a whole, 11 of the 13 districts of the province of Kerala had been affected by EUS, out of which five had been severely afflicted. The various fish types affected were murrels, *Clarias*, *Etroplus*, *Puntius*, *Wallago*, etc.

Fish farmers and fishers were estimated to have incurred a colossal economic loss ranging from Rs 20 million (per government records) to Rs 120–200 million (per newspaper reports). The value of the annual catch from the Vembanad Lake alone had been estimated at Rs 100 million.

Studies were done during the period September 1991 to April 1992 in five districts of Kerala (with particular focus on Kuttanad) to assess the socioeconomic impact of EUS on the inland fisherfolk. They found that the spread of EUS had completely paralyzed the inland fish market.

Panic-stricken consumers did not bother to eat not only EUS-affected fish, but also mussels and ducks, in spite of official announcements that unaffected fish could safely be eaten. In Alappuzha, water for domestic use had to be supplied in tankers, because people were afraid to use supposedly contaminated lake water even for washing. According to Kamonporn Tonguthai, a similar loss of confidence in freshwater fish also occurred in Thailand during the initial outbreak of EUS in 1982–83, and led to financial losses to the tune of greater than $8.7 million.

In Kerala, the immediate effect of the collapse of the market was that inland fishworkers were thrown out of work and out of gear. Hit particularly hard were women fish vendors, who had to seek alternative employment, but too often with little success, as agricultural laborers, headload and quarry workers, rubble breakers, brickmakers, and construction workers. Further, illicit brewing became the main source of livelihood in many places. In view of these difficulties, the economic consequences were severe, with earnings almost totally wiped out. Under a still worse and compelling situation, fishers often had to resort to loans with interest rates as high as 180%, while income

from other sources, notably from fishes, had been totally wiped out.

The provincial government had invited the ICAR experts to investigate the causes of the disease. At the same time, they had given free rations to the affected people, particularly fisherfolk, for a period of time. This was perhaps in contrast to other affected provinces, who possibly did not bother to show similar gestures to the affected people. Subsequently, agitations by the Kuttanad fisherfolk brought some additional relief when the government agreed to buy EUS-affected fish at Rs 2.00 per kg.

Three fishworkers' associations took up the matter of relief: the Kerala Matsya Thozhilali Aikya Vedi, the Matsya Thozhilali Union, and the Kerala Swatantra Matsya Thozhilali Federation.

Each union had its own particular opinion. The Vedi attributed the outbreak of EUS to the unscientific developmental efforts in the agricultural and fisheries sectors of Kuttanad. The Union demanded free rations for at least one month; a supply of safe drinking water; and financial aid and proper developmental works beginning at the panchayat level in order to generate employment. Meanwhile, the Federation demanded financial compensation for affected fisherfolk. Its agitation received powerful impetus from a fast by its leader Jose Kaleeckal, The Federation further tried to force the government to take necessary action, including "fixing a floor price for inland fish catches, and steps to contain the disease."

In view of the agitations indicated above, the government released Rs 3.75 million to the Fishermen's Welfare Board by November 1991 for the payment of compensation at Rs 150.00 per head. Nonetheless, though the report had been presented, the money could not possibly be distributed so promptly. However, the provincial government of its own volition proposed to sanction funds for liming in May 1992.

Clearly, the implementation of relief measures was not without its shortcomings. First, the government was slow to react, moreso as a result of being reassured by expert opinions that the impact of EUS could be over naturally with changed weather conditions. On the other hand, free rations had reached only a few thousand fisherfolk. It had reached mainly those fisherfolk who had been registered with fishers' societies. Despite the chief minister of the province admitting that he had been proved wrong on the recurrence of EUS, neither had the fisherfolk begun to receive any cash relief, nor were there enough government outlets to purchase the diseased fish even as late as October of that year.

According to the suffering mass of Kerala, but for the campaign by the unions, even this minor relief would not have possibly materialized. Interestingly enough, none of the unions had demanded prophylactic measures to combat EUS in spite of its evident recurrence.

Nevertheless, one of the useful by-products of the socioeconomic impact of the disease had been the government's fresh attention to conduct research on EUS in Kerala. Since the state experiences two monsoons and has semienclosed waters, the situation of Kerala is said to be strikingly different from others, mainly because this maritime province receives two monsoons in a year.

Notwithstanding the above, the effects of pesticides and agrochemicals on Kuttanad was debated. According to a very significant Indo-Dutch study on water conditions in Kuttanad, large doses of fertilizers (c 20,000 t) and toxic pesticides (c 500 t) had been sprayed over 66,000 ha of paddy fields. These might have entered the water body each time the fields had been drained prior to planting. Also, the "Grow More Paddy" program and the construction of the Thaneermukkom salt water barrier were said to be deleterious to the Kuttanad ecosystem because they were believed to have put an end to the erstwhile flushing out of the backwaters during monsoons.

Notwithstanding the above, a scientist from the Central Institute of Fisheries Technology (CIFT), ICAR, Mumbai, questioned the conclusion that had been drawn on pesticides. He had opined that residues of pesticides, and also heavy metals like mercury and cadmium, were within acceptable limits of toxicity. He opined that the first outbreak of EUS would have occurred in Kuttanad, and not in Wynad, if high levels of pesticides and heavy metals were the main causes of EUS.

Scientists from Thailand had a similar line of thinking. Kamonporn Tonguthai opined that there should be for more experimentation before a specific pesticide could be pinpointed as a triggering factor. She further indicated the case of Lao PDR, where EUS had been reported despite the absence of pesticide pollution. On the other hand, Supranee Chinabut indicated a study by FAO in which pesticides had been assigned as the possible cause for the outbreak of EUS. Further, studies had indicated the occurrence of EUS even in hilly ponds said to be not affected by environmental pollution. However, low alkalinity seemed to be the common factor all over SE Asia, alhough pesticides could affect water quality.

Almost similar views were put forward by the scientists from Kerala, who opined that the monsoon rains usually lower the alkalinity of water, leading to outbreaks of EUS. However, it had been pointed out that the pattern of the monsoons had been the same even prior to 1991. Further, Wynad seemed to have received fewer monsoons than Kuttanad, yet was the first site of the disease.

In addition to the above, some of the more farfetched etiological conjectures had pointed to infected fishlings brought into Kerala by private agencies, as well as to droppings of birds that had eaten diseased fish. Moreover, according to the subjective perceptions of the fisherfolk, EUS was regarded as a curse of nature and a warning against destructive human intervention into the natural aquatic environment.

Notwithstanding the above, works by different workers brought another dimension to the hypotheses on the **origin of EUS**. Highlighted was the strongly inter-linked intricate riverine network of Kerala's lotic water system. A continuity of water bodies could be established between the Kaveri and the Kaverretty Rivers, starting from Wynad in the north, which flowed through Tamil Nadu and Karnatak. Some of the prominent lotic systems in Kerala are the Pamba, Achenkoil, and Meenachil. It may be mentioned here that EUS had been reported in Tamil Nadu in early 1991 and in Karnataka in 1990. It could be probable that EUS might have spread from those neighboring provinces via the riverine network, possibly aided by floods.

2.3.4 In Assam

In India, widespread initiation of outbreak of EUS started from Barak Valley region of Assam since July 1988 (Kar, 1999, 2005, 2007, 2010, 2013, 2014; Kar and Dey, 1988a,b,c; 1990a,b,c; Kar and Das, 2004a,b; Kar et al., 1990a,b, 1993, 1994, 1995a,b,c, 1996a,b,c, 1997, 1998a,b,c,d, 1999a,b, 2000a,b,c,d, 2001a,b, 2002d,e,f, 2003a,b,c,d,e,f, 2004a,b,c; Patil et al., 2003). Outbreaks of EUS in India has been comprehensively reviewed at various levels (The Zoological Society of Assam, 1988; Jhingran and Das, 1990; National Workshop on Ulcerative Disease Syndrome in Fish, 1990); Kumar et al., 1991; ICSF, 1992; Das and Das, 1993; Mohan and Shankar, 1994; etc.).

From the Barak Valley region of Assam (July 1988), after its initiation in India, EUS has been spreading and sweeping almost unabated and more or less in a chronological manner through other regions of India, notably West Bengal (1989), Bihar (1989), Orissa (1989), UP (1990), MP (1990), Maharastra (1991−92), Karnataka (1993), Goa (1993), Tamil Nadu (1993), Andhra Pradesh (1992−93), Kerala (1994−95), until EUS had reached Pakistan during 1998−99 and is said to be progressing further.

However, in Assam, particularly in the Barak Valley region, EUS has become almost endemic, recurring almost every year between November and March. The detailed works done on EUS since its inception in India by Dr Kar and his associates of Assam (Central) University at Silchar will be presented a little later in this treatise.

2.3.5 Epizootic Ulcerative Syndrome Outbreak in Uttar Pradesh, India

Pradhan et al. (2014) have reported about the emergence of EUS and large-scale mortalities of cultured and wild fish species in Uttar Pradesh, India.

2.3.6 Epizootic Ulcerative Syndrome Outbreak in Goa, India

Rattan and Parulekar (1998) had dealt with the diseases and parasites of laboratory-reared and wild populations of banded pearl spot, *Eutroplus suratensis* (Cichlidae), in Goa.

2.3.7 Epizootic Ulcerative Syndrome Outbreak in Maharastra, India

Laharia (2013) worked on the immune reponse and status of EUS-affected fishes, *Channa punctatus* and *Clarias magur*, in Maharastra.

2.3.8 Details of the Overview of the Outbreak of Epizootic Ulcerative Syndrome, Aphanomycosis, and Studies in Different Countries of the World: Studies in Australia, Asia, the United Kingdom, the United States, and Africa

Concomitant to the occurrence and analyses of MG in Japan, there had been outbreaks of EUS in many countries outside Japan. Works on different aspects of EUS have been continuing since the early 1970s, beginning with Australia, followed by the sweeping of SE Asia and South Asia by EUS, and probably still continuing with its spread of EUS to the United States and other countries. Incidentally, the pattern of spread within and among the nations was said to be more or less consistent with progressive dissemination of a so-called single infectious agent.

2.3.9 More Information on Outbreaks and the Status of Epizootic Ulcerative Syndrome in Bangladesh

Muniruzzaman and Chowdhury (2008) had reported the ulcer diseases in cultured fish in Mymensingh and surrounding districts in Bangladesh.

The *Manual of Diagnostic Tests for Aquatic Animals 2009* has detailed about EUS and its outbreaks, and opines that **EUS** is an infection with an oomycete known as *A.invadans* or *A. piscicida.*

Hossain et al. (2011), in connection with outbreaks of EUS in Bangladesh, had worked on the isolation of some emergent bacterial pathogens recovered from capture and cultured fisheries in Bangladesh.

Islam and Chowdhury (2013) portrayed a detailed limnological status of **Trimohini Beel of** Rajshahi in Bangladesh, and reported about the outbreak of EUS.

Siddique et al. (2009) had reported about fungal disease of freshwater fishes in the Natore district of Bangladesh.

2.3.10 More Information on Outbreaks and the Status of Epizootic Ulcerative Syndrome in Pakistan

Pakistan was perhaps the most recent place on the Indian subcontinent to be invaded by EUS. EUS was confirmed among the snakeheads in Punjab province during April 1996 and among the mrigals in Sind province during January 1998 (DFID, 1998). The blotched or mud murrel was perhaps the most commonly affected species, with *Puntius* spp., *L. rohita*, and *C. reba* also reportedly affected by EUS. Approximately 20% of the fish farms in the Sialkot division of Punjab province had been affected by EUS, with the incidence being higher in ponds that were inundated by the floods during 1996 (AAHRI, ACIAR, IoA, and NACA, 1997). However, loss of fish was not very high in certain parts of the Punjab province, possibly due to the extensive use of tube well waters (with slightly higher salinity) for fish farms. EUS now seems to be quite well established in some places situated along the bank of the River Indus, and given its apparent rapid spread across the country, there are fears of serious future impacts to fisheries and aquaculture development.

2.3.11 Areas of Outbreak in Pakistan

EUS is said to have entered Pakistan during 1995 in the regions around the River Sutlej covering mainly the district of Bahawalnagar, which had borne the brunt of EUS. However, some people believed that EUS had occurred very preliminarily and briefly during 1988 as well. However, the disease at that time could not be confirmed as EUS, although it had been confirmed in the neighboring countries since the late 1980s. During 1996, the disease spread into the regions around the River Ravi (i.e., Head Balloki) and the River Dutlej (covering mainly the district of Kasur). In fact, the samples of affected murrels (notably, *Channa marulius*) collected from the backwaters at Balloki headworks (Kasur district) had been at first confirmed as EUS positive from AAHRI, Bangkok. Subsequently, a team of NACA and others visited the affected areas, collected samples in 1997, and reconfirmed the disease in the area as EUS. Subsequently, the workers at Fisheries Research and Training Institute (FRTI) in Pakistan recorded the occurrence of EUS from the five districts of the Punjab, viz., Sialkot, Gujranwala, Lahore, Kasur, and Bahawalnagar.

Later, during 1997, the disease had spread in the areas of River Chenab (covering mainly the districts of Sialkot, Gujranwala, and the ponds of Haed and Qadirabad). However, in the case of artificial ponds, the disease had been reported only from those ponds that had received floodwater, or where farmers had stocked their farms with seeds collected from natural water bodies. During 1998, interestingly, the disease spread into the regions of River Indus covering mainly Sukkur and Badin. Further, Wadahar, et al. (2012) reported further about the outbreak of EUS in Pakistan.

Anon (1998) reported about EUS in fishes of Pakistan in connection with a report regarding a mission to Punjab (Pakistan) in connection with EUS, April 22–25, 1998.

2.3.12 Status of Epizootic Ulcerative Syndrome in Nepal

Dahal et al. (2005) reported about the outbreak of EUS in Nepal.

Baidya and Prasad (2013) reported about the outbreak and prevalence of EUS among carps in Nepal.

Bhutan and the Eastern Terai of **Nepal** were first affected in 1989, and by 1993 EUS had covered almost the whole of this Himalayan region including Pokhara and Kathmandu, where even coldwater species like the *Tor* spp. were affected by EUS (Phillips, 1989; Shresta, 1994). It was estimated that *c* 20%–30% of Nepalese pond fish production (*c* 3000 mt) was lost each year as a result of EUS.

2.3.13 Status of Epizootic Ulcerative Syndrome in the UK

There has not been any possible large-scale outbreak of EUS in the UK. Like many other workers, Lilley and Roberts (1997) opined that *A. invadans* is a probable primary causative agent that could be responsible for development of the salient pathology of EUS.

2.3.14 Outbreaks and Research in Australia

Australia had also been affected by large-scale fish epidemics. The laboratory-based studies on EUS with estuarine fishes in Australia are said to have begun with various field observations. Further, Callinan et al. (1995a) made preliminary comparative studies on RSD in Australia. In addition to the above, Humphrey and Pearce (2006) revealed that EUS or "red-spot" (as it is known colloquially) is an ulcerative syndrome of fish that affects a range of native species in the Northern Territory of Australia. The disease also occurs in New South Wales, Queensland, and Western Australia, as well as in many Asian countries.

In fact, Humphrey and Pearce (2006) reported on the EUS (red-spot disease) in Australia, and had attempted to answer the question, **"what is epizootic ulcerative syndrome?"**

Choongo et al. (2009) and Boys et al. (2012) had reported about the emergence of EUS in native fish of the

Murray-Darling River System (MDRS) in Australia with a note on hosts, distribution, and possible vectors. This study was said to be the first published account of *A. invadans* in the wild fish populations of the MDRS, and was perhaps the first confirmed record of EUS in *M. ambigua, M. peelii*, and *L. unicolor*.

Anon (2011) dealt with EUS (aka RSD, MG, or UM) while preparing **Identification Field Guide** for **Aquatic Animal Diseases Significant to Australia**.

Anon (2008) reported about EUS/RSD as one of the fungal diseases among the finfishes in Australia.

Anon, under the aegis of **Gladstone Fish Health Scientific Advisory Panel,** prepared a report based on information provided to the Panel by the Queensland government and other stakeholders relating to fish health issues observed in Gladstone Harbor and surrounding areas.

A little background of the above works is given below:

In 2011, the Queensland government received reports (primarily by commercial fishers) of barramundi fish and subsequently other species being caught with obvious signs of disease, including bulging/red eyes, blindness, severe skin lesions, and skin discoloration. The government then undertook an investigation of Gladstone Harbor and surrounding areas following reports that commercial fishers were sick with what appeared to be bacterial infections on their arms, feet, and legs following contact with, or abrasions and fish spikes from, net-caught barramundi that were exhibiting evidence of disease. At that time, the Queensland government was concerned about the potential food safety issues of consuming the diseased fish, given that the type of disease remained unknown. Furthermore, there were concerns about the possibility of the transfer of the disease from affected fish to other fish and consequently its entry into the food chain.

2.3.15 More Information on Outbreaks of Epizootic Ulcerative Syndrome and Works Done in the Philippines

Callinan et al. (1995a) reported about outbreaks of EUS in the Philippines. They worked on the *Aphanomyces* species associated with EUS in the Philippines as well as RSD, which was said to be prevalent in Australia. They made preliminary comparative studies of RSD in Australia with EUS.

2.3.16 Outbreaks and Works Done in the United States

Concomitant to the account on works done on fish disease in the United States, there had possibly been some more episodes of outbreaks of EUS in the United States. Works on EUS in the United States have been done by different workers. Notably, Kiryu et al. (2002) is reported to have been able to generate the lesions of EUS by both injection and bath exposure to zoospores of *A. invadans*. In dose—response studies with menhaden (*B. tyrannus*), it was pointed out that *c* 31% of the fish specimens inoculated with as few as one zoospore are said to have developed characteristic lesions of EUS within 2 weeks of inoculation. On the other hand, LD_{50} by injection had been estimated to be only 10 zoospores per fish (Kiryu et al., 2003). These findings seemed to corroborate that oomycete is a primary pathogen and is highly virulent. These also seemed to support Koch's hypothesis.

Further, Kiryu, et al. (2002, 2003) had worked on the induction of skin ulcers in Atlantic Menhaden by injection and aqueous exposure to the zoospores of *A. invadans,* a study done possibly in connection with a probable outbreak of EUS in the United States.

Saylor et al. (2010) had reported about EUS caused by *A. invadans* in captive bullseye snakehead (*C. marulius*) collected from south Florida, United States. This pathogen is also present in estuaries and freshwater bodies of the Atlantic and Gulf coasts of the United States. There had been events of mass mortality of 343 captive juvenile bullseye snakehead *C. marulius* collected from freshwater canals in Miami-Dade County, Florida. Clinical signs appeared within the first 2 days of captivity and included petechiae, ulceration, erratic swimming, and inappetence. Histological examination revealed hyphae emanating from the skin lesions and invading deep into the musculature and internal organs. Species identification was confirmed using a species-specific PCR assay. Mortality of 100% occurred despite therapeutic attempts. This represented the first documented case of EUS in bullseye snakehead fish collected from waters in the United States. Future investigations of the distribution and prevalence of *A. invadans* within the bullseye snakehead range in south Florida might give some insight into this pathogen—host system.

2.3.17 Outbreak of Epizootic Ulcerative Syndrome in Africa

Karl and Benjamin. (2012) reported about widespread outbreaks of EUS in Africa that threatened the aquatic ecosystems of the continent. They communicated that in late 2006, an unusual ulcerative condition in wild fish was reported for the first time in Africa from the Chobe and upper Zambezi Rivers in Botswana and Namibia.

Further, Nsonga et al. (2013), in connection with outbreaks of EUS in Africa, did epidemiological works in the Zambezi River system in Zambia.

Huchzermeyer and Van der Waal (2012) had reported about the exotic fish disease called EUS threatening Africa's aquatic ecosystems.

Food and agriculture organization of the united nations (FAO) (2009) (Rome, 2009) had dealt in detail with

"**What you need to know about Epizootic ulcerative syndrome (EUS),**" in their extension brochure. It had been brought out that EUS is an infection caused by an oomycete fungi known as *A. invadans* or *A. piscicida*. **FAO** opined that *Aphanomyces* is a member of a group of organisms that were earlier commonly known as water molds. They are currently recognized as belonging to the group of heterokonts or stramenopiles (OIE, 2006). EUS is an epizootic condition that has affected wild and farmed freshwater and estuarine finfish since it was first reported in 1971. EUS is also known by other names such as RSD, MG, and UM, and in 2005 it was suggested that EUS be renamed epizootic granulomatous aphanomycosis (EGA) (Baldock et al., 2005).

Noga et al. (1996) showed that sublethal exposure to toxins produced by a recently identified "phantom" dinoflagellate *P. piscicida*, also responsible for high mortalities in the Pamlico river (Burkholder et al., 1992), can result in dermatitis and subsequent development of UM.

Munday (1985) reported the presence of severely ulcerated red cod (*P. bachus*) in the River Tamar near Launceston, Tasmania in November 1980 and 1981. Although a variety of bacteria and parasites were identified from the fish, pollution was considered to be the main cause of the disease. Munday (pers. comm.) believed that the said ulcer disease was possibly the same syndrome as EUS even though it had occurred in higher salinity waters, but he added that after Launceston's sewerage system had been improved, the disease was no longer reported.

2.4 EPIZOOTIC ULCERATIVE SYNDROME DISEASE AND DISASTER RESPONSE

Brown et al. (2014) had reported about Supporting Disaster Response and Preparedness in Aquaculture. They opined that the aquaculture sector is particularly vulnerable to natural and human-induced rapid or slow onset of disasters. This was considered to be due to the risks associated with the location of fish farm production facilities and other infrastructure, the type of activities associated with fish farming, the nature of fish in specific developmental stages (fry, fingerling, and larvae), and the limited capacity of the sector to reduce and cope with the potential consequences of a disaster.

REFERENCES

Anon, 1992. Notes on mycological procedure. In: Workshop on Mycological Aspects of Fish and Shellfish Disease, Bangkok, Thailand, January 1992. Aquatic Animal Health Research Institute.

AAHRI, ACIAR, IoA and NACA, 1997. Epizootic Ulcerative Syndrome (EUS) of fishes in Pakistan. A report of the findings of an ACIAR/ DFID-funded mission to Pakistan. pp. 9—19 March 1997

Anon, 1998. Workbook from the Workshop on the Fungal Aetiology of EUS. The Aquatic Animal Health Research Institute. Bangkok Thailand.

Anon, 2008. Disease of finfish: Fungal diseases-epizootic ulcerative syndrome/red spot disease. In Australian Government Department of Agriculture, Fisheries and Forestry, Aquatic Animal Diseases Significant to Australia: Identification Field Guide. Australian Government Department Agriculture Forestry and Fisheries, Canberra, pp. 1–3.

Anon, 2011. World Animal Health Information Database Interface: Exceptional Epidemiological Events, World Organisation for Animal Health, Paris, Viewed in 2011.

Bloch, 1785–1795. Naturgeschichte der auslandischen fish, Berlin, 9 parts in 2, and Atlas (vols. 3).

Burkholder, J.M., Noga, E.J., Hobbs, C.H., Glasgow Jr., H.B., 1992. New phantom dinoflagellate is the caustive agent of major estuarine fish kills. Nature 358, 407–410.

Balasuriya, L.K.S.W., 1994. Epizootic ulcerative syndrome in fish in Sri lanka, country status report. In: Robert, R.J., Campbell, B., Macrae, I.H. (Eds.), Proceeding of the Oda Regional Seminar on Epizootic Ulcerative. Aquatic Animal Health Research Institute, Bangkok, Thailand, pp. 325, 39–47.

Baldock, F.C., Blazer, V., Callinan, R., Hatai, K., Karunasagar, I., 2005. Outcome of a short expert consultation on epizootic ulcerative syndrome (EUS): re-examination of causal factors, case definition and nomenclature. In: Walker, P., Laster, R., Bondad-Reantaso, M.G. (Eds.), Disease in Asian Aquaculture V. Fish Health Section. Asian Fisheries Society, Manila, Phillippines, pp. 555–585.

Bondad-Reantaso, M.G., Paclibare, J.O., Lumanlan-Mayo, S.C., Catap, E.S., 1994. EUS outbreak in the Philippines: a country report: pp. 61–87. In: Roberts, R.J., Campbell, B., Mac Rae, I.H. (Eds.), Proceedings of the ODA Regional Seminar on Epizootic Ulcerative Syndrome, 25–27 January, 1994. Aquatic Animal Health Research Institute, Bangkok.

Barua, G., 1994. The status of epizootic ulcerative syndrome of fish of Bangladesh: pp. 13–20. In: Roberts, R.J., Campbell, B., Mac Rae, I.H. (Eds.), Proceedings of the ODA Regional Seminar on Epizootic Ulcerative Syndrome, 25–27 January, 1994. Aquatic Animal Health Research Institute, Bangkok.

Boys, C.A., Rowland, S.J., Gabor, M., Gabor, L., Marsh, I.B., Hum, S., Callinan, R.B., 2012. Emergence of epizootic ulcerative syndrome in native fish of the Murray–Darling River System, Australia: hosts, distribution and possible vectors. PLoS One 7, e35568. http://dx.doi.org/10.1371/journal.pone.0035568.

Baidya, S., Prasad, A., 2013. Prevalence of epizootic ulcerative syndrome (EUS) in carps. Nepal. J. Zool. 1 (1), 41–47.

Brown, D., Poulain, F., Subasinghe, R., Reantaso, M., 2014. Supporting disaster response and preparedness in aquaculture. In: Preparing for the Day after, pp. 1351, Report of the Food and Agricultural Organisation of the UN (Rome).

Callinan, R.B., Fraser, G.C., Virgona, J.L., 1989. Pathology of red spot disease in sea mullet, *Mugil cephalus* L. from Eastern Australia. J. Fish Dis. 12, 467–479.

Callinan, R.B., 1994a. Red spot disease-EUS in Australia. In: Roberts, R.J., Campbell, B., MacRae, I.H. (Eds.), Proceedings of the ODA Regional Seminar on Epizootic Ulcerative Syndrome, Aquatic Animal Health Research Institute, Bangkok, 25–27 January 1994. AAHRI, Bangkok, pp. 189–193.

Callinan, R.B., 1994b. A comparative review of *Aphanomyces* sp. associated with epizootic ulcerative syndrome, red spot disease and mycotic granulomatosis: pp. 248–252. In: Roberts, R.J., Campbell, B.,

MacRae, I.H. (Eds.), Proc. ODA Regional Seminar on Epizootic Ulcerative Syndrome, 25–27 January 1994. Aquatic Animal Health Research Institute, Bangkok.

Callinan, R.B., 1994c. Red spot disease—EUS in Australia. In: Roberts, R.J., Campbell, B., MacRae, I.H. (Eds.), Proceedings of the ODA Regional Seminar on Epizootic Ulcerative Syndrome, 25–27 January 1994. Aquatic Animal Health Research Institute, Bangkok, pp. 82–88.

Callinan, R.B., Paclibare, J.O., Reantaso, M.B., Lumanlan-Mayo, S.C., Fraser, G.C., Sammut, J., 1995a. EUS outbreaks in estuarine fish in Australia and the Philippines: associations with acid sulphate soils, rainfall and *Aphanomyces*. In: Shariff, M., Arthur, J.R., Subasinghe, R.P. (Eds.), Diseases in Asian Aquaculture II. Fish Health Section. Asian Fisheries Society, Manila, pp. 291–298.

Chinabut, S., Roberts, R.J., Willoughby, G.R., Pearson, M.D., 1995. Histopathology of snakehead, *Channa striatus* (Bloch), experimentally infected with the specific Aphanomyces fungus associated with epizootic ulcerative syndrome (EUS) at different temperatures. J. Fish Dis. 18, 41–47.

Callinan, R.B., Paclibare, J.O., Bondad-Reantaso, M.G., Chin, J.C., Gogolewski, R.P., 1995b. Aphanomyces species associated with epizootic ulcerative syndrome (EUS) in the Philippines and red spot disease (RSD) in Australia: preliminary comparative studies. Dis. Aquat. Org. 21, 233–238.

Callinan, R.B., Sammut, J., Fraser, G.C., 1996. Epizootic ulcerative syndrome (red spot disease) in estuarine fish—confirmation that exposure to acid sulfate soil runoff and an invasive aquatic fungus, Aphanomyces sp., are causative factors. In: Proceedings of the Second National Conference on Acid Sulfate Soils. Roberts J Smith and Associates and ASSMAC, Australia.

Callinan, R.B., 1997. Pathogenesis of Red Spot Disease (Epizootic Ulcerative Syndrome) in Estuarine Fish in Eastern Australia and the Philippines. University of Queensland, Australia, pp. 232.

Choonga, K., Hang'ombe, B., Samui, K.L., Syachaba, M., Phiri, H., Maguswi, C., et al., 2009. Environmental and climatic factors associated with epizootic ulcerative syndrome (EUS) in fish from the Zambezi floodplains, Zambia. Bull. Environ. Contam. Toxicol. 83, 474–478. http://dx.doi.org/10.1007/s00128-009-9799-0. PMid: 19565173.

Chulalongkorn University, 1983. The Symposium on Fresh Water Fishes Epidemic: 1982–1983. Chulalongkorn University, Bangkok, 23–24 June 1983.

Chulalongkorn University, 1985. Proceedings of the Technical Conference on Living Aquatic Resources. Chulalongkorn University, Bangkok, 7–8 March 1985.

Clark, S., 1991. Snakeheads. The species and their distribution. Aquarist and Pondkeeper 56 (8), 17–22.

Costa, H.H., Wijeyratne, M.J.S., 1989. Epidemiology of epizootic ulcerative syndrome occurring for the first time among fishes in Sri Lanka. J. Appl. Ichthyol. 1, 48–52.

Coates, D., Nunn, M.J., Uwate, K.R., 1989. Epizootic ulcerative disease of freshwater fish in Papua New Guinea. Sci. N. Guin. 15, 1–11.

Chulalongkorn University, 1987. Proceedings of the 2nd Seminar on Living Aquatic Resources, 17–18 December, 1987. Chulalongkorn University, Bangkok.

Dykstra, M.J., Noga, E.J., Levine, J.F., Moye, D.W., Hawkins, J.H., 1986. Characterization of the Aphanomyces species involved with ulcerative mycosis (UM) in menhaden. Mycologia 78 (4), 664–672.

DFID, 1998. A Report on the 2nd Mission to Investigate Epizootic Ulcerative Syndrome (EUS) in Pakistan, 19–30 April, 1998.

Das, M.K., Das, R.K., 1993. A review of the fish disease, epizootic ulcerative syndrome in India. Environ. Ecol. 11 (1), 134−145.

Dahal, S.P., Shreshta, M.K., Pradhan, S.K., Jha, D.K., 2005. Occurrence of epizootic ulcerative syndrome in pond fish of Kapilvastu district of Nepal. Dis. Asian Aquac. VI 169−178.

Egusa, S., Masuda, N., 1971. A new fungal disease of *Plecoglossus altivelis*. Fish Pathol. 6, 41−46.

FAO, 1986. Report of the Expert Consultation on Ulcerative Fish Diseases in the Asia-Pacific Region. (TCP/RAS/4508). Bangkok, August 1986. FAO, Regional Office for Asia and the Pacific, Bangkok.

Fraser, G.C., Callinan, R.B., Calder, L.M., 1992. *Aphanomyces* species associated with red spot disease: an ulcerative disease of estuarine fish of Eastern Australia. J. Fish Dis. 15, 173−181.

FAO, 2009. What You Need to Know about Epizootic Ulcerative Syndrome (EUS)—an Extension Brochure, pp. 1−36. Food and Agricultural Organisation of the UN, Rome.

Guizhen, M.J., 1990. China report: pp. 11−13. In: Regional Research Programme on Relationships between Epizootic Ulcerative Syndrome in Fish and the Environment, 13−26 August 1990. NACA, Bangkok.

Hatai, K., Egusa, S., Takashashi, S., Ooe, K., 1977a. Study on the pathogenic fungus of mycotic granulomatosis: I: Isolation and pathogenicity of the fungus from cultured ayu infected with the disease. Fish Pathol. 11 (2), 129−133.

Hatai, K., Egusa, S., Takahashi, S., Ooe, K., 1977b. Study on the pathogenic fungus of mycotic granulomatosis—I. Isolation and pathogenicity of the fungus from cultured-ayu infected with the disease. Fish Pathol. 12, 129−133.

Hatai, K., 1980a. Studies on pathogenic agents of saprolegniasis in fresh water fishes. Spec. Rep. Nagasaki. Pref. Inst. Fish 8, 1−95.

Hatai, K., 1980b. Studies on the Pathogenic Agents of Saprolegniansis in Freshwater Fishes. Special Report of Nagasaki Prefectural Inst. of Fisheries, No. 8. Matsugae-Cho, Nagasaki, Japan.

Haines, A.K., 1983. Fish fauna and ecology: pp. 367−384. In: Petr, T. (Ed.), The Purari Tropical Environment of High Rainfall River Basin. Dr.W.Junk Publishers, The Netherlands.

Hatai, K., 1994. Mycotic granulomatosis in ayu (*Plecoglossus altivelis*) due to *Aphanomyces piscicida*. In: Roberts, R.J., Campbell, B., MacRae, I.H. (Eds.), Proc. ODA Regional Sem. on Epizootic Ulcerative Syndrome, 25−27 January 1994. Aquatic Animal Health Research Institute, Bangkok.

Hatai, K., Nakamura, K., Rha, S.A., Yusa, K., Wada, S., 1994. *Aphanmyces* infection in dwarf gourami (*Colisa lalia*). Fish Pathol. 29 (2), 95−99.

Hanjavanit, C., Hiroki, S., Hatai, K., 1997. Mycotic granulomatosis found in two species of ornamental fishes imported from Singapore. Mycoscience 38, 433−436.

Humphrey, J.D., Pearce, M., 2006. Epizootic Ulcerative Syndrome (Red-spot Disease) Fishnote. Northern Tereritory Government, Australia, pp. 1−4.

Hossian, M.M.M., Rahman, M.A., Mondal, S., Shadat Mondal, A.S.M., Chowdhury, M.B.R., 2011. Isolation of some emergent bacterial pathogens recovered from capture and culture fisheries in Bangladesh. Bangladesh Res. Publ. J. 6 (1), 77−90.

Huchzermeyer, K.D.A., Van der Waal, B.C.W., 2012. Epizootic ulcerative syndrome: exotic fish disease threatens Africa's aquatic ecosystems. J. S. Afr. Vet. Assoc. 83 (1), Art. #204, 6 pages. http://dx.doi.org/10.4102/jsava.v83i1.204.

ICSF, 1992. Enigma of EUS Consultation on Epizootic Ulcerative Syndrome vis-à-vis the Environment and the People, 25−26 May 1992. International Collective in Support of Fish workers, Trivandrum (Kerala), 40 pp, Madras (India).

Islam, Md A., Chowdhury, A.H., 2013. Limnological status of Trimohini Beel of Rajshahi. Bangladesh. J. Asiat. Soc. Bangladesh, Sci. 39 (2), 173−182.

Jothy, A.A., 1981. Preliminary Report on the outbreak of Wabak Kudison in Freshwater fish in paddy-growing areas in Peninsular Malaysia. Report to the Ministry of Agriculture, Malaysia, December, 1981, 17 pp.

Jhingran, A.G., Das, M.K., 1990. Epizootic Ulcerative Syndrome in Fishes. Bulletin of the Central Inland Capture Fisheries Research Institute (No.65). CIFRI, Barrackpore, India.

Kanchanakhan, S., 1996a. Field and Laboratory Studies on Rhabdoviruses Associated with Epizootic Ulcerative Syndrome (EUS) of Fishes (Ph.D. thesis). University of Stirling, Scotland, 278 pp.

Kanchanakhan, S., 1996b. Epizootic ulcerative syndrome (EUS): a new look at the old story. AAHRI Newsletter 5, 2−3.

Kar, D., Dey, S.C., 1988a. A critical account of the recent fish disease in the Barak valley of Assam. In: Proc. Regional Symp. on Recent outbreak of Fish Diseases in North-East India, vol. 1, 8 pp.

Kar, D., Dey, S.C., 1988b. Impact of recent fish epidemics on the fishing communities of Cachar district of Assam. In: Proc. Regional Symp. On Recent outbreak of Fish Diseases in North-East India, vol. 1, 8 pp.

Kar, D., Dey, S., 1988c. Preliminary electron microscopic studies on diseased fish tissues from Barak valley of Assam. In: Proc. Annual Conference of Electron Microscopic Society of India, vol. 18, 88 pp.

Kar, D., Dey, S.C., 1990a. Fish disease syndrome: a preliminary study from Assam. Bangladesh J. Zool. 18, 115−118.

Kar, D., Dey, S.C., 1990b. A preliminary study of diseased fishes from Cachar district of Assam. Matsya 15−16, 155−161.

Kar, D., Dey, S.C., 1990c. Epizootic ulcerative syndrome in fishes of Assam. J. Assam. Sci. Soc. 32 (2), 29−31.

Kar, D., 2012. Essentials of Fish Biology. Dominant Publishers, New Delhi pp. vii + 244.

Kar, D., Das, M., 2004a. Preliminary histochemical studies with diseased fishes suffering from epizootic ulcerative syndrome in Assam. In: Proc. National Symposium for Management of Aquatic Resources for Biodiversity Maintenance and Conservation, vol. 1, 27 pp.

Kar, D., Das, M., 2004b. Preliminary enzymological studies with diseased fishes suffering from epizootic ulcerative syndrome in Fishes. In: Proc. National Conference on Fish and their Environment, vol. 1, 25 pp.

Kar, D., Dey, S.C., Michael, R.G., Kar, S., Changkija, S., 1990a. Studies on fish epidemics from Assam, India. J. Indian Fish Assoc. 20, 73−75.

Kar, D., Bhattacharjee, S., Kar, S., Dey, S.C., 1990b. An account of 'EUS' in Fishes of Cachar district of Assam with special emphasis on Microbiological studies. Souv. In: Annual Conference of Indian Association of Pathologists and Microbiologists of India (NE Chapter), vol. 8, pp. 9−11.

Khatri, S., 2004. Prevalence of Fish Diseases in Sambalpur, Orissa. M.Sc., Dissertation. G.M. College, Sambalpur. 35 pp.

Kiryu, Y., Shields, J.D., Vogelbein, W.K., Zwerner, D.E., Kator, H., Blazer, V.S., 2002. Induction of skin ulcers in Atlantic menhaden by injection and aqueous exposure to the zoospores of Aphanomyces invdans. J. Aquat. Anim. Health 14, 11−24.

Kiryu, Y., Shields, J.D., Vogelbein, W.K., Kator, H., Blazer, V.S., 2003. Infectivity and pathogenicity of the oomycete *Aphanomyces invadans* in Atlantic menhaden *Brevoortia tyrannus*. Dis. Aquat. Org. 54, 135−146.

Karl, D.A., Huchzermeyer, Benjamin, C.W., van der Wall, 2012. Epizootic ulcerative syndrome: exotic fish disease threatens Africa's aquatic ecosystems (2012), Open assess. J. S. Afr. Vet. Assoc. 6, 112−118.

Kumar, D., Day, R.K., Sinha, A., 1991. In: Sinha, V.R.P., Srivastava, H.C. (Eds.), Aquaculture Productivity. Oxford and IBH Publ. Co., New Delhi, pp. 345.

Kar, D., Dey, S.C., Kar, S., Bhattacharjee, N., Roy, A., 1993. Virus-like particles in epizootic ulcerative syndrome of fish. In: Proc. International Symp. on Virus-Cell Interaction: Cellular and Molecular Responses, vol. 1, 34 pp.

Kar, D., Dey, S.C., Kar, S., Roy, A., Michael, R.G., Bhattacharjee, S., Changkija, S., 1994. A Candidate virus in epizootic ulcerative syndrome of fish. In: Proc. National Symp. of the Indian Virological Society, vol. 1, 27 pp.

Kar, D., Roy, A., Dey, S.C., Menon, A.G.K., Kar, S., 1995a. Epizootic ulcerative syndrome in fishes of India. World Congr. In Vitro Biol. In Vitro 31 (3), 7 pp.

Kar, D., Kar, S., Roy, A., Dey, S.C., 1995b. Viral disease syndrome in fishes of North-East India. In: Proc. International Symp. Of International Centre for Genetic Engg. And Biotechnology (ICGEB) and the Univ. of California at Irvine, vol. 1, 14 pp.

Kar, D., Dey, S.C., Kar, S., 1995c. A viral disease among the fishes of North-East India. Annual Congress on Man and Environment. Natl. Environ. Sci. Acad. and Natl. Inst. Oceanogr. 10, 62.

Kar, D., Dey, S.C., Purkayastha, M., Kar, S., 1996a. An overview of the impediments in conservation of biodiversity of Lake Sone in Assam. In: Proc. Seminar on Conservation of Biodiversity in Indian Aquatic Ecosystems, vol. 1. JawaharLal Nehru University, New Delhi.

Kar, D., Dey, S.C., Roy, A., Kar, S., 1996b. Viral disease syndrome in fishes of India. Proc. Int. Congr. Virol. 10.

Kar, D., Purkayastha, M., Kar, S., 1996c. Biodiversity conservation prioritisation project: a case study from Sone Beel in Assam. In: Proc. National Workshop on Biodiversity Conservation Prioritisation Project (BCPP), World Wide Fund (WWF) for Nature-India and Centre for Ecological Sciences. Indian Institute of Science, Bangalore.

Kar, D., Saha, D., Laskar, R., Barbhuiya, M.H., 1997. Biodiversity conservation Prioritisation Project (BCPP) in barak valley region of Assam. In: Proc. National Project Evaluation Workshop on BCPP, Betla Tiger Reserve and National Park, 1 (Palamu).

Kar, D., Saha, D., Dey, S.C., 1998a. Epizootic ulcerative syndrome in Barak valley of Assam: 2—4. In: Project Report Submitted and Presented at the National Symposium of Biodiversity Conservation Prioritisation Project (BCPP) Held at WWF-India, 18—19 Jan 1998 (New Delhi).

Kar, D., Dey, S.C., Roy, A., 1998b. Present status of epizootic ulcerative syndrome (EUS) in Southern Assam. In: Proc. Regional Project Initiation Workshop for NATP-ICAR-NBFGR Project, vol. 1, 9 pp.

Kar, D., Dey, S.C., Kar, S., Roy, A., 1998c. An account of epizootic ulcerative syndrome in Assam: pp. 1—3. In: Final Project Report of Biodiversity Conservation Prioritisation Project (BCPP). Submitted and presented at the International Project Finalisation Workshop of BCPP, vol. 1, 22—28 April 1998 (New Delhi).

Kar, D., Dey, S.C., Roy, A., 1998d. Fish disease diagnosis and fish Quarantine problems in India and south-east asia with particular emphasis on epizootic ulcerative syndrome (EUS) fish disease problems. In: Proc. International Workshop on Fish Disease Diagnosis and Quarantine, Indian Council of Agricultural Research (ICAR)-Ministry of Agriculture (MoA), Govt. of India (GoI)-Network of Aquaculture Centres in Asia (NACA)-Food and Agricultural Organisation (FAO) of the United Nations (UN)-OIE, France: Held at the Central Institute of Freshwater Aquaculture (CIFA), vol. 1.

Kar, D., Rahaman, H., Barnman, N.N., Kar, S., Dey, S.C., Ramachandra, T.V., 1999a. Bacterial pathogens associated with epizootic ulcerative syndrome in freshwater fishes of India. Environ. Ecol. 17 (4), 1025—1027.

Kar, D., Mandal, M., Bhattacharjee, S., 1999b. Fungal pathogens associated with epizootic ulcerative syndrome in fishes of barak valley region of Assam. In: Proc. 1st National Conference on Fisheries Biotechnology, vol. 1. CIFE, 34 pp.

Kar, D., Dey, S.C., Roy, A., 2000a. Fish genetic resources in the principal rivers and wetlands in North-East India with special emphasis on Barak valley (Assam), in Mizoram and in Tripura, with a note on epizootic ulcerative syndrome fish disease. In: Proc. National Project Initiation Workshop of the NATP-ICAR World Bank- aided Project on Germplasm Inventory, Evaluation and Gene Banking of Freshwater Fishes, vol. 1. National Bureau of Fish Genetic Resources (NBFGR), Lucknow, 12 pp.

Kar, D., Mandal, M., Laskar, B.A., Dhar, N., Barbhuiya, M.H., 2000b. Ichthyofauna of some of the oxbow lake in Barak valley region of Assam. In: Proceedings of the National Symposium on Wetlands and Fisheries Research in the New Millennium, vol. 1, 16 pp.

Kar, D., Dey, S.C., Roy, A., Mandal, M., 2000c. Epizootic ulcerative syndrome fish disease in barak valley region of Assam, India. In: Proc. Nat. Symp. Current Trends in Wetlands and Fisheries Research in the New Millennium, vol. 1, 2 pp.

Kar, D., Dey, S.C., Mandal, M., Lalsiamliana, 2000d. Epizootic Ulcerative Syndrome among the fishes of Assam. In: Proc. National Workshop of NATP-ICAR-NEC North-East Programme, Shillong, vol. 1, 24 pp.

Kar, D., Mandal, M., Lalsiamliana, 2001a. Species composition and distribution of Fishes in the rivers in Barak valley region of Assam and in the Principal rivers of Mizoram and Tripura in relation to their habitat parameters. In: Proc. National Workshop, NATP-ICAR Project, Midterm Review, vol. 1. Central Marine Fisheries Research Institute, Cochin, 25 pp.

Kar, D., Laskar, B.A., Lalsiamliana, 2001b. Further studies on the Ichthyospecies composition and distribution of Freshwater Fishes in Barak drainage and in the Principal rivers in Mizoram and in Tripura with a note on their Feeding and Breeding biology. In: Proc. National Project Monitoring Workshop of NATP-ICAR Project, vol. 1. National Bureau of Fish Genetic Resources, Lucknow, 22 pp.

Kar, D., Dey, S.C., Roy, A., 2002d. Prevalence of epizootic ulcerative syndrome among fishes of Assam. In: Proc. Regional Symp. on Biodiversity, S.S. College, Assam (Central) University, Hailakandi (Assam).

Kar, D., Dey, S.C., Roy, A., 2002e. On the diseased fishes suffering from hitherto unknown epizootic ulcerative syndrome in fishes in India. December 2002. In: Proc. All-India Congress of Zoology, Bangalore.

Kar, D., Dey, S.C., Roy, A., 2002f. Prevalence of epizootic ulcerative syndrome (EUS) among fishes of Mizoram. In: Proc. Regional Symp. On Aquaculture, vol. 1.

Kar, D., Dey, S.C., Datta, N.C., 2003a. Welfare Biology in the New Millennium. Allied Publishers Pvt. Ltd, Bangalore pp. xx + 97.

Kar, D., Dey, S.C., Roy, A., 2003b. Capture Fishery in Lnetic Systems with some light on Sone Beel in Assam with special reference to prevalence to EUS in the Barak valley region of Assam. In: Proc. Lecture Series in UGC-sponsored Vocational Course in Industrial Fish and Fisheries. Cachar College, Assam University, Silchar. Abstracts, 1.

Kar, D., Roy, A., Dey, S.C., 2003c. Epizootic ulcerative disease syndrome in freshwater fishes with a note on its management for Sustainable fisheries. In: Proc. International Conference on Disease Management for Sustainable Fisheries, 26—29 May, 2003. Department of Aquatic Biology and Fisheries, University of Kerala, Trivandrum.

Kar, D., Lohar, M., Ngassepam, Z., Sinom, S., Tiwary, B., Bawari, M., 2003d. Fish bioresources in certain rivers of Assam and Manipur with a note on their assessment, management and conservation. In: Proc. Nat. Symp. Assessment and Management of Bioresources. North Bengal University and The Zoological Society, Calcutta, 28—30 May, 2003, 56 pp.

Kar, D., Dey, S.C., Roy, A., 2003e. Epizootic ulcerative syndrome: a virulent disease among the freshwater fishes of India. In: Proc. Guest Lecture Delivered at the Dept. of Zoology. Arunachal University, Arunachal Pradesh, p. 1.

Kar, D., Dey, S.C., Roy, A., 2003f. Biodiversity of freshwater fishes of North-East India with a note on their health status. In: Proc. DBT-sponsored Guest Lecture Series Delivered at the Dept. of Molecular Biology and Biotechnology. Tezpur (Central) University, Tezpur, p. 1.

Kar, D., Roy, A., Dey, S.C., 2004a. An overview of fish genetic diversity of North-East India, 18—19 March 2004. In: Garg, S.K., Jain, K.L. (Eds.), Proc. National Workshop on Rational use of Water Resources for Aquaculture, vol. 1. CCS Haryana Agricultural University, pp. 164—171.

Kar, D., Roy, A., Mazumder, J., Patil, P., 2004b. Biotechnological approach for defining epizootic ulcerative syndrome disease in freshwater fishes. In: Proc. Mid-term Review. DBT-sponsored Project, New Delhi, p. 1.

Kar, D., Dey, S.C., Roy, A., Mazumder, J., Patil, P., Kohlapure, R.M., 2004c. Fish disease prevalence in Assam with particular reference to epizootic ulcerative syndrome and its pathogens. In: Proc. International Conference 'Bioconvergence,' 18—20 November 2004, vol. 1. Thapar Institute of Engineering and Technology, Patiala, 220 pp.

Kar, D., 2007. Fundamentals of Limnology and Aquaculture Biotechnology. Daya Publishing House, New Delhi pp. vi + 609.

Kar, D., 1999. Microbiological and Environmental Studies in Relation to Fishes of India. Gordon Research Conference, Connecticut, USA.

Kar, D., 2005. Sustainability issues of inland fish biodiversity and fisheries in Barak drainage (Assam), in Mizoram and in Tripura. In: Proceedings of the International Symposium on Improved Sustainability of Fish Production Systems and Appropriate Technologies for Utilisation. Cochin University of Science and Technology, School of Industrial Fish and Fisheries, p. 1.

Kar, D., 2010. Biodiversity Conservation Prioritisation. Swastik Publications2, New Delhi pp. xi + 167.

Kar, D., 2013. Wetlands and Lakes of the World. pp. xxx + 687. Springer, London. Print ISBN 978-81-322-1022-1; e-Book ISBN: 978-81-322-1923-8.

Kar, D., 2014. Fish diversity and habitat parameters in the water bodies of North-east India with a note on health of Fishes. In: Proc. International Symposium, LAKE—2014. India Institute of Science, Bengaluru, p. 20.

Llobrera, A.T., Gacutan, R.Q., 1987. Aeromonas hydrophila associated with ulcerative disease epizootic in Laguna de Bay Philippines. Aquaculture 67, 273—278.

Lian, C.X., 1990. China report. In: Regional Research Programme on the Relationship between Epizootic Ulcerative Syndrome in Fish and the Environment, 13—26 August 1990. NACA, Bangkok, pp. 8—9.

Levine, J.F., Hawkins, J.H., Dykstra, M.J., Noga, E.J., Moye, D.W., Cone, R.S., 1990a. Species distribution of ulcerative lesions on finfish in the Tar-Pamlico River Estuary, North Carolina. Dis. Aquat. Org. 8 (1), 1—5.

Levine, J.F., Hawkins, J.H., Dykstra, M.J., Noga, E.J., Moye, D.W., Cone, R.S., 1990b. Epidemiology of ulcerative mycosis in Atlantic menhaden in the Tar-Pamlico River Estuary, North Carolina. J. Aquat. Animal Health 2 (3), 162—171.

Lilley, J.H., Phillips, M.J., Tonguthai, K., 1992. A Review of Epizootic Ulcerative Syndrome (EUS) in Asia. Aquatic Animal Health Research Institute and Network of Aquculture Centres in Asia-Pacific, Bangkok, pp. 73.

Lehane, R., 1996. Fish with ulcers: why and what can be done? Partners Res. Dev. Aust. 9, 13—19.

Lilley, J.H., Hart, D., Richards, R.H., Roberts, R.J., Cerenius, L., Söderhäll, K., 1997a. Pan-Asian spread of single fungal clone results in large-scale fish-kills. Vet. Rec. 140, 653—654.

Lilley, J.H., Thompson, K.D., Adams, A., 1997b. Characterization of Aphanomyces invadans by electrophoretic and Western blot analysis. Dis. Aquat. Org. 30, 187—197.

Lilley, J.H., Roberts, R.J., 1997. Pathogenicity and culture studies comparing Aphanomyces involved in epizootic ulcerative syndrome (EUS) with other similar fungi. J. Fish Dis. 20, 101—110.

Lahariya, R., 2013. Immune response and status of epizootic ulcerative syndrome (EUS) affected fishes Channa punctatus and Clarias magur. J. Appl. Nat. Sci. 5 (1), 103—107.

Miyazaki, T., Egusa, S., 1972. Studies on mycotic granulomatosis in freshwater fish I. Mycotic granulomatosis in goldfish. Fish Pathol. 7, 15—25 (In Japanese).

Miyazaki, T., Egusa, S., 1973a. Studies on mycotic granulomatosis in freshwater fish II. Mycotic granulomatosis prevailed in goldfish. Fish. Pathol. 7, 125—133 (In Japanese).

Miyazaki, T., Egusa, S., 1973b. Studies on mycotic granulomatosis in freshwater fish III. Bluegill. Mycotic granulomatosis in bluegill. Fish Pathol. 8, 41—43 (In Japanese).

Miyazaki, T., Egusa, S., 1973c. Studies on mycotic granulomatosis in freshwater fish IV. Mycotic granulomatosis in some wild fishes (In Japanese). Fish Pathol. 8, 44—47.

Mckenzie, R.A., Hall, W.T.K., 1976. Dermal ulceration of mullet (Mugil cephalus). Aust. Vet. J. 52, 230—231.

Munday, B.L., 1985. Ulcer disease in cod (Pseudophycis bachus) from the Tamar River. Tasman. Fish Res. 27, 15—18.

McGarey, D.J., Beatty, T.K., Alberts, V.A., Te Strake, D., Lim, D.V., 1990. Investigations of potential microbial pathogens associated with ulcerative disease syndrome (UDS) of Florida fish. Pathol. Mar. Sci.

Miyazaki, T., 1994. Comparison among mycotic granulomatosis, saprolegniasis and anaaki-byo in fishes: a Japanese experience. In: Roberts, R.J., Campbell, B., MacRae, I.H. (Eds.), Proceedings of the ODA Regional Seminar on Epizootic Ulcerative Syndrome. Aquatic Animal Health Research Institute, Bangkok, 25—27 January 1994. pp. 253—270. AAHRI, Bangkok.

Mohan, C.V., Shankar, K.M., 1994. Epidemiological analysis of Epizootic ulcerative syndrome of fresh and Brakishwater fishes of Karnataka, India. Curr. Sci. 66, 656—658.

Mohan, C.V., Shankar, K.M., 1995. Role of fungus in epizootic ulcerative syndrome of fresh- and brackishwater fishes of India: a histopathological assessment. In: Shariff, M., Arthur, J.R., Subasinghe, R.P. (Eds.), Diseases in Asian Aquaculture II. Fish Health Section, Asian Fisheries Society, Manila, pp. 299—305.

Muniruzzaman, M., Chowdhury, M.B.R., 2008. Ulcer diseases in cultured fish in Mymensingh and surrounding districts. Bangladesh Vet. 25 (1), 40—49.

Muthu Ramakrishnan, C., 2009. Investigation on Epizootic Ulcerative Syndrome and Prophylatic Measures in Catfish Heteropneustes

Fossilis (Ph.D. Thesis). Manonmaniam Sundaranar University, Tirunelveli, India.

Noga, E.J., Dykstra, M.J., 1986. Oomycete fungi associated with ulcerative mycosis in Menhaden, *Brevoortia tyrannus* (Latrobe). J. Fish Dis. 9, 47–53.

National Workshop on Ulcerative Disease Syndrome in Fish, 1990. In: . Proceedings of the National Workshop on Ulcerative Disease Syndrome in Fish, 6–7 March, 1990, Calcutta (India).

Noga, E.J., Wright, J.F., Levine, J.F., Dykstra, M.J., Hawkins, J.H., 1991. Dermatological diseases affecting fishes of the Tar-Pamlico Estuary, North Carolina. Dis. Aquat. Org. 10 (2), 87–92.

Noga, E.J., 1993. Fungal diseases of marine and estuarine fishes. In: Couch, J.A., Fournie, J.W. (Eds.), Pathobiology of Marine and Estuarine Organisms. CRC Press, Boca Raton, pp. 85–100.

Noga, E.J., 1994. Epidemic ulcerative diseases recently affecting estuarine fishes of the western Atlantic ocean. In: Roberts, R.J., Campbell, B., MacRae, I.H. (Eds.), Proceedings of the ODA Regional Seminar on Epizootic Ulcerative Syndrome, 25–27 January 1994. Aquatic Animal Health Research Institute, Bangkok, pp. 89–100.

Noga, E.J., Khoo, L., Stevens, J.B., Fan, Z., Burkholder, J.M., 1996. Novel toxic dinoflagellate causes epidemic disease in estuarine fish. Mar. Pollut. Bull. 32 (2), 219–224.

Nsonga, A., Mfitilodze, W., Samui, K.L., Sikawa, D., 2013. Epidermiology of epizootic ulcerative syndrome in the Zambezi River system. A case study for Zambia. Hum. Vet. Med. Int. J. Bioflux Soc. 5 (1), 1–8.

OIE, 2006. Epizootic haematopoietic necrosis. Man. Diagn Tests Aquatic Anim 2006, 85–103.

Phillips, M.J., 1989. A Report on the NACA Workshop on the Regional Research Programme on Ulcerative Syndrome in Fish and the Environment, 20–24 March, 1989. Network of Aquaculture Centres in Asia-Pacific, Bangkok.

Pearce, M., 1990. Epizootic Ulcerative Syndrome Technical Report December 1987–September 1989. Northern Territory Department of Primary Industry and Fisheries. Fisheries Report No.22. Northern Territory, Australia. 82 pp.

Pallibare, J.O., Catap, E.S., Callinan, R.B., 1994. Fungal isolation from EUS-affected fish in the Philippines: pp. 238–243. In: Roberts, R.J., Campbell, B., MacRae, I.H. (Eds.), Proceedings of the ODA Regional Seminar on Epizootic Ulcerative Syndrome, 25–27 January 1994. Aquatic Animal Health Research Institute, Bangkok.

Patil, P.R., Kohlapure, R.M., Mazumder, J., Kar, D., Roy, A., 2003. Isolation of a suspected viral agent from EUS-affected *Labeo calbasu* fish. In: Proc. International Conference on Disease Management for Sustainable Fisheries. University of Kerala, Trivandrum.

Pradhan, P.K., Rathore, G., Sood, N., Swaminathan, T.R., Yadav, M.K., Verma, D.K., Chaudhary, D.K., Abidi, R., Punia, P., Jena, J.K., 2014. Emergence of epizootic ulcerative syndrome: large-scale mortalities of cultured and wild fish species in Uttar Pradesh, India. Curr. Sci. 106 (12).

Raman, R.P., 1992. EUS strikes in the brackishwaters of Chilka Lagoon in India. Fish Health Section Newsletter 3 (2), 3–4. Asian Fisheries Society, Manila.

Roberts, R.J., Macintosh, D.J., Tonguthai, K., Boonyratpalin, S., Tayaputch, N., Phillips, M.J., Millar, S.D., 1986. Field and Laboratory Investigations into Ulcerative Fish Diseases in Asia-Pacific Region. Technical Report, FAO Project TCP/RAS/4508, pp. 214.

Roberts, R.J., Wootten, R., MacRae, I., Millar, S., Struthers, W., 1989. Ulcerative Disease Survey: Bangladesh Final Report to the Govt. Of Bangladesh and the Overseas Development Administration. Institute of Aquaculture, Stirling University, UK.

Reantaso, M.B., 1991. EUS in brackish waters of the Philippines. Fish Health Sect. Newsl. 2 (1), 8–9. Asian Fisheries Society (Manila).

Roberts, R.J., Willoughby, L.G., Chinabut, S., 1993. Mycotic aspects of epizootic ulcerative syndrome (EUS) of Asian fishes. J. Fish Dis. 16, 169–183.

Rukyani, A., 1994. Status of epizootic ulcerative disease in Indonesia. In: Proceedings of the ODA Regional Seminar on Epizootic Ulcerative Syndrome, 25–27 January, 1994. Aquatic Animal Health Research Institute, Bangkok.

Rattan, P., Parulekar, A.H., 1998. Diseases and parasites of laboratory reared and wild population of banded pearl spot *Etroplus suratensis* (Cichlidae) in Goa. I J. Mar. Sci. 27 (3–4), 407–410.

Shariff, M., Law, A.T., 1980. An incidence of fish mortality in Bekok river, Johore, Malaysia. In: Proc. International Symp. on Conservation and Input from Life Sciences, 27–30 October 1980. Universiti Kebangsaan, Bangi, Selangor (Malaysia).

Soe, U.M., 1990. Myanmar report, 13–26 August, 1990. In: Regional Research Programme on Relationships between Epizootic Ulcerative Syndrome in Fish and the Environment. NACA, Bangkok, pp. 35–38.

Sanjeevaghosh, D., 1991. EUS ravages Kerala inland fisheries. Fish Chimes 11 (9), 47–49.

Shrestha, G.B., 1994. Status of epizootic ulcerative syndrome (EUS) and its effects on aquaculture in Nepal. In: Roberts, R.J., Campbell, B., MacRae, I.H. (Eds.), Proceedings of the ODA Regional Seminar on Epizootic Ulcerative Syndrome, 25–27 January 1994. Aquatic Animal Health Research Institute, Bangkok, pp. 49–57.

Shariff, M., Saidin, T.H., 1994. Status of epizootic ulcerative syndrome in Malaysia since 1986. In: Roberts, R.J., Campbell, B., MacRae, I.H. (Eds.), Proceedings of the ODA Regional Seminar on Epizootic Ulcerative Syndrome, 25–27 January 1994. Aquatic Animal Health Research Institute, Bangkok, p. 48.

Sammut, J., White, I., Melville, M.D., 1996. Acidification of an estuarine tributary in eastern Australia due to drainage of acid sulfate soils. Mar. Freshw. Res. 47 (5), 669–684.

Saylor, R.K., Miller, D.L., Vandersea, M.W., Bevelhimer, M.S., Schofield, P.J., Bennett, W.A., 2010. Epizootic ulcerative syndrome caused by *Aphanomyces invadans* in captive bullseye snakehead (*Channa marulius*) collected from south Florida, USA. Dis. Aquat. Org. 88, 169–175.

Siddique, M.A.B., Bashar, M.A., Hussain, M.A., Kibria, A.S.M., 2009. Fungal disease of freshwater fishes in Natore district of Bangladesh. J. Bangladesh Agril. Univ. 7 (1), 157–162.

Ulcerative Fish Disease Committee, 1983. Practical Report of the Ulcerative Fish Disease Committee 1982–1983 (Bangkok).

Virgona, J.L., 1992. Environmental factors influencing the prevalence of a cutaneous ulcerative disease (red spot) in the sea mullet, MugU cephalus L., in the Clarence River, New South Wales, Australia. J. Fish Dis. 15, 363–378.

Vishwanath, T.S., Mohan, C.V., Shankar, K.M., 1997. Mycotic granulomatosis and seasonality are the consistent features of epizootic ulcerative syndrome of fresh and brackishwater fishes of Karnataka, India. Asian Fish Sci. 10, 155–160.

Vishwanath, T.S., Mohan, C.V., Shankar, K.M., 1998. Epizootic ulcerative syndrome associated with a fungal pathogen in Indian fishes-histopathology: a cause for invasiveness. Aquaculture 165, 1—9.

Widago, D., 1990. Indonesia report. In: Regional Research Programme on Relationships between Epizootic Ulcerative Syndrome in Fish and the Environment, 13—26 August 1990. NACA, Bangkok, pp. 18—22.

Wilson, K.D., Lo, K.S., 1992. Fish disease in Hong Kong and the potential role of the veterinarian. In: 23rd Annual Conference of the International Association for Aquatic Animal Medicine (IAAAM), Hong Kong, 18—22 May 1992.

Wadahar, G.M., Chandio, M.H., Mahar, M.A., 2012. Epizootic ulcerative syndrome (EUS) in freshwater fishes infected by fungal genera aphanomyces and Alternaria, sindh, Pakistan. Sindh Univ. Res. Jour. Sci. Ser. 44 (4), 663—666.

Xuan, T.T., 1990. Vietnam report. In: Phillips, M.J., Keddie, H.G. (Eds.), Regional Research Programme on Relationships between Epizootic Ulcerative Syndrome in Fish and the Environment. A Report on the 2nd Technical Workshop, 13—26 August 1990, Network of Aquaculture Centres in Asia-Pacific (Bangkok).

Zoological Society of Assam, 1988. Organised by Zoological. In: . Proceedings of the Symposium on Recent Outbreak of Fish Diseases in North Eastern India. 30 December 1988. Society of Assam, Guwahati, Assam, India, 23 pp.

Chapter 3

Efforts in Unraveling the Enigmatic Epizootic Ulcerative Syndrome and Review of Current Status on its Research and Development

3.1 INTRODUCTION

Epizootic ulcerative syndrome (EUS) is an international problem that has been studied independently and collaboratively by many different workers. The early occurrences of mycotic granulomatosis (MG) in Japan and red spot disease (RSD) in Australia more than 40 years ago are now considered to have been outbreaks of the disease that was subsequently designated as EUS. Reference to early works on MG and RSD therefore proved important to understanding later EUS outbreaks in SE and South Asia.

According to some quarters, EUS is a **fungal disease of freshwater (FW) and brackish-water fish** affecting >100 species of fishes that is caused by the fungal species *Aphanomyces invadans*. The organism requires a specific combination of factors in order to germinate within the dermis of the fish. The disease causes lesions in both the skin and the visceral organs.

The first sustained report on EUS appeared during the early 1980s. Since then, EUS has continued to spread unabated and has extended its range through different countries with devastating effects on farmed as well as wild fish populations. More is now known about EUS since those early outbreaks, but still the underlying causes are not. As such, much work remains to be done in unraveling the mysteries of this enigmatic disease. A compiled report entitled "A Review of Epizootic Ulcerative Syndrome (EUS) in Asia" was published during 1992 as a collaborative effort between the Aquatic Animal Health Research Institute (AAHRI) of the Royal Thai Government and the Network of Aquaculture Centres in Asia—Pacific (NACA). Investigations regarding possible causes of EUS and possible measures to control and contain the disease are still being conducted.

The history of serious research with the aim of trying to understand EUS could be traced back to the early 1980s, when scientists from Indonesia, and then Malaysia and Thailand, made significant studies on EUS as it swept through those countries. In 1985, FAO, through its Technical Cooperation Programme, launched a major collaborative effort between scientists of the Royal Thai Government and the University of Stirling (UK) to investigate the causes and spread of serious ulcerative diseases in the Asia—Pacific region. This project team reported the results of its study on EUS to the Expert Consultation Committee on Ulcerative Fish Diseases in the Asia—Pacific Region held in Bangkok August 5—9, 1986. Among other recommendations, the Expert Consultation Committee recommended long-term studies of the environment in relation to outbreaks of EUS. As a response to these recommendations, NACA initiated a regional study during 1987 on the relationship between the environment and outbreaks of EUS, supported by the United Nations Development Programme and the FAO through the NACA Project (RAS/86/047). This study, which started with 18 researchers located in 11 countries in the region, subsequently grew to form a network of 21 scientists in 12 countries. These scientists emphatically examined, in a quantitative manner, the interactions between EUS and environmental factors. The results obtained from these studies formed a significant basis for and component of future studies and reports. Further, an important outcome of this regional research program was the demonstrable benefits that could be obtained through collaboration of regional scientists working to resolve common regional problems—many previous efforts in fish disease research in the region had unfortunately been futile mainly due to a lack of coordinated efforts.

The "shock" of EUS also brought positive benefits. The concern generated among many Asian governments as a result of EUS led to increased awareness of the importance of fish health management in the protection of fishes in

Epizootic Ulcerative Fish Disease Syndrome. http://dx.doi.org/10.1016/B978-0-12-802504-8.00003-1

aquaculture (AQC) as well as the wild. The Workshop on Fish Disease and Fish Health Management conducted by the Asian Development Bank (ADB) in association with NACA (ADB/NACA, 1991), and collaborated by the Overseas Development Administration (ODA), demonstrated the importance of tackling fish health management problems in a coordinated and concerted manner.

There were many developments in the field of EUS research following the publication of the AAH-RI—NACA review of EUS in 1992. EUS today is a semiglobal problem that has been studied both independently and collaboratively by many different workers. The early occurrences of MG in Japan and RSD in Australia about 45 years ago are now considered to have been outbreaks of the disease that was subsequently named EUS. Reference to the earlier works on MG and RSD was therefore considered important and relevant in understanding the later outbreaks of EUS in SE and South Asia.

As indicated earlier, a number of different agencies have supported works on EUS. In summary, the initial EUS outbreaks in SE Asia were investigated by a team funded by FAO, and the name "epizootic ulcerative syndrome" was later proposed at an FAO-convened Expert Consultation in Bangkok in 1986. NACA also was integrally involved in the studies on EUS. Much of the data on environmental parameters associated with EUS outbreaks was generated by the NACA Regional Research Programme on Ulcerative Syndrome in Fish and the Environment. The Department for International Development (DFID) of the United Kingdom (formerly the ODA) and the Australian Centre for International Agricultural Research (ACIAR) subsequently funded major research projects on EUS. Both of these organizations also funded various publications through the Fisheries Programme of ACIAR and the DFID Southeast Asia Aquatic Animal Disease Control Project. The unabated spread of EUS might be due partly to the large-scale movement of fishes within the Asia—Pacific region. As such, the probability of further spread is high. Consequently, the probability of initiation of EUS in the countries that are yet unaffected is high. This has been a concern for many countries. The need for the development of effective strategies to reduce risks associated with the spread of significant aquatic animal pathogens is now widely recognized. And in Asia, it has been addressed through a cooperative FAO/NACA/OIE regional program for the development of technical guidelines on quarantine and health certification, as well as the establishment of information systems for responsible movement of live aquatic animals in Asia. As such, cooperation among countries in Asia and its neighbors is important in controlling aquatic animal diseases and in promoting sustainable aquacultural development.

3.2 REVIEW OF CURRENT STATUS ON RESEARCH AND DEVELOPMENT ON EUS

3.2.1 International Status

The syndrome now called EUS has been causing the colossal loss of fishes in many countries for more than four decades, and during that time the disease has been given several colloquial names. It was first described in **Japan** in 1971 as *Aphanomyces* infection (Egusa and Masuda, 1971). The infection was found in other fishes and named mycotic granulomatosis based on histopathological findings (Miyazaki and Egusa, 1972). Beginning in 1972, an epizootic cutaneous ulcerative syndrome in estuarine fishes in **Australia** was termed red spot disease (Mckenzie and Hall, 1976). Similar conditions with dermal ulcerations and mortalities occurred throughout SE and South Asia, and the syndrome was given its present name in 1986 at the Expert Consultation on Ulcerative Fish Diseases in Bangkok (FAO, 1986). In the United States, similar ulcerative lesions designated as ulcerative mycosis (UM) (Noga and Dykstra, 1986) occurred in estuarine fishes along the East Coast beginning in 1978.

The literature and our knowledge on EUS has been reviewed a number of times (Roberts et al., 1993; Chinabut et al., 1995; Roberts, 1997; Chinabut, 1998; Lilley et al., 1998). Egusa (1992) summarized the situation with MG in Japan, while Noga (1993) assessed the findings regarding UM through the early 1990s. In the light of this information, it is now generally accepted that EUS is characterized by the presence of ulcerative dermal lesions in which invasive fungal hyphae are believed to elicit a granulomatous response. It is also generally accepted that despite extensive investigations in many affected countries in recent decades, no naturally occurring, epidemiologically similar, but pathologically distinct cutaneous ulcerative syndrome has been found.

However, there seem to remain two key areas of dispute among scientists, both related to the pathogenesis of dermal ulcers. These differences were first brought to light by the formal definition of EUS in 1994 as "a seasonal epizootic condition of freshwater and estuarine warm water fish of complex infectious aetiology characterized by the presence of invasive *Aphanomyces* infection and necrotising ulcerative lesions typically leading to a granulomatous response" (Roberts et al., 1994a,b). Some scientists proposed that a number of invasive fungal species, not necessarily including *A. invadans* (=*Aphanomyces piscicida*) in all cases, might be involved in ulcer formation. Others believed that only *A. invadans* seemed to be consistently present and responsible for the observed tissue destruction. Also, differences existed regarding whether the viruses and bacteria recovered from EUS cases have essential or merely opportunistic roles in ulcer formation (Lio-Po, 1999; Lio-Po et al., 2002).

3.2.1.1 Geographical Distribution

Infection with the fungus A. invadans was first reported in farmed FW ayu (*Plecoglossus altivelis*) in Oita Prefecture, Kyushu Island, Japan in 1971 (Egusa and Masuda, 1971). It was later reported in estuarine fish, particularly grey mullet (*Mugil cephalus*) in eastern Australia in 1972 (Fraser et al., 1992; Mckenzie and Hall, 1976). Infection by A. invadans subsequently extended its range through Papua New Guinea, into SE and South Asia, and then into West Asia, after sweeping through Pakistan (Lilley et al., 1998; Tonguthai, 1985). Outbreaks of ulcerative disease in menhaden (*Brevoortia tyrannus*) in the United States had the same etiological agent as the disease observed in Asia (Blazer et al., 1999; Lilley et al., 1997a; Vandersea et al., 2006). The first confirmed outbreaks of infection with A. invadans on the African continent occurred in 2007 in Botswana, Namibia, and Zambia, and were linked to the Zambezi—Chobe river system (FAO, 2009). In 2010 and 2011, infection with A. invadans appeared in wild FW fish in Western Cape Province, South Africa and in wild brown bullhead fish in Lake Ontario in the province of Ontario, Canada. Infection with A. invadans was reported from >20 countries on four continents spreading across North America, Southern Africa, Asia, and Australia.

3.2.1.2 Detailed Account

The chronology of EUS briefly outlined above is discussed in more detail below.

Following the initial outbreaks of **MG and RSD**, there was a progressive spread westwards across Asia of a syndrome associated with dermal ulceration and involving large-scale mortalities in a number of FW and estuarine fish species. The syndrome was called by various names like ulcerative disease syndrome, fish disease syndrome, etc. Eventually, the name EUS was adopted in 1986 at the meeting of the Expert Consultation Committee on Ulcerative Fish Diseases in Bangkok (FAO, 1986). Outbreaks of EUS were reported from 18 countries in the Asia—Pacific region.

In 1975—1976, an ulcerative disease outbreak believed to be EUS occurred in the rivers of southern **Papua New Guinea** (Haines, 1983). During 1982—1983, there were high mortalities in the gudgeons (*Ophieleotris aporos* and *Oxyeleotris heterodon*) from inland areas, as well as mullets from estuaries in northern Papua New Guinea (Coates et al., 1989). Introduced tilapia (*Oreochromis mossambicus*) is common in these areas, but even so, proved to be resistant. Preserved affected fishes were later examined by Roberts et al. (1986) and confirmed as pathologically identical to EUS.

In 1980, outbreaks of an epizootic hemorrhagic condition occurred in **Java, Indonesia** affecting primarily cultured cyprinids and clariids, although whether this was EUS was uncertain (Roberts et al., 1986). Typically, ulcerated snakeheads and catfish were subsequently reported in the Indonesian provinces of **Sumatra**, Sulawesi, and Kalimantan (Widago, 1990). Invasive hyphae were identified from sand gobies (*Oxyeleotris marmoratus*) from eastern Kalimantan (Rukyani, 1994), and Bastiawan (1993) was said to have isolated A. invadans from an EUS-affected sand goby in Java.

Further, Roberts et al. (1986) discussed unconfirmed accounts of ulcerated walking catfish (*Clarias batrachus*) in **Singapore** in 1977. They also discussed subsequent occurrences of EUS. However, there had been no detailed reports of extensive losses to the ornamental fish industry due to EUS, in spite of Singapore being one of the biggest trading centers for ornamental fishes.

Conversely, there were reports of high mortality among fishes in southern Peninsular **Malaysia** in 1979 (Shariff and Law, 1980). The first report of large-scale fish mortality due to the outbreak of typical EUS, however, was during December 1980 among rice-field fishes in Northern Malaysia (Jothy, 1981). From that point, outbreaks occurred almost annually, although sometimes to a lesser extent (Shariff and Saidin, 1994). Major species affected were the snakeskin gourami (*Trichogaster pectoralis*), stripped snakehead (*Channa striata*), climbing perch (*Anabas testudineus*), and the walking catfish (*C. batrachus*).

In addition to the above, significant well-documented epizootics had been occurring in **Thailand** since 1981 (Ulcerative Fish Disease Committee, 1983; Chulalongkorn University, 1983, 1985, 1987). The second epidemic occurred during 1982—1983, and the third during 1983—1984. EUS outbreaks were particularly devastating, as they had affected the intensive fish culture systems in Central Thailand as well as wild fishes in waterways. Some of the most severe mortalities occurred among farmed snakeheads in rice fields. The original outbreaks started toward the end of the rainy season (September) and persisted throughout the winter until March of the following year. At that time, EUS outbreaks tended to be restricted to the cooler months of December and January. Roberts et al. (1993) described the isolation of the pathogenic fungus A. invadans from EUS-affected snakeheads collected from the Suphanburi province of Thailand.

Major outbreaks of EUS from **Cambodia (Kampuchea), Lao PDR, and Myanmar** were reported during 1983—1984 (Lilley et al., 1992). However, subsequent EUS outbreaks were less extensive—for example, EUS outbreaks affected 35 Myanmarese towns during 1984—1985 and only 11 towns during 1989—1990 (Soe, 1990). However, the impact of EUS continued to be significant because of the importance of susceptible fish species to the rural economy.

Further, several reports of EUS-affected fish came from **Vietnam, China, and Hong Kong**. The first report of an EUS outbreak among snakeheads came from the Mekong Delta during 1983 (Xuan, 1990). Ulcerated *Labeo rohita* were first observed at the Pearl River Fisheries Institute in Guangzhou, South China during 1982 (Lian, 1990). Clariid catfish were affected in the same area during 1987–1988 (Lian, 1990). Also, EUS outbreaks among *Carassius auratus* were reported from much of Eastern China during 1989 (Guizhen, 1990). Wilson and Lo (1992) reported *c* 70% mortality among the snakeheads (*Channa maculata*) in Hong Kong during summer 1988.

Laguna de Bay in **the Philippines** experienced a serious outbreak of EUS during December 1985. Approximately 5–40% of certain fish groups, notably snakeheads, gobies, gouramies, and catfishes, were found to be ulcerated, while certain other fish groups—for example, milkfish, bighead carps, and tilapia—were found to be unaffected (Llobrera and Gacutan, 1987). The disease continued to spread to at least 11 other provinces in the Philippines, affecting wild fishes in the lakes, rice fields, and swamps as well as cultured fishes in ponds (Bondad-Reantaso et al., 1994). Mullet, goatfish (*Upeneus bensasi*), croaker (*Johnius* sp.), *Psettodes* sp., and *Scatophagus argus* in a lagoon in Cagayan Province suffered an outbreak in 1990 that was confirmed as EUS through histopathological examination (Reantaso, 1991). The occurrence of EUS among these brackish-water and marine species might provide an explanation regarding the modus operandi of the spread of EUS between the islands. Several *A. invadans* isolates were observed from EUS-affected fishes in the Philippines (Paclibare et al., 1994).

A major outbreak of EUS occurred among the FW and estuarine fishes of Western Sri Lanka during December 1987 (Costa and Wijeyaratne, 1989). It was suspected that the disease may have been imported from SE Asia through shipments of imported fish that included infected angelfish (*Pterophyllum scalare*), some of which were ulcerated and suffered high mortalities (Balasuriya, 1994). Large necrotic ulcers in affected fishes—for example, snakeheads, were the most prominently visible signs of EUS. However, tilapia was found to be refractory to the disease.

For more than 15 years, EUS has been having a serious impact on fisheries throughout mainland SE Asia.

In **Bangladesh**, EUS was first reported from the Chandpur district during February 1988. The disease spread quite rapidly throughout the country, possibly aided by the flood of September 1988 (Barua, 1994). Ulceration was observed in many wild species, notably *Channa* spp., *Puntius* spp., *Clarias* sp., *Mystus* spp., *Macrognathus* spp., and *Mastacembelus* sp. The IMCs in the culture system were also widely affected, although large-scale mortalities due to EUS were probably restricted to juveniles (Roberts et al., 1989). There was a small decline in the severity of

outbreaks of EUS temporarily. However, the intensity of EUS was again on the rise beginning in 1993. *A. invadans* was isolated from farmed IMC (notably *L. rohita*) in NW Bangladesh, and from wild fishes in the wetlands of NE Bangladesh during 1993.

The earliest report of an EUS outbreak causing panic among the people because of its suspected health hazards goes back to Australia (Mckenzie and Hall, 1976), from whence it swept in more or less a chronological manner through almost all of SE Asia including Java, Borneo, Sumatra, Indonesia, Malaysia (Shariff and Law, 1980), Kampuchea, Lao PDR (Boonyaratpalin, 1985), Vietnam, Myanmar, Thailand (Tonguthai, 1985), and Bangladesh before reaching India through the Barak Valley region of Assam during July 1988 (Kar and Dey, 1990a,b,c; Kar, 2007, 2013), and has continued its progression to the current day.

Various postulates have been put forward for this disease, including physical factors such as water pollution causing changes in physicochemical characteristics and micronutrients. But these may just be the triggering factors. The origin of the disease is unknown and remains a matter of speculation even today. Attempts have been made in different laboratories to investigate the cause of EUS and to correlate outbreaks of the epizootic with environmental parameters. Rodgers and Burke (1981) stated the rapid seasonal depression of salinity and temperature as important environmental factors predisposing fish to attack by EUS due to the stress conditions created. No particular characteristic of soil fertility appeared to be related to the occurrence of EUS (Macintosh and Phillips, 1986). However, *Aeromonas hydrophila* is one of the most commonly isolated bacterium involved in hemorrhagic fish diseases in the Indo-Pacific region (Roberts, 1989). A primary viral etiology has been considered a likely possibility for the rapid and uncontrollable spread of EUS (Frerichs et al., 1986; Ahne et al., 1988).

Further, Roberts et al. (1993) reported the involvement of one or more invasive fungi and a range of opportunistic bacteria and certain viruses. Chinabut et al. (1995) reported the isolation of the invasive fungus *Aphanomyces*, from EUS-affected fishes. Willoughby et al. (1995) stated that the primary cause of EUS had not been correctly known, although there appeared to be a close correlation between outbreaks of EUS and temperature involving a range of microbes. They further stated that previous works had failed to establish a strong correlation between isolated mycotic flora and disease outbreak. Leung et al. (1994) stated that *A. hydrophila*, an opportunistic pathogen associated with hemorrhagic septicemia, is a ubiquitous and heterogeneous organism that produces disease in fish under stress conditions or in contact with infection by other pathogens. They stated that the pathogenesis of *A. hydrophila* is multifactorial, and a variety of virulence factors possibly work in concert to contribute to the overall virulence of this bacterium. Extracellular products including toxins, hemolysins, proteases, and acetyl cholinesterase appeared to contribute to the establishment of *A. hydrophila* infection in fish. Cartwright et al. (1994) reported that a virulent strain of *A. hydrophila* associated with EUS was used to produce a monoclonal antibody that identified virulent strains of *A. hydrophila.* Lilley and Frerichs (1994) compared seven rhabdoviruses isolated from fish suffering from EUS in terms of their morphology, cytopathogenicity, antigenic relatedness, and structured polypeptide composition. Roberts et al. (1994a) reported about a number of birnavirus strains, a reovirus, and at least eight rhabdovirus isolates.

According to OIE (1997), EUS is a seasonal epizootic condition of wild and farmed fresh and brackish-water fish that is of complex infectious etiology. According to them, the primary causative agent of EUS is the fungus *A. invadans*, often associated with bacterial septicemia that generally involves *A. hydrophila*. A variety of parasites had also been reported from the diseased fish, but their presence is said to be inconsistent. Associated viral infections could also occur frequently. It is believed that the fungus by itself may not normally invade fish, and it is postulated that a certain cofactor such as, epidermal damage (which could be initiated by an array of agents), severe environmental stress, or viral infection is required to in order to initiate this complex and exceedingly dangerous condition. They reported that EUS is now endemic in SE and South Asia.

In addition to the various workers in EUS research mentioned in this treatise, some other significant workers in the field of EUS research are:

Dr C. Baldock (Ausvet Animal Health Services, Australia) in the field of EUS epidemiology; Mr Graeme Fraser (NSW Agriculture, Australia); Dr L. G. Willoughby (Freshwater Biological Association, UK) with regard to EUS mycology; Prof R. J. Roberts (Stirling, UK); Dr Kim Thompson (IoA, Stirling, UK); Dr Ruth Campbell (IoA, Stirling); Mr Jes Sammut (University of New South Wales); Dr Melba B. Reantaso, Ms Susan Lumanlan-Mayo, Mr Jose Paclibare, Ms Elena Catap, and Ms Hazel Matias (Bureau of Fisheries and Aquatic Resources, Philippines); Dr Akhmad Rukyani, Mr Dayat Basiawan, and Mr Taukhid (Research Institute for Freshwater Fisheries, Indonesia); Dr C. V. Mohan (College of Fisheries, Mangalore, India); Mr R. R. Dhital (Fisheries Development Division, Nepal); Mr Masud Hossain Khan (Bangladesh Fisheries Research Institute, Bangladesh); Miss Werawan Chin-aksorn (Suphanburi Fisheries Station, Thailand); Mr Douangkham Singhanouvong (Department of Livestock and Fisheries, Ministry of Agriculture and Forestry, Lao PDR); Mr K. Subramaniam (Brackishwater Research Station, Malaysia); Mr Ing Kim Leang (Department of Fisheries, Cambodia); Mr U. Tin Myo Zaw (Department of Fisheries, Myanmar); and Mr Zafran (Gondol Research Station for Coastal Fisheries, Bali, Indonesia); and others.

3.2.2 National Status: Status of EUS in India

The records of EUS in India date back to the year 1988, when it entered India through the NE corridor from adjoining Bangladesh. The Indian Council of Agricultural Research (ICAR) started the investigation with the report of outbreaks of EUS in India along with other workers.

The fishes of the Barak Valley region of Assam received the first attack of the disease, being adjacent and quite near to Bangladesh. Dr Devashish Kar and his coworkers were among the few researchers who initiated this pioneering work in India. Details of their works are given in the following paragraphs.

In India, widespread outbreaks of EUS started in the Barak Valley region of Assam beginning in July 1988 (Kar, 1999, 2005, 2007, 2010, 2012, 2013; Kar and Dey, 1988; Kar and Dey, 1988a,b, 1990a,b,c; Kar and Das, 2004a,b; Kar et al., 1990a,b, 1993, 1994, 1995a,b,c, 1996a,b,c, 1997, 1998a,b,c,d, 1999a,b, 2000a,b,c,d, 2001a,b, 2002a,b,c, 2003a,b,c,d,e,f,g, 2004a,b,c; Patil et al., 2003). Outbreaks of EUS in India have been comprehensively reviewed at various perspectives for a comprehensive review (The Zoological Society of Assam, 1988; Jhingran and Das, 1990; National Workshop on Ulcerative Disease Syndrome in Fish, 1990; Kumar et al., 1991; ICSF, 1992; Das and Das, 1993; Mohan and Shankar, 1994; etc.).

From the Barak Valley region of Assam (July 1988), after its initiation in India, EUS spread and swept almost unabated, and more or less in a chronological manner, through other regions of India, notably West Bengal (1989), Bihar (1989), Orissa (1989), UP (1990), MP (1990), Maharastra (1991—1992), Karnataka (1993), Goa (1993), Tamil Nadu (1993), Andhra Pradesh (1992—1993), and Kerala (1994—1995), reaching Pakistan during 1996—1998 before progressing further.

However, in Assam, particularly in the Barak Valley region, EUS has become almost endemic, recurring almost every year between November and March. The detailed works done on EUS since its inception in India by Dr Kar and his associates of Assam (Central) University at Silchar will be presented a little later in this treatise.

Other Indian workers among those who worked on the EUS problem are Bhaumik et al. (1990), who worked on the impact of EUS on society. Chakrabarty and Dastidar (1991) repeatedly isolated chemoautotrophic nocardioform bacteria from fish affected by EUS. Chattopadhyay et al. (1990) conducted microbiological investigations into EUS in fishes. Das et al. (1990) prepared a comprehensive account of EUS. Goswami et al. (1988) performed studies on certain aspects of prevention and treatment of fish suffering from EUS. Jhingran (1991) worked on a strategy for containing EUS. Prasad and Sinha (1990) prepared a status paper on the occurrence of EUS in fishes of Bihar. Purkait (1990) prepared case studies on EUS of fishes in Chanditala of the Hooghly district in West Bengal. Rahman et al. (1988) studied the role of EUS-affected fish on the health of ducks in Assam. The Central Institute of Freshwater Aquaculture (ICAR) developed CIFAX, which has been claimed to control EUS. Mohan and Shankar (1994), Vishwanath et al. (1997a,b, 1998), and Mohan et al. (1999a,b) did works revealing different aspects of EUS.

Historical resume of works done on EUS globally

In addition to the above, some other significant works on EUS from different parts of the globe are briefly mentioned below as a historical resume:

Callinan et al. (1995) worked on *Aphanomyces* species associated with EUS in the Philippines, and RSD, which is said to be prevalent in Australia. They made preliminary comparative studies on RSD in Australia with EUS. Frerichs (1995) reported about the viruses associated with EUS of fish in SE Asia.

Chakrabarty et al. (1996) worked on the electron microscopic characteristics of actinomycetic agents having etiological association with human leprosy and EUS of fish.

Anon (1998) reported about EUS in fishes of Pakistan with regard to a report regarding a visit of a mission to Punjab (Pakistan) during April 2–25, 1998.

Lilley et al. (1998) published the *Epizootic Ulcerative Syndrome (EUS) Technical Handbook*. They opined that EUS is an international problem that has been studied independently and collaboratively by many different workers. The early occurrences of MG in Japan and RSD in Australia, >40 years ago, are now considered to be the initial outbreaks of the disease subsequently designated as EUS.

Chandrakant et al. (2000) reported on the characteristics and virulence of *A. hydrophila* isolates from FW fish affected by EUS.

Lio-Po et al. (2000) worked on the characterization of a virus obtained from snakeheads (*Ophicephalus striatus*) with EUS in the Philippines.

John et al. (2001) worked on the characteristics of a new reovirus isolated from EUS-infected snakehead fish.

Lilley et al. (2001) worked on the characterization of *A. invadans* isolates using pyrolysis mass spectrometry.

Miles et al. (2001) worked on the effect of macrophages and serum of fish susceptible or resistant to EUS on the EUS pathogen *A. invadans*.

Vogelbein et al. (2001) dealt with skin ulcers in estuarine fishes and made a comparative pathological evaluation of wild and laboratory-exposed fish.

Kiryu et al. (2002) worked on infectivity and the role of *A. invadans* in the etiology and induction of skin ulcers in Atlantic menhaden (*B. tyrannus*) with two laboratory challenges.

Anon (2003) worked on EUS in relation to infection with *A. invadans*. EUS has been considered the same as RSD, MG, and UM. In 2005, scientists proposed that the disease should be named epizootic granulomatous aphanomycosis (Baldock et al., 2005).

Sipaúba-Tavares et al. (2003) reported on the effect of liming management on water quality in *Colossoma macropomum* ("Tambaqui") ponds.

Anon (2003) provided excerpts of EUS in the **OIE Manual**. The account described EUS as a seasonal epizootic condition of wild and farmed FW and brackish-water fish of complex infectious etiology. It is characterized by the presence of invasive *Aphanomyces* infection and necrotizing ulcerative lesions, typically producing a granulomatous response. The disease is now endemic in SE and South Asia, and has recently extended to West Asia. EUS is indistinguishable from RSD in Australia and MG in Japan.

Zaki (2004) worked on the effect of *Saccharomyces cervisiae* on the immune status of *Oreochromis niloticus* against transportation stress and motile *Aeromonas* septicemia.

Baldock et al. (2005) reviewed the outcomes of a short expert consultation on EUS in connection with reexamination of causal factors, case definitions, and nomenclature. They opined that EUS is the same disease as MG, RSD, and UM, which is now generally accepted. Further, they posited that *A. invadans* (=*A. piscicida*) could be the only necessary infectious agent.

Callinan et al. (2005) worked on dermatitis, bronchitis, and mortality in empire gudgeon (*Hypseleotris compressa*) exposed naturally to runoff from acid sulfate soils.

Humphrey and Pearce (2006) reported on the EUS/RSD in Australia and attempted to answer the question, "What is epizootic ulcerative syndrome?"

Birgit et al. expressed concern about the likelihood of the introduction and establishment of EUS in the United Kingdom.

Mudenda (2006) reported vividly about a surveillance experience of the current state of EUS in Africa.

Zhang et al. (2007) worked on the hematological and plasma biochemical responses of crucian carp (*C. auratus*) to intraperitoneal injection of extracted microcystins, with possible mechanisms of anemia.

Hossain (2008) worked on the isolation of pathogenic bacteria from the skin of ulcerated symptomatic gourami (*Colisa lalia*) through 16S rDNA analysis.

Muniruzzaman and Chowdhury (2008) reported on ulcer diseases in cultured fish in Mymensingh and surrounding districts in Bangladesh.

Gayathri et al. (2008) worked on determination of a specific sampling site on EUS-affected fish for diagnosis of *A. invadans* by immunodot test using monoclonal antibodies

Dennis et al. (2008), while discussing **"Emerging and Endemic Aquatic Animal Diseases in Australasia,"** dealt with protozoal diseases (white spot disease and perkinsosis); OX disease; velvet disease; chilodonellosis; kudoa; other parasitic diseases (notably blood flukes, sea lice, anchor worms, etc.); fungal diseases (EUS/RSD); bacterial diseases (goldfish ulcer disease, enteric septicemia, and epitheliocystis); viral diseases (megalocytivirus, epizootic hematopoietic necrosis, viral nervous necrosis, abalone ganglioneuritis virus, and pilchard herpesvirus); and so on.

Anon (2008) reported about EUS/RSD as one of the fungal diseases among finfishes in Australia.

FAO (2009) reported about "What One May Need to Know Regarding EUS" as part of an extension brochure.

Muthu Ramakrishnan (2009) worked on EUS and its prophylactic measures in the catfish *Heteropneustes fossilis*.

Anon (2009) described EUS in fishes covering various aspects, such as its scope, disease information, agent factors, etiological agent, agent strains, stability of the agent, life cycle, host factors, and so on.

Das et al. (2009) worked on antimicrobial resistance and in vitro gene transfer in bacteria isolated from the ulcers of EUS-affected fish in India.

Choongo et al. (2009) reported about the emergence of EUS in native fish of the Murray—Darling river system (MDRS) of Australia in relation to hosts, distribution, and possible vectors.

Siddique et al. (2009) reported about fungal disease of FW fishes in the Natore district of Bangladesh.

Rodger (2010) prepared the *Fish Disease Manual* that dealt with various aspects of fish and fish disease.

Saylor et al. (2010) reported about EUS caused by *A. invadans* in captive bullseye snakehead (*Channa marulius*) collected from south Florida, USA.

Hossain et al. (2011) worked on the isolation of emergent bacterial pathogens recovered from capture and cultured fisheries in Bangladesh.

Anon (2011) dealt with distribution, signalment, clinical signs, diagnosis, treatment, control, quarantine, and health certification practices for EUS in general, and in endemic areas. The author also dealt with eradication, exclusion, management, surveillance, and treatment of the disease.

Anon (2011) dealt with EUS while preparing **Identification Field Guide** for **Aq Animal Diseases significant to Australia.**

Anon (2011) discussed the quarantine requirements for the importation of live fish and their gametes and fertilized eggs.

Anon (2012) published a report based on information provided to the panel by the Queensland government and other stakeholders, relating to fish health issues observed in Gladstone Harbour and surrounding areas under the aegis of the **Gladstone Fish Health Scientific Advisory Panel**.

Boys et al. (2012) reported about "Emergence of EUS in Native Fish of the MDRS in Australia," with a note on hosts, distribution, and possible vectors. This study is the first published account of *A. invadans* in the wild fish populations of the MDRS, and is the first confirmed record of EUS in *Macquaria ambigua*, *Maccullochella peelii*, and *Leiopotherapon unicolor*.

Huchzermeyer and van der Waal (2012) reported about the exotic fish disease called EUS threatening Africa's aquatic ecosystems.

Baruah et al. (2012) worked on interspecific transmission of the EUS pathogen *A. invadans* and associated physiological responses.

Nsonga et al. (2013) worked on the epidemiology of EUS in the Zambezi River system as a case study from Zambia in Africa.

Baidya and Prasad (2013) reported about the outbreak and prevalence of EUS among the carps in Nepal.

Islam and Chowdhury (2013) studied the **limnological status of the Trimohini Beel (wetland) of** Rajshahi in Bangladesh. A total of 38 zooplankton genera and 26 physicochemical variables were studied in Trimohini Beel. They also reported about outbreaks of EUS.

Kar (2013) has been working extensively on EUS fish disease in India with regard to its gross pathology, epidemiology, etiology, and so on.

Takuma et al. (2013) reported about two new species of *Aphanomyces*, viz. *Aphanomyces izumoensis* sp. nov. and *Aphanomyces shimanensis* sp. nov. isolated from ice fish *Salangichthys microdon*.

Adil et al. (2013) worked on the development and standardization of a monoclonal antibody-based rapid flow-through immunoassay for the detection of *A. invadans* in the field.

Brown et al. (2014) reported about supporting disaster response and preparedness in AQC. They opined that the AQC sector is particularly vulnerable to natural and human-induced rapid- or slow-onset disasters. This is due to the risks associated with the location of fish farm production facilities and other infrastructure, the type of activities associated with fish farming, the nature of fish in specific development stages (fry, fingerling, larvae), and the limited capacity of the sector to reduce and cope with the potential consequences of a disaster.

Likewise, some significant works done on different aspects of EUS fish disease in India are given below:

Anon (1992a,b) held consultation on EUS vis-à-vis the environment and the people, under the aegis of the International Collective in Support of Fishworkers.

Rattan and Parulekar (1998) dealt with the diseases and parasites of laboratory-reared and wild populations of banded pearl spot, *Etroplus suratensis* (Cichlidae), in Goa.

Mohan et al. (1999) argued that EUS-specific fungus of fishes could be a primary pathogen for the outbreak of the disease.

Riji John and Richards (1999) worked on the characteristics of a new birnavirus associated with a warmwater fish cell line.

Riji John et al. (2001) worked on the characteristics of a new reovirus isolated from EUS-infected snakehead fish.

Baldock (2002), in connection with health management issues in the rural livestock sector, advocated for certain useful lessons for consideration when formulating programs on health management in rural small-scale AQC for livelihood.

Mazid and Banu (2002) gave an overview of the social and economic impact as well as the management of fish and shrimp disease in Bangladesh, with an emphasis on small-scale AQC.

Khan and Lilley (2002) **dealt with risk factors and socioeconomic impacts associated with EUS in Bangladesh.**

Roy (2005) worked with a biotechnological approach for defining EUS in FW fishes.

Anon (2005) reported about quarantine measures adopted for a consignment of kissing gourami (*Helostoma temminckii*) in Western Australia that had been imported from Singapore, and the results thereof.

Harikrishnan et al. (2008) worked on the use of herbal concoctions in the therapy of goldfish (*C. auratus*) infected with *A. hydrophila*.

Muthukrishnan et al. (2008) worked on microbial flora from EUS-infected murrel [*C. striata* (Bloch, 1797)] in the Tirunelveli region.

Khatri et al. (2009) made **an assessment on the prevalence of fish diseases in Sambalpur, Western Orissa.**

Sunitha et al. (2010) reported that *Coleus aromaticus* **(Benth) could act as an immunostimulant in** *C. marulius* **(Hamilton).**

Chauhan (2012) studied certain fungal diseases among culturable and nonculturable species of fishes of the Upper Lake in Bhopal.

Sankar Ganesh et al. (2012) worked on isolation and identification of *Vibrio* spp. in diseased *Channa punctatus* from AQC fish farms.

Riji John and Rosalind George (2012) reported an update of the viruses associated with EUS.

Vijayakumar et al. (2013) worked on EUS in FW and brackish-water fishes.

Roy et al. (2013) did works related to isolation and identification of multiantibiotic-resistant *Aeromonas veronii* by 16S rDNA gene sequencing from the gut of FW loach *Lepidocephalichthys guntea* (Hamilton Buchanan).

Haniffa et al. (2013) studied the antimicrobial effects of *Wrightia tinctoria* (Roxb.) R.Br. against EUS in *C. striata*, both in vitro and in vivo.

Laharia (2013) worked on the immune response and status of EUS-affected fishes *C. punctatus* and *Clarias magur* in Maharastra.

Pradhan et al. (2014) reported about the emergence of EUS, and large-scale mortality of cultured and wild fishes, in Uttar Pradesh, India.

Riji John et al. (2015) worked on EUS in fishes. Sampling was done in two NE provinces of Assam and Manipur, and infected FW fish samples with different stages of development of ulcers were collected. These fish tissues were processed and analyzed for the presence of viruses by diagnostic PCR.

REFERENCES

Adil, B., Shankar, K.M., Kumar, B.T., Patil, R., Ballyaya, A., Ramesh, K.S., Poojary, S.R., Byadgi, O.V., Siriyappagouder, P., 2013. Development and standardization of a monoclonal antibody-based rapid flow-through immunoassay for the detection of *Aphanomyces invadans* in the field. J. Vet. Sci. 14 (4), 413–419.

Ahne, W., Jørgensen, P.E.V., Olesen, N.J., Wattanavijarn, W., 1988. Serological examination of a rhabdovirus isolated from snakehead fish (*Ophicephalus striatus*) in Thailand with ulcerative syndrome. J. Appl. Ichthyol. 4, 194–196.

Anon, 1992a. Enigma of EUS. Consultation of EUS vis-a-vis the environment and the people, 25–26 May 1992. Institute of Management in Government, Vikas Bhavan, Trivandrum 695 033, Kerala (India). In: Summary of Proceedings, International Collective in Support of Fish Workers, 27 C College Road, Madras 600006, India.

Anon, January 1992b. Notes on mycological procedure. In: Workshop on Mycological Aspects of Fish and Shellfish Disease. Aquatic Animal Health Research Institute, January 1992, Bangkok, Thailand.

Anon, 2005. Kissing gourami (*Helostoma tremminckii*); longitudinal section. NAAHTWG Slide of the Quarter (July–September, 2005): Epizootic Ulcerative Syndrome (EUS), Govt. of Western Australia, Dept. of Fisheries, Australia.

Anon, 2008. Diseases of Finfish: Fungal Diseases: Epizootic Ulcerative Syndrome/Red Spot Disease. Australian Govt, Dept of Agril, Fisheries and Forestry. Source: New South Wales Department of Primary Industries. Sourced further from: AGDAFF (2008) Aquatic Animal Diseases Significant to Australia: Identification Field Guide. Australian Government Department of Agriculture, Fisheries and Forestry (Canberra).

Anon, 2009. Report of the International Emergency Disease Investigation Task Force on a Serious Finfish Disease in Southern Africa, 18—26 May 2007. Food and Agriculture Organization of the United Nations, Rome. Viewed in 2011, from: http://www.fao.org/docrep/012/i0778e00.htm.

Anon, 2011a. World Animal Health Information Database Interface: Exceptional Epidemiological Events. World Organisation for Animal Health, Paris. Viewed in 2011, from: http://www.oie.int.

ADB/NACA, 1991. Fish health management system in Asia-Pacific. In: Report on a Regional Study and Workshop on Fish Disease and Fish Health Management. Network of Aquaculture Centres in Asia-Pacific, Bangkok. ADB Agriculture Department Report Series No. 1.

Anon, April 1998. Epizooic ulcerative syndrome (EUS). In: Fishes of Pakistan: A Report Regarding Visit of Mission to Punjab in Connection with Epizootic Ulcerative Syndrome (EUS), pp. 22—25.

Anon, 2003. Outbreaks of botulism in cattle associated with the spreading of poultry manure. Vet. Rec. 153 (10), 283—286.

Anon, 2012. Gladstone Fish Health Scientific Advisory Panel, pp. 1—47.

Balasuriya, L.K.S.W., 1994. Epizootic ulcerative syndrome in fish in Sri Lanka, country status report. In: Robert, R.J., Campbell, B., Macrae, I.H. (Eds.), Proceeding of the ODA Regional Seminar on Epizootic Ulcerative, 325. Aquatic Animal Health Research Institute, Bangkok, Thailand, pp. 39—47.

Bondad-Reantaso, M.G., Paclibare, J.,O., Lumanlan-Mayo, S.C., Catap, E.S., 1994. EUS outbreak in the Philippines: a country Report. In: Roberts, R.J., Campbell, B., Mac Rae, I.H. (Eds.), Proceedings of the ODA Regional Seminar on Epizootic Ulcerative Syndrome, 25—27 January, 1994. Aquatic Animal Health Research Institute, Bangkok, pp. 61—87.

Blazer, V.S., Vogelbein, W.K., Densmore, C.L., May, E.B., Lilley, J.H., Zwerner, D.E., 1999. Aphanomyces as a cause of ulcerative skin lesions of menhaden from Chesapeake Bay tributaries. J. Aquat. Anim. Health 11, 340—349.

Baldock, F.C., Blazer, V., Callinan, R., Hatai, K., Karunasagar, I., Mohan, C.V., Bondad-Reantaso, M.G., 2005. Outcomes of a short expert consultation on epizootic ulcerative syndrome (EUS): reexamination of causal factors, case definition and nomenclature. In: Diseases in Asian Aquaculture V. Fish Health Section, Asian Fisheries Society, Manila, pp. 555—585.

Baidya, S., Prasad, A., 2013. Prevalence of epizootic ulcerative syndrome (EUS) in carps. Nepal. J. Zool. 1 (1), 41—47.

Brown, D., Poulain, F., Subasinghe, R., Reantaso, M., 2014. Supporting Disaster Response and Preparedness in Aquaculture. Report of the Food and Agricultural Organisation of the UN, Rome. Preparing for the day after, pp. 1351.

Baldock, C., 2002. Health management issues in the rural livestock sector: useful lessons for consideration when formulating programmes on health management in rural, small-scale aquaculture for livelihood. In: Arthur, J.R., Phillips, M.J., Subasinghe, R.P., Reantaso, M.B., MacRae, I.H. (Eds.), Primary Aquatic Animal Health Care in Rural. Small-scale, AQC Development, Rome, pp. 7—19, pp. 54, FAO Fisheries Tech. Pap No. 406.

Bloch, 1797. Naturgeschichte der auslandischen fish. Berlin, 9 parts in 2, and Atlas (3 vols).

Boonyaratpalin, S., 1985. Fish Disease Outbreak in Burma. FAO Special Report, TCP/BUR/4402: 7 pp.

Bastiawan, D., Taukhid, 1993. Innate immunity on fish seed catfish (Clarias sp.) results spawning parent that vaccinated anti Aeromonas hydrophila. In: Proceedings Seminar on Freshwater Fisheries.

Barua, G., 1994. The status of epizootic ulcerative syndrome of fish of Bangladesh. In: Roberts, R.J., Campbell, B., Mac Rae, I.H. (Eds.), Proceedings of the ODA Regional Seminar on Epizootic Ulcerative Syndrome, 25—27 January, 1994. Aquatic Animal Health Research Institute, Bangkok, pp. 13—20.

Bhaumik, U., Pandit, P.K., Chatterjee, J.G., 1990. Impact of epizootic ulcerative syndrome on the society. In: National Workshop on Ulcerative Disease Syndrome in Fish, 6—7 March, 1990 (Calcutta).

Boys, C.A., Rowland, S.J., Gabor, M., Gabor, L., Marsh, I.B., Hum, S., Callinan, R.B., 2012. Emergence of epizootic ulcerative syndrome in native fish of the Murray—Darling River System, Australia: hosts, distribution and possible vectors. PLoS One 7, e35568. http://dx.doi.org/10.1371/journal.pone.0035568.

Baruah, A., Saha, R.K., Kamilya, D., 2012. Inter-species transmission of the epizootic ulcerative syndrome (EUS) pathogen, Aphanomyces invadans, and associated physiological responses. Israeli J. Aquacult. Bamidgeh 696—699.

Chandrakant, W.H.S., et al., 2000. Characteristics and virulence of Aeromonas hydrophila isolates from freshwater fish with epizootic ulcerative syndrome (EUS). J. Natl. Sci. Found. Sri Lanka 28 (1), 29—42, 200.

Chulalongkorn University, June 23—24, 1983. In: The Symposium on Fresh Water Fishes Epidemic: 1982—1983. Chulalongkorn University, Bangkok.

Chulalongkorn University, March 7—8, 1985. In: Proceedings of the Technical Conference on Living Aquatic Resources. Chulalongkorn University, Bangkok.

Chulalongkorn University, 1987. In: Proc. of the 2nd Seminar on Living Aquatic Resources, 17—18 December, 1987. Chulalongkorn University, Bangkok.

Costa, H.H., Wijeyaratne, M.J.S., 1989. Epidemiology of epizootic ulcerative syndrome occurring for the first time among fishes in Sri Lanka. J. Appl. Ichthyol. 1, 48—52.

Coates, D., Nunn, M.J., Uwate, K.R., 1989. Epizootic ulcerative disease of freshwater fish in Papua New Guinea. Sci. New Guinea 15, 1—11.

Chakrabarty, A.N., Dastidar, S.G., 1991. Repeated isolation of chemoautotrophic nocardioform bacteria from fish epizootic ulcerative syndrome. Indian J. Exp. Biol. 29, 623—627.

Chinabut, S., Roberts, R.J., Willoughby, G.R., Pearson, M.D., 1995. Histopathology of snakehead, Channa striatus (Bloch), experimentally infected with the specific Aphanomyces fungus associated with epizootic ulcerative syndrome (EUS) at different temperatures. J. Fish Dis. 18, 41—47.

Callinan, R.B., Paclibare, J.O., Reantaso, M.B., Lumanlan-Mayo, S.C., Fraser, G.C., Sammut, J., 1995. EUS outbreaks in estuarine fish in Australia and the Philippines: associations with acid sulphate soils, rainfall and Aphanomyces. In: Shariff, M., Arthur, J.R., Subasinghe, R.P. (Eds.), Diseases in Asian Aquaculture II. Fish Health Section. Asian Fisheries Society, Manila, pp. 291—298.

Chinabut, S., 1998. Epizootic ulcerative syndrome: information up to 1997. Fish Pathol. 33, 321—326.

Choongo, K., Hang'ombe, B., Samui, K.L., Syachaba, M., Phiri, H., Maguswi, C., et al., 2009. Environmental and climatic factors associated with epizootic ulcerative syndrome (EUS) in fish from the Zambezi floodplains, Zambia. Bull. Environ. Contam. Toxicol. 83, 474–478. PMID: 19565173. http://dx.doi.org/10.1007/s00128-009-9799-0.

Chauhan, R., 2012. Study on Certain Fungal Diseases in Culturable and Non-culturable Species of Fishes of Upper Lake, Bhopal. Barkatullah University, Bhopal.

Chattopadhyay, U.K., Pal, D., Das, M.S., Das, S., Pal, R.N., 1990. Microbiological investigations into epizootic ulcerative syndrome (EUS) in fishes. In: The National Workshop on Ulcerative Disease Syndrome in Fish, 6–7 March, 1990 (Calcutta).

Cartwright, G.A., Chen, D., Henna, P.J., Gudkovs, N., Tajima, K., 1994. Immunodiagnosis of virulent strains of *Aeromonas hydrophila* associated with epizootic ulcerative syndrome (EUS) using monoclonal antibody. J. Fish Dis. 17, 123–133.

Chakrabarty, A., Mukherjee, M., Chakraborty, A.N., Dastidar, S.G., Basak, P., Saha, B., 1996. Electronmicroscopic characteristics of actinomycetic agents having aetiological association with human leprosy and epizootic ulcerative syndrome of fish. Indian J. Exp. Biol. 34, 810–812.

Callinan, R.B., Sammut, J., Fraser, G.C., 2005. Dermatitis, branchitis and mortality in empire gudgeon *Hypseleotris compressa* exposed naturally to runoff from acid sulfate soils. Dis. Aquat. Org. 63, 247–253.

Das, M.K., Das, R.K., 1993. A review of the fish disease, epizootic ulcerative syndrome in India. Environ. Ecol. 11 (1), 134–145.

Das, M.K., Pal, R.N., Ghosh, A.K., Das, R.K., Joshi, H.C., Mukhopadhyaya, M.K., Hajra, A., 1990. Epizootic ulcerative syndrome: a comparative account. In: Proc. National Workshop on Ulcerative Disease Syndrome in Fish, 6–7 March, 1990 (Calcutta).

Das, A., Saha, D., Pal, J., 2009. Antimicrobial resistance and in vitro gene transfer in bacteria isolated from the ulcers of EUS-affected fish in India. Lett. Appl. Microbiol. 49 (4), 497–502.

Dennis, M., Yoshida, T., Atsuta, S., Kobayashi, M., 2008. Enhancement of resistance to vibriosis in rainbow trout, *Oncorhynchus mykiss* (Walaum), by oral administration of Clostridium butyricum bacterin. J. Fish Dis. 18, 187–190.

Egusa, S., Masuda, N., 1971. A new fungal disease of *Plecoglossus altivelis*. Fish Pathol. 6, 41–46.

Egusa, S., 1992. Mycotic granulomatosis. In: Infectious Diseases of Fish. A.A. Balkema, Rotterdam, pp. 392–396.

FAO, 1986. Report of the Expert Consultation on Ulcerative Fish Diseases in the Asia-Pacific Region. (TCP/RAS/4508). Bangkok, August 1986. FAO, Regional Office for Asia and the Pacific, Bangkok.

Frerichs, G.N., Millar, S.D., Roberts, R.J., 1986. Ulcerative rhabdovirus in fish in Southeast Asia. Nature 322, 216.

Fraser, G.C., Callinan, R.B., Calder, L.M., 1992. *Aphanomyces* species associated with red spot disease: an ulcerative disease of estuarine fish of eastern Australia. J. Fish Dis. 15, 173–181.

Frerichs, G.N., 1995. Viruses associated with the epizootic ulcerative syndrome (EUS) of fish in South-East Asia. Vet. Res. 26, 449–454.

FAO, 2009. What You Need to Know about Epizootic Ulcerative Syndrome (EUS) − an Extension Brochure. Food and Agricultural Organisation of the UN, Rome, pp. 1–36.

Gayathri, D., Shankar, K.M., Mohan, C.V., Devaraja, T.N., 2008. Determination of specific sampling site on EUS affected fish for diagnosis of *Aphanomyces invadans* by immunodot test using monoclonal antibodies. Res. J. Agric. Biol. Sci. 4 (6), 757–760. © 2008, INSInet Publication.

Guizhen, M.J., 1990. China report. In: Regional Research Programme on Relationships between Epizootic Ulcerative Syndrome in Fish and the Environment, 13–26 August 1990. NACA, Bangkok, pp. 11–13.

Goswami, U.C., Datta, A., Sarma, D.K., Sarma, G., 1988. Studies on certain aspects of prevention and treatment of fish suffering from epizootic ulcerative syndrome. In: Proc. Symp, on Recent Outbreak of Fish Diseases in North-Eastern India, 30 December, 1988, Guwahati (Assam), pp. 16–17.

Harikrishnan, R., Balasundaram, C., Moon, Y.G., Kim, M.C., Kim, J.S., Heo, M.S., 2008. Use of Herbal Concoction in the Therapy of Goldfish *(Carassius auratus)* Infected with *Aeromonas hydrophila*. Bharathidasan University, Tiruchirapalli.

Hossain, M., 2008. Isolation of pathogenic bacteria from the skin ulcerous symptomatic gourami (*Colisa lalia*) through 16S rDNA analysis. Univ. J. Zool. Rajshahi Univ. 27, 21–24.

Haniffa, M.A., Jeya Sheela, P., James Milton, M., Britto, J. de, 2013. In vitro and in vivo antimicrobial effects of *Wrightia tinctoria* (Roxb.) R. Br. against epizootic ulcerative syndrome in *Channa striatus*. Int. J. Pharm. Pharm. Sci. 5 (Suppl. 3), 219.

Humphrey, J.D., Pearce, M., 2006. Epizootic Ulcerative Syndrome (Red-Spot Disease). Fishnote, Northern Territory Government, Australia, 1–4.

Haines, A.K., 1983. Fish fauna and ecology. In: Petr, T. (Ed.), The Purari Tropical Environment of High Rainfall River Basin. Dr. W. Junk Publishers, The Netherlands, pp. 367–384.

Hossain, M.M.M., Rahman, M.A., Mondal, S., Shadat Mondal, A.S.M., Chowdhury, M.B.R., 2011. Isolation of some emergent bacterial pathogens recovered from capture and culture fisheries in Bangladesh. Bangladesh Res. Publ. J. 6 (1), 77–90.

Huchzermeyer, K.D.A., van der Waal, B.C.W., 2012. Epizootic ulcerative syndrome: exotic fish disease threatens Africa's aquatic ecosystems. J. S. Afr. Vet. Assoc. 83 (1) http://dx.doi.org/10.4102/jsava.v83 i1.204. Art. # 204, 6 pages.

ICSF, 1992. Enigma of EUS. Consultation on Epizootic Ulcerative Syndrome vis-a-vis the Environment and the People. 25–26 May, 1992. Trivandrum, Kerala. International Collective in Support of Fishworkers, Madras, India, 40 pp.

Islam, Md A., Chowdhury, A.H., 2013. Limnological status of Trimohini Beel of Rajshahi, Bangladesh. J. Asiat. Soc. Bangladesh Sci. 39 (2), 173–182.

Jothy, A.A., 1981. Preliminary Report on the Outbreak of Wabak Kudison in Freshwater Fish in Paddy-Growing Areas in Peninsular Malaysia. Report to the Ministry of Agriculture, Malaysia, December, 1981, pp. 17.

Jhingran, A.G., Das, M.K., 1990. Epizootic Ulcerative Syndrome in Fishes. Bulletin of the Central Inland Capture Fisheries Research Institute (No. 65). CIFRI, Barrackpore, India.

Jhingran, V.G., 1991. Fish and Fisheries of India. Hindustan Publishing Corporation, New Delhi, pp. xxiii + 727.

John, K.R., George, M.R., Richards, R.H., Frerichs, G.N., 2001. Characteristics of a new reovirus isolated from epizootic ulcerative syndrome infected snakehead fish. Dis. Aquat. Org. 46 (2), 83–92.

Kar, D., Dey, S., 1988. Preliminary electron microscopic studies on diseased fish tissues from Barak valley of Assam. In: Proc. Annual Conference of Electron Microscopic Society of India, vol. 18, 88 pp.

Kar, D., Dey, S.C., 1988a. A critical account of the recent fish disease in the Barak valley of Assam. In: Proc. Regional Symp. on Recent Outbreak of Fish Diseases in North-East India, vol. 1, 8 pp.

Kar, D., Dey, S.C., 1988b. Impact of recent fish epidemics on the fishing communities of Cachar district of Assam. In: Proc. Regional Symp. on Recent Outbreak of Fish Diseases in North-East India, vol. 1, 8 pp.

Kar, D., Dey, S.C., 1990a. Fish disease syndrome: a preliminary study from Assam. Bangladesh J. Zool. 18, 115—118.

Kar, D., Dey, S.C., 1990b. A preliminary study of diseased fishes from Cachar district of Assam. Matsya 15—16, 155—161.

Kar, D., Dey, S.C., 1990c. Epizootic ulcerative syndrome in fishes of Assam. J. Assam Sci. Soc. 32 (2), 29—31.

Kar, D., Dey, S.C., Michael, R.G., Kar, S., Changkija, S., 1990a. Studies on fish epidemics from Assam. J. Indian Fish. Assoc. 20, 73—75.

Kar, D., Bhattacharjee, S., Kar, S., Dey, S.C., 1990b. An account of 'EUS' in fishes of Cachar district of Assam with special emphasis on microbiological studies. Souv. Annu. Conf. Indian Assoc. Pathol. Microbiol. India (NE Chapter) 8, 9—11.

Kumar, D., Dey, R.K., Sinha, A., 1991. Outbreak of epizootic ulcerative syndrome of fish in India. In: Sinha, V.R.P., Srivastava, H.C. (Eds.), Aquaculture Productivity. Oxford and IBH Publishing Company, New Delhi, pp. 345—356.

Kar, D., Dey, S.C., Kar, S., Bhattacharjee, N., Roy, A., 1993. Virus-like particles in epizootic ulcerative syndrome of fish. In: Proc. International Symp. on Virus-Cell Interaction: Cellular and Molecular Responses, vol. 1, 34 pp.

Kar, D., Dey, S.C., Kar, S., Roy, A., Michael, R.G., Bhattacharjee, S., Changkija, S., 1994. A candidate virus in epizootic ulcerative syndrome of fish. In: Proc. National Symp. of the Indian Virological Society, vol. 1, 27 pp.

Kar, D., Roy, A., Dey, S.C., Menon, A.G.K., Kar, S., 1995a. Epizootic ulcerative syndrome in fishes of India. World Congr. In Vitro Biol. In Vitro 31 (3), 7.

Kar, D., Kar, S., Roy, A., Dey, S.C., 1995b. Viral disease syndrome in fishes of North-East India. In: Proc. International Symp. of International Centre for Genetic Engg. And Biotechnology (ICGEB) and the Univ. of California at Irvine, vol. 1, 14 pp.

Kar, D., Dey, S.C., Kar, S., 1995c. A viral disease among the fishes of North-East India. In: Annual Congress on Man and Environment, National Environment Sci. Acad. And National Institute of Oceanography, vol. 10, 62 pp.

Kar, D., Dey, S.C., Purkayastha, M., Kar, S., 1996a. An overview of the impediments in conservation of biodiversity of Lake Sone in Assam. In: Proc. Seminar on Conservation of Biodiversity in Indian Aquatic Ecosystems. Jawaharlal Nehru University, New Delhi, p. 1.

Kar, D., Dey, S.C., Roy, A., Kar, S., 1996b. Viral disease syndrome in fishes of India. Proc. Int. Congr. Virol. 10.

Kar, D., Purkayastha, M., Kar, S., 1996c. Biodiversity conservation prioritisation project: a case study from Sone Beel in Assam. In: Proc. National Workshop on Biodiversity Conservation Prioritisation Project (BCPP), World Wide Fund (WWF) for Nature-India and Centre for Ecological Sciences. Indian Institute of Science, Bangalore.

Kar, D., Saha, D., Laskar, R., Barbhuiya, M.H., 1997. Biodiversity conservation prioritisation project (BCPP) in barak valley region of Assam. In: Proc. National Project Evaluation Workshop on BCPP, vol. 1. Betla Tiger Reserve and National Park, Palamu.

Kar, D., Saha, D., Dey, S.C., 1998a. Epizootic ulcerative syndrome in Barak valley of Assam. In: Project Report Submitted and Presented at the National Symposium of Biodiversity Conservation Prioritisation Project (BCPP) Held at WWF-India, 18—19 January 1998 (New Delhi), pp. 2—4.

Kar, D., Dey, S.C., Roy, A., 1998b. Present status of epizootic ulcerative syndrome (EUS) in southern Assam. In: Proc. Regional Project Initiation Workshop for NATP-ICAR-NBFGR Project, vol. 1, 9 pp.

Kar, D., Dey, S.C., Kar, S., Roy, A., 22—28 April 1998c. An account of epizootic ulcerative syndrome in Assam. In: Final Project Report of Biodiversity Conservation Prioritisation Project (BCPP). Submitted and Presented at the International Project Finalisation Workshop of BCPP, vol. 1, pp. 1—3 (New Delhi).

Kar, D., Dey, S.C., Roy, A., 1998d. Fish disease diagnosis and fish quarantine problems in India and South-East Asia with particular emphasis on epizootic ulcerative syndrome (EUS) fish disease problems. In: Proc. International Workshop on Fish Disease Diagnosis and Quarantine, Indian Council of Agricultural Research (ICAR)-Ministry of Agriculture (MoA), Govt. of India (GoI)-Network of Aquaculture Centres in Asia (NACA)-Food and Agricultural Organisation (FAO) of the United Nations (UN)-OIE, vol. 1. Held at the Central Institute of Freshwater Aquaculture (CIFA), France.

Kar, D., 1999. Microbiological and environmental studies in relation to fishes of India. In: Gordon Research Conference, Connecticut, USA.

Kar, D., Rahaman, H., Barnman, N.N., Kar, S., Dey, S.C., Ramachandra, T.V., 1999a. Bacterial pathogens associated with epizootic ulcerative syndrome in freshwater fishes of India. Environ. Ecol. 17 (4), 1025—1027.

Kar, D., Mandal, M., Bhattacharjee, S., 1999b. Fungal pathogens associated with epizootic ulcerative syndrome in fishes of barak valley region of Assam. In: Proc. 1st National Conference on Fisheries Biotechnology, vol. 1. CIFE, 34 pp.

Kar, D., Dey, S.C., Roy, A., 2000a. Fish genetic resources in the principal rivers and wetlands in North-East India with special emphasis on Barak valley (Assam), in Mizoram and in Tripura, with a note on epizootic ulcerative syndrome fish disease. In: Proc. National Project Initiation Workshop of the NATP-ICAR World Bank- Aided Project on 'Germplasm Inventory, Evaluation and Gene Banking of Freshwater Fishes', vol. 1. National Bureau of Fish Genetic Resources (NBFGR), Lucknow, 12 pp.

Kar, D., Mandal, M., Laskar, B.A., Dhar, N., Barbhuiya, M.H., 2000b. Ichthyofauna of some of the oxbow lake in Barak valley region of Assam. In: Proceedings of the National Symposium on Wetlands and Fisheries Research in the New Millennium, vol. 1, 16 pp.

Kar, D., Dey, S.C., Roy, A., Mandal, M., 2000c. Epizootic ulcerative syndrome fish disease in Barak valley region of Assam, India. In: Proc. Nat. Symp. Current Trends in Wetlands and Fisheries Research in the New Millennium, vol. 1, 2 pp.

Kar, D., Dey, S.C., Mandal, M., Lalsiamliana, 2000d. Epizootic ulcerative syndrome among the fishes of Assam. In: Proc. National Workshop of NATP-ICAR-NEC North-East Programme, Shillong, vol. 1, 24 pp.

Kar, D., Mandal, M., Lalsiamliana, 2001a. Species composition and distribution of fishes in the rivers in Barak valley region of Assam and in the principal rivers of Mizoram and Tripura in relation to their habitat parameters. In: Proc. National Workshop, NATP-ICAR Project, Mid-term Review, vol. 1. Central Marine Fisheries Research Institute, Cochin, 25 pp.

Kar, D., Laskar, B.A., Lalsiamliana, 2001b. Further studies on the ichthyospecies composition and distribution of freshwater fishes in Barak drainage and in the principal rivers in Mizoram and in Tripura with a note on their feeding and breeding biology. In: Proc. National Project Monitoring Workshop of NATP-ICAR Project, vol. 1. National Bureau of Fish Genetic Resources, Lucknow, 22 pp.

Kiryu, Y., Shields, J.D., Vogelbein, W.K., Zwerner, D.E., Kator, H., Blazer, V.S., 2002. Induction of skin ulcers in Atlantic menhaden by injection and aqueous exposure to the zoospores of *Aphanomyces invdans*. J. Aquat. Anim. Health 14, 11–24.

Kar, D., Dey, S.C., Roy, A., 2002a. Prevalence of epizootic ulcerative syndrome among fishes of Assam. In: Proc. Regional Symp. on Biodiversity. S.S. College, Assam (Central) University, Hailakandi (Assam).

Kar, D., Dey, S.C., Roy, A., 2002b. On the diseased fishes suffering from hitherto unknown epizootic ulcerative syndrome in fishes in India. In: Proc. All-India Congress of Zoology, December 2002 (Bangalore).

Kar, D., Dey, S.C., Roy, A., 2002c. Prevalence of epizootic ulcerative syndrome (EUS) among fishes of Mizoram. In: Proc. Regional Symp. on Aquaculture, vol. 1.

Kar, D., Dey, S.C., Datta, N.C., 2003a. Welfare Biology in the New Millennium. Allied Publishers Pvt. Ltd, Bangalore pp. xx + 97.

Kar, D., Dey, S.C., Roy, A., 2003b. Capture fishery in lnetic systems with some light on Sone Beel in Assam with special reference to prevalence to EUS in the Barak valley region of Assam. In: Proc. Lecture Series in UGC-sponsored Vocational Course in Industrial Fish and Fisheries, Cachar College, Assam University, Silchar, Abstracts, 1.

Kar, D., Roy, A., Dey, S.C., 2003c. Epizootic ulcerative disease syndrome in freshwater fishes with a note on its management for sustainable fisheries. In: Proc. International Conference on Disease Management for Sustainable Fisheries, 26–29 May, 2003. Department of Aquatic Biology and Fisheries, University of Kerala, Trivandrum.

Kar, D., Lohar, M., Ngassepam, Z., Sinom, S., Tiwary, B., Bawari, M., May 28–30, 2003d. Fish bioresources in certain rivers of Assam and Manipur with a note on their assessment, management and conservation. In: Proc. Nat. Symp. Assessment and Management of Bioresources. North Bengal University and The Zoological Society, Calcutta, 56 pp.

Kar, D., Dey, S.C., Roy, A., 2003e. Epizootic ulcerative syndrome: a virulent disease among the freshwater fishes of India. In: Proc. Guest Lecture delivered at the Dept. of Zoology, vol. 1. Arunachal University, Arunachal Pradesh.

Kar, D., Dey, S.C., Roy, A., 2003f. Biodiversity of freshwater fishes of North-East India with a note on their health status. In: Proc. DBT-sponsored Guest Lecture Series delivered at the Dept. of Molecular Biology and Biotechnology, vol. 1. Tezpur (Central) University, Tezpur.

Kar, D., Dey, S.C., Roy, A., 2003g. Epidemiology of epizootic ulcerative syndrome disease among the freshwater fishes of India. In: Proc. Regional Symp. on World Bank-aided Programme Entitled, 'Assam Rural Infrastructure and Agricultural Services Project (ARIASP)', Silchar, 1.

Kar, D., Das, M., 2004a. Preliminary histochemical studies with diseased fishes suffering from epizootic ulcerative syndrome in Assam. In: Proc. National Symposium for Management of Aquatic Resources for Biodiversity Maintenance and Conservation, vol. 1, 27 pp.

Kar, D., Das, M., 2004b. Preliminary enzymological studies with diseased fishes suffering from epizootic ulcerative syndrome in fishes. In: Proc. National Conference on Fish and their Environment, vol. 1, 25 pp.

Kar, D., Roy, A., Dey, S.C., March 18–19, 2004. An overview of fish genetic diversity of North-East India. In: Garg, S.K., Jain, K.L. (Eds.), Proc. National Workshop on Rational use of Water Resources for Aquaculture, vol. 1. CCS Haryana Agricultural University, pp. 164–171.

Kar, D., Roy, A., Mazumder, J., Patil, P., 2004. Biotechnological approach for defining epizootic ulcerative syndrome disease in freshwater fishes. In: Proc. Mid-term Review, vol. 1. DBT- sponsored Project, New Delhi.

Kar, D., 2005. Sustainability issues of inland fish biodiversity and fisheries in barak drainage (Assam), in Mizoram and in Tripura. In: Proceedings of the International Symposium on Improved Sustainability of Fish Production Systems and Appropriate Technologies for Utilisation. Cochin University of Science and Technology, School of Industrial Fish and Fisheries, p. 1.

Kar, D., Dey, S.C., Roy, A., Mazumder, J., Patil, P., Kohlapure, R.M., 2004c. Fish disease prevalence in Assam with particular reference to epizootic ulcerative syndrome and its pathogens. In: Proc. International Conference 'Bioconvergence', 18–20 November 2004, Thapar Institute of Engineering and Technology, Patiala, vol. 1, 220 pp.

Kar, D., 2007. Fundamentals of Limnology and Aquaculture Biotechnology. Daya Publishing House, New Delhi vi + 609.

Khatri, S., Sahu, J., Peasad, M.M., 2009. An assessment on prevalence of fish diseases in Sambalpur, Western Orissa. Asian Fish. Sci. 22, 569–581.

Khan, M.H., Lilley, J.H., 2002. Risk factors and socio-economic impacts associated with EUS in Bangladesh. In: Arthur, J.R., Phillips, M.J., Subasinghe, R.P., Reantaso, M.B., MacRae, I.H. (Eds.), Primary Aquatic Animal Health Care in Rural. Small-scale, AQC Development, Rome, pp. 27–39, pp. 54, FAO Fisheries Tech. Pap No. 406.

Kar, D., 2010. Biodiversity Conservation Prioritisation. pp. xi + 167. Swastik Publications 2, New Delhi.

Kar, D., 2012. Essentials of Fish Biology. pp. vii + 244. Dominant Publishers, New Delhi, ISBN 978-93-82007-16-6.

Kar, D., 2013. Wetlands and Lakes of the World. pp. xxx + 687. Springer, London. Print ISBN: 978-81-322-1022-1; e-Book ISBN: 978-81-322-1923-8.

Laharia, R., 2013. Immune reponse and status of epizootic ulcerative syndrome (EUS) affected fishes, *Channa punctatus* and *Clarias magur*. J. Appl. Nat. Sci. 5 (1), 103–107.

Lian, C.X., 1990. China report. In: Regional Research Programme on the Relationship between Epizootic Ulcerative Syndrome in Fish and the Environment, 13–26 August 1990. NACA, Bangkok, pp. 8–9.

Lilley, J.H., Callinan, R.B., Chinabut, S., Khanchanakhan, S., MacRae, I.H., Phillips, M.J., 1998. Epizootic Ulcerative Syndrome (EUS) Technical Handbook. The Aquatic Animal Health Research Institute, Bangkok, p. 88.

Lilley, J.H., Chinabut, S., Miles, J.C., 2001. Applied Studies on Epizootic Ulcerative Syndrome. Institute of Aquaculture, University of Stirling, Scotland. Aquatic Animal Health Research Institute, Department of Fisheries, Thailand, Stirling.

Llobrera, A.T., Gacutan, R.Q., 1987. *Aeromonas hydrophila* associated with ulcerative disease epizootic in Laguna de Bay Philippines. Aquaculture 67, 273–278.

Lio-Po, G.D., 1999. Epizootic ulcerative syndrome (EUS): recent developments. In: Book of Abstracts, 4th Symposium on Diseases in Asian Aquaculture: Aquatic Animal Health for Sustainability. Cebu City, Philippines. Fish Health Section, Asian Fisheries Society, Manila, Philippines. OP-77.

Lio-Po, G.D., Traxler, G.S., Albright, L.J., Leaño, E.M., 2000. Characterization of a virus obtained from snakeheads *Ophicephalus striatus* with epizootic ulcerative syndrome (EUS) in the Philippines. Dis. Aquat. Org. 43, 191–198.

Lio-Po, G.D., Albright, L.J., Traxler, G.S., Leano, E.M., 2002. Experimental pathogenicity of the epizootic ulcerative syndrome (EUS)-associated rhabdovirus, *Aeromonas hydrophila* and *Aphanomyces* sp. in snakehead *Channa striata*. In: Book of Abstracts, World Aquaculture 2002, Beijing. World Aquaculture Society, Baton Rouge USA, p. 434.

Lilley, J.H., Chinabut, S., Miles, J.C., 2001a. Applied Studies on Epizootic Ulcerative Syndrome. Institute of Aquaculture, University of Stirling, Scotland; Aquatic Animal Health Research Institute, Department of Fisheries, Thailand, Stirling.

Lilley, J.H., Phillips, M.J., Tonguthai, K., 1992. A review of epizootic ulcerative syndrome (EUS) in Asia. Aquatic Animal Health Research Institute and Network of Aquaculture Centres in Asia-Pacific, Bangkok, pp. 73.

Lilley, J.H., Frerichs, G.N., 1994. Comparison of rhabdoviruses associated with epizootic ulcerative syndrome (EUS) with respect to their structural proteins, cytopathology and serology. J. Fish Dis. 17, 513−522.

Leung, K.Y., Yeap, I.V., Lam, T.J., Sin, Y.M., 1994. Serum resistance as a good indicator for virulence in *Aeromonas hydrophila* strains isolated from diseased fish in South-East Asia. J. Fish Dis. 18, 511−518.

Lilley, J.H., Hart, D., Richards, R.H., Roberts, R.J., Cerenius, L., Söderhäll, K., 1997a. Pan-Asian spread of single fungal clone results in large-scale fish-kills. Vet. Rec. 140, 653−654.

Lilley, J.H., Petchinda, T., Panyawachira, V., 2001b. *Aphanomyces invadans* zoospore physiology: 4. In vitro viability of cysts. AAHRI Newsl. 10, 1−4.

Mazid, M.A., Banu, A.N.H., 2002. An overview of the social and economic impact and management of fish and shrimp disease in Bangadesh, with an emphasis on small-scale aquaculture. In: Arthur, J.R., Phillips, M.J., Subasinghe, R.P., Reantaso, M.B., MacRae, I.H. (Eds.), Primary Aquatic Animal Health Care in Rural. Small-Scale, AQC Development, Rome, pp. 21−25, pp. 54, FAO Fisheries Tech. Pap No. 406.

Mohan, C.V., Shankar, K.M., Ramesh, K.S., 1999a. Is epizootic ulcerative syndrome (EUS)-specific fungus of fishes a primary pathogen?: an opinion. Naga ICLARM Q. 22 (1), 15−18.

Mohan, C.V., Callinan, D., Fraser, G., Shankar, K.M., 1999b. Initiation and progression of epizootic ulcerative syndrome (EUS) in mullets in brackishwater ponds of Karnataka, India. In: Fourth Symposium on diseases in asian aquaculture: aquatic animal health for sustainability, 22−26 November. Cebu International Convention Center, Waterfront Cebu City Hotel, Cebu City.

Muthukrishnan, D., Haniffa, M.A.K., Muthu Ramakrishnan, C., Vijayakesava, S., Singh, A., 2008. Microbial flora from the epizootic ulcerative syndrome (EUS)-infected murrel [*Channa striatus* (Bloch, 1797)] in Tirunelveli region. Turk. J. Vet. Anim. Sci. 32 (3), 221−224.

Muthu Ramakrishnan, C., 2009. Investigation on Epizootic Ulcerative Syndrome and Prophylatic Measures in Catfish Heteropneustes Fossilis (Ph.D. thesis). Manonmaniam Sundaranar University, Tirunelveli, India.

Mckenzie, R.A., Hall, W.T.K., 1976. Dermal ulceration of mullet *(Mugil cephalus)*. Aust. Vet. J. 52, 230−231.

Mohan, C.V., Shankar, K.M., 1994. Epidemiological analysis of epizootic ulcerative syndrome of fresh and brackishwater fishes of Karnataka, India. Curr. Sci. 66, 656−658.

Miyazaki, T., Egusa, S., 1972. Studies on mycotic granulomatosis in freshwater fish I. Mycotic granulomatosis in goldfish. Fish Pathol. 7, 15−25 (In Japanese).

Mohan, C.V., Callinan, D., Fraser, G., Shankar, K.M., 1999b. Initiation and progression of epizootic ulcerative syndrome (EUS) in mullets in brackishwater ponds of Karnataka, India. In: Fourth Symposium on Diseases in Asian Aquaculture: Aquatic Animal Health for Sustainabilitzy, 22−26 November. Cebu International Convention Center, Waterfront Cebu City Hotel, Cebu City, 1999.

Miles, D.J.V., Polchana, J., Lilley, J.H., Kanchanakhan, S., Thompson, K.D., Adams, A., 2001. Immunostimulation of striped snakehead *Channa striata* against epizootic ulcerative syndrome. Aquaculture 195, 1−15.

Macintosh, D.J., Phillips, M.J., 1986. The contribution of environmental factors to the ulcerative fish diseases condition in south-east asia. In: Field and Laboratory Investigations into Ulcerative Fish Diseases in the Asia-Pacific Region, pp. 175−208. Technical Report of the FAO Project TCP/RAS/4508 Bangkok (Thailand).

Mudenda, C.G., 2006. Economic Perspectives of Aquaculture. Development Strategy of Zambia. Consultant report, Development Consultant, Lusaka Zambia.

Muniruzzaman, M., Chowdhury, M.B.R., 2008. Ulcer diseases in cultured fish in Mymensingh and surrounding districts. Bangladesh Vet. 25 (1), 40−49.

National Workshop on Ulcerative Disease Syndrome in Fish, 1990. In: . Proceedings of the National Workshop on Ulcerative Disease Syndrome in Fish, 6−7 March, 1990, Calcutta (India).

Noga, E.J., Dykstra, M.J., 1986. Oomycete fungi associated with ulcerative mycosis in Menhaden, *Brevoortia tyrannus* (Latrobe). J. Fish Dis. 9, 47−53.

Noga, E.J., 1993. Fungal diseases of marine and estuarine fishes. In: Couch, J.A., Fournie, J.W. (Eds.), Pathobiology of Marine and Estuarine Organisms. CRC Press, Boca Raton, pp. 85−100.

Nsonga, A., Mfitilodze, W., Samui, K.L., Sikawa, D., 2013. Epidermiology of epizootic ulcerative syndrome in the Zambezi river system. A case study for Zambia. Hum. Vet. Med. Int. J. Bioflux Soc. 5 (1), 1−8.

OIE, 1997. Epizootic Ulcerative Syndrome: Diagnostic Manual for Aquatic Animal Diseases. Office International Des Epizootics, Paris, pp. 127−129.

Patil, P.R., Kohlapure, R.M., Mazumder, J., Kar, D., Roy, A., 2003. Isolation of a suspected viral agent from EUS-affected *Labeo calbasu* fish. In: Proc. International Conference on Disease Management for Sustainable Fisheries. University of Kerala, Trivandrum.

Paclibare, J.O., Catap, E.S., Callinan, R.B., January 25−27, 1994. Fungal isolation of EUS-affected fish in the Philippines. In: Roberts, R.J., Campbell, B., MacRae, I.H. (Eds.), Proceedings of the ODA Regional Seminar on Epizootic Ulcerative Syndrome, Aquatic Animal Health Research Institute, Bangkok. AAHRI, Bangkok, pp. 238−243.

Pradhan, P.K., Rathore, G., Sood, N., Swaminathan, T.R., Yadav, M.K., Verma, D.K., Chaudhary, D.K., Abidi, R., Punia, P., Jena, J.K., 2014. Emergence of epizootic ulcerative syndrome: large-scale mortalities of cultured and wild fish species in Uttar Pradesh, India. Curr. Sci. 106 (12), 1711−1718.

Purkait, M.C., 1990. Case studies on the epizootic ulcerative syndrome of fishes in Chanditala:I of Hooghly district. In: National Workshop on Ulcerative Disease Syndrome in Fish, 6−7 March, 1990 (Calcutta).

Prasad, P.S., Sinha, J.P., 1990. Status paper on the occurrence of ulcerative disease syndrome in fishes of Bihar. In: National Workshop on Ulcerative Disease Syndrome in Fish, 6−7 March, 1990 (Calcutta).

Riji John, K., Richards, R.H., 1999. Characteristics of a new birnavirus associated with a warm-water fish cell line. J. General Virol. 80, 2061–2065.

Riji John, K., Rosalind George, M., 2012. Viruses associated with epizootic ulcerative syndrome: an update. Indian J. Virol. 23 (2), 106–113.

Riji John, K., Rosalind George, M., Richards, R.H., Frerichs, G.N., 2001. Characteristics of a new reovirus isolated from epizootic ulcerative syndrome-infected snakehead fish. Dis. Aquat. Org. 46, 83–92.

Riji John, K., Rosalind George, M., Mansoor, Singha, R., Mahesh, Selva Mahesh, Kar, D., 2015. Molecular Characterisation of Pathogens Associated with Fish Diseases in Assam. Project Report. Department of Biotechnology, Govt. of India-Funded, New Delhi.

Rodgers, L.J., Burke, J.B., 1981. Seasonal variation in the prevalence of red spot disease in estuarine fish with particular reference to the sea mullet, *Mugil cephalus* L. J. Fish Dis. 4, 297–307.

Roberts, R.J., Willoughby, L.G., Chinabut, S., 1993. Mycotic aspects of epizootic ulcerative syndrome (EUS) of Asian fishes. J. Fish Dis. 16, 169–183.

Roberts, R.J., Macintosh, D.J., Tonguthai, K., Boonyratpalin, S., Tayaputch, N., Phillips, M.J., Millar, S.D., 1986. Field and Laboratory Investigations into Ulcerative Fish Diseases in Asia-Pacific Region. Technical Report, FAO Project TCP/RAS/4508, pp. 214.

Roberts, R.J., 1997. Epizootic ulcerative syndrome (EUS): progress since 1985. In: Flegel, T.W., Macrae, I.H. (Eds.), Diseases in Asian Aquaculture III. Fish Health section, Asian Fisheries Society, Manila, pp. 125–128.

Roberts, R.J., Wootten, R., MacRae, I., Millar, S., Struthers, W., 1989. Ulcerative Disease Survey: Bangladesh Final Report to the Govt. of Bangladesh and the Overseas Development Administration. Institute of Aquaculture, Stirling University, UK.

Roberts, R.J., Campbell, B., MacRae, I.H., 1994a. Proceedings of the Regional Seminar on Epizootic Ulcerative Syndrome. 25–27 January 1994. The Aquatic Animal Health Research Institute, Bangkok.

Roberts, R.J., Frerichs, G.N., Tonguthai, K., Chinabut, S., 1994b. Epizootic ulcerative syndrome of farmed and wild fishes. In: Muir, J.F., Roberts, R.J. (Eds.), Chapter 4. Recent Advances in Aquaculture V. Blackwell Science, pp. 207–239.

Rukyani, A., 1994. Status of epizootic ulcerative disease in Indonesia. In: Proceedings of the ODA Regional Seminar on Epizootic Ulcerative Syndrome, 25–27 January, 1994. Aquatic Animal Health Research Institute, Bangkok.

Rahman, T., Choudhury, B., Barman, N., 1988. Role of UDS affected fish in the health of ducks in Assam: an experimental study. In: Proceedings of the Symposium on Recent Outbreak of Fish Diseases in North Eastern India. 30 December 1988, p. 10 Guwahati, Assam, India.

Roberts, R.J., Campbell, B., MacRae, I.H., 1994a. In: Proceedings of the Regional Seminar on Epizootic Ulcerative Syndrome. 25–27 January 1994. The Aquatic Animal Health Research Institute, Bangkok.

Roberts, R.J., Frerichs, G.N., Tonguthai, K., Chinabut, S., 1994b. Epizootic ulcerative syndrome of farmed and wild fishes. In: Muir, J.F., Roberts, R.J. (Eds.), Chapter 4. Recent Advances in Aquaculture V. Blackwell Science, pp. 207–239.

Roy, A., 2005. Biotechnological approach for defining epizootic ulcerative syndrome in freshwater fishes. In: Annual Report 2004–2005. National Institute of Virology, Pune.

Roy, R.P., Bahadur, M., Barat, S., 2013. Isolation and identification of multi-antibiotic resistant *Aeromonas veronii* by 16s rDNA gene sequencing from the gut of fresh Water Loach, *Lepidocephalycthys guntea* (Hamilton Buchanan). Int. J. Bioresour. Stress Manage. 4 (4), 496–499.

Rattan, P., Parulekar, A.H., 1998. Diseases and parasites of laboratory reared and wild population of banded pearl spot *Etroplus suratensis* (Cichlidae) in Goa. Indian J. Mar. Sci. 27 (3–4), 407–410.

Roberts, R.J. (Ed.), 1989. Fish Pathology. Bailliere Tindall, London x + 318.

Reantaso, M.B., 1991. EUS in brackish waters of the Philippines. Fish Health Sect. Newsl. 2 (1), 8–9. Asian Fisheries Society (Manila).

Rodger, H.D., 2010. Fish Disease Manual.

Sankar Ganesh, P., Saravaanan, J., Krishnamurthy, P., Chandrakaala, N., Rajendran, K., 2012. Isolation and identification of *Vibrio* spp. in diseased *Channa punctatus* from aquaculture fish farm. Indian J. Geo Mar. Sci. 41 (2), 159–163.

Shariff, M., Law, A.T., 1980. An incidence of fish mortality in Bekok river, Johore, Malaysia. In: Proc. International Symp. on Conservation and Input from Life Sciences, 27–30 October 1980. Universiti Kebangsaan, Bangi, Selangor (Malaysia).

Shariff, M., Saidin, T.H., 1994. Status of epizootic ulcerative syndrome in Malaysia since 1986. In: Roberts, R.J., Campbell, B., MacRae, I.H. (Eds.), Proceedings of the ODA Regional Seminar on Epizootic Ulcerative Syndrome, 25–27 January 1994. Aquatic Animal Health Research Institute, Bangkok, pp. 48.

Soe, U.M., August 13–26, 1990. Myanmar report. In: Regional Research Programme on Relationships between Epizootic Ulcerative Syndrome in Fish and the Environment. NACA, Bangkok, pp. 35–38.

Saylor, R.K., Miller, D.L., Vandersea, M.W., Bevelhimer, M.S., Schofield, P.J., Bennett, W.A., 2010. Epizootic ulcerative syndrome caused by *Aphanomyces invadans* in captive bullseye snakehead (*Channa marulius*) collected from south Florida, USA. Dis. Aquat. Org. 88, 169–175.

Sipaúba-Tavares, L.H., Gomes, J.P.F., Dos, S., Braga, F.M., De, S., 2003. Effect of liming management on the water quality in *Colossoma macropomum* ("Tambaqui"), ponds. Acta Limnol. Bras. 15 (3), 95–103.

Siddique, M.A.B., Bashar, M.A., Hussain, M.A., Kibria, A.S.M., 2009. Fungal disease of freshwater fishes in Natore district of Bangladesh. J. Bangladesh Agric. Univ. 7 (1), 157–162.

Sunitha, K.S., Haniffa, M.A., James Milton, M., Manju, A., 2010. *Coleus aromaticus* Benth act as an immunostimulant in *Channa marulius* Hamilton. IJBT 1 (2), 55.

Tonguthai, K., 1985. A Preliminary Account of Ulcerative Fish Diseases in the Indo-Pacific Region (A Comprehensive Study Based on Thai Experiences). National Inland Fisheries Institute, Bangkok, Thailand, pp. 39.

Takuma, D., Sano, A., Wada, S., Kurata, O., Hatai, K., 2010. A new species, *Aphanomyces salsuginosus* sp. nov., isolated from ice fish Salangichthys microdon. Mycoscience 51 (in press).

Takuma, D., Sano, A., Hatai, K., 2013. Two new species, *Aphanomyces izumoensis* sp. nov. and *Aphanomyces shimanensis* sp. nov. isolated from Ice Fish *Salangichthys microdon*. Int. J. Res. Pure Appl. Microbiol. 3 (3), 67–76.

Ulcerative Fish Disease Committee, 1983. Practical Report of the Ulcerative Fish Disease Committee 1982–1983. Bangkok.

Vishwanath, T.S., Mohan, C.V., Shankar, K.M., 1998. Epizootic ulcerative syndrome associated with a fungal pathogen in Indian fishes-histopathology: a cause for invasiveness. Aquaculture 165, 1–9.

Vogelbein, W.K., Shields, J.D., Haas, L.W., Reece, K.S., Zwerner, D.E., 2001. Skin ulcers in estuarine fishes: a comparative pathological evaluation of wild and laboratory-exposed fish. Environ. Health Perspect. 109 (Suppl. 5), 687–693.

Vishwanath, T.S., Mohan, C.V., Shankar, K.M., 1997a. Mycotic granulomatosis and seasonality are the consistent features of EUS of fresh and brackish water fishes of Karnataka, India. Asian Fish. Sci. 10, 155–160.

Vishwanath, T.S., Mohan, C.V., Shankar, K.M., 1997b. Chemical and histopathological characterization of different types of lesions associated with epizootic ulcerative syndrome (EUS). J. Aquacult. Tropics 12 (1), 35–42.

Vandersea, M.W., Litaker, R.W., Yonnish, B., Sosa, E., Landsberg, J.H., Pullinger, C., Moon-Butzin, P., Green, J., Morris, J.A., Kator, H., Noga, E.J., Tester, P.A., 2006. Molecular assays for detecting Aphanomyces invadans in ulcerative mycotic fish lesions. Appl. Environ. Microbiol. 72, 1551–1557.

Vijayakumar, R., Raja, K., Sinduja, K., Gopalakrishnan, A., 2013. Epizootic ulcerative syndrome on fresh water and brackish water fishes. Afr. J. Basic Appl. Sci. 5 (4), 179–183.

Willoughby, L.G., Roberts, R.J., Chinabut, S., 1995. Aphanomyces invaderis sp. nov., the fungal pathogen of freshwater tropical fish affected by epizootic ulcerative syndrome. J. Fish Dis. 18, 273–275.

Wilson, K.D., Lo, K.S., 1992. Fish disease in Hong Kong and the potential role of the veterinarian. In: 23rd Annual Conference of the International Association for Aquatic Animal Medicine (IAAAM), Hong Kong, 18–22 May 1992.

Widago, D., 1990. Indonesia report. In: Regional Research Programme on Relationships between Epizootic Ulcerative Syndrome in Fish and the Environment, 13–26 August 1990. NACA, Bangkok, pp. 18–22.

Xuan, T.T., 1990. Vietnam report. In: Phillips, M.J., Keddie, H.G. (Eds.), Regional Research Programme on Relationships between Epizootic Ulcerative Syndrome in Fish and the Environment, A Report on the 2nd Technical Workshop, 13–26 August 1990. Network of Aquaculture Centres in Asia-Pacific, Bangkok.

Zoological Society of Assam, 1988. In: Proceedings of the Symposium on Recent Outbreak of Fish Diseases in North Eastern India. 30 December 1988. Organised by Zoological Society of Assam, Guwahati, Assam, India, 23 pp.

Zaki, V.H., 2004. Effect of Saccharomyces cervisiae on the immune status of Oreochromis niloticus against transportation stress and motile Aeromonas septicemia. In: First Scientific Conference of Fac. Vet. Med.; Moshtohor, September 1–4, Benharas Sedr.

Zhang, X., Xie, P., Li, D., Shi, Z., 2007. Hematological and plasma biochemical responses of crucian carp (Carassius Auratus) to intraperitoneal injection of extracted microcystins with the possible mechanisms of anemia. Toxicon 49 (8), 1150–1157.

FURTHER READING

Archetti, I., Horsfall Jr., F.L., 1950. Persistent antigenic variation of influenza A viruses after incomplete neutralization in ovo with heterologous immune serum. J. Exp. Med. 92, 441–462.

Avtalion, R.R., Wichkovsky, A., Katz, D., 1980. Regulatory effect of temprature on specific supression and enhancement of the humoral response in fish. In: Manning, M.J. (Ed.), Phylogeny of Immunological Memory. Elsevier, Amsterdam, pp. 113–121.

Alderman, D.J., Polglase, J.L., 1988. Pathogens, parasites and commensals. In: Holdich, D.M., Lowery, R.S. (Eds.), Freshwater Crayfish: Biology, Management and Exploitation, pp. 167–212.

Austin, B., Austin, D.A., 1993. In: Horwood, E. (Ed.), Bacterial Fish Pathogens: Disease in Farmed and Wild Fish, second ed. Chichester, United Kingdom.

Ahmed, M., Rab, M.A., 1995. Factors affecting outbreaks of epizootic ulcerative syndrome in farmed and wild fish in Bangladesh. J. Fish Dis. 18, 263–271.

AAHRI, ACIAR, IoA and NACA, March 9–19, 1997. Epizootic ulcerative syndrome (EUS) of fishes in Pakistan. In: A Report of the Findings of an ACIAR/DFID-Funded Mission to Pakistan.

Ahmed, G.U., Hoque, A., 1998. Mycotic involvement in epizootic ulcerative syndrome of freshwater fishes of Bangladesh: a histopathological study. In: Proceedings of the Fifth Asian Fisheries Forum, Chiang Mai, Thailand. Asian Fisheries Society, Manila, pp. 11–14.

Al Harbi, A.H., 2003. Faecal coliforms in pond water, sediments and hybrid tilapia Oreochromis niloticus × Oreochromis aureus in Saudi Arabia. Aquacult. Res. 34 (7), 517–524.

Ashton, P.J., 2007. Riverine biodiversity and conservation in South Africa: current situation and future prospects. Aquat. Conserv. Mar. Freshwater Ecosyst. 17, 441–445.

Andrew, T.G., Huchzermeyer, K.D., Mbeha, B.C., Nengu, S.M., 2008. Epizootic ulcerative syndrome affecting fish in the Zambezi river system in southern Africa. Vet. Rec. 163, 629–631.

Anon, 2010. Manual of Diagnostic Tests for Aquatic Animals. World Organisation for Animal Health, Paris. Viewed in 2011, from: http://www.oie.int/international-standard-setting/aquatic-manual/access-online/.

Blazevic, D.J., Ederer, G.M., 1975. Principles of Biochemical Tests in Diagnostic Microbiology. John Wiley & Sons, Inc., New York.

Bennet, R.W., 1984. Bacteriological Analytical Manual, sixth ed. Association of Official Analytical Chemists, Arlington.

Bielenin, A., Jeffers, S.N., Wilcox, W.F., Jones, A.L., 1988. Separation by protein electrophoresis of six species of Phytophthora associated with deciduous fruit crops. Phytopathology 78, 1402–1408.

Barua, G., August 13–26, 1990. Bangladesh report. In: Regional Research Programme on Relationships between Epizootic Ulcerative Syndrome in Fish and the Environment. NACA, Bangkok, pp. 2–4.

Bhaumik, U., Pandit, P.K., Chatterjee, J.G., 1991. Impact of epizootic ulcerative syndrome on the fish yield, consumption and trade in West Bengal. J. Inland Fish. Soc. India 23 (1), 45–51.

Bondad-Reantaso, M.G., Lumanlan, S.C., Natividad, J.M., Phillips, M.J., 1992. Environmental monitoring of the epizootic ulcerative syndrome (EUS) in fish from Munoz, Nueva Ecija in the Philippines. In: Shariff, M., Subasinghe, R.P., Arthur, J.R. (Eds.), Diseases in Asian Aquaculture I, Fish Health Section. Asian Fisheries Society, Manila, The Philippines, pp. 475–490.

Bruno, D.W., Wood, B.P., 1994. Saprolegnia and other Oomycetes. In: Wood, P.T.K., Bruno, D.W. (Eds.), Fish Diseases and Disorder, Viral, Bacterial and Fungal Infections, vol. 3. CABI publishing, Wallingford, Oxon, United Kingdom, pp. 599–659.

Boyd, C.E., 1995. Bottom Soils, Sediment, and Pond Aquaculture. Chapman and Hall, New York, 348 pp.

Boyd, C.E., 1997. Practical aspects of chemistry in pond aquaculture. Prog. Fish Cult. 59 (2), 85–93.

Bondad-Reantaso, M.G., Hatai, K., Kurata, O., 1999a. Aphanomyces from EUS-infected fish in the Philippines and Bangladesh and MG-infected fish from Japan: I. Comparative cultural, biological and biochemical characterization. In: Book of Abstracts, Fourth Symposium on Diseases in Asian Aquaculture, November 22–26, 1999, Cebu City, Philippines. Fish Health Section, Asian Fisheries Society, pp. 35.

Bondad-Reantaso, M.G., Hatai, K., Kurata, O., 1999b. Aphanomyces from EUS-infected fish in the Philippines and Bangladesh and MG-infected fish from Japan: II. Pathogenicity studies. In: Book of Abstracts, Fourth Symposium on Diseases in Asian Aquaculture, November 22–26, 1999, Cebu City, Philippines. Fish Health Section, Asian Fisheries Society, pp. 36.

Bondad-Reantaso, M.G., Hatai, K., Kurata, O., 1999c. Aphanomyces from EUS-infected fish in the Philippines and Bangladesh and MG-infected fish from Japan: III. Presence of galactose-binding protein. In: Book of Abstracts, Fourth Symposium on Diseases in Asian Aquaculture, November 22–26, 1999, Cebu City, Philippines. Fish Health Section, Asian Fisheries Society, pp. 37.

Bondad-Reantaso, M.G., McGladdery, S., East, I., Subasinghe, R.P. (Eds.), 2001. Asia Diagnostic Guide to Aquatic Animal Diseases. FAO Fisheries Technical Paper No. 402. Supplement 2. Rome, FAO. 240 p. Also available online at: http://www.enaca.org/NACA-Publications/ADG-complete.pdf.

Blazer, V.S., Lilley, J., Schill, W.B., Kiryu, Y., Densmore, C.L., Panyawachira, V., Chinabut, S., 2002. Aphanomyces invadans in Atlantic menhaden along the east coast of the United States. J. Aquat. Anim. Health 14, 1–10.

Callinan, R.B., Keep, J.A., 1989. Bacteriology and parasitology of red spot disease in sea mullet, Mugil cephalus L., from eastern Australia. J. Fish Dis. 12, 349–356.

Callinan, R.B., Fraser, G.C., Virgona, J.L., 1989. Pathology of red spot disease in sea mullet, Mugil cephalus L. from eastern Australia. J. Fish Dis. 12, 467–479.

Chen, W., Hoy, J.W., Schneider, R.W., 1991. Comparisons of soluble proteins and isozymes for seven Pythium species and applications of the biochemical data to Pythium systematics. Mycol. Res. 95, 548–555.

Chen, M.M., Kou, G.H., Chen, S.N., 1993. Establishment and characterization of a cell line persistently infected with infectious pancreatic necrosis virus (IPNV). Bull. Inst. Zool. Acad. Sin. 32, 265–272.

Chinabut, S., Roberts, R.J., 1999. Pathology and Histopathology of Epizootic Ulcerative Syndrome (EUS). Aquatic Animal Health Research Institute, Department of Fisheries, Royal Thai Government, Bangkok, Thailand, ISBN 974-7604-55-8, 33 p.

Chowdhury, M.B.R., 1993. Research priorities for microbial fish disease and its control in Bangladesh for fish health. In: Tollervey, A. (Ed.), Disease Prevention and Pathology, pp. 8–11.

Callinan, R.B., 1994. Red spot disease-EUS in Australia. In: Roberts, R.J., Campbell, B., MacRae, I.H. (Eds.), Proceedings of the ODA Regional Seminar on Epizootic Ulcerative Syndrome, Aquatic Animal Health Research Institute, Bangkok, 25–27 January 1994. AAHRI, Bangkok, pp. 189–193.

Callinan, R.B., 1994a. A comparative review of Aphanomyces sp. associated with epizootic ulcerative syndrome, red spot disease and mycotic granulomatosis. In: Roberts, R.J., Campbell, B., MacRae, I.H. (Eds.), Proc. ODA Regional Seminar on Epizootic Ulcerative Syndrome, 25–27 January, 1994. Aquatic Animal Health Research Institute, Bangkok, pp. 248–252.

Cruz-Lacierda, E.R., Shariff, M., 1995. Experimental transmission of epizootic ulcerative syndrome (EUS) in snakehead, Ophicephalus striatus. In: Shariff, M., Arthur, J.R., Subasinghe, R.P. (Eds.), Diseases in Asian Aquaculture II. Fish Health Section, Asian Fisheries Society, Manila. Philippines, pp. 327–336.

Callinan, R.B., Sammut, J., Fraser, G.C., 1996. Epizootic ulcerative syndrome (red spot disease) in estuarine fish: confirmation that exposure to acid sulphate soil runoff and an invasive fungus, Aphanomyces sp., are causative factors. In: Proceedings of the National Conference on Acid Sulphate Soils. Robert J. Smith and Associates and ASSMAC, Australia.

Callinan, R.B., 1997. Pathogenesis of Red Spot Disease (Epizootic Ulcerative Syndrome) in Estuarine Fish in Eastern Australia and the Philippines. University of Queensland, Australia, pp. 232.

CAMP, 1998. In: . Report of the Workshop on 'Conservation, Assessment and Management Plan (CAMP) for Freshwater Fishes of India'. Zoo Outreach Organization and NBFGR, Lucknow, pp. 156, September 22–26, 1997.

Catap, E.S., Munday, B.L., 1998. Effects of variations of water temperature and dietary lipids on the expression of experimental epizootic ulcerative syndrome (EUS) in sand whiting, Sillago ciliata. Fish Pathol. 33, 327–335.

Catap, E.S., Munday, B.L., 2002. Development of a method for reproducing epizootic ulcerative syndrome using controlled doses of Aphanomyces invadans in species with different salinity requirements. Aquaculture 209, 35–47.

Duncan, R., Dobos, P., 1986. The nucleotide sequence of infectious pancreatic necrosis virus (IPNV) dsRNA segment A reveals one large ORF encoding a precursor polyprotein. Nucleic Acids Res. 14, 5934.

Dykstra, M.J., Noga, E.J., Levine, J.F., Moye, D.W., Hawkins, J.H., 1986. Characterization of the Aphanomyces species involved with ulcerative mycosis (UM) in menhaden. Mycologia 78 (4), 664–672.

Dobos, P., Nagy, E., Duncan, R., 1991. Birnaviridae. In: Kurstak, E. (Ed.), Viruses of Invertebrates. Marcel Dekker, New York, pp. 301–314.

Dobos, P., Berthiaume, L., Leong, J.A., Kibenge, F.S.B., Muller, H., Nicholson, B.L., 1995. Family Birnaviridae. In: Murphy, F.A., Fauquet, C.M., Bishop, D.H.L., Ghabrial, S.A., Jarvis, A.W., Martelli, G.P., Mayo, M.A., Summers, M.D. (Eds.), Virus Taxonomy. Sixth Report of the International Committee on Taxonomy of Viruses. Springer-Verlag, Vienna & New York, pp. 240–244.

DFID, April 19–30, 1998. A Report on the 2nd Mission to Investigate Epizootic Ulcerative Syndrome (EUS) in Pakistan.

Drancourt, M., Bollet, C., Carlioz, A., Martelin, R., Gayral, J.P., Raoult, D., 2000. 16S ribosomal DNA sequence analysis of a large collection of environmental and clinical unidentifiable bacterial isolates. J. Clin. Microbiol. 38 (10), 3623–3630.

Daskalov, H., 2006. The importance of Aeromonas hydrophila in food safety. Food Cont. 6, 474–483.

Egusa, 1978. Studies on the pathogenic fungus of mycotic granulomatosis – II. Some note on the MG-fungus. Fish Pathol. 13, 85–89 (In Japanese, English abstract).

Evenberg, D., de Graff, P., Fleuren, W., van Muiswinkel, W.B., 1986. Blood changes in carp (Cyprinus carpio) induced by ulcerative Aeromonas salmonicida infections. Vet. Immunol. Immunopathol. 12, 321–330.

EFSA, 2007. Scientific Opinion of the Panel on Animal Health and Welfare on a request from the European Commission on possible vector species and live stages of susceptible species not transmitting disease as regards certain fish diseases. EFSA J. 584, 1–163.

Enviro-Fish Africa, 2007. Report of a Preliminary Investigation into the Fish Disease Outbreak, Chobe River, Botswana. Report of the Department of Wildlife and National Parks, Botswana.

FAO, 2004. The State of World Fisheries and Aquaculture. FAO Fisheries Department, Rome, 153 pp.

Fowles, B., 1976. Factors affecting growth and reproduction in selected species of *Aphanomyces*. Mycologia 68, 1221–1232.

Frerichs, G.N., Hill, B.J., Way, K., 1989. Ulcerative disease rhabdovirus: cell line susceptibility and serological comparison with other fish rhabdoviruses. J. Fish Dis. 12, 51–56.

Frerichs, G.N., Millar, S.D., Chinabut, S., 1993. Clinical response of snakeheads (*Ophicephalus striatus*) to experimental infection with snakehead fish rhabdovirus and snakehead cell line retrovirus. Aquaculture 116, 297–301.

Funge-Smith, S., Dubeau, P., 2002. Aquatic animal health management issues in rural AQC development in Lao PDR. In: Arthur, J.R., Phillips, M.J., Subasinghe, R.P., Reantaso, M.B., MacRae, I.H. (Eds.), Primary Aquatic Animal Health Care in Rural. Small-scale, AQC Development, Rome, pp. 41–-54, pp. 54, FAO Fisheries Tech. Pap No. 406.

Green, I.C., Wagner, D.A., Glogowski, J., Skipper, P.L., Wishnok, J.S., Tannenbaum, S., 1982. Analysis of nitrate, nitrite and (15N) nitrate in biological fluids. Anal. Biochem. 126, 131–138.

Greuter, W., McNeill, J., Barrie, F.R., Burdet, H.M., Demoulin, V., Filgueiras, V.T.S., Nicolson, D.H., Silva, P.C., Skog, J.E., Trehane, P., Turland, N.J., Hawksworth, D.L., 2000. International Code of Botanical Nomenclature (Saint Louis Code). In: Regnum Vegetabile, 138. Koeltz Scientific Books, Königstein, Germany, 474 pp.

Garcia, M.J.A., Gonzalez, F.A.N., Pineda, O.C., Rodriguez, R.H.A., Hernandez, M.R.E., 2003. Microbiological quality of the culture of rainbow trout (*Oncorhynchus mykiss*) and water quality in the north-western region of Chihuahua, Mexico. Hidrobiologica (Iztapalapa) 13, 111–118.

Golas, I., Zmyslowska, I., Harnisz, M., Teodorowicz, M., Lewandowsca, D., Szarek, J., 2004. Isolation of heterotrophic bacteria in the water, fish feed and fish skin in Poland, during the pond rearing of European catfish (*Silurus glanis* L.) fry. Bull. Sea Fish. Inst. Gdynia 161, 3–14.

Hoshima, T., Sano, T., Sunamayo, M., 1960. Studies on the Saprolegniasis of eel. J. Tokyo Univ. Fish. 47, 59–79.

Hatai, K., Egusa, S., Takashashi, S., Ooe, K., 1977a. Study on the pathogenic fungus of mycotic granulomatosis: I: Isolation and pathogenicity of the fungus from cultured ayu infected with the disease. Fish Pathol. 11 (2), 129–133.

Hatai, K., Egusa, S., Takahashi, S., Ooe, K., 1977b. Study on the pathogenic fungus of mycotic granulomatosis – I. Isolation and pathogenicity of the fungus from cultured-ayu infected with the disease. Fish Pathol. 12, 129–133.

Hatai, K., Egusa, S., 1979. Studies on pathogenic fungus of mycotic granulomatosis III. Development of the medium for MG-fungus. Fish Pathol. 13, 147–152.

Hatai, K., Egus, S., 1979. Studies on the pathogenic fungus of mycotic granulomatosis - III. Development of the medium for MG-fungus. Fish Pathol. 13, 147–152 (In Japanese, English abstract).

Hatai, K., 1980a. Studies on pathogenic agents of saprolegniasis in fresh water fishes. Spec. Rep. Nagasaki. Pref. Inst. Fish. 8, 1–95.

Hatai, K., 1980b. Studies on the Pathogenic Agents of Saprolegniansis in Freshwater Fishes. Special Report of Nagasaki Prefectural Inst. of Fisheries, No. 8. Matsugae-Cho, Nagasaki, Japan.

Hatai, K., Takahashi, S., Egusa, S., 1984. Studies on the pathogenic fungus of mycotic granulomatosis - IV. Changes of blood constituents in both ayu, *Plecoglossus altivelis*, experimentally inoculated and naturally infected with *Aphanomyces piscicida*. Fish. Pathol. 19, 17–23.

Humphrey, J.D., Langdon, J.S., 1986. Ulcerative Disease in Northern Territory Fish. Internal Report. Australian Fish Health Reference Laboratory, Benalia, Victoria, Australia.

Hedrick, R.P., Eaton, W.D., Fryer, J.L., Croberg, W.G., Boonyaratpalin, S., 1986. Characteristics of a birnavirus isolated from cultured sand goby *Oxyeleotris marmoratus*. Dis. Aquat. Org. 1, 219–225.

Havelaar, A.H., During, M., Versteegh, J.F.M., 1987. Ampicillin dextrin agar medium for the enumeration of *Aeromonas* species in water by membrane filtration. J. Appl. Bacteriol. 62, 279–287.

Hussenot, M., Martin, J.L.M., 1995. Assessment of the quality of pond sediment in aquaculture using simple, rapid techniques. Aquacult. Int. 3 (2), 123–133.

Hatai, K., 1994. Mycotic granulomatosis in ayu *(Plecoglossus altivelis)* due to *Aphanomyces piscicida*. In: Roberts, R.J., Campbell, B., MacRae, I.H. (Eds.), Proc. ODA Regional Sem. on Epizootic Ulcerative Syndrome, 25–27 January 1994. Aquatic Animal Health Research Institute, Bangkok.

Hatai, K., Nakamura, K.,, Rha, S.A., Yusa, K., Wada, S., 1994. *Aphanmyces* infection in dwarf gourami *(Colisa lalia)*. Fish Pathol. 29 (2), 95–99.

Hill, B.J., Way, K., 1996. Serological classification of infectious pancreatic necrosis (IPN) virus and other aquatic birnaviruses. Annu. Rev. Fish Dis. 5, 55–77.

Hanjavanit, C., Hiroki, S., Hatai, K., 1997. Mycotic granulomatosis found in two species of ornamental fishes imported from Singapore. Mycoscience 38, 433–436.

Haniffa, M.A., Marimuthu, K., 2004. Seed production and culture of snakehead. INFOFISH Int. 2, 16–18.

Iqbal, M.M., Tajima, K., Ezura, Y., 1998. Phenotypic identification of motile *Aeromonas* isolated from fishes with epizootic ulcerative syndrome in Southeast Asian countries. Bull. Fac. Fish. Hokkaido Univ. 49, 131–141.

Isonhood, J.H., Drake, M., 2002. *Aeromonas* species in foods. J. Food Prot. 65, 557–582.

Janda, J.M., Abbott, S.L., 1996. Human pathogens. In: Austin, B., Attwegg, M., Golsin, P.J., Joseph, S. (Eds.), The Genus Aeromonas. John Wiley and Sons, New York, pp. 151–173.

Janda, J.M., Abbott, S.L., 1998. Evolving concepts regarding the genus *Aeromonas*: an expanding panorama of species, disease presentations, and unanswered questions. Clin. Infec. Dis. 27, 332–344.

Johnson Jr., T.W., Seymour, R.L., Padgett, D.E., 2002. Biology and the systematics of the Saprolegniaceae. http://dl.uncw.edu/digilib/biology/fungi/taxonomy%20and%20systematics/padgett%20book/.

Jewel, M.A.S., Affan, M.A., 2003. Epizotic ulcerative syndrome (EUS) in fishes of small-scale farmers' ponds in Bogra district of Bangladesh. Univ. J. Zool. Rajshahi Univ. 22, 11–18.

Kumamaru, A., 1973. A fungal disease of fishes in lake Kasumigiura and lake Kitaura. Report Stan. Ibaraki Prefect. 11, 129–142 (In Japanese).

Kasornchandra, J., Engelking, H.M., Lannan, C.N., Rohovec, J.S., Fryer, J.L., 1992. Characteristics of three rhabdoviruses from snakehead fish *Ophicephalus striatus*. Dis. Aquat. Org. 13, 89–94.

Karunasagar, I., Karunasagar, I., January 25–27, 1994. Bacteriological studies on ulcerative syndrome in India. In: Roberts, R.J., Campbell, B., Macrae, I.H. (Eds.), Proceedings of the ODA Regioanl Seminar on Epizootic Ulcerative Syndrome. Aquatic Animal Health Research Institute, Bangkok, pp. 158–170.

Karunasagar, I., Sugumar, G., Karunasagar, I., 1995. Virulence characters of *Aeromonas* spp. isolated from EUS-affected fish. In: Shariff, M., Arthur, J.R., Subasinghe, R.P. (Eds.), Diseases in Asian Aquaculture II. Fish Health Section, Asian Fisheries Society, Manila, pp. 307–314.

Khulbe, R.D., Joshi, C., Bisht, G.S., 1995. Fungal diseases of fish in NanakSagar, Nainai Tal, India. Mycopathologia 130, 71–74.

Kanchanakhan, S., 1996b. Field and Laboratory Studies on Rhabdoviruses Associated with Epizootic Ulcerative Syndrome (EUS) of Fishes (Ph.D. thesis). University of Stirling, Scotland, 278 pp.

Kar, D., Dey, S.C., Roy, A., Mandal, M., 2000e. Epizootic ulcerative syndrome in fish at Barak Valley, Assam, India. In: Ramachandra, T.V., Murthy, C.R., Ahalya, N. (Eds.), Lake 2000: Int. Symp. Restoration of Lakes and Wetlands. Indian Inst. Sci, Bangalore, India, pp. 317–319.

Kühn, I., Allestam, G., Huys, G., Janssen, P., Kersters, K., Krovacek, K., Stenstrom, T.X., 1997a. Diversity, persistence and virulence of *Aeromonas* strains isolated from drinking water distribution systems in Sweden. Appl. Environ. Microbiol. 63, 2708–2715.

Kühn, I., Albert, M.J., Ansaruzzaman, M., Bhuiyan, N.A., Alabi, S.A., Islam, M.S., Neogi, P.K., Huys, G., Janssen, P., Kersters, K., Mollby, R., 1997b. Characterization of *Aeromonas* spp. isolated from humans with diarrhea, from healthy controls, and from surface water in Bangladesh. J. Clin. Microbiol. 35, 369–373.

Kanchanakhan, S., Saduakdee, U., Areerat, S., 1998. Virus isolation from epizootic ulcerative syndrome-diseases fishes. In: The 5th Asian Fisheries Forum, November 11–14, 1998. Chiangmai, Thailand (Abstract, p. 99).

Khan, M.H., Marshall, L., Thompson, K.D., Campbell, R.E., Lilley, J.H., 1998. Susceptibility of five fish species (nile tilapia, rosy barb, rainbow trout, stickleback and roach) to intramuscular injection with the oomycete fish pathogen, *Aphanomyces invadans*. Bull. Eur. Assoc. Fish Pathol. 18 (6), 192–197.

Kanchanakhan, S.S., Saduakdee, U., Areerat, S., 1999. Virus isolation from epizootic ulcerative syndrome-diseased fishes. Asian Fish. Sci. 4, 327–335.

Kurata, O., Kanai, H., Hatai, K., 2000. Hemagglutinating and hemolytic capacities of *Aphanomyces piscicida*. Fish. Pathol. 35, 29–33.

Kasai, H., Nakamura, N., Yoshimizu, M., 2002. Observations on bacteria collected in the Shibetsu, Ichani and Kotanuka rivers in Eastern Hokkaido, Japan. Bull. Fish. Sci. Hokkaido Univ. 53 (2), 75–82.

Kanchanakhan, S., Chinabut, S., Tonguthai, K., Richards, R.H., 2002. Epizootic ulcerative syndrome of fishes: rhabdovirus infection and EUS induction experiments in snakehead fish. In: Lavilla-Pitogo, C.R., Cruz-Lacierda, E.R. (Eds.), Diseases in Asian Aquaculture IV. Fish Health Section, Asian Fisheries Society, Manila, pp. 383–394.

Kurata, O., Hatai, K., 2002. Activation of carp leukocytes by a galactose-binding protein from *Aphanomyces piscicida*. Dev. Comp. Immunol. 26, 461–469.

Kurata, O., Sanpei, K., Hikijii, K., Hatai, K., 2002. A galactose-binding protein revealed as a hemagglutinin in *Aphanomyces piscicida*. Fish Pathol. 37, 1–6.

Kiryu, Y., Shields, J.D., Vogelbein, W.K., Kator, H., Blazer, V.S., 2003. Infectivity and pathogenicity of the oomycete *Aphanomyces invadans* in Atlantic menhaden *Brevoortia tyrannus*. Dis. Aquat. Org. 54, 135–146.

Katoch, R.E.C., Sharma, M., Pathania, D., Verma, S., Chahota, R., Arvind Mahajan, L., 2003. Recovery of bacterial and mycotic fish pathogen from carp and other fish in Himachal Pradesh. Ind. J. Microbiol. 43, 65–66.

Khatri, S., 2004. Prevalence of Fish Diseases in Sambalpur, Orissa (M.Sc., dissertation). G.M. College, Sambalpur, 35 pp.

Leano, E.M., Lio-Po, G.D., Dureza, L.A., 1995. Siderophore detection among bacteria associated with the epizootic ulcerative syndrome (EUS). In: Shariff, M., Arthur, J.R., Subasinghe, R.P. (Eds.), Diseases in Asain Aquaculture II. Fish Health Section, Asian Fisheries Society, Manila, pp. 315–325.

Lilley, J.H., Roberts, R.J., 1997. Pathogenicity and culture studies comparing *Aphanomyces* involved in epizootic ulcerative syndrome (EUS) with other similar fungi. J. Fish Dis. 20, 101–110.

Lilley, J.H., Thompson, K.D., Adams, A., 1997b. Characterization of *Aphanomyces invadans* by electrophoretic and Western blot analysis. Dis. Aquat. Org. 30, 187–197.

Lilley, J.H., Hart, D., Panyawachira, V., Kanchanakhan, S., Chinabut, S., Soderhall, K., Cerenius, L., 2003. Molecular characterization of the fish-pathogenic fungus *Aphanomyces invadans*. J. Fish Dis. 26, 263–275.

Lio-Po, G.D., Albright, L.J., Alapide-Tendencia, E.V., 1992. *Aeromonas hydrophila* in the epizootic ulcerative syndrome (EUS) of snakehead, *Ophicephalus striatus*, and catfish, *Clarias batrachus*: quantitative estimation in natural infection and experimental induction of dermo-necrotic lesion. In: Shariff, M., Subasinghe, R.P., Arthur, J.R. (Eds.), Diseases in Asian Aquaculture I. Fish Health Section, Asian Fisheries Society, Manila, The Philippines, pp. 461–474.

Lio-Po, G.D., Albright, L.J., Traxler, G.S., Leano, E.M., 2001. Pathogenicity of the epizootic ulcerative syndrome (EUS)-associated rhabdovirus to snakehead *Ophicephalus striatus*. Fish Pathol. 36, 57–66.

Lumanlan-Mayo, S.C., Callinan, R.B., Paclibare, J.O., Catap, E.S., Fraser, G.C., 1997. Epizootic ulcerative syndrome (EUS) in rice-fish culture systems: an overview of field experiments 1993–1995. In: Flegel, T.W., MacRae, I.H. (Eds.), Diseases in Asian Aquaculture III. Fish Health Section. Asian Fisheries Society, Manila, The Philippines, pp. 129–138.

Levine, J.F., Hawkins, J.H., Dykstra, M.J., Noga, E.J., Moye, D.W., Cone, R.S., 1990a. Species distribution of ulcerative lesions on finfish in the Tar-Pamlico River Estuary, North Carolina. Dis. Aquat. Org. 8 (1), 1–5.

Levine, J.F., Hawkins, J.H., Dykstra, M.J., Noga, E.J., Moye, D.W., Cone, R.S., 1990b. Epidemiology of ulcerative mycosis in Atlantic menhaden in the Tar-Pamlico River Estuary, North Carolina. J. Aquat. Anim. Health 2 (3), 162–171.

Lilley, J.H., Inglis, V., 1997. Comprative effects of various antibiotics, fungicides and disinfectants on *Aphanomyces invaderis* and other saprolegniceous fungi. Aquacult. Res. 28, 461–469.

Limsuwan, C., Chinabut, S., 1983. Histological changes of some freshwater fishes during 1982-83 disease outbreak. In: Proceedings of the Symposium on Freshwater Fish Epidemic 1982–1983, Bangkok, 23–24 June 1983. Chulalongkorn University, Bangkok, p. 255.

Lio-Po, G.D., Albright, L.J., Traxler, G.S., Lean~o, E.M., 2003. Horizontal transmission of epizootic ulcerative syndrome (EUS)-associated virus in the snakehead *Ophicephalus striatus* under simulated natural conditions. Dis. Aquat. Org. 57, 213–220.

Lee, C., Cho, J.C., Lee, S.H., Lee, D.G., Kim, S.J., 2002. Distribution of *Aeromonas* spp. as identified by 16S rDNA restriction fragment length polymorphism analysis in a trout farm. J. Appl. Microbiol. 93, 976–985.

Laemmli, U.K., 1970. Cleavage of structural proteins during assembly of the head of bacteriophage T4. Nature 227, 680—685.

Macdonald, R.D., Gower, D.A., 1981. Genomic and phenotypic divergence among three serotypes of aquatic birnaviruses (infectious pancreatic necrosis virus). Virology 114, 187—195.

Magyar, G., Dobos, P., 1994. Expression of infectious pancreatic necrosis virus polyprotein and VP1 in insect cells and the detection of the polyprotein in purified virus. Virology 198, 437—445.

Mahy, B.W.J., Kangro, H.O., 1996. Virology Methods Manual. Academic Press, London.

Morgan, K.L., 2001. Epizootic ulcerative syndrome: an epidemiological approach. In: Rodgers, C.J. (Ed.), Risk Analysis in Aquatic Animal Health. Proceedings of an International Conference Held in Paris, France, 8—10 February 2000, pp. 209—214.

Miyazaki, T., Egusa, S., 1973a. Studies on mycotic granulomatosis in freshwater fish II. Mycotic granulomatosis prevailed in goldfish. Fish Pathol. 7, 125—133 (In Japanese).

Miyazaki, T., Egusa, S., 1973b. Studies on mycotic granulomatosis in freshwater fish III. Bluegill. Mycotic granulomatosis in bluegill. Fish Pathol. 8, 41—43 (In Japanese).

Miyazaki, T., Egusa, S., 1973c. Studies on mycotic granulomatosis in freshwater fish IV. Mycotic granulomatosis in some wild fishes. Fish Pathol. 8, 44—47 (In Japanese).

Macintosh, D.J., 1986. Environmental background to the ulcerative disease conditionin South-east Asia. In: Roberts, R.J., Macintosh, D.J., Tonguthai, K., Boonyaratpalin, S., Tayaputch, N., Phillips, M.J., Millar, S.D. (Eds.), Field and Laboratory Investigations into Ulcerative Fish Diseases in the Asia-Pacific Region. Technical Report of FAO Project TCP/RAS/4508. Bangkok, pp. 61—113.

McGarey, D.J., Beatty, T.K., Alberts, V.A., Te Strake, D., Lim, D.V., 1990. Investigations of potential microbial pathogens associated with ulcerative disease syndrome (UDS) of Florida fish. Pathol. Mar. Sci. 65—75.

Munday, B.L., 1985. Ulcer disease in cod (Pseudophycis bachus) from the Tamar river. Tasman. Fish. Res. 27, 15—18.

Markosova, R., Jezek, J., 1994. Indicator bacteria and limnological parameters in fish ponds. Water Res. 28 (12), 2477—2485.

McGarey, D.J., Milanesi, L., Foley, D.P., Reyes, B.J., Frye, L.C., Lim, D.V., 1991. The role of motile aeromonads in the fish disease, ulcerative disease syndrome (UDS). Exp. Rev. 47, 441—444.

Nakamura, K., Yuasa, K., Simnuk, S., Hatai, K., Hara, N., 1995. The ubiquinone system in oomycetes. Mycoscience 36, 121—123.

Noga, E.J., Wright, J.F., Levine, J.F., Dykstra, M.J., Hawkins, J.H., 1991. Dermatological diseases affecting fishes of the Tar-Pamlico Estuary, North Carolina. Dis. Aquat. Org. 10 (2), 87—92.

Niewolak, S., Tucholski, S., 2000. Evaluation of bacteria identified on broth agar in 20 and 37 degree C, coli forms, fecal coli forms, fecal streptococci, anaerobic sulfite-reducing Clostridium perfringens, Pseudomonas aeruginosa, Aeromonas sp., and Salmonella sp. in the muscles, skin, and digestive tract content of common carp (Cyprinus carpio). Archiwum rybactwa polskiego/Arch polish fish. Olsztyn 8:63—74. Asian Fish. Sci. 22 (2009), 569—581.

Nakatsu, C.H., Torsvik, V., Øvreås, L., 2000. Soil community analysis using DGGE of 16S rDNA polymerase chain reaction products. Soil Sci. Soc. Am. J. 64, 1382—1388.

Nyblom, H., Bjornsson, E., Simren, M., Aldenborg, F., Almer, S., Olsson, R., 2006. The AST/ALT ratio as an indicator of cirrhosis in patients with PBC. Liver Int. 26, 840—845.

Ogbulie, J.N., Cobiajuru, I.O., 2003. The impact of seasonal changes on the bacteriological water quality and substrate specificity of isolates in a Clarias lazera culture system at Obirikom, Rivers State, Nigeria. J. Aquac. Sci. 18, 105—112.

Ogbonna, C.I.C., Alabi, R.O., 1991. Studies on species of fungi associated with mycotic infections of fish in a Nigerian freshwater fish pond. Hydrobiologia 220, 131—135.

OIE, 2006. Epizootic haematopoietic necrosis. Man. Diagn. Tests Aquat. Anim. 2006, 85—103.

OIE, 2003. Manual of Diagnostic Tests for Aquatic Animals. Office International des Epizooties, Paris, France.

Oidtmann, B., Steinbauer geiger, S., Hoffmann, R.W., 2008. Experimental infection and detection of Aphanomyces invadans in European catfish, rainbow trout and European eel. Dis. Aquat. Org. 82, 185—207.

Palu, A.P., Gomes, L.M., Miguel, M.A.L., Balassiano, I.T., Queiroz, M.L.P., Freitas-Almeida, A.C., de Oliveira, S.S., 2005. Antimicrobial resistance in food and clinical Aeromonas isolates. Food Microbiol. 23, 504—509.

Prasad, M.M., Rao, C.C.P., Surendran, P.K., 1998. Motile aeromonads associated with epizootic ulcerative syndrome affected Channa striatus. In: Balachandran, K.K., Iyer, T.S.G., Madhavan, P., Joseph, J., Perigreen, P.A., Raghunath, M.R., Varghese, M.D. (Eds.), Advance and Priorities in Fisheries Technology. Society of Fisheries Technologists of India, Kochi, pp. 374—377.

Pathiratne, A., Epa, U.P.K., Jayasinghe, R.P.P.K., 2002. Hematological changes in snakeshkm gourami Trichogaster pectoralis effected by epizootic ulcerative syndrome. In: Lavilla-Pitogo, C.R., Cruz-Lacierda, E.R. (Eds.), Diseases in Asian Aquaculture IV. Fish Health Section, Asian Fisheries Society, Manila.

Phillips, M.J., March 20—24, 1989. A Report on the NACA Workshop on the Regional Research Programme on Ulcerative Syndrome in Fish and the Environment. Network of Aquaculture Centres in Asia-Pacific, Bangkok.

Phiri, H., 2007. Report on the Field Expedition to Assess the Possible Causes of the General Fish Kill in Western Province from 18th September to 1st October 2007. School of Veterinary Medicine, University of Zambia, 29 pp.

Pal, J., Pradhan, K., 1990. Bacterial involvement in ulcerative condition of air-breathing fish from India. J. Fish Biol. 36, 833—839.

Palisoc, F.P., 1990. Histopathology of the epizootic ulcerative syndrome (EUS) positive snakehead, Ophicephalus striatus, from Laguna Lake, Philippines. In: Abstracts from the Symposium on Diseases in Asian Aquaculture. 26—29 November 1990. Bali, Indonesia. Fish Health Section, Asian Fisheries Society, Manila, p. 22.

Pearce, M., 1990. Epizootic Ulcerative Syndrome Technical Report December 1987—September 1989. Northern Territory Department of Primary Industry and Fisheries. Fisheries Report No. 22. Northern Territory, Australia, 82 pp.

Phillips, M.J., Keddie, H.G., 1990. Regional research programme on relationships between epizootic ulcerative syndrome in fish and the environment. In: A Report on the Second Technical Workshop, 13—26 August 1990. Network of Aquaculture Centres in Asia-Pacific, Bangkok, 133 pp.

Pichyangkura, S., Tangtrongpiros, J., 1985. The relationship between microscopic exam of Achlya sp. infection and characteristic of lesions Ophicephalus striatus. In: Proceedings of the Living Aquatic Resources. 7—8 March 1985. Chulalongkorn University, Bangkok, pp. 19—23 (In Thai, English abstract).

Plumb, J.A., Grizzle, J.M., Defiguriedo, J., 1976. Necrosis and bacterial infection in channel catfish (Ictaluris punctatus) following hypoxia. J. Wildl. Dis. 12, 247–253.

Poonsuk, K., Saitanu, K., Navephap, O., Wongsawang, S., 1983. Antimicrobial susceptibility testing of Aeromonas hydrophila: strain isolated from fresh water fish infection. In: The Symposium on Fresh Water Fishes Epidemic: 1982–1983. 23–24 June 1983. Chulalongkorn University, Bangkok, pp. 436–438 (In Thai, English abstract).

Pathiratne, A., Rajapakshe, W., 1998. Hematological changes associated with epizootic ulcerative syndrome in the Asian cichlid Fish, Etroplus suratensis. Asian Fish. Sci. 11, 203–211.

Parveen, R., Parulekar, A.H., 1998. Diseases and parasites of laboratory-reared and wild population of banded pearl spot, Eutroplus suratensis (Cichlidae) in Goa. Indian J. Mar. Sci. 27, 407–410.

Qureshi, T.A., Chouhan, R., Prasad, Y., Mastan, S.A., 1995. Mycological studies on EUS affected catfish, Mystus cavasius. In: Abstracts of the Fourth Asian Fisheries Forum. 16–20 October 1995. Asian Fisheries Society and China Society Fisheries, Beijing, p. 38.

Rahman, M., Colque-Navarro, P., Kühn, I., Huys, G., Swings, J., Möllby, R., 2002. Identification and characterization of pathogenic Aeromonas veronii biovar sobria associated with epizootic ulcerative syndrome in fish in Bangladesh. Appl. Environ. Microbiol. 68, 650–655.

Rha, S.A., Sinmuk, S., Wada, S., Yuasa, K., Nakamura, K., Hatai, K., Ishii, H., 1996. Pathogenicity to ayu (Plecoglossus altivelis) of Aphanomyces sp. isolated from dwarf gourami (Colisa lalia). Bull. Nippon Vet. Anim. Sci. Univ. 45, 9–15.

Raman, R.P., 1992. EUS strikes in the brackishwaters of Chilka Lagoon in India. Fish Health Sect. Newsl. 3 (2), 3–4 (Asian Fisheries Society, Manila).

Rattanaphani, R., Wattanavijarn, W., Tesprateep, T., Wattanodorn, S., Sukolopong, V., Vetchagarun, S., Eongpakornkeaw, A., 1983. Preliminary report on fish virus in snakehead fish. Thai J. Vet. Med. 13, 44–50.

Riji John, K., 1997. Characterisation of Reovirus-Like Agents Associated with Snakehead Fish and Cell Culture (Ph.D. thesis), University of Stirling, Scotland.

Rovozzo, G.C., Burke, C.N., 1973. A Manual of Basic Virological Techniques. Prentice Hall, Englewood Cliffs, NJ, pp. 126–163.

Royo, F., Andersson, G., Bangyeekhun, E., Muzquiz, J.L., Soderhall, K., Cerenius, L., 2004. Physiological and genetic characterisation of some new Aphanomyces strains isolated from freshwater crayfish. Vet. Microbiol. 104, 103–112.

Sambrook, J., Fritsch, E.F., Maniatis, T., 1989. Molecular Cloning. A Laboratory Manual. Cold Spring Harbor Laboratory, Cold Spring Harbor, NY.

Shrestha, G.B., 1994. Status of epizootic ulcerative syndrome (EUS) and its effects on aquaculture in Nepal. In: Roberts, R.J., Campbell, B., MacRae, I.H. (Eds.), Proceedings of the ODA Regional Seminar on Epizootic Ulcerative Syndrome, 25–27 January 1994. Aquatic Animal Health Research Institute, Bangkok, pp. 49–57.

Saitanu, K., Wongsawang, S., Sunyascotcharee, B., Sahaphong, S., 1986. Snakehead fish virus isolation and pathogenicity studies. In: Maclean, J.L., Dizon, L.B., Hosillos, L.V. (Eds.), The First Asian Fisheries Forum. Asian Fisheries Society, Manila, pp. 327–330.

Sammut, J., White, I., Melville, M.D., 1996. Acidification of an estuarine tributary in eastern Australia due to drainage of acid sulfate soils. Mar. Freshwater Res. 47 (5), 669–684.

Sanjeevaghosh, D., 1991. EUS ravages Kerala inland fisheries. Fish. Chimes 11 (9), 47–49.

Shanor, L., Saslow, H.B., 1944. Aphanomyces as a fish parasite. Mycologia 36, 413–415.

Sharifpour, I., 1997. Histology of the Inflammatory Response of the Carp (Cyprinus carpio L.) to Various Stimuli (Ph.D. thesis). University of Stirling, Scotland.

Srivastava, R.C., 1979. Aphanomycosis - a new threat to fish population. Mykosen 22 (1), 25–29.

Subasinghe, R.P., 1993. Effects of controlled infections of Trichodina sp on transmission of epizootic ulcerative syndrome (EUS) to naive snakehead, Ophicephalus striatus Bloch. J. Fish Dis. 16, 161–164.

Subramaniam, S., Chew-lim, M., Chong, S.Y., Howe, J., Ngoh, G.H., Chan, Y.C., 1993. Molecular and electron microscopic studies of infectious pancreatic necrosis virus from snakehead. In: IXth International Congress of Virology. Glasgow, UK (Abstract), p. 368.

Suthi, G., 1991. Pathogenicity of motile aeromonads for Puntius schwanfeldi and Oreochromis niloticus with particular reference to the ulcerative disease syndrome (EUS) (M.Sc. thesis). University of Stirling, Scotland, 71 pp.

Shireman, J.V., Cichra, C.E., 1994. Evaluation of aquaculture effluents. Aquaculture 123, 53–68.

Sivakami, R., Premkishore, G., Chandran, M.R., 1996. Occurrence and distribution of potentially pathogenic Enterobacteriaceae in carps and pond water in Tamil Nadu, India. Aquacult. Res. 27 (5), 375–378.

Sen, K., Rodgers, M., 2004. Distribution of six virulence factors in Aeromonas species isolated from US drinking water utilities: a PCR identification. J. App. Microbiol. 97, 1077–1086.

Snieszko, S.F., 1974. The effects of environmental stress on outbreaks of infectious disease of fishes. J. Fish Biol. 6, 197–208.

Secombes, C.J., 1990. Isolation of salmonid macrophages and analysis of their killing activity. In: Stolen, J.S., Fletcher, T.C., Anderson, D.P., Robertson, B.S., van Muiswinkel, W.B. (Eds.), Techniques in Fish Immunology. SOS Publ., Fair Haven, pp. 137–154.

Skelton, P.H., 1993. A Complete Guide to the Freshwater FIshes of Southern Africa. Southern Book Publishers, Halfway House.

Skelton, P.H., Cambray, J.A., Lombard, A., Benn, G.A., 1995. Patterns of distribution and conservation status of freshwater fishes in South Africa. S. Afr. J. Zool. 30, 71–81.

Sosa, E.R., Landsberg, J.H., Kiryu, Y., Stephenson, C.M., Cody, T.T., Dukeman, A.K., et al., 2007a. Pathogenicity studies with the fungi Aphanomyces invadans, Achlya bisexualis, and Phialemonium dimorphosporum: Induction of skin ulcers in striped mullet. J. Aquat. Anim. Health 19, 41–48. http://dx.doi.org/10.1577/H06–013.1. PMID:18236631.

Sosa, E.R., Landsberg, J.H., Stephenson, C.M., Forstchen, A.B., 2007b. Aphanomyces invadans and ulcerative mycosis in estuarine and freshwater fish in Florida. J. Aquat. Anim. Health 19, 14–26. http://dx.doi.org/10.1577/H06–012.1. PMID:18236628.

Scott, W.W., 1961. A monograph of the genus Aphanomyces. Va. Agric. Exp. Stn. Tech. Bull. 151.

Shah, K.L., Jha, B.C., Jhingran, A.G., 1977. Observations on some aquatic phycomycetes pathogenic to eggs and fry of freshwater fish and prawn. Aquaculture 12, 141–147.

Subasinghe, R.P., Jayasinghe, L.P., Balasuriya, K.S., Kulathilake, M., 1990. Preliminary investigations into the bacterial and fungal pathogens associated with the ulcerative fish disease syndrome in Sri Lanka. In: Hirano, R., Hanyu, I (Eds.), The Second Asian Fisheries Forum. Asian Fishenes Society, Manila, pp. 655–665.

Twiddy, D.R., Reilly, P.J.A., 1994. The occurrence of antibiotic-resistant human pathogenic bacteria in integrated fish farms in a Southeast Asian country. In: Research Contributions Presented at the 9th Session of the Indo-Pacific Fisheries Commission Working Party on Fish Technology and Marketing, Cochin, and Kerala, India, pp. 23−37. FAO Technical papers 514.

Thompson, K.D., Lilley, J.H., Chinabut, S., Adams, A., 1997. The antibody response of snakehead, *Channa striata* Bloch, to *Aphanomyces invaderis*. Fish Shellfish Immunol. 7 (5), 349−353.

Thompson, K.D., Lilley, J.H., Chen, S.-C., Adams, A., Richards, R.H., 1999. The immune response of rainbow trout (*Oncorhynchus mykiss*) against *Aphanomyces invadans*. Fish Shellfish Immunol. 9 (3), 195−210.

Thune, R.L., Stanley, L.A., Cooper, R.K., 1993. Pathogenesis of gram-negative bacteria infections in warm-water fish. Annu. Rev. Fish Dis. 3, 37−68.

Torres, J.L., Shariff, M., Tajima, K., 1992. Serological relationships among motile *Aeromonas* spp. associated with healthy and epizootic ulcerative syndrome-positive fish. In: Shariff, M., Subasinghe, R.P., Arthur, J.R. (Eds.), Diseases in Asian Aquaculture I. Fish Health Section. Asian Fisheries Society, Manila, Philippines, pp. 451−460.

Takuma, D., Sano, A., Wada, S., Kurata, O., Hatai, K. A new species, *Aphanomyces salsuginosus* sp. nov., isolated from ice fish *Salangichthys microdon*. Mycoscience 51, in press.

Unestem, T., 1972. On the host range and origin of the crayfish plague fungus. Rep. Inst. Freshwater Res. Drottningholm 52, 192−198.

Valairatana, W., Willoughby, L.G., 1994. The aquatic fungi *Aphanomyces* and *Pythium*, as wound pathogens on a soft shell turtle (*Trionyx cartilogineus*). AAHRI Newsl. 3 (1), 2.

Van Helden, L., Grewar, J., 2011. 'Epizootic ulcerative syndrome − Eerste river system', epidemiology report, Veterinary services. West. Cape Gov. 3 (11), 1−2.

Virgona, J.L., 1992. Environmental factors influencing the prevalence of a cutaneous ulcerative disease (red spot) in the sea mullet *Mugil cephalus* L., in the Clarence River, New South Wales, Austalia. J. Fish Dis. 15, 363−378.

Vishwanath, T.S., Mohan, C.V., Shankar, K.M., 1997. Mycotic granulomatosis and seasonality are the consistent features of epizootic ulcerative syndrome of fresh and brackishwater fishes of Karnataka, India. Asian Fish. Sci. 10, 155−160.

World Organisation for Animal Health OIE, 2009. Manual of Diagnostic Tests for Aquatic Animals. OIE, Paris.

Wada, S., Yuasa, K., Rha, S., Nakamura, K., Hatai, K., 1994. Histopathology of *Aphanomyces* infection in dwarf gourami (*Colisa lalia*). Fish Pathol. 29, 229−237.

Wada, S., Rha, S.A., Kondoh, T., Suda, H., Hatai, K., Ishii, H., 1996. Histopathological comparison between ayu and carp artificially infected with *Aphanomyces piscicida*. Fish Pathol. 31, 71−80.

Wattanavijarn, W., Tangtrongpiros, J., Wattanodorn, S., Hunnak, P., 1985. Detection of viruses from cell cultures by using electron microscope and induction of disease by infected cell culture fluid. Thai J. Vet. Med. 36, 143−151.

Wattanavijarn, W., Ousavaplangchai, L., Rattanaphani, R., Sukolapong, V., Tesaprateep, T., Eongpakornkeaw, A., Tangtrongpiros, J., Vetchangarun, S., Thirapatsakun, T., 1983a. Virus-like particles in the skeletal muscle, capillaries and spleen of sick snakehead fish (*Ophicephalus striatus*) during a disease epidemic. Thai J. Vet. Med. 13 (2), 122−130.

Wattanavijarn, W., Tesaprateep, T., Wattanodorn, S., Sukolapong, V., Vetchangarun, S., Eongpakornkeaw, A., 1983b. Virus-like particles in the liver snakehead fish. Thai J. Vet. Med. 13 (1), 51−57.

Wattanavijarn, W., Tangtrongpiros, J., Rattanaphani, R., Sukolapong, V., Eongpakornkeaw, A., Vetchangarun, S., 1984. Examination of sick catfish by scanning and transmission electron microscopy. Thai J. Vet. Med. 14 (1), 31−38.

Wattanavijarn, W., Torchy, C., Tangtrongpiros, J., de Kinkelin, P., 1988. Isolation of a birnavirus belonging to Sp serotype, from southeast Asia fishes. Bull. Eur. Assoc. Fish Pathol. 8 (5), 106−108.

Wolf, K., 1988. Fish Viruses and Fish Viral Diseases. Cornell University Press, Ithaca.

Wolf, K., Quimby, M.C., 1969. Fish cell line and tissue culture. In: Hoar, W.S., Randall, D.J. (Eds.), Fish Physiology, vol. 3. Academic Press, New York, pp. 253−305.

Wolf, K., Gravell, M., Malsberger, R.G., 1966. Lymphocystis virus: isolation and propagation in centrarchid fish cell lines. Science 151, 1004−1005.

Wedemeyer, G.A., Meyer, F., Smith, L., 1999. Environmental Stress and Fish Diseases. Narendra Publishing House, New Delhi, 181 pp.

Yuasa, K., Hatai, K., 1996. Some biochemical characteristics of the genera *Saprolegnia*, *Achlya* and *Aphanomyces* isolated from fishes with fungal infection. Mycoscience 37, 477−479.

Chapter 4

Epidemiological Study

Epidemiology

Epidemiology is the study of the distribution and determinants (i.e., causes) of disease in populations. Epidemiologists typically take a wide view of causal factors, defining them as "any event, condition, or characteristic that plays an essential role in producing an occurrence of disease." In contrast to this, many pathologists and microbiologists may consider, for example, a particular infectious agent to be the cause of a disease, and relegate all other contributions to "contributing" or "predisposing" factors.

The epidemiological approach to outbreak investigations is based on the premise that cases of a disease are not distributed randomly but occur in patterns within the at-risk population. It is the role of the epidemiologist to record and analyze these patterns to help meet the primary objective.

From an epidemiologic viewpoint, it is important to characterize the outbreak in terms of the above three variables. This characterization must be done in such a way that hypotheses could be developed regarding the source, mode of transmission, and duration of the outbreak. The information is organized in an attempt to find answers to the kinds of aspects mentioned above.

4.1 DETAILS OF EPIDEMIOLOGY

4.1.1 The Early Status

Epidemiology literally means "the study of what is upon the people." The word is derived from the Greek *epi*, meaning "upon, among," *demos*, meaning "people, district," and *logos*, meaning "study, word, discourse." It suggests that it perhaps applies only to human populations. Nevertheless, the term is widely used in studies of zoological populations including veterinary epidemiology. On the other hand, the term "epizoology" is applied to studies of plant populations—that is, botanical or plant disease epidemiology. In addition, epidemiologists are said to deal with the interaction of diseases in a population, a condition known as a "syndemic." Moreover, epidemiology is often viewed as an aggregation of statistical tools that could be used to discuss the associations of exposures to health outcomes. However, a thorough appreciation of this science is that of discovering *causal* relationships.

The Greek physician Hippocrates (*c* 200 BC) is considered the father of medicine. He is also regarded as a pioneer for drawing a distinction between "epidemic" and "endemic." He had attempted to distinguish between diseases that are "visited upon" a population, called "epidemic," from those that "reside within" a population, termed "endemic." In other words, the term *endemic* could be used for diseases usually found in some places but not in others, while the word *epidemic* may be used for diseases that are seen sometimes, but not every time (Anon, 2009b).

Hippocrates perhaps also sought logic to sickness. He is also regarded as a pioneer for examining the relationships between the occurrence of diseases and environmental influences. Hippocrates had believed that illness of the human body could be caused by an imbalance of the four humors, viz., air, fire, water, and earth "atoms." The cure to the sickness could be achieved by either removing or adding the humor in question in order to balance the body. This idea led to the applications of bloodletting and dieting in medicine. Also, the "Ayurveda" in ancient India had considered disease to be a manifestation of imbalance in three bodily humors, called "Doshas." Systems of diagnosis were said to have been based around this theory.

In addition, one of the earliest theories of the origin of disease advocated that diseases are caused primarily by faults in human luxury. This was expressed by philosopher such as Rousseau (1754), as well as social critics like Jonathan Swift (2008).

Nevertheless, history seemed to portray the term "epidemiology" as appearing at first to describe the study of epidemics in 1802 by the Spanish physician Villalba in *Epidemiología Española*. Concomitantly, a doctor from Verona named Girolamo Fracastoro, during the middle of the sixteenth century, is often regarded as the pioneer who proposed the theory that the very small and unobservable particles that cause diseases are alive. They were considered able to spread by air, having the capacity of self-multiplication, and could be destroyed by fire. In this way, he is believed to have refuted Galen's miasma theory (poison gas in sick people). He had written a book entitled *De contagione et contagiosis morbis* in 1543. He was considered the pioneer in promoting personal and environmental hygiene to prevent disease through this treatise. Subsequently, Anton van Leeuwenhoek had invented a

Epizootic Ulcerative Fish Disease Syndrome. http://dx.doi.org/10.1016/B978-0-12-802504-8.00004-3

sufficiently powerful microscope in 1675. This provided visual evidence of living particles consistent with the germ theory of disease.

Notwithstanding the above, Thomas Sydenham (1624–1689) is considered another pioneer, who distinguished the fevers of Londoners during the later 1600s. His theories on cures of fevers met with much resistance from traditional physicians at that time. He was also unable to find the initial causes of the smallpox fever he researched and treated. Concomitantly, John Graunt was a haberdasher and amateur statistician. He published a number of "Natural and Political Observations upon the Bills of Mortality" in 1662. In these advocacies, he analyzed the mortality rolls in London before the Great Plague. He also presented one of the first life tables and reported time trends for many new and old diseases. Further, he provided statistical evidence for many theories on disease and also criticized some of the widespread ideas about them.

4.1.2 The Modern Era

Concomitant to the above, John Snow is considered famous for his investigations into the causes of the cholera epidemics during the nineteenth century. He is also often regarded as the father of modern epidemiology. His exposition of the Broad Street pump as the cause of the Soho epidemic is regarded as a classic example of epidemiology. Snow used chlorine and also removed the handle in an attempt to clean the water. This ended the outbreak. Further, this has therefore been perceived as a major event in the history of public health, and has been considered the founding event of the science of epidemiology, having assisted in shaping public health policies around the globe. However, results of Snow's research on preventive measures to avoid further outbreaks was not completely accepted or put into practice until after he had expired. In addition to the above, Danish physician Peter Anton Schleisner is also regarded as a pioneer, who related his work on the prevention of the epidemic of neonatal tetanus on the Vestmanna Islands in Iceland in 1849 (Garðarsdóttir and Guttormsson, 2008, 2009).

In addition to the above, another significant pioneer was the Hungarian physician Ignaz Semmelweis, who in 1847 brought down infant mortality at a Vienna hospital by instituting a disinfection procedure. His findings were published in 1850, but were said to be ill received by his colleagues, who had discontinued the procedure. Disinfection did not become widely practiced until British surgeon Joseph Lister discovered antiseptics in 1865 in light of the work of Louis Pasteur.

Notwithstanding the above, mathematical methods were introduced into epidemiology by Ronald Ross, Janet Lane-Claypon, Anderson Gray McKendrick, and others during the early twentieth century (Karl et al., 1900–1945; Kermack and McKendrick, 1927).

With the advancement of research in the biomedical sciences during the late twentieth century, a number of molecular markers in blood, other biospecimens, and environment were identified as predictors of development or risk of certain diseases. Thus, "molecular epidemiology" broadly emanated out of epidemiological research in order to examine the relationship between these biomarkers analyzed at the molecular level and the disease. Specifically, "genetic epidemiology" was used for the epidemiology of germ line genetic variation and disease. Genetic variation is typically determined by using DNA from peripheral blood leucocytes. Genome-wide association studies have been commonly done to identify genetic risk factors for many diseases and health conditions since the 2000s.

It has been increasingly recognized that disease evolution represents an inherently heterogeneous process that differs from person to person, while most molecular epidemiology studies still possibly use the conventional disease diagnosis and classification system. Conceptually, each individual is said to possess a unique disease process, called the "unique disease principle," that is said to be different from that of any other individual (Ogino et al., 2012a, 2013), considering the uniqueness of the exposome (which is said to be a totality of endogenous and exogenous/environmental exposures, and its unique influence on molecular pathologic process in each individual). In this regard, it could be noted that studies related to the examination of the relationship between an exposure and molecular pathologic signature of disease became increasingly common throughout the 2000s. However, the use of molecular pathology in epidemiology has posed unique challenges such as the lack of research guidelines and standardized statistical methodologies, and the paucity of interdisciplinary experts and training programs (Ogino et al., 2012b). Still further, the concept of disease heterogeneity appeared to be in conflict with the long-standing premise in epidemiology that individuals with the same disease name have similar etiologies and disease processes. In this connection, it may be mentioned here that "molecular pathology" and "epidemiology" were integrated into a new interdisciplinary field of "molecular pathological epidemiology" (MPE) (Ogino and Stampfer, 2010; Ogino et al., 2011), in order to resolve these issues and advance population health science in the era of molecular precision medicine (MPE). MPE may be defined as "epidemiology of molecular pathology and heterogeneity of disease." In MPE, investigators analyze the relationships among the following: (1) environmental, dietary, lifestyle, and genetic factors; (2) alterations in cellular or extracellular molecules; and (3) evolution and progression of disease. However, a better understanding of heterogeneity of disease pathogenesis will further contribute to elucidate etiologies of disease. Incidentally, the MPE approach could be applied to both neoplastic and nonneoplastic diseases (Field et al., 2013). The concept and paradigm of MPE are said to have become widespread in the 2010s (Curtin et al., 2011, Hughes et al., 2012; Ku et al., 2012; Chia et al., 2012; Ikramuddin and Livingston, 2013).

4.1.2.1 Epidemiology and Bradford Hill Criteria

Austin Bradford Hill (1965) had advocated a series of considerations to help in assessing the evidence of causation (Hill, 1965), which are commonly called the "Bradford Hill criteria." However, Hill's considerations are now sometimes considered a checklist to be implemented for assessing causality (Phillips and Karen, 2004). Hill himself had asserted that none of his viewpoints could bring indisputable evidence for or against the cause-and-effect hypothesis and none could also be required sine qua non (Hill, 1965).

Concomitant to the above, "correlation may not imply causation." It is a common theme for much epidemiological literature. For epidemiologists, the key is in the term inference. Epidemiologists use gathered data and a broad range of biomedical and psychosocial theories in an iterative way to generate or expand theory, test hypotheses, and make educated, informed assertions about what relationships are causal—they also try to furnish explanations for exactly how they are causal.

Notwithstanding the above, epidemiologists like Rothman and Greenland have emphasized that the "**one cause—one effect**" understanding is a simplistic misbelief. Most outcomes, whether disease or death, are caused by a chain or web consisting of many component reasons. Causes could further be distinguished as necessary, sufficient, or probabilistic conditions. If a necessary condition could be identified and controlled—for example, antibodies to a disease agent—the harmful outcome could be avoided.

Some related terminologies are briefly dealt with below:

1. **Strength**: A weak association may not mean that there is no causal effect. However, the larger the association, the more likely that it is causal (Hill, 1965).
2. **Consistency**: Consistent findings observed by different persons in different places with different samples are said to strengthen the likelihood of an effect (Hill, 1965).
3. **Specificity**: Causation is likely for a very specific population, at a specific site, that has a disease but no other likely explanation. The more specific the association between a factor and an effect, the bigger the probability of a causal relationship (Hill, 1965).

4. **Temporality**: The effect has to occur after the cause, and if there is an expected delay between the cause and expected effect, then the effect may occur after that delay (Hill, 1965).
5. **Biological gradient**: Greater exposure should generally lead to greater incidence of the effect. However, in some cases, the mere presence of the factor may trigger the effect. In other cases, an inverse proportion is observed: greater exposure leads to lower incidence (Hill, 1965).
6. **Plausibility**: A plausible mechanism between cause and effect is helpful. However, Hill noted that knowledge of the mechanism is limited by current knowledge (Hill, 1965).
7. **Coherence**: Coherence between epidemiological and laboratory findings increase the likelihood of an effect. However, Hill noted that lack of such (laboratory) evidence may not nullify the epidemiological effect on association (Hill, 1965).
8. **Experiment**: Occasionally it is possible to appeal to experimental evidence (Hill, 1965).
9. **Analogy**: The effect of similar factors may be considered (Hill, 1965).

4.1.2.2 Legal Interpretation

Epidemiology is said to be more related to the incidence of disease in "populations" than to addressing the reasons for an individual's disease. The aspect of individual ailment is often referred to as **specific causation** and is beyond the regime of the science of epidemiology. Epidemiology has its limitations at the point where a conclusion can be drawn that the relationship between an agent and a subject disease is causal (general causation), and the magnitude of excess risk attributed to the agent has been determined. This could mean that epidemiology emphasizes whether an agent could cause a disease and not whether an agent did cause a specific plaintiff's disease (Green et al., 2008). Epidemiological studies can only prove that an agent could have caused, not that it did cause, an effect in any particular case.

In the legal scenario of the United States, epidemiology alone may not be enough to outweigh the fact that causal association does not generally exist. Conversely, in certain individual cases, some US courts have allowed the conclusion that a causal association does exist based upon a balance of probability.

Concomitant to the above, the subdiscipline of forensic epidemiology is said to be directed toward the investigation of specific causation of an ailment or injury in individuals or groups of individuals, in instances in which causation is disputed or is unclear, for presentation in legal settings.

Epidemiological evidence is sometimes used to plead, as in a public health discipline, for both personal measures like diet change and corporate measures like removal of junk food advertising. The findings of such studies are often disseminated to the people. This could help them to make informed decisions about their health. Often uncertainties about these findings are not well communicated. News items and articles often prominently report the latest result of one study with little mention of its limitations, caveats, or context. Nevertheless, epidemiological tools have proved to be effective in establishing major causes of diseases like cholera and lung cancer (Hill, 1965). However, such tools may often experience difficulties when dealing with more subtle health issues where causation is more complex. Nevertheless, it may be noted here that inferences drawn from observational studies may be reconsidered, as later data, from the randomized controlled trials once they become available, as was the case with the association between the use of hormone replacement therapy and cardiac risk (Gabriel et al., 2005).

Potential health hazards to men, which could happen due to availability and consumption of sick fishes, could also be dealt with under the considerations indicated above. Attempts could therefore be made in that direction.

4.1.2.3 Epidemiology and Population-Based Health Management

Further, epidemiological practice and the results of epidemiological analysis possibly make a significant contribution to the emerging population-based health management infrastructure.

However, population-based health management encompasses the capacity to:

1. adjudicate the health status and health requirements of the target population;
2. materialize and evaluate interventions that are designed to upkeep the health of that population; and
3. provide care for members of that population efficiently, effectively, and in a way consistent with the community's cultural, policy, and health-resource values and ethos.

Notwithstanding the above, the modern population-based health management system is complex and requires a multiple set of skills, which include medical, political, technological, and mathematical. Nevertheless, among these, epidemiological practice and analysis are considered core components that collaborate with management science to provide efficient and effective health care and health guidance to a population. Such a task needs the ability to look forward to modern risk management approaches that could assist in transforming health risk factors, incidence, prevalence, and mortality statistics (derived from epidemiological analysis) into management matrices. This not only could guide how a health system might respond to current population health

issues, but also could give some clues to the issues of how a health system could be managed to better respond to future potential population health issues (Neil and Debra, 2011).

In this regard, it may be mentioned here that organizations use population-based health management practices, which leverage the work. Some examples of epidemiological practice are said to include the Canadian Strategy for Cancer Control, Health Canada Tobacco Control Programs, Rick Hansen Foundation, Canadian Tobacco Control Research Initiative (Smetanin and Kobak, 2005a,b, 2006), and so on. Such initiatives are also expected to become available among fish health management workers.

In this connection, it may be pointed out that each of these organizations is believed to use a population-based health management framework called life at risk. This combines epidemiological quantitative analysis with demographics, health agencies, operational research, and economics to perform the following functions:

Simulation of impacts of disease on life of the population: It is said to aim at measurement of the future potential impact of a disease upon a population with regard to new cases of disease prevalence, premature death, and potential years of life lost from disability and death.

Simulation of impacts of disease on the life of the labor force: It is said to aim at measurement of the future potential impact of a disease upon the labor force with regard to new cases, disease-prevalence, premature death, and potential years of life lost from disability and death.

Simulation of impacts of disease on the economy: It is said to aim at measurement of the future potential impact of a disease upon disposable income of the private sector (e.g., wages, corporate profits, and private health care costs) and public sector (e.g., personal income tax, corporate income tax, consumption taxes, and publicly funded health care costs).

4.1.2.4 Types of Studies Related to Epidemiology

Case studies, sometimes supported by statistical techniques, may lead to formulation of new hypotheses. Using the data, analytical studies could be done to reveal possible causal factors. These could include case—control and prospective studies. The former would involve matching comparable controls without the disease to the cases in the series. A prospective study would involve pursuing the case series over time to evaluate the disease's natural history (Hennekens and Julie, 1987). On the other hand, the latter type, more formally denoted as self-controlled case-series studies, splits individual patient follow-up time into exposed and unexposed periods, and then uses fixed-effects Poisson regression processes to compare the incidence rate of a given outcome between exposed and unexposed periods. This technique has been extensively used in the study of adverse reactions to vaccination, and has been shown in some circumstances to provide statistical power comparable to that available in cohort studies. Attempts could be made to obtain tangible results in the fisheries sector through the application of such techniques, maybe after some necessary modification.

4.1.2.5 Case–Control Studies

Case–control studies usually choose subjects based on their disease status. A group of individuals who are disease positive (the "case" group) is compared with a group of individuals that are disease negative (the "control" group). The control group is supposed to come from the same population, ideally, that yielded the cases. Further, the case–control study looks back through time at potential exposures that both the groups (cases and controls) might have encountered. It may be pointed out here that case--control studies are usually faster and more cost-effective than cohort studies, but are sensitive to bias (both recall bias and selection bias). The main task is to identify the appropriate control group. The distribution of exposure among the members of the control group should preferably be representative of the distribution in the population that gave rise to the cases. However, there are certain drawbacks in case–control studies.

4.1.2.6 Cohort Studies

Cohort studies attempt to choose subjects based on their exposure status. The subjects of the study are usually at risk of the outcome under investigation at the starting point of the cohort study. This usually means that they should preferably be disease-free when the cohort study starts. The cohort is followed through time to adjudge later outcome status.

4.1.2.7 Validity: Precision and Bias Related to Epidemiology

Notwithstanding the above, it may be noted here that different arenas in epidemiology have different levels of validity. One of the means to estimate the validity of findings is the ratio of false positives (claimed effects that are not correct) to false negatives (studies that fail to support a true effect). If we consider the field of genetic epidemiology, candidate gene studies may produce >100 false-positive findings for each false negative. By contrast, genome-wide association is nearly the reverse, with only one false positive for every 100 or more false negatives (Ioannidis et al., 2011). This ratio is said to have improved over time in genetic epidemiology, because the field has adopted stringent criteria. Conversely, other epidemiological arenas are usually not required to have such rigorous reporting. Hence, they are said to be much less reliable (Ioannidis et al., 2011).

In this connection, the different kinds of errors are briefly discussed below.

Random Error (RE)

Random error is the outcome or result of fluctuations around a true value because of sampling variability. It may occur during data collection, coding, transfer, or analysis. RE could be of different kinds; for example, poorly worded questions, misunderstanding in interpreting an individual answer from a particular respondent, or a typographical error during coding. RE may affect measurement in a transient mainly because of inconsistent manners, and it may be impossible to correct RE. Further, RE is said to occur in all sampling procedures. This could be regarded as sampling error.

In epidemiological variables, precision is a measure of RE, and also is said to be inversely related. Thus, RE could be reduced by increasing precision. Moreover, confidence intervals are computed in order to demonstrate the precision of relative risk estimates. Thus, the narrower the confidence interval, the more precise the relative risk estimate could be.

Concomitant to the above, there are two basic ways of reducing RE in epidemiological studies. The first could be to increase the sample size of the study—that is, to add more subjects to our objectives of study. The second could be to lessen the variability in measurement in the study. This might be accomplished by using a more precise measuring device or by increasing the number of observations in measurements. But it may be noted here that if sample size or number of observations in measurements are increased, or a more precise measuring tool is purchased, the costs of the study are also supposed to increase. There is generally an uneasy balance between the requirement for an adequate precision and the practical issue of study cost.

Systematic Error or Systematic Bias

Systematic error (SE) or systematic bias (SB) is said to occur when there is a difference between the true value (in the population) and the observed value (in the study) due to any factor other than sampling variability. A mistake in coding, which may affect *all* responses for that particular question, is another example of SB. Further, the validity of a study is dependent on the degree of SB. Validity, however, is usually separated into two components as follows:

1. Internal validity: It is said to be dependent on the number of errors in measurement, including exposure, disease, and associations between these variables. Good internal validity could indicate a lack of errors in measurement. It also suggests that conclusions may be

drawn, at least as they pertain to the subjects under study.

2. External validity: It is said to pertain to the process of generalizing the findings of the study to the population from which the sample was drawn. It could even extend beyond that population to inferences that are more universal. This requires an understanding of which conditions are relevant to the generalization. In fact, internal validity is clearly said to be a prerequisite for external validity.

Concomitant to the above, "bias" plays a significant role in sampling exercises. Three different types of bias are briefly discussed below:

1. Selection Bias
   Selection bias is one of three types of bias that could impair the validity of a study. It is said to occur when the subjects of study are selected or become part of the study as a result of a third, unmeasured variable that may be associated with both the exposure and the outcome of interest (Hernán et al., 2004).
2. Information Bias (IB)
   Information bias is said to arise from systematic error in the assessment of a variable. An example of this is recall bias.
3. Confounding
   It has been traditionally defined as bias arising from the co-occurrence or mixing of effects of extraneous factors, referred to as confounders, with the main effects of interest (Greenland and Morgenstern, 2001). Some epidemiologists prefer to think of confounding separately from common categorizations of bias. This is mainly because unlike selection and information bias, confounding stems from real causal effects.

Notwithstanding the above, **epidemiology**, while dealing with other aspects, could be said to be that branch of science that deals with the study of the patterns, causes, and effects of disease conditions on health in defined populations. It is the principal item of study in public health, and conveys policy decisions and evidence-based practices by identifying risk factors of diseases and also fixes the goals for preventive healthcare. Epidemiologists further assist in designing a study through exploration and statistical analysis of the data. Epidemiology also does not lag behind in the interpretation and dissemination of results, including conducting peer and systematic reviews. Epidemiology is said to have played a key role in developing methodologies to be used in clinical research, public health studies, and to a lesser extent basic research in the biological sciences. Concomitantly, some of the major areas of epidemiological study include etiology of the disease, its transmission and propagation, investigation of the outbreak, surveillance of the disease, as well as screening,

biomonitoring, and comparing the effects of treatment in clinical trials, etc. Moreover, epidemiologists often rely on other scientific disciplines like the biological sciences in order to understand disease processes in better ways. It may be noted here that statistics helps to make efficient use of the data and assists in drawing appropriate conclusions. The role of social sciences is significant because it provides a better understanding of proximate and distal causes. Incidentally, engineering is said to play a key role in exposure assessment.

Concomitant to the above, epidemiologists are said to apply a range of study designs from the observational to the experimental stage. These could generally be categorized into three aspects, viz., (1) descriptive; (2) analytical, which usually aims to examine known associations further or to try to hypothesize relationships; and (3) experimental, which attempts to make clinical or community trials of treatments and other interventions. Further, it may be noted here that in observational studies, nature is allowed to take its course as epidemiologists observe from the sidelines. Conversely, in experimental studies, the epidemiologist is the one in control of all factors entering a certain case study. Furthermore, epidemiological studies are aimed, where possible, at revealing unbiased relationships between exposures such as alcohol or smoking, biological agents, stress, or chemicals to mortality or morbidity. The identification of causal relationships between these exposures and outcomes is an important aspect of epidemiology. Modern epidemiologists use informatics as a tool. Interestingly, observational studies have two components, viz., (1) descriptive and (2) analytical. Descriptive observations generally pertain to aspects like the "who, what, where, and when" of health-related states of occurrence. Incidentally, analytical observations deal more with the "how" of a health-related event.

Conversely, experimental epidemiology contains three case types, viz., (1) a randomized control trial, which is often used for new medicine or drug testing; (2) a field trial, which is conducted on those at a high risk of conducting a disease; and (3) a community trial, which is the research on social implication of diseases (Refere, 2009). Incidentally, the term "epidemiologic triad" is used to describe the interaction of *host, agent*, and *environment* in analyzing an outbreak.

Notwithstanding the above, epidemiology is the study of the distribution and determinants (i.e., causes) of disease in populations. Epidemiologists typically take a wide view of causal factors, defining them as "any event, condition, or characteristic that plays an essential role in producing an occurrence of the disease." By contrast, many pathologists and microbiologists may consider, for example, a particular infectious agent to be the cause of a disease and may relegate all other contributions to "contributing" or "predisposing" factors.

Notwithstanding the above, the term "epidemiology" is widely applied today to cover the description and causation of not only epidemic diseases, but also diseases in general, and even many nondisease health-related conditions such as high blood pressure and obesity. Therefore, the term "epidemiology" in a broader sense is based upon how the pattern of the disease cause changes in the function of everyone.

4.2 MAJOR SPECIES AFFECTED

In India as well as many other countries, systematic study since 1988 has revealed wide-scale attack among four species of fishes. These are:

Channa punctata
Macrognathus aral
Mystus vittatus
Puntius conchonius

Study revealed a periodicity in the magnitude of the disease among the four most widely affected species as detailed below by periods studied.

1. From around mid-July to mid-August 1988:
   M. vittatus > *C. punctata* > *M. aral* > *P. conchonius*
2. From around mid-August to the first week of September 1988:
   M. vittatus > *M. aral* > *P. conchonius* > *C. punctata*
3. From around first week of September to around the last week of September 1988:
   M. vittatus > *P. conchonius* > *C. punctata* > *M. aral*
4. From around the last week of September to around the third week of October 1988:
   P. conchonius > *C. punctata* > *M. aral* > *M. vittatus*

Other species affected by epizootic ulcerative syndrome (EUS) during the same period, but not widely, included *Sperata aor*, *Amblypharyngodon mola*, *Catla catla*, *Cirrhinus mrigala*, *Heteropneustes fossilis*, *Labeo rohita*, *Lepidocephalus guntea*, *Notopterus notopterus*, and *Salmostoma bacaila*.

Our continued study revealed that during the period 1992–1994, the following species were found to be severely affected by EUS: *Parambassis ranga*, *Chanda nama*, *Nandus nandus*, and *Glossogobius giuris* (Kar et al., 1994). Some of the other species affected by EUS to a lesser extent during this period, included *Mastacembelus armatus*, *Macrognathus pancalus*, *Xenentodon cancila*, *Colisa fasciatus*, etc.

Our recent studies (Kar et al., 1997, 1998a,b,c, 1999, 2000, 2001, 2004) indicated that the following species of fishes have been severely affected by EUS since 1995, particularly during the period November–February, causing large-scale mortality among them: *Channa striata*, *C. punctata*, *Anabas testudineus*, and *Clarias batrachus*.

Thus, there has been a differential pattern of spread of EUS among different fish species during different seasons.

With regard to the global scenario, more than 100 fish species have been reported to be affected by EUS (Lilley et al., 1992), but relatively few reports have been confirmed by demonstrating the presence of MG in histological section or by isolation of the pathogenic fungus *Aphanomyces* sp. from tissues underlying ulcers.

Some commercially important species are considered to be particularly resistant to EUS. But not many studies have been done to confirm these observations and to investigate the mechanism of resistance. Species reported to be unaffected by EUS outbreaks include the Chinese carps, tilapias, and milkfish (*Chanos chanos*). Hatai (1994) experimentally injected catfish (*Parasilurus asotus*), loach (*Misgurnus anguillicaudatus*), and eel (*Anguilla japonica*) with hyphae of *Aphanomyces invadans* and found them to be refractory to infection. Wada et al. (1996) and Sharifpour (1997) experimentally injected common carp (*Cyprinus carpio*) with zoospores of *Aphanomyces* from MG and EUS outbreaks respectively, and demonstrated that fungal growth was suppressed by an intense inflammatory response.

Humphrey and Pearce had reported that in the Northern Territory of Australia, EUS had been reported in archer fish (*Toxotes chartareus*), barramundi (*Lates calcarifer*), bony bream (*Nematolosa erebi*), chanda perch (*Ambassis agassiz*), fork-tailed catfish (*Arius* sp.), long tom (*Strongylura krefftii*), mangrove jack (*Lutjanus argentimaculatus*), mouth almighty (*Glossamia aprion*), mullet (*Liza diadema*), primitive archer fish (*Toxotes lorentzi*), red scat (*Scatophagus argus*), saratoga (*Scleropages jardini*), rainbow fish (*Melanotaenia splendida*), sleepy cod (*Oxyeleotris lineolatus*), spangled perch (*Leiopotherapon unicolor*), striped grunter (*Amniataba percoides*), and nursery fish (*Kurtus gulliveri*).

Notwithstanding the above, in Pakistan, EUS at the moment appears to be confined to the regions of Rivers Chenab and Ravi and their associated irrigation canals in the districts of Kasur, Lahore, Gujranwala, and Sialkot. A wide range of infections and large-scale mortalities had occurred among some of the ichthyospecies, mainly *C. punctata*, *Channa marulius*, *Wallago attu*, and *Puntius* sp. Surprisingly, there had possibly been no mortalities among *L. rohita*, *C. catla*, and *C. mrigala*—although they became infected, no mortality was said to have been observed, as reported by Anon (1998): "Epizootic Ulcerative Syndrome (EUS) in Fishes of Pakistan: A report regarding visit of Mission to Punjab in connection with Epizootic Ulcerative Syndrome (EUS)," April 22–25, 1998. However, the common, silver, and grass carps had been found to be refractory to EUS. Further, as reported by a number of other workers and mentioned earlier in this treatise, EUS is said to be caused by a fungus called *A. invadans*. EUS appears to occur with distinct lesions in the temperature range of 15–30 °C, and sometimes at temperatures <15 °C.

Concomitant to the above, in Bangladesh (Baruah et al., 1991), the most affected ichthyospecies had been recorded as those that were said to be highly susceptible to EUS, notably the *Puntius* spp., *Channa* spp., *Mastacembelus* spp., and so on. Moreover, there had been reports of variations in the status of susceptibility at the species level among the different genera. This had been revealed during selection of candidate species for aquaculture. For example, *Puntius phutunio* got priority over *Puntius terio* with regard to selection of candidate species for aquaculture. It was mainly because the former was refractory to EUS attack (with perhaps none of the fishes affected by EUS infection), whereas the latter was severely affected by EUS. Furthermore, variations had also been noted in the occurrence of lesions among different fishes, which might not be accounted for by flooding or other "risk factors." Differences in the occurrence of lesions among different habitats had supported the findings of Khan and Lilley (2001), who are said to have opined that EUS is less likely to occur in the lotic water bodies. Notwithstanding the above, it had been further reported that fishes collected by little uncommon fish catching devices, like the seine nets, seemed to have a lower prevalence of lesions compared with fishes collected using more common fishing gears, like the scooping nets, spears, etc. The latter often damaged the body of the fish and could make it more susceptible to disease.

Notwithstanding the above, some authors have commented that the most severely affected species in natural outbreaks are generally bottom dwellers (Llobrera and Gacutan, 1987; Chondar and Rao, 1996) or fishes that possess air-breathing organs.

In the case of snakeheads, no particular size group appears to be more susceptible, with affected fish usually ranging from 40 to 900 g (Cruz-Lacierda and Shariff, 1995). However, there is a possibility that size or age may be significant in other species. For example, the IMCs suffer high mortalities as fingerlings, but larger fish, although they appear ulcerated, are not reported as dying in large numbers (AAHRI, ACIAR, IoA, NACA, 1997).

Notwithstanding the above, some of the EUS-susceptible species seem to have a wide range of geographical distribution, which could sometimes be beyond the current limits of EUS outbreaks. For example, several snakehead and clariid catfish species occur in Africa and Central Asia. This suggests that there is potential for further spread of the disease to these areas, and the EUS did spread.

4.3 EPIDEMIOLOGY AND PATTERN OF SPREAD OF EPIZOOTIC ULCERATIVE SYNDROME

Concomitant to the above, epidemiological studies had also been done with the disease-related episodes in fishes. As indicated earlier, the hitherto unknown virulent and enigmatic epizootic ulcerative condition in fish, known as mycotic granulomatosis (MG), was first described from Japan in 1971 (Egusa and Masuda, 1971). Involvement of an invasive oomycete, subsequently named *A. invadans*, had been widely postulated (Baldock et al., 2005). The disease had since been reported from an increasing number of countries, where it had become widespread in both cultured and wild fish populations (Blazer et al., 2002). The pattern of spread to distinctly separate geographic locations within a relatively short period of time had been considered by some quarters as consistent with the progressive dissemination of a single infectious agent, probably a mycotic element (Baldock et al., 2005; Lilley and Roberts, 1997; Kar, 2007, 2010, 2013).

Concomitant to the above, the disease had been spreading unabated from Australia since long. In Australia, the infection had been recognized since 1972 and was known as red spot disease, as indicated earlier. It was regarded as endemic in many Australian rivers and estuaries (Baldock et al., 2005). Soon after, the disease, variously called EUS, was reported from the Philippines (Callinan et al., 1995), other East Asian countries (Lilley and Roberts, 1997) and India (Kar and Dey, 1988, 1990a,b,c; Kar, 1999, 2007, 2010, 2013; Kar et al., 1990a,b, 1993, 1994, 1995a,b,c; Das and Das, 1993), EUS had been spread to other countries beyond India. Likewise, UM had been reported in estuarine fish on the East Coast of the United States since the early 1980s. This had been ascribed to the same cause according to some workers (Blazer et al., 2002). Further, the epidemic outbreak of ulcerative conditions in fish in the Chobe and Zambezi Rivers in Botswana and Namibia (near the Chobe Game Reserve) in 2006 represented the first reports of the disease on the African continent (Andrew et al., 2008; Anon, 2009a). Further, data were collected in the Caprivi Region of Namibia during regular biological surveys using experimental gill nets, inspection of fishermen's catches, and so on. Many of the juvenile and smaller floodplain species were affected. Surveys had indicated that *Clarias gariepinus* and *Clarias ngamensis* had depicted the highest prevalence of lesions. Further data from angling competitions indicated a high prevalence of lesions in larger cichlids. In this regard, the report about the outbreak, obtained from the fishing headmen during 2006, while working in the Caprivi Region, Namibia, for the Integrated Management of the

Zambezi/Chobe River System Fishery Resource Project, was also a significant one from Africa (Nsonga et.al., 2013). There was also reported outbreak on the Zambian side of the Zambezi River above Victoria Falls. EUS-infected fishes have been observed seasonally at the end of subsequent summers, but with a lower prevalence. Fishes with typical EUS lesions could be found during surveys of fishes brought to the Katima Mulilo Open Market during the initial period of the outbreak, between February and August 2007. However, subsequent to August 2007, the presence of infected fish in the market had been rare. Since fishermen were still reporting about catching fish with lesions, it was assumed that fishermen or traders had removed the infected fish before offering them for sale.

The ulcerative condition was subsequently confirmed as EUS and the causative agent A. invadans was identified by histopathology (Andrew et al., 2008; Anon, 2009a), culture and polymerase chain reaction (PCR) (Anon, 2009b). The annual cycle of flooding had brought about changes in water quality that were thought to favor the infectivity of A. invadans, with diseased fish appearing soon after the plains had been flooded. However, it could be noted that the optimal temperatures for vegetative growth in vitro for A. invadans are in the range of 20–30 °C (Fraser et al., 1992; Lilley and Roberts, 1997), and probably, for this reason, natural outbreaks of EUS, to date, have been found to be restricted to the latitudes between 35°N and 35°S. Experimental injection challenges of native European and American fish species have shown that the pathogenic fungus A. invadans is capable of causing lesions in rainbow trout at 18 °C, but is seems to be less infective in stickleback (Gasterosteus aculeatus) and roach (Rutilus rutilus) at 11–16 °C (Khan et al., 1998). Further, according to Humphrey and Pearce (2004), infection occurs when motile spores of A. invadans in the water are attracted to the skin of fish. The spores penetrate the skin and germinate, forming fungal filaments or hyphae. The hyphae invade widely into the surrounding skin and deeply into underlying muscle tissues, resulting in extensive ulceration and destruction of tissues.

Concomitant to the above, it may be noted here that the upper Zambezi floodplain at the confluence with the Chobe River spans the four countries of Botswana, Namibia (Reantaso and Subasinghe, 2012), Zambia, and Zimbabwe, making disease control a challenge. The floodplain ecosystem was considered to support a high fish diversity of around 80 species, and was regarded as an important breeding and nursery ground.

Since 2006, the disease had spread rapidly upstream along the upper Zambezi and its tributaries. Reports from Zambia had also indicated a rapid spread upstream in the Zambezi River and some of its tributaries, along with large-scale fish mortality in some cases (Anon, 2009a; Choonga et al., 2009). So far, however, EUS had not been reported from the Zambezi River downstream of Victoria Falls. Nevertheless, EUS had of late been reported from the Kafue River in Zambia (Mudenda, 2006). The lower Zambezi River has two large impoundments, Lake Kariba and Lake Cahora Bassa, both of which support significant fisheries, while the former has one of the largest aquaculture facilities in sub-Saharan Africa. By 2010, the disease was reported from the Okavango Delta in Botswana and in 2011 from the Western Cape Province of South Africa. It is often opined that EUS has the potential to disrupt floodplain ecosystems elsewhere in Africa where high fish diversity forms the basis of subsistence fisheries and local economies, and is a direct threat to freshwater aquaculture.

Concomitant to the above, a detailed epidemiological investigation on the EUS fish disease was conducted from January to December 2011 by Albert Nsonga et al. (2013) in the Zambezi River System (ZRS) of Sesheke District in the Western Province of Zambia. The study had aimed at determining: (1) factors associated with outbreaks of EUS; (2) the rate of infection; and (3) the distribution of the disease. The disease outbreak had occurred during 2006 causing high mortality in various fish species including Barbus and Clarias spp. As a result, the biodiversity of the fishery was threatened, thus posing food insecurity to more than 700,000 people in 2000 villages, with Zambia affected most. Incidentally, EUS had been regarded as a newly confirmed disease in Southern Africa, said to be caused by a fungal pathogen, A. invadans. Active surveillance was conducted where a total of 4800 fishes had been examined during February, June, and October for gross EUS-like lesions. The disease was subsequently confirmed using histopathology. Environmental cues were also assessed monthly for one year in order to determine their association with the disease outbreaks. The spread of EUS was also monitored by feedbacks through a questionnaire. The Geographic Information System (GIS) had helped to map the distribution of the disease. Results of the study had implicated a number of predisposing environmental factors. Heavy rains preceded outbreaks resulting in excess flooding that had resulted in a rise in the water level to 2 m higher than normal. Further, the predominantly gleysol and arenosol soils of the ZRS resulted in low water pH (4.53–6.5). Thus, acidic soils of the plains had reduced water pH and were implicated as a risk factor associated with EUS outbreak. However, disease outbreaks were subsequently reported in nonacidic agroecological zones as well. Other factors significantly had predisposed the fishes to disease conditions. An active surveillance coupled with participatory epidemiological study had therefore being carried out to further investigate the risk factors, prevalence, severity, and spatial distribution of the disease.

4.3.1 Current African Situation

In 2006, fishes caught in the Chobe—Zambezi River were found with clinical signs of ulcers and focal areas of skin inflammation. In 2007, the disease was confirmed as EUS, by the OIE reference Laboratory in Thailand with the support of FAO. EUS was earlier reported to be confined only to Asia, Australia, and America.

As from 2007, the disease was being reported in Namibia and Zambia. In Namibia, the disease was being reported from fish farms, while in Zambia, the wild fisheries were affected.

By 2008—2009, the entire Zambezi River was affected along with its tributaries on the upper part.

In 2010, the disease was reported in the Kafue River.

In 2011, the disease was confirmed in some tributaries of the Kafue River (Chongwe River). Zambia had been mostly affected.

Currently in Zambia, there reportedly has been enzootic stability in disease occurrence. Some species have depicted signs of healing.

4.3.2 Brief Historical Resume of the Spread of Epizootic Ulcerative Syndrome in Africa

In the African region, the disease had been reported in Zambia, Namibia, and Botswana, and subsequently in South Africa, Botswana, and Namibia. Disease confirmation was made up to 2010. The Okavango Delta had reported clinical signs of the disease in 2010. Further, in South Africa, the disease had affected the wild fish such as the bluegill sunfish (*Lepomis macrochirus*) and the cichlids, but trouts in contact inside cages in the dam were said to be not affected. Moreover, it was reported that fishes found in a semiclosed freshwater pond were believed to have developed the first incidence of the disease in South Africa. Another report on EUS infection was on the freshwater sharptooth catfish.

Concomitantly, Kamilya and Baruah (2014) had reviewed the history and current status of understanding of EUS disease in fishes. According to them, >100 species of wild and cultured finfish have been affected by EUS. EUS was first reported in farmed ayu (*Plecoglossus altivelis*) from Japan in 1971 and had since then swept across different continents including Asia, Australia, North America, and Africa as mentioned above. Epidemiological studies showed that the spread of the disease, especially in the Asia—Pacific region and Africa, had resulted in substantial damage and enormous losses in fish resources and to the livelihoods of fish farmers. Perhaps no reports are available as yet that confirm outbreak of the disease in Europe and South America. The latest outbreak of EUS had

been reported from Canada. A new susceptible species of brown bullhead, *Ameiurus nebulosus*, seemed to have been affected by EUS. It appeared that the disease has the potential to spread further because of its epizootic nature and broad susceptible host range.

In addition to the above, Humphrey and Pearce (2006) had dealt with EUS, or "red spot," as it is known colloquially, and attempted to answer **"what is epizootic ulcerative syndrome?"** They had opined that it is an ulcerative syndrome of fish that had affected a range of native species in the Northern Territory of Australia. The disease had also occurred in New South Wales, Queensland, and Western Australia, and had been occurring in many Asian countries as mentioned earlier. They had opined that EUS begins as a small area of reddening over a single scale, which subsequently spreads to involve a number of adjacent scales. This is the characteristic "red spot." As the condition progresses, the "red spot" expands and deepens, giving a deep ulcer that sometimes extends into the abdominal cavity. Some fish, especially "barramundi" (*L. calcarifer*) are said to develop unilateral or bilateral cloudiness of the cornea. These changes in the eye may or may not be accompanied by lesions in the skin. Some cases of EUS may heal spontaneously, but many affected fish, especially juveniles, were reported as having died. Like many others, they (2004) also reported that a pathogenic fungus *A. invadans* causes EUS. In the Northern Territory of Australia, the factors that initiate infection are not clear. Low water temperature depresses the immune system of fish, which may partially explain the apparent dry season occurrence of EUS. There appeared to be no definite relationship between dissolved oxygen (DO) levels and EUS. Affected fish had been found to occur in water with DO levels between 2 and 8 mg/L. Further, while dealing with the epidemiology of EUS, they (2004) further pointed out that EUS was first seen in the Northern Territory of Australia during 1986 on mullet from the Mary River. Subsequently, it was reported to affect fish from many other river systems in the Top End. In 1986, fish of all ages had appeared to be affected by EUS. As reported, the disease had continued to appear in the subsequent years and had occurred particularly in the early dry season, around May and June. Fish younger than one year had appeared to be most seriously affected, while only sporadic cases had been reported in matured fish. There had been no confirmed reports of EUS during the wet season from December to April, and the syndrome seemed to be restricted to fish in freshwater and in water of low salinity in the upper estuarine reaches of the rivers.

With regard to implications for wild fish stocks in aquaculture, it had been opined that the overall impact of EUS on wild stocks of fish is uncertain, although high losses in juvenile stocks are known to occur. Sometimes, restrictions are placed in areas where EUS occurs on translocation of live fish from river basins and estuaries in order to prevent further spread of the disease. If EUS invades aquaculture farms, it may cause unsightly injuries in fish, market rejection, and fish mortality.

Concomitant to the above, certain programs in Bangladesh, notably the Bangladesh Flood Action Plan 17; the Fisheries Studies and Pilot Project (FAP 17 1995), are said to have accumulated significant information on the occurrence of lesions on about 35,000 wild FW fishes (Lilley et al., 2002). Analysis of the data depicted a surprisingly high prevalence of lesions on the fishes. Of the fishes examined, fishes with lesions made up greater than 50% of the entire population of fishes in the water body. Thus, it is said to eliminate the danger of inadvertently selecting subsamples of less healthy fishes. In fact, it appeared that the population of samples had only a slightly lower percentage of lesions (c 24%) than the population of subsamples (c 28%). In addition, a greater occurrence of lesions during the winter and in some species that are usually considered more EUS-susceptible seems to provide further evidence that lesions are predominantly the result of EUS infections. Further, the FAP 17 database is said to include comments on the severity of the infections in each sample. These had been coded with X (mild) to XXXXX (severe lesion). Further, the most affected ichthyospecies had been recorded as those that were highly susceptible to EUS, notably the *Puntius* spp., *Channa* spp., and *Mastacembelus* spp. Moreover, there had been reports of variations in the status of susceptibility at the species level among the different genera. This had been revealed during selection of candidate species for aquaculture. For example, *P. phutunio* got priority over *P. terio* with regard to selection of candidate species for aquaculture, mainly because the former had been refractory to EUS attacks (with perhaps none of the fishes affected by EUS infection), whereas the latter was severely affected by EUS. Furthermore, variations had also been noted in the occurrence of lesions among different fishes that might not be accounted for by flooding or other "risk factors." Differences in the occurrence of lesions among different habitats supported the findings of Khan and Lilley (2001), who are said to have opined that EUS is less likely to occur in lotic water bodies.

Concomitant to the above, a study on the prevalence of EUS in three floodplain areas of Bangladesh had been undertaken by Subasinghe and Hossain (1997). They used histological technique for diagnosis of the disease. It was revealed from their work that prevalence of EUS was generally lower in artificially stocked fishponds than in the wild fish population. It seems quite unlikely that fry, which had been reared in hatcheries within Bangladesh, could pose a significant risk of EUS to wild fishes, mainly because there are few accounts of EUS in carp hatcheries and EUS seemed to have become endemic in natural waterways throughout Bangladesh (Khan and Lilley, 2001).

Furthermore, water bodies that had been artificially stocked with fishes did not display any significant association with occurrences of EUS. The above-mentioned account seems to be quite different from the situation in Northeast (NE) India, where, EUS had been significantly affecting both farmed and wild fishes since its inception in 1988 (Kar and Dey, 1988a,b,c, 1990a,b,c).

Notwithstanding the above, reports revealed that there had been a decreasing trend of occurrence of EUS in both wild and farmed fish populations in Bangladesh since last 15 years approximately. The reduced severity of outbreaks had been even more evident. EUS-affected ponds, netted during initial outbreaks, had usually revealed that certain ichthyospecies had been 100% infected with high rates of mortality (Ahmed and Rab, 1995). However, these susceptible species now usually depict lower rates of infection. Further, lesions often heal as temperature gradually rises. Moreover, an ADB/NACA (1991) questionnaire survey of carp farmers in Bangladesh had portrayed that c 17% of the extensive farmers and c 53% of the intensive and semi-intensive farmers reported somewhat little respite from the EUS problem since 1992–1995. Such was not the status, however, in neighboring India. Similarly, Hossain (1998) reported that c 85% of the Thana Fisheries officers (TFOs) had indicated that the general aquatic animal disease situation had improved from 1994 to 1996; and that c 91% of the TFOs had indicated an improvement from 1996 to 1998. Concomitant to the above, survey of wild EUS fishes in Bangladesh revealed the following:

1. Approximately 26% of 34,451 fishes examined had lesions as noted during the period October 1992 to March 1994 (FAP 17 data).
2. Approximately 0–23% of the stocked fishes and 2–37% of the wild fishes had been affected by EUS in three floodplains, as recorded during the period December 1992 to March 1993 (Subasinghe and Hossain, 1997).
3. Approximately 0–9% of the stocked fishes and 0–11% of the wild fishes were said to be affected by EUS, as recorded during the period December 1993 to March 1994 (Subasinghe and Hossain, 1997).
4. Approximately 16% of the c 6433 wild fishes had lesions, as noted during the period November 1998 to March 1999 (Khan and Lilley, 2001).

4.4 CAUSAL HYPOTHESIS AND MODEL FOR THE DISEASE

For most diseases, including EUS, there happens to be strong evidence that outbreaks generally occur only when a number of causal factors combine together. Many of the causal factors that have been identified or suggested for EUS, based on reasonable evidence, may be represented in a causal web. It may be noted here that there are several levels within the web, and that several factors may act at the same level but not necessarily at the same time or intensity. It may further be noted that for EUS to occur, combinations of causal factors must ultimately lead to exposure of dermis, attachment to it by *A. invadans,* and subsequent invasion of the dermis and muscle by the fungus. The resulting mycotic granulomatous dermatitis and myositis may be considered expression due to EUS (Lilley et al., 1998).

The **multifactorial nature** of the causes of EUS could also be represented by using the concepts of "necessary cause, component cause, and sufficient cause." Each combination of various causal factors ("component causes") that together cause a disease could collectively be regarded as a "sufficient cause" for that disease to be initiated. It may be important to note here that different combinations of "component causes" may result in a sufficient degree of cause for the initiation of a disease, which could be different, however, under different circumstances. Moreover, all sufficient causes for a particular disease are said to have in common at least one component cause, known as a "necessary cause," that must always be present as expected for that disease to prevail.

For EUS, studies have suggested a number of sufficient causes, and each could make up its component causes. It may be noted here that each of these sufficient causes includes, among its component causes, one of the recognized necessary causes, viz., propagules of *A. invadans* (Lilley et al., 1998).

Notwithstanding the above, a unique feature of an epidemiologist is to test the etiological hypothesis and to identify the underlying causes (or risk factors) for the diseases. A number of factors have been hypothesized as either factors or determinants of EUS outbreaks. So, in order to find out the exact causative organism, or the infectious etiology, it is the need of the hour to study the epidemiological aspects of the disease.

A standard pro forma has been designed (Kar, 2007). This pro forma has been used to collect different information related to epidemiology of EUS in the Barak Valley region of Assam, where EUS had made its initiation in India. It has been covered from the angles of socioeconomic impact of it, epidemiology of the disease, and its impact on aquaculture farms.

Since 1988, EUS has been considered to be one of the most serious diseases affecting freshwater fishes. Here, according to our (Kar, 2007) survey, the highest mortality rate has been found among *C. mrigala, C. striata, Puntius ticto, Labeo gonius,* and *L. guntea.* When analysis is used to throw light on etiology, it is essential to use a specific

death rate (SDR). Therefore, we have attempted to determine the SDR due to EUS:

SDR due to EUS

$$= \frac{\text{Number of deaths during a year}}{\text{Midyear population (total number of species)}} \times 100$$

To represent the killing power of the disease, we have studied the case fatality ratio (CFR), which is calculated as follows:

$$\frac{\text{Total number of deaths due to EUS}}{\text{Total number of cases due to EUS}} \times 100$$

Concomitant to the above, it may be mentioned here that when EUS spreads into a fish culture pond, such as a snakehead fishpond, high morbidity (>50%) and high mortality (>50%) might be observed in those years that have a long cold season, with water temperatures between 18 and 22 °C. However, some infected fishes may recover when the cold period is over.

4.4.1 Characterize the Outbreak in Terms of Time, Affected/Unaffected Fishes, and Place

To describe patterns of disease by fish categories, it is first necessary to outline what measures of disease frequency are used in outbreak investigations. The basic measure of disease frequency in outbreaks is the "attack rate" (AR). An attack rate is a special form of incidence rate where the period of observation is relatively short. An attack rate is the number of cases of the disease divided by the number of animals at risk at the beginning of the outbreak. Where different risk factors for the disease are to be evaluated, attack rates specific for the particular factor must be calculated. For example, suppose there were deaths due to suspected EUS in a pond and it appeared that small fish were at greater risk of having EUS than larger fish. We might make the following calculations:

For small fish, $AR1 = \dfrac{\text{Number with EUS}}{\text{Total small fish}}$

For large fish, $AR2 = \dfrac{\text{Number with EUS}}{\text{Total large fish}}$

With EUS, of course, the total numbers of fish in different putative risk categories would have to be estimated through representative sampling. Another problem could be the estimation of the number of mortalities and when they died. When investigating an outbreak of EUS, it is postulated that such information could be necessary.

It may also be noted that in studies with different models, it had been understood that small fish (<30 cm in length) could be three times more likely to have EUS infection than medium-sized fish and six times more likely than large fish. Also, medium-sized fish were at twice the risk of larger fish. This dose–response phenomenon, when relating size to attack rate, would lend support to a hypothesis that nutritional stress, as manifested in size, could be a "component cause" of EUS.

4.4.2 Prepare a Working Hypothesis

Based on analyses of time, place, and fish data, working hypotheses are usually developed for further investigation. These may concern one or more of the following:

1. Whether the outbreak is common source or propagating
2. If common source, whether it is point or multiple exposure
3. The mode of transmission—contact, vehicle, or vector. Any hypothesis should be compatible with all the facts
4. Corrective action that could be taken based on the more realistic hypotheses

4.4.3 Intensive Follow-Up

This generally includes clinical, pathological, and microbiological examinations together with examinations of water quality data and recent meteorological data. Epidemiological follow-up will include detailed analyses of these data as well as the search for additional cases on other premises. Flowcharts of the management and movements of fish, water, and equipment, for example, may be required as part of this process. Transmission trials may be necessary where additional infectious agents, such as bacteria or ectoparasites, are suspected as component causes of the outbreak.

Further, a **model** related to the study of factors affecting the outbreak of EUS in farmed fish in Bangladesh, carried out by Ahmed and Rab (1995), in collaboration with ICLARM and the Department of Economics, Islamic University of Bangladesh at Kusthia, was replicated by us in the field in Assam University at Silchar (Kar, 2007). In this model, three experimental ponds (viz., ponds A, B, C) were prepared of sizes $15' \times 15' \times 3'$ each. Lime was applied during the poststocking, generally in proportions of 0.027, 0.032, and 0.043 kg, respectively. Pond A was seeded with five species of fishes, viz., *C. mrigala, L. rohita, Ctenopharyngodon idella, Oreochromis mossambicus,* and *C. punctata.* Pond B was seeded with only two species of fishes, viz., *C. mrigala* and *C. punctata.* Pond C was seeded with only a single species of fish, viz., the *C. mrigala.* No EUS-outbreak had been found to occur in any of the experimental ponds

(viz., A, B, C) during the period from November 2003 to March 2004.

Our epidemiological study conducted during 2002–2004 (Kar et al., 2003, 2004) revealed that the highest mortality rate was found among *C. mrigala, Channa* spp., *Puntius* spp., *Labeo* spp., and *C. batrachus.* SDR due to EUS = (number of deaths during a year/midyear population (total number of species)) $\times 100 = (22/38) \times 100 = 57.89\%$. In the same study and period, the killing power of EUS was represented by CFR = (total number of deaths due to EUS/total number of cases due to EUS) $\times 100$. CFR was calculated species-wise and found to be high among *C. mrigala* (83.72%), followed by *P. ticto* (75%), *C. marulius* (70%), *M. armatus* 69.23%), and *A. testudineus* (53.85%).

Details of the outbreak of EUS, including inter- and intracontinental states, were elucidated earlier.

4.4.4 Some More Information on the Epidemiology of Epizootic Ulcerative Syndrome

Notwithstanding the above, some more information on the epidemiology of EUS is given below:

1. Callinan et al. (1995) had done preliminary comparative studies on the *Aphanomyces* species associated with EUS in the Philippines and RSD in Australia. On the basis of preliminary comparative studies, they proposed that the patterns of spread of both EUS and RSD suggest involvement of one or more primary infectious agents (Rodgers and Burke, 1981; Roberts et al., 1992). A number of viruses and bacteria have been recovered inconsistently from fishes with EUS (Frerichs et al., 1986; Hedrick et al., 1986; Llobrera and Gacutan, 1987; Roberts et al., 1990; Torres et al., 1992) and RSD (Burke and Rodgers, 1981; Callinan and Keep, 1989; Pearce, 1990)—and as had been opined, no etiological agent has been conclusively identified for either condition.
2. Anon (2009a), while dwelling upon EUS, dealt with this fish disease as an exotic fish disease that threatens Africa's aquatic ecosystems.
3. Anon (2009b) described EUS in fishes by covering various aspects such as scope; disease information; agent factors; etiological agent, agent strains; stability of the agent; life cycle; host factors; and so on.
4. In the event of disease outbreak, government agency disease information as well as additional surveys revealed that on the Indian subcontinent in general, *c* 57.7% (75 out of 130) of affected persons had consulted a government fishery extension officer for guidance and help. Concomitantly, affected farmers also contacted other farmers (*c* 44.6%, n = 58), other hatchery owners (*c* 52.3%, n = 68), and drug and

chemicals salespersons (c 11.5%, n = 15). Ranking of information support revealed that the government extension officers had provided the affected farmers with perhaps the most useful information (c 42.3%, n = 55), followed by other hatchery owners (c 27.6%, n = 36), other farmers (c 16.9%, n = 22), and drug and chemicals salespersons (c 6.2%, n = 8).

5. Anon (2012), under the aegis of the **Gladstone Fish Health Scientific Advisory Panel**, had prepared a report that is based on information provided to the panel by the Queensland government and other stakeholders relating to the fish health issues observed in Gladstone Harbor and surrounding areas. In 2011, the Queensland government received reports (primarily by commercial fishers) of barramundi fish and subsequently other species being caught with obvious signs of disease, including bulging/red eyes, blindness, severe skin lesions, and skin discoloration. The government then undertook an investigation of Gladstone Harbor and surrounding areas following reports that commercial fishers were sick with what appeared to be bacterial infections on their arms, feet, and legs following contact with or abrasions and fish spikes from net-caught barramundi that were exhibiting evidence of disease. At that time, the Queensland government was concerned about the potential food safety issues of consuming the diseased fish, given that the type of disease remained unknown. Furthermore, there were concerns about the possibility of the transfer of the disease from affected fish to other fish and consequently its entry into the food chain.

From the initial testing of nine diseased barramundi, two conditions were identified that were affecting barramundi in the Gladstone area:

a. Red-spot disease (EUS), a fungus endemic to finfish species of mainland Australia. This condition was only confirmed in one fish from Port Alma.

b. An external parasitism due to the fluke *Neobenedenia* sp. was affecting the eyes and skin, particularly in the barramundi in Gladstone Harbor. This parasite has previously been found in Queensland waters in Hinchinbrook Channel between Hinchinbrook Island and mainland Queensland where barramundi are in high densities.

6. Mohan and Shankar (1994) had made epidemiological analysis of EUS of fresh and brackish water fishes of Karnataka in India. According to them, EUS had been causing large-scale mortality among the fresh and brackish water fishes in various parts of South and SE Asia since the early 1980s. It is an infectious disease with mixed etiology. They had further stated that EUS had first surfaced in India during 1989 in the NE provinces, and then, had extended its range to become a serious menace to

fresh and brackish water fishes in different parts of the country.

7. Lilley et al. (1998) published *Epizootic Ulcerative Syndrome (EUS) Technical Handbook*. They had opined that EUS is an international problem that has been studied independently, and collaboratively, by many different workers. The early occurrences of MG in Japan and the RSD in Australia more than 35 years ago are now considered to have been outbreaks of the disease subsequently designated as EUS. It had further been opined that the spread of EUS might be due partly to the large-scale movement of fish within the Asia—Pacific region. The potential for further spread is high. Consequently, the risk of introducing EUS should be a matter of concern for countries that are as yet unaffected. The need for development of effective strategies to reduce risks associated with the spread of important aquatic animal pathogens had been widely recognized, and in Asia had been addressed through a cooperative FAO/NACA/OIE Regional Program for the Development of Technical Guidelines on Quarantine and Health Certification and Establishment of Information Systems for the Responsible Movement of Live Aquatic Animals in Asia development.

8. Anon (2011) had dealt with EUS (aka RSD, MG, and UM) while preparing *Identification Field Guide for Aq Animal Diseases Significant to Australia.*

9. Anon (1992) had dealt with the spread of EUS across Asia from Japan and Australia where it was first identified in the early 1970s, and then to Pakistan in 1996 and to southern Africa in 2006. It was regarded as a major epizootiological phenomenon.

10. Humphrey and Pearce (2006) had dealt with EUS (red-spot disease) in Australia.

4.5 EPISODES OF FISH KILLS

Notwithstanding the above, there have been episodes of fish kills from many causes that could be in addition to EUS. The matter is briefly discussed below:

4.5.1 Fish Kills Due to Environmental Degradation

Large-scale fish deaths occur worldwide and can be attributed to a variety of factors including infectious diseases, algal blooms, jellyfish blooms, toxins, brushfire ash, water temperature extremes, stranding of school fish, lifecycle events (spawning or migration), deranged water pH, and depletion of dissolved oxygen (DO). Notably, some of these events are completely natural occurrences whereas others are the direct result of human activities. Toxins may be introduced from dino-flagellate or algal blooms, industrial effluents, agricultural spraying, precipitation from atmosphere, or spillage/transport accidents. Lethal agricultural chemicals include those that may derange water pH (fertilizer or lime), pesticides (organophosphates, chlorinated hydrocarbon), and herbicides. Industrial chemicals that may be lethal to fish include elemental phosphorus, phenols, cyanide, petrochemicals, and polychlorinated biphenyls. Water pH may also be deranged by overflow of human or animal waste matter.

4.5.2 Fish Kills Due to Algal Blooms

Algal blooms are especially a threat to fish that are contained in freshwater, estuarine, or coastal aquaculture systems, but could also affect free-living fish that swim into affected waters. Algal blooms may be noticed by discoloration of water or identification of scum accumulation on the water surface. Some algae are also thought to result in fish death through the production of neurotoxins. This is the case with blooms of the dinoflagellate, *Gymnodinium breve* (commonly referred to as "red tide") in the Gulf of Mexico. In Australia, blooms of *Gymnodinium* sp. have been associated with fish kills in New South Wales, Tasmania, and Victoria. However, algal blooms could result in fish death without neurotoxins. Some species of algae, especially diatoms, have spiny surface architecture and may lodge in the gills, mechanically causing microtrauma and inciting bronchitis that may result in death. The diatom *Cerataulina pelagica* was associated with fish kills in New Zealand in 1983. Similarly, a bloom of green microalgae, *Tetraselmis* sp., that adheres to the gill surface was held responsible for a massive die-off of pilchards in Wellington Harbor in 1993. Another important mechanism in which algal blooms result in fish death is referred to as "algal crash." In this case, the unsettling of nutrient-rich sediment or provision of excess nutrients from anthropogenic or natural sources contributes to the occurrence of an algal bloom. The new algal population, often comprising cyanobacteria, is unsustainable (has a "boom and bust" cycle) and massive death of algae ensues. Dying algae liberally consume oxygen, and thus fish within affected water die from hypoxia. The water containing dying algae often has brown discoloration and smells due to algal decomposition. "Algal crash" could also be induced without an algal bloom when various algicidal chemicals (e.g., copper sulfate and formalin) are used to treat fish disease in confined water systems with large algal populations; for example, in ponds. It may be of interest to note here that cyanobacteria that are toxic to mammals (some causing hepatocellular necrosis) apparently have not yet been found to cause toxicosis in fish.

4.5.3 Fish Kills Affecting River Systems

Large-scale mortalities affecting multiple species and wide age ranges of wild fishes often represent particularly disconcerting phenomena having profound potential ecological impact. In the last decade in Australia, two such fish kills affecting river systems in northern New South Wales gained considerable public attention. These involved the Richmond River catchment in 2001 and 2008. During these periods, this area experienced widespread severe rainfall and flooding followed by the deaths of millions of finfish and shellfish. Although natural fish kills were known to occur in Australian coastal rivers prior to European settlement, fish kills of this scale had not been previously documented (with the exception of the massive pilchard kills occurring in southern Australian oceans during 1995 and 1998 due to pilchard herpesvirus). As a response to the fish kills, river and near shore areas were closed to all forms of fishing for an extended period to allow ecosystem repair. Although supported by the community and industry, this had profound economic impact on the community, especially commercial and recreational fisheries, etc.

The Richmond River catchment fish kills are an excellent example of kills that occur as a result of low dissolved oxygen (DO). In aquaculture, low DO is often simply the result of power failure and loss of aeration, or overcrowding. However, in natural water bodies consumption of DO is a complex multifactorial process. Water conditions that lead to low DO may include low water flow, high water sediment or organic load (i.e., decomposing biota), high water temperature, and algal growth. In coastal rivers such as the Richmond River catchment, land use practices in areas surrounding catchments are often to blame for episodes of low DO and low pH. These areas typically have acid sulfate soils. In these areas, floodgates and drainage systems have been placed to remove water from developed wetlands and floodplains during rains and high tides. These changes have increased drainage flow and depth, resulting in lower groundwater and surface water levels of wetlands and floodplains, thereby drying and oxidizing the acid sulfate soils. Developed wetlands and floodplains had also been altered to replace native inundation-tolerant vegetation with inundation-intolerant vegetation (i.e., pasture for cattle grazing, etc.). As a result of these land use practices, during subsequent rainfalls, topsoil containing soluble acid and iron is readily mobilized, acidifying the water and consuming DO. Inundated vegetation readily decomposes via microbial assistance, contributing to water sedimentation and DO consumption. "Black water" forms on floodplains, reflecting the accumulation of monosulfide black ooze (organic sediment that has been enriched with iron monosulfides); this material potentiates water acidification and consumption of DO. This toxic oxygen-depleted backwater is flushed rapidly through drainage systems, overwhelming river ways and killing all fish swimming in the supplied waterway. Interestingly, these water conditions (a result of agriculture land use) had led to fish kills of a magnitude that surpass those that had occurred elsewhere in the world as a result of millions of liters of raw human sewage and concentrated swine waste being diverted into waterways following hurricane-associated flooding.

4.6 MANAGEMENT OF EPIDEMIOLOGICAL EPISODES

4.6.1 Management

Epidemiological evidence suggests that EUS outbreaks in farmed fish are more severe when stocking densities are high. During high-risk periods; for example, when EUS prevalence is high in adjacent wild fish populations, stocking densities should be kept as low as possible and farmed populations subjected to minimal stress. In particular, fish could be monitored to ensure that bacterial and parasitic skin pathogens do not cause problems during high-risk periods. Similarly, the pond environment could be monitored to ensure that abiotic factors that may induce skin damage; for example, low DO concentrations, are kept within acceptable limits. A simple and effective form of prevention, which may be acceptable in some endemic areas, is to farm species that are resistant to EUS. For example, EUS has never been reported in tilapia or milkfish, and very rarely in Chinese or European carps. Incidentally, as a guide one could note that epidemiological studies conducted by Kumamaru (1973) from 1971 to 1972 in Lake Kasumigaura revealed that freshwater fish such as yellowfin goby (*Acanthogobius flavimanus*), gray mullet, oily gudgeon (*Sarcocheilichthys variegatus*), trident goby, and slender bitterling (*Acheilognathus lanceolatus*) were highly susceptible to MG, while carp (*C. carpio*), silver carp (*Hypophthalmichthys molitrix*), and eel (*A. japonica*) were not affected during the outbreaks.

As mentioned earlier, from an epidemiological perspective to accommodate the apparently multifactorial nature of EUS, Lilley et al. (1998) used the concepts of "necessary cause," "component cause," and "sufficient cause." Each combination of various "component causes" that results in disease is known (collectively) as a "sufficient cause" for that disease. However, it is important to recognize that under different circumstances, different combinations of "component causes" may constitute "sufficient cause" for a disease and these "sufficient causes" for a particular disease have in common at least one "component cause," known as a "necessary cause," that must always be present for that disease to occur.

It should also be noted that proponents of "EUS as an aphanomycosis" usually consider *A. invadans* (=*Aphanomyces piscicida*) the only necessary infectious cause for the disease. By contrast, proponents of "EUS as a polymicrobial disease" generally suggest that the condition has three necessary infectious causes—a virus (in some cases a rhabdovirus), a fungus (in many cases *A. invadans*), and a bacterium (such as an *Aeromonas* sp. or a *Vibrio* sp.)—that are believed to act in sequence or in concert to induce lesions.

Both groups propose that one or more additional component causes are usually involved in outbreak causation. These include environmental insults that increase the probability that the necessary infectious causes could infect the host and induce lesions. As examples, exposure to acidified water, trauma-induced epidermal damage, and conditions that could lead to immunosuppression have been shown to increase susceptibility of fish to EUS.

REFERENCES

ADB/NACA, 1991. Fish Health Management in Asia-Pacific. Report on a Regional Study and Workshop on Fish disease and Fish Health Management. ADB Agriculture Department Report Series, No.1. Network of Aquaculture Centres in Asia-Pacific, Bangkok.

Anon, 1992. Notes on mycological procedure. In: Workshop on Mycological Aspects of Fish and Shellfish Disease, Bangkok, Thailand, January 1992. Aquatic Animal Health Research Institute.

Ahmed, M., Rab, M.A., 1995. Factors affecting outbreaks of epizootic ulcerative syndrome in farmed and wild fish in Bangladesh. J. Fish Dis. 18, 263–271.

AAHRI, ACIAR, IoA and NACA, 1997. Epizootic Ulcerative Syndrome (EUS) of Fishes in Pakistan. A Report of the Findings of an ACIAR/DFID-Funded Mission to Pakistan. 9–19 March 1997.

Anon, April 1998. Epizooic Ulcerative Syndrome (EUS) in Fishes of Pakistan: A Report Regarding Visit of Mission to Punjab in Connection with Epizootic Ulcerative Syndrome (EUS), pp. 22–25.

Andrew, T.G., Huchzermeyer, K.D., Mbeha, B.C., Nengu, S.M., 2008. Epizootic ulcerative syndrome affecting fish in the Zambezi river system in southern Africa. Vet. Rec. 163, 629–631.

Anon, 2009a. Report of the International Emergency Disease Investigation Task Force on a Serious Finfish Disease in Southern Africa, 18–26 May 2007. Food and Agriculture Organization of the United Nations, Rome. Viewed in 2011, from: http://www.fao.org/docrep/012/i0778e00.htm.

Anon, 2009b. Manual of Diagnostic Tests for Aquatic Animals.

Anon, 2011. World Animal Health Information Database Interface: Exceptional Epidemiological Events. World Organisation for Animal Health, Paris. Viewed in 2011, from: http://web.oie.int/wahis/public.php?page=country_reports.

Anon, 2012. Gladstone Fish Health Scientific Advisory Panel, pp. 1–47.

Barua, G., Banu, A.N.H., Khan, M.H., 1991. An investigation into the prevalence of fish disease in Bangladesh during 1988–1989. Bangladesh J. Aquac. 11–13 (For 1989–1991): 27:29.

Baldock, F.C., Blazer, V., Callinan, R., Hatai, K., Karunasagar, I., Mohan, C.V., Bondad-Reantaso, M.G., 2005. Outcomes of a short expert consultation on epizootic ulcerative syndrome (EUS): re-examination of causal factors, case definition and nomenclature. In: Wlaker, P., Lester, R., Bondad-Reantaso, M.G. (Eds.), Diseases in Asian Aquaculture V. Fish Health Section, Asian Fisheries Society, Manila, p. 555585.

Blazer, V.S., Lilley, J., Schill, W.B., Kiryu, Y., Densmore, C.L., Panyawachira, V., Chinabut, S., 2002. *Aphanomyces invadans* in Atlantic menhaden along the east coast of the United States. J. Aquat. Anim. Health 14, 1–10.

Callinan, R.B., Keep, J.A., 1989. Bacteriology and parasitology of red spot disease in sea mullet, *Mugil cephalus* L., from eastern Australia. J. Fish Dis. 12, 349–356.

Callinan, R.B., Paclibare, J.O., Bondad-Reantaso, M.G., Chin, J.C., Gogolewski, R.P., 1995. *Aphanomyces* species associated with epizootic ulcerative syndrome (EUS) in the Philippines and red spot disease (RSD) in Australia: preliminary comparative studies. Dis. Aquat. Org. 21, 233–238.

Cruz-Lacierda, E.R., Shariff, M., 1995. Experimental transmission of epizootic ulcerative syndrome (EUS) in snakehead, *Ophicephalus striatus*. In: Shariff, M., Arthur, J.R., Subasinghe, R.P. (Eds.), Diseases in Asian Aquaculture II. Fish Health Section, Asian Fisheries Society, Manila, Philippines, pp. 327–336.

Chia, W.K., Ali, R., Toh, H.C., 2012. Aspirin as adjuvant therapy for colorectal cancer-reinterpreting paradigms. Nat. Rev. Clin. Oncol. 2012 (9), 561–570.

Chondar, S.L., Rao, P.S., January 29–February 2, 1996. Epizootic Ulcerative Syndrome Disease to Fish and Its Control: A Review. World Aquaculture 1996, Book of Abstracts. World Aquaculture Society, Bangkok, p. 77.

Choongo, K., Hang'ombe, B., Samui, K.L., Syachaba, M., Phiri, H., Maguswi, C., Muyangaali, K., Bwalya, G., Mataa, L., 2009. Environmental and climatic factors associated with epizootic ulcerative syndrome (EUS) in fish from the Zambezi floodplains, Zambia. Bull. Environ. Contam. Toxicol. 83, 474–478. Springer, New York.

Curtin, K., Slattery, M.L., Samowitz, W.S., 2011. CpG island methylation in colorectal cancer: past, present and future. Pathol. Res. Int. 2011, 902674.

Das, M.K., Das, R.K., 1993. A review of the fish disease, epizootic ulcerative syndrome in India. Environ. Ecol. 11 (1), 134–145.

Egusa, S., Masuda, N., 1971. A new fungal disease of *Plecoglossus altivelis*. Fish Pathol. 6, 41–46.

Field, A.E., Camargo, C.A., Ogino, S., 2013. The merits of subtyping obesity: one size does not fit all. JAMA 2013 (310), 2147–2148.

Frerichs, G.N., Millar, S.D., Roberts, R.J., 1986. Ulcerative rhabdovirus in fish in Southeast Asia. Nature 322, 216.

Fraser, G.C., Callinan, R.B., Calder, L.M., 1992. *Aphanomyces* species associated with red spot disease — an ulcerative disease of estuarine fish from eastern Australia. J. Fish Dis. 15, 173–181.

Gabriel, S.R., Sanchez, G.L.M., Carmona, L., Roqué i Figuls, M., Bonfill Cosp, X., 2005. Hormone replacement therapy for preventing cardiovascular disease in post-menopausal women. Cochrane Database Syst. Rev. 2005 (2). Art. No: CD002229.

Garðarsdóttir, O., Guttormsson, L., 2008. An Isolated Case of Early Medical Intervention. The Battle against Neonatal Tetanus in the Island of Vestmannaeyjar (Iceland) during the 19th Century. Instituto de Economía y Geografía (Retrieved 19.04.11).

Garðarsdóttir, O., Guttormsson, L., 2009. Public health measures against neonatal tetanus on the island of Vestmannaeyjar (Iceland) during the 19th century. Hist. Fam. 14 (3), 266–279.

Green, M.D., Freedman, M.D., Gordis, L., 2008. Reference Guide on Epidemiology (PDF). Federal Judicial Centre (Retrieved 03.02.08).

Greenland, S., Morgenstern, H., 2001. Confounding in health research. Annu. Rev. Public Health 2.

Hatai, K., 1994. Mycotic granulomatosis in ayu (*Plecoglossus altivelis*) due to *Aphanomyces piscicida*. In: Roberts, R.J., Campbell, B., MacRae, I.H. (Eds.), Proc. ODA Regional Sem. on Epizootic Ulcerative Syndrome, 25–27 Jan 1994. Aquatic Animal Health Research Institute, Bangkok.

Hedrick, R.P., Eaton, W.D., Fryer, J.L., Croberg, W.G., Boonyaratapalin, S., 1986. Characteristics of a birnavirus isolated from cultured sand goby *Oxyeleotris marmoratus*. Dis. Aquat. Org. 1, 219–225.

Hennekens, C.H., Julie, E.B., 1987. In: Mayrent, S.L. (Ed.), Epidemiology in Medicine. Lippincott, Williams and Wilkins, ISBN 978-0-316-35636-7.

Hernán, M.A., Hernández-Díaz, S., Robins, J.M., 2004. A structural approach to selection bias. Epidemiology (Cambridge, MA) 15 (5), 615–625.

Hill, A.B., 1965. The environment and disease: association or causation? Proc. R. Soc. Med. 58 (5), 295–300. PMC 1898525. PMID 14283879.

Hossain, M.A., 1998. Survey of aquatic animal disease: Bangladesh 1994–1998. In: MacRae, I.H. (Ed.), The Proceedings of the Fourth DFID-SEAADCP Network Meeting. Aquatic Animal Health Research Institute, Bangkok.

Hughes, L.A., Khalid-de Bakker, C.A., Smits, K.M., van den Brandt, P.A., Jonkers, D., Ahuja, N., Herman, J.G., Weijenberg, M.P., van Engeland, M., 2012. The CpG island methylator phenotype in colorectal cancer: progress and problems. Biochim. Biophys. Acta 2012 (1825), 77–85.

Humphrey, J.D., Pearce, M., 2006. Epizootic Ulcerative Syndrome (Red-spot Disease). Fishnote,Northern Territory Government, Australia, pp. 1–4.

Ikramuddin, S., Livingston, E.H., 2013. New insights on bariatric surgery outcomes. JAMA 2013 (310), 2401–2402.

Ioannidis, J.P.A., Tarone, R., McLaughlin, J.K., 2011. The false-positive to false-negative ratio in epidemiologic studies. Epidemiology 22 (4), 450–456.

Khan, M.H., Marshall, L., Thompson, K.D., Campbell, R.E., Lilley, J.H., 1998. Susceptibility of five fish species (Nile tilapia, rosy barb, rainbow trout, stickleback and roach) to intramuscular injection with the oomycete fish pathogen, *Aphanomyces invadans*. Bull. Eur. Assoc. Fish Pathol. 18, 192–197.

Kar, D., Dey, S., 1988a. Preliminary electron microscopic studies on diseased fish tissues from Barak valley of Assam. In: Proc. Annual Conference of Electron Microscopic Society of India, vol. 18, p. 88.

Kar, D., Dey, S.C., 1988b. A critical account of the recent fish disease in the Barak valley of Assam. In: Proc. Regional Symp. on Recent outbreak of Fish Diseases in North-East India, vol. 1, p. 8.

Kar, D., Dey, S.C., 1988c. Impact of recent fish epidemics on the fishing communities of Cachar district of Assam. In: Proc. Regional Symp. on Recent outbreak of Fish Diseases in North-East India, vol. 1, p. 8.

Kar, D., Dey, S.C., 1990. Fish disease syndrome: a preliminary study from Assam. Bangladesh J. Zool. 18, 115–118.

Kar, D., Dey, S.C., Michael, R.G., Kar, S., Changkija, S., 1990a. Studies on fish epidemics from Assam, India. J. Indian Fisheries Assoc. 20, 73–75.

Kar, D., Bhattacharjee, S., Kar, S., Dey, S.C., 1990b. An Account of 'EUS' in Fishes of Cachar District of Assam with Special Emphasis on Microbiological Studies. Souv. Ann. Conf. Indian Association of Pathologists and Microbiologists of India (NE Chapter) 8, 9–11.

Kar, D., Dey, S.C., Kar, S., Bhattacharjee, N., Roy, A., 1993. Virus-like Particles in Epizootic Ulcerative Syndrome of Fish. Proc. Int. Symp. On Virus-Cell Interaction: Cellular and Molecular Responses 1, 34.

Kar, D., Dey, S.C., Kar, S., Roy, A., Michael, R.G., Bhattacharjee, S., Changkija, S., 1994. A candidate virus in Epizootic Ulcerative Syndrome of Fish. Proc. Natl. Symp. Indian Virological Society 1, 27.

Kar, D., Dey, S.C., Kar, S., 1995a. Lake Sone in Assam and its Biodiversity. Xth Annual Congress on Man and Environment. National Environment Science Academy and National Institute of Oceanography 10, 52.

Kar, D., Roy, A., Dey, S.C., Menon, A.G.K., Kar, S., 1995b. Epizootic Ulcerative Syndrome in Fishes of India. World Congress of *In Vitro* Biology. In Vitro 31 (3), 7.

Kar, D., Kar, S., Roy, A., Dey, S.C., 1995c. Viral Disease Syndrome in Fishes of North-East India. Proc. Int. Symp. Int. Centre for Genetic Eng. and Biotechnology (ICGEB) and the Univ, California at Irvine, 1, 14pp.

Kar, D., Saha, D., Laskar, R., Barbhuiya, M.H., 1997. Biodiversity Conservation Prioritisation Project (BCPP) in barak valley region of Assam. In: Proc. National Project Evaluation Workshop on BCPP, Betla Tiger Reserve and National Park, vol. 1 (Palamu).

Kar, D., Saha, D., Dey, S.C., 1998a. Epizootic ulcerative syndrome in Barak valley of Assam: 2–4. In: Project Report Submitted and Presented at the National Symposium of Biodiversity Conservation Prioritisation Project (BCPP) Held at WWF-India, 18–19 Jan 1998 (New Delhi).

Kar, D., Dey, S.C., Roy, A., 1998b. Present status of epizootic ulcerative syndrome (EUS) in southern Assam. In: Proc. Regional Project Initiation Workshop for NATP-ICAR-NBFGR Project, vol. 1, p. 9.

Kar, D., Dey, S.C., Kar, S., Roy, A., 1998c. An account of epizootic ulcerative syndrome in Assam. In: Final Project Report of Biodiversity Conservation Prioritisation Project (BCPP). Submitted and Presented at the International Project Finalisation Workshop of BCPP, 1, 22–28 April 1998 (New Delhi), pp. 1–3.

Kar, D., Rahaman, H., Barnman, N.N., Kar, S., Dey, S.C., Ramachandra, T.V., 1999. Bacterial pathogens associated with epizootic ulcerative syndrome in freshwater fishes of India. Environ. Ecol. 17 (4), 1025–1027.

Kar, D., Dey, S.C., Roy, A., 2000. Fish Genetic resources in the principal rivers and wetlands in North-East India with special emphasis on Barak valley (Assam), in Mizoram and in Tripura, with a note on epizootic ulcerative syndrome fish disease. In: Proc. National Project Initiation Workshop of the NATP-ICAR World Bank- aided Project on 'Germplasm Inventory, Evaluation and Gene Banking of Freshwater Fishes', vol. 1. National Bureau of Fish Genetic Resources (NBFGR), Lucknow, p. 12.

Kar, D., Mandal, M., Lalsiamliana, 2001. Species composition and distribution of fishes in the rivers in Barak valley region of Assam and in the principal rivers of Mizoram and Tripura in relation to their habitat parameters. In: Proc. National Workshop, NATP-ICAR Project, Mid-term Review, vol. 1. Central Marine Fisheries Research Institute, Cochin, p. 25.

Kar, D., Dey, S.C., Datta, N.C., 2003. Welfare Biology in the New Millennium. Allied Publishers Pvt. Ltd, Bangalore pp. xx + 97.

Kar, D., Roy, A., Dey, S.C., March 18–19, 2004. An overview of fish genetic diversity of North-East India. In: Garg, S.K., Jain, K.L. (Eds.), Proc. National Workshop on Rational use of Water Resources for Aquaculture, vol. 1. CCS Haryana Agricultural University, pp. 164–171.

Kar, D., 2007. Fundamentals of Limnology and Aquaculture Biotechnology. Daya Publishing House, New Delhi pp. vi + 609.

Kar, D., 2010. Biodiversity Conservation Prioritisation. Swastik Publications2, New Delhi pp. xi + 167.

Kar, D., 2013. Wetlands and Lakes of the World. pp. xxx + 687. Springer, London. Print ISBN: 978-81-322-1022-1; e-Book ISBN: 978-81-322-1923-8.

Kar, D., 1999. Microbiological and environmental studies in relation to fishes of India. In: Gordon Research Conference, Connecticut, USA.

Kar, D., Dey, S.C., 1990b. A preliminary study of diseased fishes from Cachar district of Assam. Matsya 15-16, 155–161.

Kar, D., Dey, S.C., 1990c. Epizootic ulcerative syndrome in fishes of Assam. J. Assam Sci. Soc. 32 (2), 29–31.

Kamilya, D., Baruah, A., 2014. Epizootic ulcerative syndrome (EUS) in fish: history and current status of understanding. Rev. Fish Biol. Fish. 24, 369–380.

Karl, P., Ronald, R., Major, G., Hill, A.B., 1900–1945. Statistical Methods in Epidemiology. Trust Centre for the History of Medicine at UCL, London.

Kermack, W.O., McKendrick, A.G., 1927. A contribution to the mathematical theory of epidemics. Proc. R. Soc. Lond. Ser. A, Containing Papers of a Mathematical and Physical Character 115 (772), 700–721.

Khan, M.H., Lilley, J.H., 2001. Risk factors and socio-economic impacts associated with epizootic ulcerative syndrome (EUS) in Bangladesh. In: Proceedings of DFID/FAO/NACA Asia Regional Scoping Workshop "Primary Aquatic Animal Health Care in Rural, Small-scale Aquaculture Development in Asia". Dhaka, 26–30 September 1999.

Ku, C.S., Cooper, D.N., Wu, M., Roukos, D.H., Pawitan, Y., Soong, R., Iacopetta, B., 2012. Gene discovery in familial cancer syndromes by exome sequencing: prospects for the elucidation of familial colorectal cancer type X. Mod. Pathol. 2012 (25), 1055–1068.

Kumamaru, A., 1973. A fungal disease of fishes in lake Kasumigiura and lake Kitaura. Report Stan. Ibaraki Prefecture 11, 129–142.

Llobrera, A.T., Gacutan, R.Q., 1987. *Aeromonas hydrophila* associated with ulcerative disease epizootic in Laguna de Bay Philippines. Aquaculture 67, 273–278.

Lilley, J.H., Philhps, M.J., Tonguthai, K., 1992. A Review of Epizootic Ulcerative Syndrome (EUS) in Asia. Aquatic Animal Health Research Institute and Network of Aquaculture Centres in Asia-Pacific, Bangkok.

Lilley, J.H., Roberts, R.J., 1997. Pathogenicity and culture studies comparing *Aphanomyces* involved in epizootic ulcerative syndrome (EUS) with other similar fungi. J. Fish Dis. 20, 101–110.

Lilley, J.H., Callinan, R.B., Chinabut, S., Kanchanakhan, S., MacRae, I.H., Phillips, M.J., 1998. EUS Technical Handbook. AAHRI, Bangkok, 88 p. McKenzie, R.A., Hall, W.T.K., 1976. Dermal ulceration of mullet (*Mugil cephalus*). Aust. Vet. J. 52, 230–231.

Lilley, J.H., Callinan, R.B., Khan, M.H., 2002. Social, economic and biodiversity impacts of epizootic ulcerative syndrome (EUS). In: Arthur, J.R., Phillips, M.J., Subasinghe, R.P., Reantaso, M.B., MacRae, I.H. (Eds.), Primary Aquatic Animal Health Care in Rural, Small-scale, Aquaculture Development, pp. 127–139. FAO Fish. Tech. Pap. No. 406.

Mohan, C.V., Shankar, K.M., 1994. Epidemiological analysis of epizootic ulcerative syndrome of fresh and Brakishwater fishes of Karnataka, India. Curr. Sci. 66, 656–658.

Mudenda, C.G., 2006. Economic Perspectives of Aquaculture Development Strategy of Zambia. Consultant Report, Development Consultant, Lusaka Zambia.

Neil, M., Debra, J., 2011. Measuring Health and Disease I: Introduction to Epidemiology (Retrieved 16.12.11).

Nsonga, A., Mfitilodze, W., Samui, K.L., Sikawa, D., 2013. Epidemiology of epizootic ulcerative syndrome in the Zambezi river system. A case study for Zambia. Hum. Vet. Med. Int. J. Bioflux Soc. 5 (1), 1–8.

Ogino, S., Stampfer, M., 2010. Lifestyle factors and microsatellite instability in colorectal cancer: the evolving field of molecular pathological epidemiology. J. Natl. Cancer Inst. 2010 (102), 365–367.

Ogino, S., Chan, A.T., Fuchs, C.S., Giovannucci, E., 2011. Molecular pathological epidemiology of colorectal neoplasia: an emerging transdisciplinary and interdisciplinary field. Gut 2011 (60), 397–411.

Ogino, S., Fuchs, C.S., Giovannucci, E., 2012a. How many molecular subtypes? Implications of the unique tumor principle in personalized medicine. Expert Rev. Mol. Diagn. 2012 (12), 621–628.

Ogino, S., King, E.E., Beck, A.H., Sherman, M.E., Milner, D.A., Giovannucci, E., 2012b. Interdisciplinary education to integrate pathology and epidemiology: towards molecular and population-level health science. Am. J. Epidemiol. 2012 (176), 659–667.

Ogino, S., Lochhead, P., Chan, A.T., Nishihara, R., Cho, E., Wolpin, B.M., Meyerhardt, J.A., Meissner, A., Schernhammer, E.S., Fuchs, C.S., Giovannucci, E., 2013. Molecular pathological epidemiology of epigenetics: emerging integrative science to analyze environment, host, and disease. Mod. Pathol. 2013 (26), 465–484.

Pearce, M., 1990. Epizootic Ulcerative Syndrome Technical Report December 1987–September 1989. Northern Territory Department of Primary Industry and Fisheries. Fisheries Report No. 22. Northern Territory, Australia. 82pp.

Phillips, C.V., Karen, J.G., 2004. The missed lessons of Sir Austin Bradford Hill. Epidemiol. Perspect. Innov. 1 (3), 3.

Reantaso, M.B., Subasinghe, R.P., 2012. Field observations of fish species susceptible to epizootic ulcerative syndrome in the Zambezi River basin in Sesheke District of Zambia. Trop. Anim. Health Prod. 44, 179–183. VA.

Refere, C., 2009. Principles of Epidemiology. In: Key Concepts in Public Health, vol. 2009. Sage, London, UK.

Roberts, R.J., Frerichs, M.G., Miller, S.D., 1990. Studies on the Epizootic Ulcerative Disease in South and South-East Asia. In: Symposium on Diseases in Asian Aquaculture, 26–29 Nov., 1990, Bali (Indonesia). Asian Fisheries Society, Manila, The Philippines, p. 20.

Rodgers, L.J., Burke, J.B., 1981. Seasonal variation in the prevalence of red spot disease in estuarine fish with particular reference to the sea mullet, *Mugil cephalus* L. J. Fish Dis. 4, 297–307.

Rousseau, J.J., 1754. A Dissertation on the Origin and Foundation of the Inequality of Mankind and Is It Authorised by Natural Law. France.

Sharifpour, I., 1997. Histology of the Inflammatory Response of the Carp (*Cyprinus carpio* L.) to Various Stimuli (Ph.D. thesis). University of Stirling, Scotland.

Subasinghe, R.P., Hossain, M.S., 1997. Fish health management in Bangladesh in floodplains. In: Tsai, C., Ali, M.Y. (Eds.), Openwater Fisheries of Bangladesh, vol. 4. The University Press Ltd., Dhaka, pp. 65–74.

Smetanin, P., Kobak, P., 2005a. Interdisciplinary cancer risk management: Canadian life and economic impacts. In: First International Cancer Control Congress.

Smetanin, P., Kobak, P., 2005b. Selected Canadian Life and Economic Forecast Impacts of Lung Cancer.

Smetanin, P., Kobak, P., 2006. A population-based risk management framework for Cancer control (PDF). In: The International Union against Cancer Conference.

Swift, J., 2008. Gulliver's Travels: Part IV. A Voyage to the Country of the Houyhnhnms (Retrieved 03.02.08).

Torres, J.L., Shariff, M., Tajima, K., 1992. Serological relationships among motile *Aeromonas* spp. associated with healthy and epizootic ulcerative syndrome-positive fish. In: Shariff, M., Subasinghe, R.P., Arthur, J.R. (Eds.), Diseases in Asian Aquaculture I. Fish Health Section. Asian Fisheries Society, Manila, Philippines, pp. 451–460.

Wada, S., Rha, S.A., Kondoh, T., Suda, H., Hatai, K., Ishii, H., 1996. Histopathological comparison between ayu and carp artificially infected with *Aphanomyces piscicida*. Fish Pathol. 31, 71–80.

Chapter 5

Pathology of Epizootic Ulcerative Syndrome

5.1 PATHOLOGY OF EPIZOOTIC ULCERATIVE SYNDROME

Epizootic ulcerative syndrome (EUS) is a seasonal epizootic condition of great importance in wild and farmed freshwater and estuarine fishes. It has a complex infectious etiology, and according to some workers, it is clinically characterized by the presence of invasive *Aphanomyces* infection and necrotizing ulcerative lesions, typically leading to a granulomatous response. EUS is also known as RSD, MG, and UM. In 2005, scientists had proposed that EUS be named epizootic granulomatous aphanomycosis or EGA. However, the term EUS is being used by many scientists. The oomycete that causes EUS is known as *Aphanomyces invadans* or *Aphanomyces piscicida*. So far, only one genotype of the EUS oomycete has been recognized, and is said to be the cause of the widespread outbreaks of EUS from the first one in 1971 in Japan to the outbreak in 2008 in Zambia, and after. Further, 24 countries in four continents, viz., North America, Southern Africa, Asia, and Australasia, have recorded occurrences of EUS. In addition, parasites and viruses (notably rhabdoviruses) had also been associated with particular outbreaks, while secondary Gram-negative bacteria are usually associated with lesions due to EUS (FAO, 2009).

5.2 DIAGNOSIS

In order to distinguish EUS fish disease from other ulcerative conditions, correct diagnosis of EUS is a prerequisite. NACA's general findings on the progressive diagnostic symptoms of EUS are summarized below:

1. **Presumptive diagnosis** of EUS could be based on gross appearance (open dermal ulcers) (Bondad-Reantaso et al., 2001) and the observation of aseptate hyphae in squashed preparations of the muscle underlying the gross lesions.
2. **Confirmatory diagnosis** requires histological demonstration of the typical granulomatous inflammation (OIE, 2006) around invasive hyphae or the isolation of *A. invadans* from the underlying muscle (OIE, 2006).

In other words, **positive diagnosis** of EUS is made by portraying the presence of mycotic granulomas in histological sections followed by isolation of *A. invadans* from internal tissues. Putting in a different form, positive diagnosis is made by **analysis of histological sections demonstrating mycotic granulomas and isolation of the causal fungus**.

Diagnosis is based on **clinical signs** and **histological** evidence of the typical aggressive invasiveness of the nonseptate fungal hyphae, within the context of high mortality. In fact, isolation of the fungus allows its characteristic growth profile to be used as an aid to identification.

5.2.1 Differential Diagnosis

The list of similar diseases may be usually available in standard books dealing with fish diseases. However, gross pathological signs may be representative of many diseases. This may not always be helpful and conclusive in leading to the correct diagnosis of an ailment. Thus, such information may not always be used as a tool for a definitive diagnosis, but rather as a tool to help identify the listed diseases in standard guides.

5.2.2 Progressive Diagnostic Symptoms of Epizootic Ulcerative Syndrome

Behavior: (1) reduced appetite; (2) swimming with head out of the water; (3) floating lethargically; and (4) death.

5.3 GENERAL SIGNS OF DISEASE

It may be noted here that animals with disease may show one or more of the signs mentioned below, but the pathogen may still be present even in the absence of any signs. Disease signs at the farm, tank, or pond level generally are:

1. loss of appetite;
2. dark body color;
3. mass mortality;
4. erratic swimming;
5. rubbing on the surfaces of tanks; and
6. increased respiration and respiratory effort.

Epizootic Ulcerative Fish Disease Syndrome. http://dx.doi.org/10.1016/B978-0-12-802504-8.00005-5

5.4 EXTERNAL SIGNS OF EPIZOOTIC ULCERATIVE SYNDROME

1. Appearance of small amber-colored, red, or gray erosions/lesions
2. Expansion of lesions into large ulcers, loss of scales, hemorrhage, and formation of oedema
3. Snakeheads may show severe erosion of the head or body cavity

5.5 GROSS PATHOLOGICAL SIGNS OF EPIZOOTIC ULCERATIVE SYNDROME

1. Lesions on the body: red spots, black burn-like marks, or deeper ulcers with red centers and white rims
2. Progressive lesions: lesions start as reddening under a single scale but quickly spread to involve adjacent scales. They continue to widen and deepen, forming ulcers that erode underlying tissues to expose skeletal musculature, vertebrae, the brain, or visceral organs, cause unilateral or bilateral clouding of the eye and so on.

5.6 CLINICAL SIGNS OF EPIZOOTIC ULCERATIVE SYNDROME

In the early stage of the disease, **red spots or small hemorrhagic lesions** are generally found on the surface of the fish. The spots/lesions then progress to become **ulcers and eventually large necrotic erosions.** Fungal **mycelium** is often visible on the surface of ulcers. **Death** usually follows rapidly as a result of **visceral granulomata,** septicemia, and failure of osmoregulatory balance.

Notwithstanding the above, the pathology in fish collected from the Chobe and Zambezi Rivers in Africa had been described by Andrew et al. (2008). Early lesions had appeared as raised hyperemic foci in the skin several millimeters in diameter. Advanced lesions consisted of large, deep skin ulcers exposing the underlying musculature. The tissues around the ulcers were characterized histologically by a granulomatous inflammatory response surrounding aseptate branched oomycete hyphae. Histological lesions in the musculature typically consisted of necrotic muscle remnants with an infiltration of large number of macrophages and other mixed inflammatory cells. Within these areas, macrophages and fibroblasts formed dense granulomatous sheaths around hyphae. In severe cases, hyphae were found to penetrate internal organs, and could be demonstrated in the kidney and to a lesser extent in the liver and mesentery of affected fish.

5.7 CLINICAL SIGNS AND PATHOLOGY

Studies on the pathology of EUS in Asia tended to focus on the striped snakeheads (*Channa striata*), as this is the species that is by and large said to be the most commonly and severely affected. However, significant differences with other species have been noted. In general, lesions on EUS-affected fishes can be distinguished into three groups based on gross appearance (Viswanath et al., 1997). Clinical signs in the early stages of the disease are said to be quite similar. Appetite is reduced or absent and fish becomes lethargic, either floating just beneath the surface or swimming with the head out of water.

Concomitant to the above, there may be differences in clinical signs and pathology among the affected fishes under different situations. The initial signs could involve mass mortality associated with distinct dermal lesions including ulcers. The surviving fishes may typically have lesions of varying degrees of severity. These may appear as red-spots, blackish burn-like marks, or deeper ulcers with red centers and white rims. Some fish, especially snakeheads may survive for a longer time with such ulcers, which may erode so deeply as to expose the vertebrae, brain, and viscera.

Histological studies may portray necrotizing, granulomatous dermatitis and myositis associated with invasive, nonseptate fungal hyphae, 10–20 μm in diameter. The fungus may penetrate visceral organs, such as the kidney and liver, after it has penetrated the musculature.

5.8 GENERAL SYMPTOMS AND GROSS PATHOLOGY

EUS is characterized by the occurrence of large hemorrhagic or necrotic ulcerative lesions on the bases of the fins and other parts of the body that becomes larger inflamed areas with acute degeneration of epidermal tissues (Kar and Dey, 1988, 1990a,b,c).

Initially, pinhead-sized red spots usually develop on the body surface, head and fins, caudal peduncle, dorsum, or operculum with no noticeable hemorrhages or ulcers. In the early stages, these may simply be areas of acute dermatitis forming rosacea.

Intermediate stage Lesions are represented by small (2–4 cm) dermal ulcers, with associated loss of scales, hemorrhage, and oedema. Roberts et al. (1989) had noted that in *Puntius* spp., gouramies, and other mid-water fishes, ulcers are particularly dark and usually circular; often only one large superficial lesion occurs on the flank or dorsum. Most species other than snakeheads and mullets may die at this stage.

Advanced stage Lesions appear on other parts of the fish's body and expand into large necrotic open ulcers, resulting eventually in death. Some affected species, for example, stripped snakeheads, may survive with much more severe chronic lesions that may have completely destroyed the caudal peduncle or eroded deep into the cranium or abdominal cavity, sometimes even exposing the swim bladder. Erosion of the cephalic tissue is a particularly common feature of the diseased stripped snakeheads, and specimens had been found with exposed optic nerves or loosened articular bones such as maxillae and mandibles. However, diseased stripped snakeheads, with moderately advanced lesions, placed in improved water-quality conditions, may often recover. Similarly, lesions on estuarine fish such as mullets appear to heal quickly when the fish move into brackish or marine environments. Healing ulcers are characterized by a conspicuous dark color caused by increased number of melanophores (Lilley et al., 1992, 1998).

Notwithstanding the above, in certain countries, notably in Pakistan, EUS-infected fishes had displayed severe ulcers on the skin, but the distribution of ulcers had shown variations in different species. The ulcerations were more pronounced and found to be more severe on the head and caudal regions in ichthyospecies like *Channa marulius, Channa punctata, Wallago attu, Sperata seenghala, Heteropneustes fossilis,* and *Puntius* spp. In some cases, the ulcers had been deep enough to expose the cranium, thereby resulting in grayish or red necrotic areas. In extreme cases, the caudal lesions had eroded the affected areas to such an extent that there had been total degeneration of the caudal peduncle portion, and in very severe cases, the degeneration even had extended up to posterior part of the abdominal cavity. However, hemorrhagic ulcerations were pronounced on the lateral sides of the body, the tail, and the head region in fishes such as *C. marulius, W. attu,* and *S. seenghala.* Nevertheless, *Catla catla* had depicted ulcerations only in the head region. On the other hand, *Labeo rohita* had displayed less intense ulcers on the body.

5.9 GROSS PATHOLOGY AND LIGHT MICROSCOPY

Egusa (1992) had summarized the clinical signs of MG based on numerous published reports from Japanese studies from naturally infected fishes (Egusa and Masuda, 1971; Miyazaki and Egusa, 1972, 1973a,b,c; Kumamaru, 1973; Hatai et al., 1977). These include localized swellings on the body surface, protruding scales, hemorrhage, scale loss, skin disintegration, exposure of underlying musculature and ulceration. Ulcers spread over a broad area and develop into a wide ulcer with exposed scarlet granuloma extending from several mm below the skin. A nonseptate, branching form of fungal hyphae with a diameter of 10—25 μm and fungal granuloma was easily and consistently observed from wet mount slide of the tissue under light microscopy.

More detailed pathology and the suggestion of a fungal etiology were subsequently reported from naturally infected cultured goldfish (Miyazaki and Egusa, 1972) and ayu

(Miyazaki and Egusa, 1973a). In goldfish, details of morphological changes of granuloma progression were described in tissues (e.g., in gills, skin, skeletal muscle, brain, and spinal cord) and internal organs (e.g., liver, spleen, pancreas, kidney, small intestine, gonads, and mesenteries). Observations on multinucleate giant cells at the lesion were also seen and varied depending on species affected and there appeared to be some geographic differences. The multinucleated cells were classified as two types: those that engulfed the fungal hyphae, which were larger in size with a maximum diameter of 90 μm and containing about 47 nuclei in the plane of the observed section; and those that did not. Fish from Oita, Miyazaki, and Tokyo areas exhibited multinucleate giant cells that engulfed hyphae (80—90% prevalence), while fish from Tokushima, Shiga, Nagano, and Tochigi Prefectures exhibited multinucleate giant cells without hyphae (80—100% prevalence).

5.10 CLINICAL PATHOLOGY

The initial features of an EUS outbreak in a new location are dramatic, whether in wild fishes in rivers, estuaries, lakes, and rice fields (where heavy losses often occur), or in farmed fishes (Chinabut and Roberts, 1999). A significant proportion of the fish stocks had been found to be floating on the surface: most of them already dead, with large red or gray shallow ulcers, often with a brown necrotic, fungus-covered center. Generally, the ulcers are on the side, though they can occur on any part of the head and body. There is often remarkable symmetry of lesions, with the ulcer located in the same position on each fish of a particular species. This effect is very much noticeable among major and minor carps and among the barbs such as *Puntius* spp., all of which are highly susceptible to the disease. EUS is seen in both wild and farmed fishes. Often such lesions appear as brand marks or burn marks.

Some fish species such as the snakeheads (Channidae), although highly susceptible, nevertheless, take much longer time to succumb. Their lesions are therefore more extensive and could lead to complete erosion of the posterior portion of the body, or to necrosis of both soft and hard tissues of the cranium, so that the brain is exposed in the living animal.

When an outbreak occurs, although dead and dying fish will appear to have similar, extensive lesions, by examining fish from neighboring ponds or rice fields, it could be possible to build up a clinical picture of the stages in the pathogenesis of the disease.

The first signs of affected fish are that they become darker and do not feed. They may float just below the surface, or in some species, with the head just breaking the surface. Sometimes, they may be hyperactive with a very jerky pattern of movement.

However, the initial lesions could vary. In IMCs and *Puntius* spp., which in some regions appear particularly susceptible, a small area of darkened skin may be seen before they become moribund and die. In other fishes, there may be one or more reddish or amber-colored areas of acute dermatitis, which gradually ulcerate to produce small (2–3 mm) ulcers that often displace a scale. Subsequently, these enlarge and develop a gray necrotic center, with a white rim of exposed dermal collagen. Around this is usually a wide zone of darkly pigmented skin. In the center of such larger lesions, there will generally be necrotic adherent mat, consisting of fungal hyphae of secondary species such as *Achlya*, hemorrhagic tissue and saprophytic protozoa such as *Epistylis* sp., and certain opportunistic bacteria.

Notwithstanding the above, our studies have revealed the following with regard to initiation of EUS:

EUS in carps begins with a small tubercle-like swelling in which bacteria and fungi could not be detected and isolated. Gradually, this tubercle grew to a tuber-like size and the fungi could be isolated. This faintly indicates that bacteria and fungi are not the primary pathogens for the initiation of EUS.

There could be further information on progressive diagnostic symptoms of EUS in India (Viswanath et al., 1997).

Type I: Early lesions
Pinhead-sized reddish or amber-colored spots on the body surface. No noticeable hemorrhage or ulceration. Skin around the spots is normal with no discoloration. Sections show focal inflammatory changes. There are several nodular structures associated with fungal hyphae. Dermis and skeletal muscle are normal, without evidence of fungal invasion.

Type II: Moderately advanced lesions
Moderately advanced lesions are raised, circular, and discolored areas on the body surface that are approximately 2–4 cm in diameter. These areas are soft with relatively intact skin and scales. In sections, mycotic granulomas are seen in epidermis, dermis, and skeletal musculature, associated with many nonseptate fungal hyphae. Significant necrotizing dermatitis and myositis due to fungal invasion are also seen. In most cases of these lesions, however, the scales and epidermis are not completely lost.

Type III: Advanced lesions
Advanced lesions are circular or oval, open dermal ulcers extending to the skeletal musculature. They are characterized by large hemorrhagic and necrotic open ulcers on the body surface, devoid of epidermis and scales, with loss of dermis at the site of the ulcer. In most cases, the underlying musculature is exposed and largely replaced by fungal granulomas and host inflammatory tissue. Fungal hyphae extend to all directions from the focus of the dermal ulcer. Necrotic muscle fibers and fungal hyphae often found within granulomas.

Notwithstanding the above, EUS has been described in more than 30 species of FW fishes. Also, some species of brackish water fishes have been affected by EUS. As might be expected, there is great variation in the clinical picture between the different species, and thus also in the underlying pathology.

Concomitant to the above, some workers (Chinabut and Roberts, 1999) had conveniently attempted to classify the clinical pattern based on their histopathology (HP) into five types as follows:

Type (a)
Peracute infection with minimal host response or secondary infection and rapid death

Type (b)
Acute infection with some amount of host response and often a secondary infection before death

Type (c)
Chronic slowly progressive infection with considerable amount of host response; the outcome is generally fatal but invariably with secondary bacterial and/or fungal involvement

Type (d)
Chronic infection with host response sufficient to allow recovery unless there is secondary infection

Type (e)
Superficial response to adventitious infection in fish species that are generally resistant to infection. Unless there is a significant amount of secondary infection, such fish would almost invariably recover.

The above-mentioned classifications of clinical picture could only be considered general in their application. Specific circumstances may affect the outcome, but in general, the response of a particular species and age group could be consistent. This may allow a clear clinical picture for an outbreak to be defined and the basic diagnosis to be made based on the specific histopathology that underlies it.

5.11 SPECIES AFFECTED

It is believed that more than 100 species of fishes have been affected by EUS (Lilley et al., 1992), but only relatively few reports have been confirmed by demonstrating the presence of mycotic granulomas in histological section or by isolation of the pathogenic *Aphanomyces* fungus from tissues underlying ulcers.

In addition to the above, >50 species of both farmed and wild fishes (freshwater and estuarine), are said to be susceptible to EUS. These include, for example, the barbs, the breams, the catfishes, the cichlids, and the churchills in Africa; the eels, the gobies, the gouramies, the IMCs (catia, mrigal, rohu), the mullets, the perches, the sea basses, the sea breams, and the snakeheads in Asia; ayu in Asia, but said to be particularly in Japan; menhaden in the United States of America; the tilapias, the perches in Africa, and so on.

However, ulcerative skin lesions are generally common in freshwater and estuarine fishes.

As reported, EUS had affected some nonculturable but economically important species, notably *C. punctata, C. marulius, W. attu, S. seenghala,* etc. The culturable affected fishes (mainly cyprinids) had included *C. catla, Cirrhinus mrigala,* etc. The Chinese carps (notably, the silver carp and the grass carp) had been refractory to EUS, as usual. However, interestingly, *L. rohita* had been found to be intermediate between susceptible and nonsusceptible (i.e., could be mildly affected) although its IMC brethren had been affected by EUS as mentioned above. As further reported, the presence of *A. invadans* had been confirmed in *L. rohita.*

Notwithstanding the above, Assam is the province where EUS made its initiation in India. The present author of this treatise, Professor D. Kar (Kar and Dey, 1988) began his study of the ichthyological survey of the lentic water bodies in the Barak Valley region of Assam (24°10′ N, 93°15′ E) with the identification of the four species of fishes most widely affected by EUS during July 1988. These were:

1. *Macrognathus aral,* which usually portrays lesions on the skin, in the abdomen and occasionally in the caudal region
2. *Channa punctata,* which generally shows lesions on the skin, at the base of the caudal peduncle and sometimes show reddish swollen eyes (exophthalmos)
3. *Mystus vittatus,* which usually depicts amber-colored lesions at the base of the rayed dorsal fin and sometimes at the base of the caudal fin
4. *Puntius conchonius,* which generally exhibits amber-colored lesions at the base of the dorsal and caudal fins and sometimes in other parts of the body

Other species affected (Kar and Dey, 1990a,b,c) by the present epidemic, but not widely, include the following: *Mastacembelus armatus, Macrognathus pancalus, Salmophasia bacaila, Gudusia chapra, Badis badis, Glossogobius giuris, Ailia coila, Lepidocephalichthys guntea, Clarias batrachus, C. mrigala, Cirrhinus rebZFa, Parambassis ranga,* and others. *W. attu* and *Tenualosa ilisha* have not been found to be widely affected by the prevailing EUS epidemics. Among the IMC during

July 1988, barring a few *C. catla* and *C. mrigala,* none were found to be affected by EUS. Nevertheless, *Amblypharyngodon mola, Ctenopharyngodon idella, Rasbora daniconius, Puntius ticto,* and *C. marulius* have been found to be affected only in certain areas of the district.

More information on EUS revealed from the continued study of Dr Kar from 1988 to 2015 is presented in Chapter 6 of this treatise.

Studies conducted by workers in other countries, notably Pakistan (FRTI, Lahore) and missions by foreign experts, revealed that EUS had affected both wild and farmed fishes. *Puntius* spp. and the *Channa* spp. were the most affected species in the wild, whereas IMCs were possibly the most affected species in the aquaculture farms. It had also been reported that EUS was first recorded in the lentic water bodies followed by its record in the lotic water bodies with *Channa* spp. being the most affected ones.

Concomitant to the above, Africa has a unique aquatic biodiversity, which in many areas forms the basis of important subsistence fisheries. The Zambezi River supports at least 134 species of freshwater fishes (Skelton, 1993). Serious regional biosecurity shortcomings were identified in Botswana and Namibia at the time of the EUS outbreak during 2006 (Andrew et al., 2008; Anon, 2009) and a number of urgent recommendations were made (Anon, 2009). Subsequently, however, in 2010, the disease was reported from the pristine wilderness area of the Okavango Delta in Botswana (Anon, 2011) and in February 2011 from South Africa (Anon, 2011).

Notwithstanding the above, an outbreak of EUS in Africa during October 2011 caused serious mortality among the sharptooth catfish (*Clarias gariepinus*) in a farm on the Eerste River near Stellenbosch in the Western Cape (Anon, 2011; Van Helden and Grewar, 2011). The Western Cape Province was known to be home to 11 endangered endemic species of freshwater fishes (Skelton, 1993; Skelton et al., 1995), with small populations in a few rivers. It was not known how the disease was introduced into South Africa.

Concomitant to the above, studies further conducted in Africa revealed that a wide range of freshwater and brackish-water wild and cultured fishes, involving >100 species, had been affected by EUS. Snakeheads were by far the most seriously affected species, even though species like *Puntius* spp., catfishes (*H. fossilis, C. batrachus*), IMCs (*C. catla, C. mrigala, L. rohita*), climbing perch (*Anabas testudineus*), mullet (*Mugil cephalus*), gobies (*Glossogobius giurus, Oxyeleotris marmoratus*), spiny eel (*Mastacembalus armatus*), swamp eel (*Fluta alba*), and gouramis (*Trichogaster pectoralis*) were also among those seriously affected. Among the snakeheads, *Channa striatus* was the most severely affected, while *C. marulius* and *C. punctata* were affected to a slightly lesser extent (Scholin et al., 1994). A few commercially important freshwater and brackish water species including milkfish (*Chanos chanos*), tilapias, and Chinese carps were found to be consistently resistant to EUS (Dykstra et al., 1986).

From the list of susceptible fish species (Anon, 2010), it was clear that *A. invadans* had shown little host specificity. Broad species susceptibility had been reported from Australia (Anon, 2008), the United States (Sosa et al., 2007), and elsewhere (Anon, 2010, 2011). The disease had spread by spores released into the water. These might have been disseminated through water, contaminated equipment, and transport of live and dead fish.

In this connection, it may be noted here that the presence of lesions may often indicate contaminated or stressed aquatic environments and may be associated with a variety of infections including parasites, bacteria, viruses, and fungi, as well as noninfectious causes such as the toxic algae.

5.12 LESS SUSCEPTIBLE SPECIES

Notwithstanding the above, a few fish species had been known to be less susceptible to EUS. Both the common carp (*Cyprinus carpio*) and Nile tilapia (*Oreochromis niloticus*) had been believed to be resistant (Anon, 2010). At the expense of reiteration, it may be mentioned here that *W. attu* and *T. ilisha* have not been found to be widely affected by the prevailing EUS epidemics in India. Moreover, the role of *O. niloticus* in the spread of EUS in Africa had not been very well known. However, one outbreak of EUS had been noted in a tributary of the Kabompo River, which is said to be a part of the upper Zambezi system in western Zambia, just downstream from a farm holding *O. niloticus*. Other indigenous tilapia species were, however, believed to be susceptible (Anon, 2010). Both Nile tilapia and carps are known to be generally and actively farmed in many parts of the world, including the southern regions of Namibia. Although neither species was actively farmed along the upper Zambezi and Chobe Rivers, they had been introduced into the Kafue River in Zambia and elsewhere in Southern Africa, and the Nile tilapia formed the basis of a major aquaculture fishery in Lake Kariba. Fishes like the common carp (*Cyprinus capio*), Nile tilapia (*O. niloticus*), and milkfish (*C. chanos*) are believed to be resistant to EUS.

5.13 SPECIES IN RELATION TO EUS

Some of the commercially significant species are considered to be particularly resistant to EUS. Species reported to be unaffected by EUS outbreaks include the Chinese carps, tilapias, and milkfish (*C. chanos*). But few studies had been undertaken to confirm these observations, as well as to

investigate the mechanism of resistance. Further, Hatai (1994) had injected catfish (*Parasilurus asotus*), loach (*Misgurnus anguillicausatus*), and eel (*Anguilla japonica*) experimentally with hyphae of *A. invadans* and found them to be refractory to infection. Further, Wada et al. (1996) and Shariffpour (1997) had experimentally injected common carp (*C. carpio*) with zoospores of *Aphanomyces* from MG and EUS outbreaks respectively. They had demonstrated that fungal growth was suppressed by an intense inflammatory response.

Some authors have commented that the most severely affected species in natural outbreaks are generally bottom dwellers (Llobrera and Gacutan, 1987; Chondar and Rao, 1996) or the fishes that possess air-breathing organs (Roberts et al., 1994), but this may not be always true. Further, in the case of snakeheads, no particular size group appears more susceptible than others do, with affected fish ranging from 40 to 900 g (Cruz-Lacierda and Shariff, 1995). However, there is a possibility that size or age may be significant in other species. For example, IMC suffer high mortalities as fingerlings (Roberts et al., 1989) but larger fish, although appearing ulcerated, were not reported as dying in large number (AAHRI, ACIAR, IoA, and NACA, 1997).

Some of the EUS-susceptible species may have a wide geographical distribution, beyond the current limits of EUS outbreaks. For example, several snakehead and clariid catfish species occur in Africa as well as in Central Asia. This suggests that there is potential for further spread of the disease to these areas. However, it may be noted here that optimal temperatures for vegetative growth in vitro for *A. invadans* are in the range 20–30 °C (Fraser et al., 1992; Lilley and Roberts, 1997). Probably, for this reason, natural outbreaks, to date, have been limited to latitudes between 35° N and 35° S. Experimental injection challenges of native European and American fish species have shown that the pathogenic fungus *A. invadans* is capable of causing lesions in rainbow trout at 18 °C, but is less infective in stickleback (*Gasterosteus aculeatus*) and roach (*Rutilus rutilus*) at 11–16 °C (Khan et al., 1998).

5.14 HOST FACTORS

5.14.1 Susceptible Host Species

EUS causes disease and mortality among both farmed and wild fishes, almost worldwide. Around 76 species of fishes have been confirmed by histological diagnosis to be naturally affected by EUS. Suspect cases of natural infection with *A. invadans*, in species other than those known, may immediately be referred to the appropriate OIE Reference Laboratory, whether or not clinical signs are associated with the findings.

5.14.2 Susceptible Stages of the Host

The susceptible life stages of the fishes are usually juvenile and young adults. There had perhaps been no report of EUS having attacked any fish fry or fish larvae.

Notwithstanding the above, an experimental injection of *A. invadans* into the yearling life stage of the IMC catla, rohu, and mrigal revealed resistance to EUS, even though they are naturally susceptible species. Experimental infections demonstrated that goldfish are susceptible, but common carp and Nile tilapia are said to be resistant.

5.14.3 Species or Subpopulation Predilection (Probability of Detection)

EUS could be readily detected in diseased fish specimens collected from EUS-affected areas using histological techniques. However, *Aphanomyces* could be isolated only from fish with mild or moderate clinical signs of EUS, exhibiting red spots or small ulcers.

5.14.4 Target Organs and Infected Tissue

The motile zoospore plays an important role in the spread of the disease. Once the motile spore attaches to the skin of the fish, the spore will germinate under suitable conditions and its hyphae will invade the fish skin, muscular tissue and reach the internal organs. Fish skeletal muscle is the target organ and exhibits major EUS clinical signs with mycotic granulomas.

5.14.5 Host Range

More than 100 estuarine and freshwater species of fishes are known to be susceptible to EUS. Concomitant to the EUS-affected fish species in Asia, etc., mentioned above, known susceptible species likely to be encountered in Australia are the Australian bass (*Macquaria novemaculeata*).

5.15 PATHOGENS INVOLVED

As reported, dead and live fishes had been examined by different workers in order to ascertain the presence of pathogens, parasites, etc., along with the fungus *A. invadans*. Parasites like *Gyrodacrylus, Epistylis, Trichodina, Costia, Lernea,* and *Ichthyophthirius* had been reported from the EUS-affected fishes. In addition, Anon (1998) had recorded ectoparasites like *Lernea* sp., *Costia* sp., and *Trichodina* sp., in EUS-affected fishes.

5.15.1 Disease Agent

According to a section of scientists, EUS is caused by infection with the oomycete *A. invadans*. Although previously regarded as a fungus, the genus *Aphanomyces* is now classified with the diatoms and brown algae in a group called Stramenopiles or Chromista.

5.16 POSSIBLE CAUSES (ETIOLOGY)

A number of predisposing "stress factors" leading to initiation of EUS-infection had been identified (Baldock et al., 2005; Choonga et al., 2009). In Asia, notably in India, the sudden abrupt fall in total alkalinity (TA) of water had been found to act as a predisposing stress factor for the initiation of EUS outbreak (Kar, 2007, 2010, 2013; Kar and Dey, 1990a,b,c; Kar et al., 1993, 1994, 1995, 2000). In other locations, the temperature could act as predisposing stress factor. When the water temperature drops, EUS disease has often manifested (Baldock et al., 2005), possibly as a consequence of a retarded immune response in the fish. Low soil and water pH had been said to increase susceptibility of the fish to infection with *A. invadans* (Baldock et al., 2005). The low soil pH might be because they are naturally acidic, or disturbed by agricultural or residential development. Either might lead to a lowering of soil and water pH. Heavy rainfall might sometimes also accelerate the process (Baldock et al., 2005; Choonga et al., 2009). This might have also partly explained the relatively high seasonal prevalence of infection in fish in the floodplains.

Concomitant to the above, severe floods in the upper Zambezi river had followed several years of low rainfall during 2007. Changes in water quality, resulting from flooding of previously dry land, had been suspected to have played a role in increasing susceptibility of fish to infection by *A. invadans*. The natural acidification of groundwater following periods of drought and subsequent contamination of surface waters of floodplains was identified as one of the main predisposing factors for EUS outbreaks in Zambia (Choonga et al., 2009). Low water pH might have played a similar role in the Okavango Delta outbreaks, where seasonal flooding was considered an integral driver of the delta ecosystem. Other causes of skin trauma, providing a portal of entry for infective zoospores, could not be discounted in an aquatic environment where predation is believed to be common.

The role of the pathogenic fungi, *A. invadans* had been found to be largely acceptable in the spread of EUS (Hargis, 1985), However, according to some quarters, the agent, by itself, had not been hitherto shown to induce the disease. Hence, the role of different agents, including infectious and noninfectious, had been implicated in the onset of the disease. Further, bacterial pathogens had been consistently found associated with EUS-infected fishes (Noga et al., 1996; Burkholder and Glasgow, 1997; Friedland et al., 1996). Experimental infections using *Aeromonas hydrophila*, isolated from EUS-infected fishes had been said to induce EUS-like lesions in fish (Noga et al., 1988). A further report was said to have suggested possible entry of the zoospores of *A. invadans* through intact skin of Atlantic menhaden (*Brevoortia tyrannus*). However, the study had indicated that the entry could have been through the epidermal cells.

Notwithstanding the above, it had remained unclear how EUS was introduced into the Southern African region and how it had spread between the catchments. It had been suggested that *A. invadans* might have spread locally by flood events, but that regional and international spreads might have occurred through movements of fish for aquaculture or ornamental purposes (Lilley et al., 1997). A number of tropical ornamental fish are known to be susceptible to EUS (Anon, 2010), as is the goldfish (*Carassius auratus*) (Miyazaki and Egusa, 1972). Goldfish, as well as species of certain other tropical fish had been farmed in the warmer regions of Southern Africa, often in open water systems close to natural water sources containing wild fish. The Okavango Delta and Chobe and upper Zambezi Rivers are popular resorts of sport fishing. Both sport and subsistence fishing might be associated with movements of potentially infected fishes and equipment.

Environmental factors such as high rainfall, poor water quality, and low pH in the induction of EUS have been documented (Brazer et al., 1999). Low salinity and sudden rainfall were implicated in EUS induction in mullets in estuaries. On the other hand, it had been found that an increase in salinity increases healing of ulcers. Diseased striped snakeheads, with moderately advanced lesions, placed in improved water quality conditions, may often recover, as had been possibly indicated earlier. Similarly, lesions on estuarine fish such as mullet appear to heal quickly when fish move into brackish or marine environments. Healing ulcers are characterized by a conspicuous dark color caused by increased number of melanophores.

5.17 HOW EPIZOOTIC ULCERATIVE SYNDROME IS SAID TO SPREAD

Spread of *A. invadans* is via zoospores in the water. Secondary zoospores enter the skin following a breach of the epidermal barrier by physical or environmental causes (Baldock et al., 2005; Lilley and Roberts, 1997). The zoospores germinate and the hyphae invade the skin and musculature, causing a focal necrotizing granulomatous dermatitis and myositis. Hyphae may invade more deeply, causing granulomatous inflammation in the internal organs (Baldock et al., 2005). The tissue necrosis in early lesions is associated with an intense inflammatory reaction that is observed as a characteristic focal reddening and swelling of the skin; hence the name "red spot disease" used in some countries (Anon, 2008). More advanced lesions are characterized by large and deep ulcers surrounded by a raised rim of inflamed tissue. EUS's full descriptive name *epizootic ulcerative syndrome* has become the internationally accepted term for infection with *A. invadans*.

5.17.1 Transmission Mechanisms

Notwithstanding the above, EUS is said to be transmitted horizontally. The *Aphanomyces* zoospores may be transmitted horizontally, from one fish to another, through the water. It is believed that only the zoospores are capable of attaching to the damaged skin of fish, and then germinate into hyphae. If the zoospores cannot find the susceptible species, or if they encounter unfavorable conditions, they may form secondary zoospores. The secondary zoospores may encyst in the aquatic environment and may wait for conditions that favor the activation of the spores. How the *Aphanomyces* pathogen or its spores survive after the outbreak is still unclear as outbreaks usually occur about the same time every year in the endemic areas (OIE, 1997).

Anon (1992) had held "Consultation on Epizootic Ulcerative Syndrome vis-à-vis the environment and the people" under the aegis of International Collective in Support of Fishworkers. It had been opined that the spread of EUS shows a certain pattern. Outbreaks are typically cyclical, with the first occurrence being particularly severe and recurrences over the next two to three years less so. There is, however, no uniformity to this pattern. While the disease spread rapidly in some areas like Malaysia and Thailand, in other areas like Indonesia its progression was slow. Moreover, in Malaysia there was a one-year gap between outbreaks.

The mechanism of spread is also not clear. The disease had been said to spread rapidly northward, where the rivers flow from east to west, and equally rapidly westward in areas where the rivers are oriented from north to south. It would thus not be possible to attribute the transport of the pathogens to, say, monsoonal floods alone. Also mysterious is the spread of EUS to areas like Sri Lanka and some islands of the Philippines. The unrestricted trade in live fish could be a mode of transmission.

5.18 PERSISTENT INFECTION WITH LIFELONG CARRIERS

There is no information to indicate that fish could be lifelong carriers of *A. invadans*. Generally, most of the infected fishes die during an outbreak. Some mild or moderately EUS-infected fishes could recover, but they are unlikely to be lifelong carriers.

Vectors: Not much information is perhaps available about the possible vectors of EUS.

Known or suspected wild aquatic animal carriers: Not much information is perhaps available about the known or possible wild aquatic animal carriers of EUS.

5.18.1 Prevalence

The prevalence of EUS in the wild and in aquaculture farms is high in the endemic areas that is said to share the same water way or system. Uncontrolled water exchange in fish farms in endemic areas could result in EUS outbreaks in most of the farms that culture susceptible fish species.

Notwithstanding the above, the following are some other works on EUS pathology:

Anon (2009) described EUS in fishes covering various aspects such as scope, disease information, agent factors, etiological agent, agent strains, stability of the agent, life cycle, host factors, and so on.
Pradhan et al. (2014) reported about the emergence of EUS and large-scale mortalities of cultured and wild fish species in Uttar Pradesh, India.

HP analysis was carried out according to the method described by Chinabut and Egusa. The tissue sections were stained with hematoxylin and eosin (H&E) for general pathological investigation, and with Grocott's methenamine silver nitrate for the presence of *A. invadans* hyphae. **Gross pathology** of the affected fishes varied from red hemorrhagic spots and scale losses to severe ulceration and exposure of underlying tissues.

5.19 SIMILAR DISEASES

Aeromonas salmonicida: atypical strains, koi herpesvirus disease, viral hemorrhagic septicemia. Diagnosis is based on clinical signs and histopathology. No specific diagnostic tests are currently available.

The fungus may be isolated and cultured without difficulty, provided measures are taken to exclude bacterial and other fungal contaminants. Besides their typical vulnerability to temperature above 30 °C, *A invadans* and *A. piscicida* are said to be very similar to nonpathogenic opportunistic *Aphanomyces* spp. that readily contaminate the surface of affected fishes and often interfere with isolation attempts. In the laboratory, the fungus had also been shown to be pathogenic to a wide range of fishes, inducing similar pathology and mortality under various predisposing experimental conditions.

REFERENCES

Anon, 1992. Consultation on Epizootic Ulcerative Syndrome *vis-à-vis* the Environment and the People, International Collective in Support of Fishworkers, 25–26 May 1992. Institute of Management in Government, Vikas Bhavan, Trivandrum 695 033, Kerala (India).

AAHRI, ACIAR, IoA, and NACA, 1997. Epizootic Ulcerative Syndrome (EUS) of Fishes in Pakistan. A Report of the Findings of an ACIAR/DFID-Funded Mission to Pakistan. 9–19 March 1997.

Anon, 1998. Epizooic Ulcerative Syndrome (EUS) in Fishes of Pakistan: A Report Regarding 'visit of Mission to Punjab in Connection with Epizootic Ulcerative Syndrome (EUS), 22–25 April, 1998.

Anon, 2008. Off. Int. Epizoot. Dis. Info. Bluetongue, Spain, pp. 21, 3.

Andrew, T.G., Huchzermeyer, K.D., Mbeha, B.C., Nengu, S.M., 2008. Epizootic ulcerative syndrome affecting fish in the Zambezi river system in southern Africa. Vet. Rec. 163, 629–631.

Anon, 2009. Report of the International Emergency Disease Investigation Task Force on a Serious Finfish Disease in Southern Africa, 18–26 May 2007. Food and Agriculture Organization of the United Nations, Rome.

Anon, 2010. Manual of Diagnostic Tests for Aquatic Animals. World Organisation for Animal Health, Paris.

Anon, 2011. World Animal Health Information Database Interface: Exceptional Epidemiological Events. World Organisation for Animal Health, Paris viewed in 2011.

Burkholder, J.M., Glasgow Jr., H.B., 1997. *Pfiesteria piscicida* and other *Pfiesteria*-like dinoflagellates: behavior, impacts, and environmental controls. Limnol. Oceanogr. 42, 1052–1075.

Blazer, V.S., Vogelbein, W.K., Densmore, C., Zwerner, D.E., May, E.B., 1999. *Aphanomyces* as a cause of ulcerative skin lesions in menhaden from Chesapeake Bay. J. Aquat. Anim. Health 11, 340–349.

Bondad-Reantaso, M.G., McGladdery, S., East, I., Subasinghe, R.P. (Eds.), 2001. Asia Diagnostic Guide to Aquatic Animal Diseases. FAO Fisheries Technical Paper No. 402. Supplement 2. FAO, Rome, p. 240.

Baldock, F.C., Blazer, V., Callinan, R., Hatai, K., Karunasagar, I., 2005. Outcome of a short expert consultation on epizootic ulcerative syndrome (EUS): re-examination of causal factors, case definition and nomenclature. In: Walker, P., Laster, R., Bondad-Reantaso, M.G. (Eds.), Disease in Asian Aquaculture V. Fish Health Section. Asian Fisheries Society, Manila, Phillipines, pp. 555–585.

Cruz-Lacierda, E.R., Shariff, M., 1995. Experimental transmission of epizootic ulcerative syndrome (EUS) in snakehead, *Ophicephalus striatus*. In: Shariff, M., Arthur, J.R., Subasinghe, R.P. (Eds.), Diseases in Asian Aquaculture II. Fish Health Section, Asian Fisheries Society, Manila. Philippines, pp. 327–336.

Chondar, S.L., Rao, P.S., 1996. Epizootic Ulcerative Syndrome Disease to Fish and its Control: A Review. World Aquaculture 1996, Book of Abstracts. Bangkok, 29 January–2 February, 1996, p. 77. World Aquaculture Society.

Chinabut, S., Roberts, R.J., 1999. Pathology and Histopathology of Epizootic Ulcerative Syndrome (EUS). Aquatic Animal Health Research Institute, Department of Fisheries, Royal Thai Government, Bangkok, Thailand, ISBN 974-7604-55-8, 33 p.

Choonga, K., Hang'ombe, B., Samui, K.L., Syachaba, M., Phiri, H., Maguswi, C., et al., 2009. Environmental and climatic factors associated with epizootic ulcerative syndrome (EUS) in fish from the Zambezi floodplains, Zambia. Bull. Environ. Contam. Toxicol. 83, 474–478.

Dykstra, M.J., Noga, E.J., Levine, J.F., Moye, D.W., Hawkins, J.H., 1986. Characterization of the *Aphanomyces* species involved in ulcera-tive mycosis (UM) in menhaden. Mycologia 78, 664–672.

Egusa, S., Masuda, N., 1971. A new fungal disease of *Plecoglossus altivelis*. Fish. Pathol. 6, 41–46.

Egusa, S., 1992. Mycotic granulomatosis. In: Infectious Diseases of Fish. A.A. Balkema, Rotterdam, 392–336.

Friedland, K.D., Ahrenholz, D.W., Guthrie, J.F., 1996. Formation and seasonal evolution of Atlantic menhaden juvenile nurseries in coastal estuaries. Estuaries 19, 105–114.

FAO, 2009. What You Need to Know about Epizootic Ulcerative Syndrome (EUS) – an Extension Brochure. Food and Agricultural Organisation of the UN, Rome, pp. 1–36.

Fraser, G.C., Callinan, R.B., Calder, L.M., 1992. *Aphanomyces* species associated with red spot disease: an ulcerative disease of estuarine fish of Eastern Australia. J. Fish Dis. 15, 173–181.

Hatai, K., Egusa, S., Takahashi, S., Ooe, K., 1977. Study on the pathogenic fungus of mycotic granulomatosis—I. Isolation and pathogenicity of the fungus from cultured-ayu infected with the disease. Fish. Pathol. 12 (11), 129–133.

Hargis Jr., W.J., 1985. Quantitative effects of marine diseases on fish and shellfish populations. Trans. N. Am. Wildl. Nat. Resour. Conf. 50, 608–640.

Hatai, K., 1994. Mycotic granulomatosis in ayu (*Plecoglossus altivelis*) due to *Aphanomyces piscicida*. In: Roberts, R.J., Campbell, B., MacRae, I.H. (Eds.), Proc. ODA Regional Sem. on Epizootic Ulcerative Syndrome, 25–27 January 1994. Aquatic Animal Health Research Institute, Bangkok.

Kumamaru, A., 1973. A fungal disease of fishes in lake Kasumigiura and lake Kitaura. Rep. Stan. Ibaraki Prefect. 11, 129–142 (In Japanese).

Kar, D., Dey, S.C., 1988. A critical account of the recent fish disease in the Barak valley of Assam. Proc. Reg. Symp. Recent Outbreak Fish Dis. North East India 1, 8.

Kar, D., Dey, S.C., 1990a. Fish disease syndrome: a preliminary study from Assam. Bangladesh J. Zool. 18, 115–118.

Kar, D., Dey, S.C., 1990b. A preliminary study of diseased fishes from Cachar district of Assam. Matsya 15–16, 155–161.

Kar, D., Dey, S.C., 1990c. Epizootic ulcerative syndrome in fishes of Assam. J. Assam. Sci. Soc. 32 (2), 29–31.

Kar, D., Dey, S.C., Kar, S., Bhattacharjee, N., Roy, A., 1993. Virus-like particles in epizootic ulcerative syndrome of fish. Proc. Int. Symp. Virus Cell Interact. Cell. Mol. Responses 1, 34.

Kar, D., Dey, S.C., Kar, S., Roy, A., Michael, R.G., Bhattacharjee, S., Changkija, S., 1994. A candidate virus in epizootic ulcerative syndrome of fish. Proc. Natl. Symp. Indian Virol. Soc. 1, 27.

Kar, D., Roy, A., Dey, S.C., Menon, A.G.K., Kar, S., 1995. Epizootic ulcerative syndrome in fishes of India. World Congr. In Vitro Biol. In Vitro 31 (3), 7.

Khan, M.H., Marshall, L., Thompson, K.D., Campbell, R.E., Lilley, J.H., 1998. Susceptibility of five fish species (nile tilapia, rosy barb, rainbow trout, stickleback and roach) to intramuscular injection with the oomycete fish pathogen, *Aphanomyces invadans*. Bull. Eur. Assoc. Fish Pathol. 18, 192–197.

Kar, D., Dey, S.C., Roy, A., Mandal, M., 2000. Epizootic ulcerative syndrome in fish at Barak Valley, Assam, India. In: Ramachandra, T.V., Murthy, C.R., Ahalya, N. (Eds.), Lake 2000: Int. Symp. Restoration of Lakes and Wetlands. Indian Inst. Sci, Bangalore, India, pp. 317–319.

Kar, D., 2007. Fundamentals of Limnology and Aquaculture Biotechnology. Daya Publishing House, New Delhi pp. vi + 609.

Kar, D., 2013. Wetlands and Lakes of the World. Springer, London. Print ISBN: 978-81-322-1022-1; e-Book ISBN: 978-81-322-1923-8, pp. xxx + 687.

Kar, D., 2010. Biodiversity Conservation Prioritisation. Swastik Publications2, New Delhi pp. xi + 167.

Lilley, J.H., Phillips, M.J., Tonguthai, K., 1992. A Review of Epizootic Ulcerative Syndrome (EUS) in Asia. Aquatic Animal Health Research Institute and Network of Aquculture Centres in Asia-Pacific, Bangkok, 73.

Lilley, J.H., Roberts, R.J., 1997. Pathogenicity and culture studies comparing *Aphanomyces* involved in epizootic ulcerative syndrome (EUS) with other similar fungi. J. Fish Dis. 20, 101–110.

Lilley, J.H., Callinan, R.B., Chinabut, S., Kanchanakhan, S., MacRae, I.H., Phillips, M.J., 1998. Epizootic Ulcerative Syndrome (EUS) Technical Handbook. The Aquatic Animal Health Research Institute, Bangkok, Thailand.

Llobrera, A.T., Gacutan, R.Q., 1987. *Aeromonas hydrophila* associated with ulcerative disease epizootic in Laguna de Bay Philippines. Aquaculture 67, 273–278.

Miyazaki, T., Egusa, S., 1972. Studies on mycotic granulomatosis in freshwater fish I. Mycotic granulomatosis in goldfish. Fish. Pathol. 7, 15–25 (In Japanese).

Miyazaki, T., Egusa, S., 1973a. Studies on mycotic granulomatosis in freshwater fish II. Mycotic granulomatosis prevailed in goldfish. Fish. Pathol. 7, 125–133 (In Japanese).

Miyazaki, T., Egusa, S., 1973b. Studies on mycotic granulomatosis in freshwater fish III. Bluegill. Mycotic granulomatosis in bluegill. Fish. Pathol. 8, 41–43 (In Japanese).

Miyazaki, T., Egusa, S., 1973c. Studies on mycotic granulomatosis in freshwater fish IV. Mycotic granulomatosis in some wild fishes. Fish. Pathol. 8, 44–47 (In Japanese).

Noga, E.J., Levine, J.F., Dykstra, M.J., Hawkins, J.H., 1988. Pathology of ulcerative mycosis in Atlantic menhaden. Dis. Aquat. Org. 4, 189–197.

Noga, E.J., Khoo, L., Stevens, J.B., Fan, Z., Burkholder, J.M., 1996. Novel toxic dinoflagellate causes epidemic disease in estuarine fish. Mar. Pollut. Bull. 32 (2), 219–224.

OIE, 1997. Epizootic Ulcerative Syndrome: Diagnostic Manual for Aquatic Animal Diseases. Office International Des Epizootics, Paris, pp. 127–129.

OIE, 2006. Epizootic haematopoietic necrosis. Man. Diagn. Tests Aquat. Anim. 2006, 85–103.

Pradhan, P.K., Rathore, G., Sood, N., Swaminathan, T.R., Yadav, M.K., Verma, D.K., Chaudhary, D.K., Abidi, R., Punia, P., Jena, J.K., 2014. Emergence of epizootic ulcerative syndrome: large-scale mortalities of cultured and wild fish species in Uttar Pradesh, India. Curr. Sci. 106 (12), 1710–1717.

Roberts, R.J., Wootten, R., MacRae, I., Millar, S., Struthers, W., 1989. Ulcerative Disease Survey: Bangladesh Final Report to the Govt. of Bangladesh and the Overseas Development Administration. Institute of Aquaculture, Stirling University, UK.

Roberts, R.J., Frerichs, G.N., Tonguthai, K., Chinabut, S., 1994. Epizootic ulcerative syndrome of farmed and wild fishes. In: Muir, J.F., Roberts, R.J. (Eds.), Chapter 4. Recent Advances in Aquaculture V. Blackwell Science, pp. 207–239.

Skelton, P.H., 1993. A Complete Guide to the Freshwater FIshes of Southern Africa. Southern Book Publishers, Halfway House.

Scholin, C.A., Herzog, M., Sogin, M., Anderson, D.M., 1994. Identification of group- and strain-specific genetic markers for globally distributed *Alexandrium* (Dinophyceae). II: Sequence analysis of a fragment of the LSU rRNA gene. J. Phycol. 30, 999–1011.

Skelton, P.H., Cambray, J.A., Lombard, A., Benn, G.A., 1995. .Patterns of distribution and conservation status of freshwater fishes in South Africa'. South Afr. J. Zool. 30, 71–81.

Sharifpour, I., 1997. Histology of the Inflammatory Response of the Carp (*Cyprinus carpio* L.) to Various Stimuli (Ph.D. thesis), University of Stirling, Scotland.

Sosa, E.R., Landsberg, J.H., Stephenson, C.M., Forstchen, A.B., 2007. *Aphanomyces invadans* and ulcerative mycosis in estuarine and freshwater fish in Florida. J. Aquat. Anim. Health 19, 14—26.

Vishwanath, T.S., Mohan, C.V., Shankar, K.M., 1997. Mycotic granulomatosis and seasonality are the consistent features of epizootic ulcerative syndrome of fresh and brackishwater fishes of Karnataka, India. Asian Fish. Sci. 10, 155—160.

Van Helden, L., Grewar, J., 2011. Epizootic ulcerative syndrome — Eerste river system, epidemiology report, veterinary services. West. Cape Gov. 3 (11), 1—2.

Wada, S., Rha, S.A., Kondoh, T., Suda, H., Hatai, K., Ishii, H., 1996. Histopathological comparison between ayu and carp artificially infected with *Aphanomyces piscicida*. Fish. Pathol. 31, 71—80.

Chapter 6

Aspects of Investigation of Epizootic Ulcerative Syndrome Outbreaks

6.1 INTRODUCTION

A living body is prone to suffer from disease due to attacks from pathogens or parasites. This is true also with a fish body. Often, a disease may be so virulent that it could sweep unabated in an epidemic dimension. In this connection, it may be said that patterns and long-term trends are important when deciding whether an epidemic exists in the present period and in predicting future epidemics.

The temporal pattern of an outbreak is described in terms of its epidemic curve. The epidemic curve is a graph showing the onset of cases of the disease as either a bar graph or a frequency polygon. The first case identified for a particular outbreak is referred to as the "index" case. For infectious diseases such as epizootic ulcerative syndrome (EUS), information about the index case could be valuable in identifying the source of the outbreak.

In general, an epidemic curve has four and sometimes five segments:

1. The endemic level
2. An ascending branch
3. A peak or plateau
4. A descending branch
5. A secondary peak

The slope of the ascending branch could indicate the exposure (propagating or common source) or the mode of transmission and the incubation period.

If transmission is rapid and the incubation period is short, then the ascending branch will be steeper than if transmission is slow or the incubation period is long.

The length of the plateau and slope of the descending branch are related to the availability of susceptible animals, which in turn depends on many factors, such as stocking densities, the changing importance of different mechanisms of transmission, and the proportion of resistant or immune fish in the population at risk.

The interval of time chosen for graphing cases is important to the subsequent interpretation of the epidemic curve. The time interval could be selected based on the incubation or latency period and the period over which cases are distributed. A common error in this regard is the selection of a time interval that is too long and thus may obscure subtle differences in temporal patterns. A general rule is to make the interval between one-eighth and one-quarter of the incubation period. Accordingly, for EUS, which has an incubation period of about 10 days, the incidence/prevalence in the population could be measured every two days.

Snieszko (1974) had stated that an overt infectious disease could occur when a susceptible host is exposed to a virulent pathogen under stress. The influence of each subset could be variable, and disease outbreak (e.g., EUS in this case) may occur only if there is sufficient relationship among them. Although a filterable biological infectious agent is thought to be the primary cause of EUS outbreak, it is generally accepted that certain abiotic factors that result in sublethal stress of fish are also important in initiation of disease outbreaks. Snieszko cited temperature, eutrophication, sewage, metabolic products of fishes, industrial pollution, and pesticides as potential sources of stressful environmental conditions.

Life and disease processes of fishes are similar in many ways to those of other vertebrates, in that most animals have muscles, skeletons, skins, and internal organs that function in approximately similar ways. However, there is one major difference: "fish live in water." As such, all their physiological structures and functions are influenced by this fact. Thus, to understand what water means to fish is to lay a foundation for a more complete understanding of what fish needs for health and how disease processes are related to physiological and environmental requirements.

In order to understand the influence of environmental factors on fish disease, one must realize that there is much greater chemical and physical variability in aquatic environments than in terrestrial ones. In very large bodies of water, such as oceans, conditions may be relatively stable, but in the coastal and estuarine areas, there is a greater magnitude of environmental change. In small bodies of water, this variability is even greater, while in fish hatchery operations (Kar et al., 2006), many human-initiated stresses are added. Therefore, in the aquatic environment, life goes on under dynamic and unstable circumstances and fishes must continually adapt to changes in population density

Epizootic Ulcerative Fish Disease Syndrome. http://dx.doi.org/10.1016/B978-0-12-802504-8.00006-7

pressure, temperature, dissolved gases, light, pH, etc. However, it may be noted here that the effects of these parameters could be far more severe than those ever faced by terrestrial animals. Such environmental changes could impose stresses of considerable magnitude on the somewhat limited homeostatic mechanisms of fishes.

EUS is a hitherto unknown, dreadful, virulent, and enigmatic disease among freshwater (FW) fishes that has swept through water bodies in an epidemic dimension almost semiglobally causing large-scale mortality among FW fishes. This has thus rendered many of them endangered and thrown the lives of fishers out of gear though the loss of avocation, while causing difficulties for fish consumers through the scarcity of fish flesh as a protein-rich source of nutrition. This dreaded fish disease has thus been a major concern in several countries of the world, particularly in the Asia—Pacific region.

As a historical résumé, in Queensland, Australia, an epizootic of marine and estuarine fishes, characterized by shallow hemorrhagic ulcers, had occurred in 1972 with recurrences in subsequent years. The disease had been named "red spot disease" (RSD). A similar disease, characterized by dermal ulcers, had been reported from Papua New Guinea from rivers of the south during 1975—76, and from the north during 1982—88. Concomitantly, Indonesia had also reported a similar disease in Bogor during 1980, which subsequently had spread to West, Central, and East Java. This disease was named infectious dropsy or "hemorrhagic septicemia." During 1981—83, the disease had been reported from Malaysia. Infected fishes had depicted amber-colored necrotic areas of ulceration over their whole body. Such symptoms were, possibly, called 'Webak Kudes' in local vernacular language. The disease was later reported from Malaysia during 1981—83. Affected fishes had red or necrotic areas of ulceration over their entire bodies and were called "Webak Kudes." The disease was subsequently reported from fishing areas of Kampuchea during early 1984, along with a significant decrease in the natural fish stock. A similar disease was reported from southern and central parts of Lao PDR during 1984. Myanmar had experienced the outbreak of EUS during 1984—85 affecting both wild and cultured fish stocks. In Thailand, the disease epizootic was first reported in 1980 in the natural water systems and the disease had recurred somewhat regularly almost every year from 1980 to 1985 in different water bodies. In Sri Lanka, the disease was first reported in 1988 in the Kelani River, Dandugan Oya, and in the nearby streams, causing extensive fish mortality. In Bangladesh, the first outbreak of EUS had occurred during February—March 1988 in the rivers Meghna, Padma, and Jamuna, as well as in adjoining water bodies, resulting in colossal losses of the commercial fish stock. In India, the outbreak of the virulent EUS fish disease had been first encountered during July 1988 among fishes in freshwater bodies of Northeast (NE) India. (Kar, 2000a,b, 2003b) In 1989, Nepal was also affected by EUS (Das and Das, 1993; Kar, 2007a,b,c, 2013).

EUS had been a hitherto unknown enigmatic virulent, epidemic among fishes for about four decades. It had been designated by various names, some of which are often colloquial. Egusa and Masuda (1971) first described the disease in Japan as *Aphanomyces* infection. The infection had also possibly occurred among other fishes and was named mycotic granulomatosis (MG). Such nomenclature was mainly based on histopathological (HP) results (Miyazaki and Eugusa, 1972). Concomitantly, a hitherto unknown EUS condition had been encountered since 1972, mainly on the skin of estuarine fishes in Australia. Such an element had been named RSD (McKenzie and Hall, 1976). Subsequently, similar conditions had occurred among FW fishes throughout Southeast (SE) and South Asia since the late 1970s. The main symptoms included dermal ulcerations, and there had been large-scale mortalities among fishes. The condition was designated "epizootic ulcerative syndrome" (EUS) at the consultation of Experts on Ulcerative Fish Diseases held in Bangkok during 1986 (FAO, 1986). In addition to the above, a similar ulcerative condition was also reported among the estuarine fishes in east coast of the United States since around 1978. The condition was termed ulcerative mycosis (UM) (Noga and Dykstra, 1986).

Severe periodic outbreaks of EUS have affected wild and cultured FW fishes, and wild estuarine fishes, in many countries of South and SE Asia since the late 1970s (Lilley et al., 1992). Fishes, affected typically by EUS, are characterized in having one or more large dermal ulcers with varying degrees of destruction of underlying tissues. Further, mortality rates are often high. Periodic outbreaks of EUS have occurred in the Philippines since 1985, affecting a number of fish genera including *Mugil, Anabas,* and so on (Leobrera and Gaeutan, 1987; Reantaso, 1991).

The last two decades have witnessed the transformation of aquaculture (AQC) and fisheries from their traditional natures to important economic activities in India and many other countries in Asia (Kar, 2001, 2002, 2003c). This has been possible because of the development and adoption of various scientific technologies in this field. However, anthropogenic pressure and intensification in AQC has created some problems in the form of environmental stress (Kurhalyuk and Tkachenko, 2011) resulting in outbreak of diseases. Pollutants (industrial, pesticidal, and intensive fish-farming effluents) discharged in various water bodies had been creating stress to fishes and endangering their lives. Recent outbreaks of EUS in fishes throughout India showed that fish disease could be a major limiting factor for enhancing fish production. It also brought to focus the lack of sufficient knowledge on fish disease and lack of required infrastructure for fish health management in India. There are many types of fish diseases so far known to affect fishes externally and internally. They are caused by various types of organisms (fungi, bacteria, protozoans, crustaceans, helminthes, etc.), causing damage to fish from minor to major magnitude. Whereas therapeutic measures for some diseases are still not well known, there are even such diseases known whose causative organisms are yet to be ascertained. As such, though considerable know-how in this field is available, much more is needed. Even many fish health management measures known to scientists have not reached functionaries in the field. These gaps are due to many factors, including the lack of a trained workforce and inadequate infrastructure.

Asian AQC is characterized by a bewildering diversity of species (Kar and Dey, 1982a,b, 1986, 1987), with innumerable marine and FW species being farmed (Kar, 2004a,b, 2005b). During the past two decades, EUS has been a serious condition affecting a large number of wild and cultured FW finfish species (Kar, 2004c, 2005c). The etiological agent, according to some workers, is *Aphanomyces invadans* (also called *Aphanomyces invaderis* and *Aphanomyces piscicida*); a variety of syndromes have been attributed, namely (MG), EUS, RSD, and UM (APHA, 1995), in both estuarine and FW fish (wild and cultured). The geographical distribution of these *Aphanomyces* infections includes the eastern United States, Australia, Japan, South and SE Asia (Barua, 1994), and India (Blazer et al., 2002). The actual economic losses in the AQC industry worldwide are estimated to be >US$9.0 billion per year, equal to c 15% of the value of the world's farmed fish and shellfish production.

Disease has becomes a major problem in fish production both in culture system (Tamuli et al., 1995) and in the wild in Asian nations, notably India (Kar et al., 1998d) and Bangladesh (Khulbe and Sati, 1981). Extreme environmental changes may be fatal or cause stress to fish leading to secondary infections. Fish remaining in an unfavorable environment loaded with innumerable agents such as chemicals, pollutants, bacteria, viruses, parasites, and fungi, either individually or in combination, can experience damage to body tissues or systems that produce several kinds of disease. Moreover, the external and internal biology of these fishes are also altered by other physical, chemical, and biological factors of the environment (Khulbe, 2001).

Common diseases of FW fishes in Asian nations, notably India, Bangladesh, etc., are ulcer type disease including EUS, septicemia, tail and fin rot, bacterial gill rot, dropsy, and various fungal, protozoan, parasitic, and nutritional diseases, tumors, etc. (Chauhan and Qureshi, 1994; Chinabut, 1994). In most cases, hemorrhages, septicemia, various lesions, gill damage, etc. are common symptoms of the affected fish. Carps are also affected by a wide range of diseases and parasites. These are dermatomycosis, saprolegniasis, dropsy, fish pox or epidermal epithelioma, bronchomycosis, dactylogryposis, furunculosis, columnaris, tail and fin rot, bacterial gill disease, bacterial kidney disease, white spot disease, trichodiniasis, myxosporidiosis, argulosis etc. In fish, the most obvious external clinical signs are inflammation, erythremia, and hemorrhage of fins, skin or head, frayed or eroded fins,

hemorrhaged opaque eye, open necrotic and ulcerative lesions at any location on the body, lepidorthosis of scales, and excessive mucus production. A total lack of mucus; edema; an enlarged abdomen; presence of yellow, white, or black spots on the skin; prolapsed anus; and exophthalmia are all clinical sings of fish disease (Khulbe, 1992). The clinical signs, combined with parasitic investigation and histopathology, may be very helpful in diagnosing fish disease, which has been widely used throughout the world. In Bangladesh, limited HP studies have been done on carp species. So, the present work was undertaken for identifying the occurrence of diseases in *Cyprinus carpio*.

EUS had been a new occurrence at the time of its first initiation (during the late 1980s) in this region, and it is still a hitherto unknown enigmatic virulent fish disease in the Indian subcontinent (Kar et al., 2001b, 2002c,g,h). Because of this, fish consumers, particularly at the initial period of outbreak, apprehended a health disaster in the event of consumption of EUS-affected fish; this was a little later proven to be unfounded (Kar, 1991, 2007a,b,c, 2013; Kar and Dey, 1988a,b; 1990a,b,c; Kar et al., 1990a,b, 1993, 1995a,b,c, 2007(a,b)).

Nevertheless, there had been an alarming decline in fish prices and sales during the initial period of EUS outbreak. This had given a severe blow to the economy of fish farmers and fishers at large. Moreover, this situation might also have had detrimental effects on the nutritional status of fish consumers, because fish serves as a staple food and source of animal protein in their diets. As reported, c 250 million families in SE Asia cultivate paddy (rice) as their principal crop, often accompanied by fish culture—that is, paddy-cum-fish culture. Incidentally, much of the fish harvested from this paddy-cum-fish culture goes into the diets of growers as nutritional supplements (Macintosh, 1986). However, economic losses due to the ongoing EUS outbreak had been estimated as 118.3 million taka in Bangladesh (US$ 4.3 million; 1 US$ = 35 taka at that time) during 1988—89. Similar colossal losses occurred in other Asian nations. The disease occurred with less severity and magnitude during 1989—90, and economic losses were estimated at 88.2 million taka (US$ 2.2 million). Incidentally, the price of fishes dropped to 25—40% of the pre-disease level during the first outbreak (Barua, 1994).

Likewise, EUS was considered a virulent disease in epidemic dimension that had affected FW fishes on the Indian subcontinent (including India and Bangladesh) since 1988. However, evidence suggests that the initiation of EUS outbreaks happens when a number of determinants or causal precipitating and predisposing stress factors amalgamate to cause the effect. The principal effect is said to be the exposure of the dermis. These exposed sites could serve as potential points of entry for primary and secondary etiological agents, viz. bacteria, fungi, and viruses—not only primarily initiating EUS, but also secondarily causing later havoc for fishes. There had been several hypotheses about the dynamics of initiation of EUS. According to Lilley et al. (1998), the exposed sites could provide points of attachment and entry for spores of the fungus *A. invadans.* According to Callinan et al. (1996), outbreaks of EUS in estuarine fishes of Australia are said to be due to acid-sulfate soils, and they were said to have reproduced EUS in estuarine fish by exposing susceptible fish to acid water and spores of *A. invadans.* Alternatively, Kanchanakhan (1996a,b) had shown that EUS could be reproduced when susceptible snakeheads (*Channa* sp.) were injected with a particular strain of rhabdovirus and bathed in spores of *A. invadans.* Thus, it appeared that there could be a number of sufficient causes for the initiation (outbreak) of EUS, although every set of sufficient causes could be different from another. However, each combination might have the common result of exposing the dermis and allowing the entry of etiological agents. Nevertheless, the presence of the fungus *A. invadans* is said to be common in almost all cases of EUS outbreak, whether as a primary or secondary etiological agent. The highly invasive ability of *A. invadans* to enter the tissues of bones, gizzards, and spinal cords had been reported and could indicate that under certain circumstances, the fungus is able to invade the healthy skin of fish (Vishwanath et al., 1998). During the peak SW monsoon period (July–August) of 1991, the outbreak of a severe fish disease was reported in FW fishes of Kerala province. The nature and spread of the disease indicated that it was similar to the ulcerative disease syndrome reported from various parts of SE Asia. The main target of EUS was the snakehead (*Channa* spp.). In 1990, the disease reached Sri Lanka. In February 1991, the disease was reported from various parts of Tamil Nadu (TN), particularly in the Kaveri Delta region. In July 1991, the disease appeared in Pookodu Lake and the Padinjarathara Reservoir of the Wyanad district. Simultaneously the disease started appearing in various parts of Kuttanadu paddy fields and the canals adjoining Vembanad Lake. The areas affected were mainly within the districts of Kottayam and Alappuzha, in Kerala.

Everywhere, snakeheads (*Channa striata*) were the first victims of the disease, followed by other bottom-dwelling fishes like *Heteropneustes fossilis, Wallago attu, Glossogobius giuris, Etroplus suratensis, Etroplus maculatus*, and *Puntius filamentosus.* It was observed that fishes stopped feeding, came to the surface, and remained vertical. The most remarkable feature of this disease was the formation of deep oozing ulcers all over the body, including the tail and fins, giving an ugly appearance to fish. In areas where the disease surfaced initially, it disappeared from bottom-dwelling fishes within a week, but spread to small pelagic fishes. Most necrotic patches and erosions were initially covered with the cotton wool-like growth of the fungus *Saprolegnia* sp. Histological studies showed inflammatory, necrotic, and proliferative lesions in a number of tissues/organs like the

intestine, spleen, and pancreas. Experimental inoculation of *Aeromonas salmonicida* produced disease symptoms in goldfish. Microbiological studies revealed heavy infection by fungi and Gram-negative bacteria such as *A. salmonicida, Aeromonas hydrophila*, and *Pseudomonas* sp. Among them, the role of *A. hydrophila* as a major pathogen is doubtful because of its opportunistic nature. In this direction, Kerala Agricultural University, Trichur, has identified a paramyxovirus as a possible primary causative agent of EUS.

Notwithstanding the above, the epidemiology of EUS seemed not to have been studied in much detail. A number of factors had been hypothesized as "risk factors," predisposing and precipitating "stress factors," or "determinants" of EUS outbreaks. These factors were based on observations of the mode of disease transmission, species, habitats, systems of culture, and so on. Identification of the true "risk factors" of EUS could allow rational control measures to be developed and adopted. According to recommendations of the FAO (1986), research requirements of EUS include the need for a greater understanding of the influence of environmental factors and pollutants on the disease and the identification of causative agents. Roberts et al. (1989), during an EUS survey in Bangladesh, stressed the need for an epidemiological study of individual water bodies to collect information on disease transmission, relative species susceptibility, mortality and recovery rates for different species and ages of fish, fish losses, and economic impacts. The present cross-sectional survey was aimed at quantifying the risk factors that affect outbreaks. Fish farmers and fishers were interviewed and the information was used to measure the strength of association between EUS and hypothesized risk factors.

While emphasizing India's AQC potential with regard to her enormous natural biotic and abiotic resources in the form of lentic and lotic water bodies, their stock and utilization, and AQC and fish breeding technologies, it has been opined that India is a country bestowed with an amazing diversity of enormous abiotic and biotic resources. In India, tanks and ponds occupy 2.85 million hectares of land; lakes and reservoirs occupy 2.05 million hectares; derelict water bodies such as swamps, beels, and oxbow lakes make up 0.78 million hectares; rivers and canals account for 0.17 ha km; and brackish-water land, lakes, and estuaries make up 1.42 million hectares. These resources have great potential. The estimated potential for fish production in the country, both marine and inland, is about 8.4 million tonnes comprising 3.9 million tonnes of marine and 4.5 million tonnes of inland fish production. Against this, the present fish production is about 6.18 million tonnes, of which 3 million is inland and 3.18 million is marine.

Asia contributes >90% of the world's AQC production. Like other farming systems, AQC is plagued with disease problems resulting from its intensification and commercialization. This paper describes various factors, providing specific examples, that have contributed to the current disease problems faced by what is now the fastest-growing food-producing sector globally. These include increased globalization of trade and markets; intensification of fish-farming practices through the movement of broodstock, post larvae, fry, and fingerlings; introduction of new species for AQC development; expansion of the ornamental fish trade; enhancement of marine and coastal areas through the stocking of aquatic animals raised in hatcheries; unanticipated interactions between cultured and wild populations of aquatic animals; poor or lacking biosecurity measures; slow awareness of emerging diseases; misunderstanding and misuse of specific pathogen-free stocks; climate change; and other human-mediated movements of AQC commodities. Data on the socioeconomic impacts of aquatic animal diseases are also presented, including estimates of industry failures and losses in production, direct and indirect income/employment, market access to or share of investment, consumer confidence, and food availability. Examples of costs of investment in aquatic animal health-related activities, including national strategies, research, surveillance, control, and other health management programs, are also provided. Finally, strategies currently being implemented in the Asian region to deal with transboundary diseases affecting the AQC sector are highlighted. These include compliance with international codes and the development and implementation of regional guidelines and national aquatic animal health strategies; new diagnostic and therapeutic techniques and new information technology; new biosecurity measures including risk analysis, epidemiology, surveillance, reporting, and planning for emergency response to epizootics; targeted research; institutional strengthening; and workforce development—for example, education, training, and extension research and diagnostic services (Melba et al., 2005).

There is a common Bengali proverb *"Machhe Vate Bangalee."* It means that Bengalees may not be able to live without fish in two square meals a day. It is believed that *c* 1.4 million people earn their livelihood from the fisheries sector. Further, *c* 11.0 million people are involved in fishing and its related and other ancillary activities (Mazid, 1995). Furthermore, *c* 80% of the population is rural and generally catches wild fishes from various small and big lentic (lakes, wetlands [beels, haors, baors, anuas, etc.], ponds, marshes, fens, ditches, etc.), and lotic (rivers, streams, brooks, canals, etc.) water bodies. Although fish is the staple food of the people for nutrition, many poor people may not be able to buy fish for every meal in a day.

Fisheries have been a lucrative sector in Asian countries, notably India and Bangladesh, in view of burgeoning population pressure with increasing demand for finfish and shellfish. In fact, pisciculture in the country seemed to have been progressing toward semi-intensive AQC. Conversely, the shrimp culture seems to have been moving toward an improved traditional system. In fact, uncalculated,

unrestricted, and indiscriminate use of fertilizers and feed has caused severe stress on fishes due to changes in water quality parameters in the aquatic domain, thereby rendering fishes susceptible to pathogens in foul water. At the same time, not many rich databases are available on diseases and health problems of finfish and shellfish, although mortality and economic losses are inevitable consequences of disease problems. Mass mortality of juvenile carps due to protozoan and metazoan parasites is often reported. Nevertheless, EUS has been one of the most virulent, enigmatic, hitherto unkown diseases among fishes and has been sweeping through FW fishes semiglobally since around 1972. Concomitantly, white spot disease (WSD), said to be caused by white spot syndrome virus—(WSSV) reported as a systematic epidermal and mesodermal baculovirus (SEMBV)—is said to have caused a loss of 44.4% of the shrimp production alone during the mid-1990s in Asia in general and the Indian subcontinent in particular. It is believed that the shrimp culture industry has suffered a setback since then. About 350,000 people associated with the shrimp culture were thrown out of employment. Consequently, it appears that diseases among finfishes and shellfishes have had severe economic impacts on the lives of those in the low-income group (LIG).

Production through AQC is said to be growing by leaps and bounds in developing countries in Asia in order to meet nutritional and avocational needs. Burgeoning population pressure vis-a-vis shortages of alternative employment opportunities in these countries has increased the attraction of fisheries and AQC to people, mainly for avocation. But these countries, notably Bangladesh and India, have faced many difficulties in the appropriate management of AQC, mainly due to a lack of up-to-date knowledge of management practices and a lack of awareness among growers (producers). Information pertaining to appropriate management of emerging problems is to be disseminated among the fisheries and AQC sectors.

6.1.1 Freshwater Fisheries Sector

There seems to be widespread appreciation that the fisheries and AQC sectors play significant roles in meeting national goals in social, economic, and nutritional arenas. As already mentioned, indiscriminate use of fertilizers and feed has rendered fishes susceptible to diseases due to foul water, and consequently the effects of diseases on improved systems of AQC are said to be significant. At the same time, outbreaks of disease had been insignificantly reported and poorly documented.

It may be a common notion that there could be a number of similarities between a smallholder of livestock and a big AQC system. In both systems, similar outcomes may result after the application of same principle of health management. Thus, it may appear appropriate that lessons learned from the livestock sector could be applied to AQC sector and vice versa. However, improvement in the situation of disease prevention and control may happen only when there is a clear understanding of the etiology of various disease syndromes and their epidemiology. In this regard, effective measures could be adopted only when the main reasons are found through epidemiological surveys.

Differences exist in capabilities of governments at different levels for improving the health of animals in the smallholder sector. The responsibilities of livestock and aquatic animal health are entrusted with the same governmental department in countries like Lao PDR and the Philippines. Thus, marvelous opportunities exist for sharing experiences, expertise, and resources. Further, livestock health surveillance and control systems for smallholders are quite well developed in some Asian nations. This could be expanded to other nations through network collaboration.

Concomitant to the above, governments in different nations are now seem to be under increasing pressure to supply better-quality information about the health status of livestock and aquatic animals. This could become more increasingly important, over the coming decades, for countries that export aquatic animal products, particularly live animals and their products. The *International Aquatic Animal Health Code* of The Office International des Epizooties has guidelines about the modus operandi of surveillance to be conducted and the concomitant reporting of diseases, if any. Improvements in the reporting system could be an ongoing process, with inputs in the form of tables, figures, maps, plates, etc., so that targeted goals could be achieved. Nevertheless, it may be appreciated that the modus operandi of disease surveillance for livestock, fisheries, and AQC may not be the same, and thus may call for different strategies.

It is often appreciated that there are similarities between smallholder livestock and AQC systems with regard to the application of similar principles of health management and their outcomes. Similar problems often occur with increased stocking densities coupled with commercial pressures for making production efficiency gains continually. This often may lead to enhanced chances of infectious diseases.

Concomitant to the above, global production from AQC is said to be increasing approximately 13%/y, and thus production is expected to double every six years if this rate of production is sustained along with uniform rates of production from different countries. Conversely, fast growth may also help the potential pathogens to find a greater number of susceptible hosts. Also, rapid unchecked growth might boost unhealthy competition among the neighboring farms, thereby, might lead to spread of infectious diseases in epidemic dimension,

leading to devastation among not only fishes, but also both large-scale and small-scale aquafarms.

In this communication, certain issues from the livestock sector have been outlined with some discussion regarding how issues of the livestock sector might suitably be applied to the rural small-scale AQC sector.

Small-scale fish farming has become increasingly popular among unemployed youths as a potential source of income in view of the burgeoning unemployment problem. Some small-scale farmers own ponds, but instead of culturing, they lease their ponds to others in a kind of ownership—tenant business. In view of this, diseases like EUS, in the event of outbreak, cause colossal losses to AQC, mainly because the tenant farmer may not always be competent enough to tackle and manage grievous situations posed by the outbreak of disease.

Indian major carps (IMCs) occupy an important position among the cultivable fish species of Bangladesh. At present, both major and exotic carps are commonly cultured in our inland AQC because they are very delicious and contain high levels of nutrients. They are easy to culture because they are fast-growing species with minimum mortality due to disease-resistant capacity and the ability to tolerate a wide range of environmental conditions. However, natural populations of these fishes are rapidly decreasing due to a lack of scientific management, and also the draining of beels, marshes, ditches, and similar natural habitats in order to meet excessive market demand and the production needs of paddy and other agricultural crops (Blazer et al., 2002).

Diseases in AQC are common phenomena today. Disease-control strategies could form strong databases for present and future works on fish health management. Such databases, for each country, should invariably be based on competent diagnosis, well-designed disease reporting systems, and specific field surveys. The collection of high-quality representative basic data from farms and villages forms the prerequisite need for building a sound database.

With innovative practices like high stocking densities, water circulation and aeration, and provision of feed, production rates of 8—10 t/ha/y have been achieved by Andhra Pradesh, as against the national average of 2 t/ha/y. Thus, provinces like Andhra Pradesh, Punjab, Maharashtra, Haryana, and Gujarat are taking up fish culture on a large scale, mainly with the objective of exporting the produce to fish-consuming provinces and elsewhere.

With regard to **AQC and fish breeding technologies,** the different carp culture systems that have been standardized with minimum achievable production rates are composite carp culture (4—6 t/ha/y), sewage-fed fish culture (3—5 t/ha/y), weed-based polyculture (3—5 t/ha/y), biogas slurry-fed fish culture (3—5 t/ha/y), integrated fish farming with poultry, pigs, ducks, horticulture, etc. (3—5 t/ha/y), intensive pond culture with feeding and aeration (10—15 t/ha/y), pen culture (4 t/ha/y), cage culture (10—15 kg/m2/y), and running-water fish culture (20—25 kg/m2/y). Intensification of culture systems is being standardized in pond culture, cage culture, pen culture, and running-water fish culture/flow-through systems. A significant achievement in this direction has been the production of 15 tonnes ha y in carp culture with high stocking densities, biofertilization, supplementary feeding, aeration, and water replenishment. Techniques for breeding and seed rearing catfish species— viz. magur (*Clarias batrachus*) and singhi (*H. fossilis*)— having been standardized, the culture of these species is increasingly being adopted.

EUS and crayfish plague are two fungal diseases that affect FW fish and crayfish, respectively. EUS is caused by *A. invadans*, and crayfish plague is due to *Aphanomyces astaci*. Members of the *Aphanomyces* genus belong to a class of water molds (Oomycetes) from the family Saprolegniaceae, which comprise several pathogens of fish, crustaceans, and plants. *A. invadans* is the causative agent of Australian RSD, synonymous with EUS in SE Asia, MG from Japan (Egusa and Masuda, 1971; McKenzie and Hall, 1976; Callinan et al., 1995a,b) and UM in North America (Dykstra et al., 1986). EUS continued to spread through SE Asia into India and Pakistan, and was implicated in mortalities of menhaden in the United States (Blazer et al., 2002). In Australia, *A. invadans* affects ornamental, FW, and estuarine fish, in particular silver perch, black bream, and barramundi. Work by Dr. Richard Callinan (NSW) and others fulfilled Henle-Koch's postulates, and thus established that the fungus *A. invadans* is the primary pathogen in the syndrome, following an initial impact usually in acid sulfate soil runoff. The acidified water damages the skin increasing the fish's susceptibility to infection with *A. invadans* (Callinan et al., 1995a,b; Sammut et al., 1996). The disease causes high mortality, and up to 80% of fish may be affected. The large red ulcerated lesions make fish unsalable. Affected fish (both wild caught and AQC) accounted for the production of 6546 tonnes at a value of $288 million in 2002—03 (Australian Fisheries Statistics, 2003).

Fishes are considered one of the important food sources for human beings because their flesh contains a high percentage of proteins, calcium, and phosphorus. Therefore, there is an increased attention given to fish farms and their diseases. Pathogens such as viruses, bacteria, and fungi are all causes for concern to aquaculturists. In fish farms, many of these pathogens cause widespread mortality. The use of antibiotics to treat bacterial infections has resulted in a global increase of resistance. For this reason, studies are conducted on immunostimulants that represent a modern and promising tool in AQC as they enhance the resistance of cultured fish to disease and stress. It comprises a group of biological or synthetic compounds that enhance the

hormonal and cellular response both in specific and nonspecific way. Immunostimulants also stimulate the natural killer cells, complement, and lysozyme antibody responses of fish (Sakai, 1999). Investigations regarding the health status of a fish could be assessed by hematological parameters (Hickey, 1976). By analyzing blood cells, characteristic clues for diagnosis and prognosis of diseases may be found (Anderson, 2003). Ascorbic acid is an indispensable and multifunctional micronutrient. It plays important roles in improving immune function, improving growth, providing good health, feed conversion, survival, stress resistance, and oxidation. Most fish species are not capable of vitamin C biosynthesis due to the absence of the enzyme L-gulonolactone oxidase necessary for ascorbic acid synthesis. Vitamin C also plays an important role in animal health as antioxidants by inactivating damaging free radicals produced through normal cellular activity and from various stress. It has been suggested that the antioxidant function of these micronutrients could enhance immunity by preserving the functional and structural integrity of immune cells. In this respect, the need for specific nutrients may be increased during infection that could require the feeding on diets formulated for optimal immune competence rather than growth and survival. Supplementation with vitamin C improved antibody production against *Edwardsiella ictaluri* in channel cat fish (Li and Lovell, 1985).

Notwithstanding the above, Baldock et al. (2005) had displayed results of a short expert consultation on EUS that had taken place during the 5th Symposium on Diseases in Asian Aquaculture held at Gold Coast in Australia during November 2002. The workshop had aimed to review the accumulation of information and knowledge on EUS. To provide a scope for experts to present mainstream and dissenting views on causal pathways and reexamine issues pertaining to case definitions, syndrome names and fungal nomenclature were some additional aims of the workshop. Corroborating with the symposium, it is now generally believed that EUS could be the same disease as MG, RSD, and UM. Further, as a joint declaration developed after the symposium in the presence of experts and participants, the Japanese works on MG were reviewed, and compared with results of works done on EUS in Australia, SE Asia, South Asia, and the United Kingdom, as well as UM in the United States. The mainstream view of the symposium/workshop was that EUS could essentially be an "aphanomycosis" with *A. invadans* (=*A. piscicida*) as the only necessary infectious cause; this was deliberated upon by most participants and experts, coupled with support by the weight of published evidence and ongoing research findings. Their deliberations had been juxtaposed with those of the minority of participating experts who had asserted that EUS is a polymicrobial infection involving specific viral, bacterial, and mycotic pathogens with concomitant reflection of primary and secondary etiologies. A number of case definitions had been proposed, which could be appropriate for use in field surveys or for laboratory diagnosis. Further, many participants had suggested a new name for the disease, viz. "epizootic granulomatous aphanomycosis" (EGA). It was further proposed that the name *A. invadans* (=*A. piscicida*) might be used in any initial reference to the putative causal fungal pathogen, other than in taxonomic contexts. The name *A. invadans* may be used thereafter. Efforts had been made to identify the key issues with an aim to unify the currently divergent views. Further, there had been recommendations for further works.

6.2 REVIEW OF LITERATURE

Concomitant to the above, literature on EUS and other possible related diseases in fishes had been reviewed by different workers since long ago (Roberts et al., 1993; Chinabut, 1995, 1998; Lilley et al., 1998; Kar, 1990, 1999a, b, 2003, 2004, 2007a,b, 2013; Kar and Dey, 1988; Kar and Dey, 1988a,b, 1990a,b,c; Kar et al., 1990, 1993, 1994, 1995a,b,c, 1996a,b,c; 1997, 1998a,b,c, 1999a,b, 2000a,b,c,d, 2002d,e,f, 2003a,b,c,d, 2004a,b,c, 2006, 2007a,b; Kar and Das, 2004a,b; Kar and Upadhyaya, 1998; Ahne et al., 1988; Barua, 1990, 1994; Bondad-Reantaso et al., 1990, 1994; Boonyaratpalin et al., 1983; Boonyaratpalin, 1985; Cartwright et al., 1994; Chattopadhyay et al., 1990; Das and Das, 1993; FAO, 1986; Frerichs et al., 1986, 1989; Hatai, 1980, 1994; Hatai et al., 1977, 1994; Miyazaki and Egusa, 1972; Mohan et al., 1999; Mohan and Shankar, 1994; OIE, 1997; Phillips, 1989; Phillips and Keddie, 1990; Poonsuk et al., 1983; Prasad and Sinha, 1990; Priyadi, 1990; Purkait, 1990; Quddus, 1982; Rahman et al., 1988; Raina et al., 1986; 1991; Reichenbach-Klinke and Elkan, 1965; Reinprayoon et al., 1983; Richards and Pickering, 1978; Roberts, 1989; Roberts et al., 1986, 1989, 1992, 1993, 1994a,b; Rodgers and Burke, 1981; Saitanu and Poonsuk, 1983; Saitanu et al., 1986; Shariff and Law, 1980; Shariff and Saidin, 1994; Shrestha, 1990, 1994; Snieszko, 1974; Soe, 1990; Tachushong and Saitanu, 1983; 1985; Temeyavanich, 1983; Tesprateed et al., 1983; Tiffney, 1939; Tonguthai, 1985; Viswanath et al., 1997a,b; 1998; Wattanavijarn, 1983; Wattanavijarn et al., 1985; Widago, 1990; Wikramanayake and Moyle, 1989; Williams and Amend, 1976; Wilson, 1909; Wilson and LoK, 1992; Willoughby, 1968, 1970; Willoughby et al., 1995; Wood, 1968; Wolf et al., 1960; Wolf and Darlington, 1971; Wolf et al., 1972).

A disease outbreak is said to occur as an outcome of the interaction between the organism and the pathogens involving the environment. Stress factors influence the initiation of the disease.

Further, some stress-inducing parameters, including the underlying concept of stress, are given below:

6.3 STRESS-INDUCING PARAMETERS AND THE UNDERLYING CONCEPT OF STRESS

Mortality of fish or decline in a fish population in a water body is, at present, the sole indicator that the effects of environmental stress factors are exceeding the acclimation tolerance limit of fish. However, several physiological and whole animal changes occur that can be used to provide prior information that the effect of stress will exceed acclimation tolerance limit of fish and lead to dysfunction, such as impaired fish health, growth, or survival. These changes are a direct or indirect result of the physiological response to environmental stress and could be quantified and used as predictive indices.

6.3.1 Definition of Stress

The term "stress," "stressor," or "stress factor" could be defined as a force or challenge in response to which there is a compensatory physiological change in fish. Thus, an environmental or biological stress is of significance, if it requires a compensating response by a fish population or ecosystem.

Some stresses and their associated parameters are briefly discussed below:

6.3.2 Stress Due to Metal Toxicity

In natural water bodies, the concentration of metals is generally low. However, if the water reaction is acidic, the concentrations of Fe, Mn, and Al increase significantly and may cause "stress" to aquatic animals. (Kurhalyuk and Tkachenko, 2011) In West Bengal (Jalpaiguri) and Assam, where soil and water reactions are generally acidic, the growth and survival of fish in ponds are usually poor. Further, industrial effluents containing significantly high quantities of Cr, Pb, Hg, Zn, etc., when discharged to the river, may cause considerable harm to fish and fish food organisms. A tentative idea of toxic and safe levels of the following elements reveal that aluminum is toxic at pH 5.0–5.5, safe level = 0.1 ppm; cadmium: safe level = 0.004 ppm in hard waters and 0.0004 ppm in soft water; copper: safe level = 0.005–0.04 ppm; iron: safe level = 1 ppm and a level of 1.2–1.5 ppm is lethal to fishes; lead: safe level = 0.005 ppm in soft water, and a level of 0.05 ppm is toxic to the nervous systems of fishes; mercury: safe level = 0.0002 ppm; zinc: safe level = 0.01–0.05 ppm depending on the TH of water; and chromium: permissible criteria = 0.05 ppm, desirable =should be absent. However, some metals may be either beneficial or toxic depending on their concentrations. At slightly alkaline pH (8.0—8.5), metal toxicity is usually low. While deliberating upon **pesticides,** it should be noted that they are complex organic compounds that enter water bodies from adjacent catchment areas such as agricultural fields and forests. There are three types of pesticides, and in natural water, they are typically present in very low concentrations. However, organochlorine pesticides are very stable and may bioaccumulate in living organisms. Given this, the damage caused to biological structures by the accumulation of pesticides may be considerable. Even in sublethal concentrations, the pesticides DDT and ethyl parathion are known to significantly affect the standard (basal) metabolism and activity of fishes.

Safe levels of some common pesticides in water are as follows—DOT: safe level = 0.002 ppm; dieldrin: safe level =0.005 ppm; endrin: safe level = 0.002 ppm; and malathion: safe level = 0.008 ppm. It may be further noted that the joint action of a mixture of pesticides may be more lethal to fish than effect of a single pesticide. The toxicity of pesticides varies with species and size of fish.

Notwithstanding the above, some other stress-related information and associated syndromes and phenomena are briefly discussed below:

6.3.3 Conceptual Framework of Stress Response

The conceptual framework is to consider the stress response in terms of primary secondary and tertiary changes.

1. Primary response: Following the perception of a stressful stimulus by the central nervous system, the stress hormones viz. cortisol and epinephrine are synthesized and released into the bloodstream.
2. Secondary response: Changes in the blood and tissue chemistry and in the hematology occur, such as elevated blood sugar levels and reduced clotting time. Diuresis begins followed by blood electrolyte losses and osmoregulatory dysfunction. Tissue changes include depletion of liver glycogen and interrenal vitamin C hypertrophy of the interrenal body.
3. Tertiary response: Manifestation in reduction of growth, resistance to diseases, reproductive health, and survival. These may decrease recruitment to succeeding life stages, and as a result, population decline occurs.

Use of physiological responses as indicators: Several changes that occur as responses to stress could be used as measurable indices of the severity of stress on fish. These changes could be quantified and used as predictive indices.

Methods for stress diagnosis: Several biochemical and physiological procedures have been developed to assess the severity of the physiological effects resulting from stress. The physiological parameters of importance for assessing stress in fish at the primary, secondary, and tertiary levels are discussed below.

Plasma cortisol: A relatively direct assessment of the severity and duration of the primary stress response could be obtained by monitoring the rise and fall of plasma cortisol or catecholamines (epinephrine and norepinephrine) concentrations.

Secondary stress response: Secondary changes, which occur mainly in the blood chemistry, also characterize the severity of stress in fishes, viz. blood glucose, chloride, and lactic acid. They are frequently used for assessing stress response. Hyperglycemia for blood glucose and hypochloremia for blood chloride is the physiological effect of concern during stress response. Accumulation of lactic acid in muscle is also an indicator of stress due to severe exhaustion. The hematological parameters also provide useful information about animals' tolerance to stress.

Hemoglobin/Hematocrit: It increases or decreases following acute stress, and could indicate whether hemodilution or hemo-concentration has occurred.

Leukocyte: It generally decreases (leucopenia), commonly occur during the physiological response to acute stressors. The blood clotting time and changes in the WBC count are among the most sensitive parameters indicating stress response.

Histopathology: Many biochemical changes that occur in response to stress are the result of cellular pathology. Moreover, histological examinations could frequently provide information on the effects of stress factors on fish. For example, interrenal hypertrophy, atrophy of the gastric mucosa, and cellular changes in gills are indicative of stress response (Jobling et al., 1996).

Tertiary stress response: Experiences had depicted that several tertiary stress responses, including changes in metabolic rate, health, behavior, growth, survival, and reproductive success could indicate unfavorable environmental conditions that have exceeded acclimation tolerance limits of fish.

Metabolic rate: This is a fundamental aspect of animal performance and is affected by stress.

Reproduction: Detrimental effects on reproduction, as manifested by oocyte atresia, spawning inhibition, and decreased fecundity and hatching success, are taken into consideration for assessing stress response.

Disease: Incidence of fish disease is an important indicator of environmental stress. Fish disease is actually the outcome of the interactions among fish, their pathogens, and the environment. If the environment deteriorates, stressed fish are unable to resist pathogens that they normally can resist. Certain diseases sometimes may prove to be useful indicators of the fact that tolerances of adverse environmental condition might have exceeded. Thus, it is apparent that knowledge of the tolerance limits for acclimation to the single or cumulative effects of various biotic and abiotic stress factors is an important part of the database for the species—habitat relationship that could be required for effective management of fisheries. Such information could solve many problems ranging from the prediction of the tolerance level the fish could have for proposed habitat alterations to evaluation of the effects on fish health exerted by modem intensive fish culture.

6.4 GENERAL ADAPTATION SYNDROME

The various physiological changes that occur, as a fish responds to stressful stimuli are compensatory, or in other words it is adaptive in nature and are required for acclimation. Collectively, these phenomena have been termed general adaptation syndrome (GAS).

6.5 CLIMATE VIS-À-VIS SEASONALITY

The globe is faced today with large-scale unpredictable climate change. As such, there has been considerable erratic behavior among climatic parameters. Many new ailments have never been heard of before. Farmers, who had reported the occurrence of EUS in their ponds during the previous season, were faced with a higher risk of EUS occurrence even during the current year (RR = 3.00). Unusually low air temperature and/or incessantly heavy rainfall for approx. 3–15 days (prior to the present survey had begun), as opined by fish farmers and fishers, might also be correlated with occurrences of EUS.

Concomitant to the above, occurrences of EUS had generally been associated with low temperature, and had often occurred after periods of heavy rain. It had been observed by Phillips and Keddie (1990) from the data from 1988 to 1989 that EUS outbreaks had occurred during months when the average daily temperature was less than the annual average temperature in India, Bangladesh, China, and Lao PDR. Nevertheless, EUS outbreaks in the Philippines and Thailand were also recorded during the warmer months. In addition, Chinabut et al. (1995) had challenged stripped snakehead (*C. striata*) by injecting zoospores of *A. invadans,* and had found a weak inflammatory response. There had been higher rates of mortality and more extensive fungal invasion among fishes maintained at 19 °C , compared with fishes held at 26 °C and 31 °C.

Some important environmental parameters of significance to fish health are discussed below (Kar, 2007a,b,c).

6.6 ENVIRONMENTAL PARAMETERS OF SIGNIFICANCE TO FISH HEALTH

Maintenance of a healthy aquatic environment and production of sufficient fish food organisms in a water body are two very important factors for fish production. To keep the water body conducive for fish growth, physical and chemical parameters like temperature, transparency, color,

odor, pH, dissolved oxygen (DO), and free carbon dioxide (FCO_2); toxic gases like ammonia (NH_3) and hydrogen sulfide (H_2S); and nutrient elements like nitrogen, phosphorus, and organic matter may be monitored regularly. When the physicochemical factors are at the optimum level, the water body is usually said to be productive, but when they are present in quantities above or below the normal range, fishes and other aquatic organisms may be under stress that may lead to fish disease and fish mortality. It may be mentioned here that "stress" is the sum of the physiological response by which an animal tries to maintain or reestablish a normal metabolism in the face of a physical or chemical force. Brett (1958) correlated stress with the fish disease situation; that is, when the normal functioning is reduced significantly and finally may result in death. In an aquatic environment, there is a profound and inverse relationship between environmental quality and fish disease. As environmental conditions deteriorate, severity of infectious diseases increases. Thus, proper health maintenance practices could play a major role in maintaining a suitable environment where healthy fish could be produced. With regard to environmental stress and fish disease, a fish prefers an optimum environmental condition for its growth and reproduction. Any change in environmental conditions causes stress on fish. If such a change occurs toward the increasing side at an arithmetic progression rate, the stress on fish may increase at a geometric progression rate. Productivity of the ecosystem that supplies food to fish is important for their growth and maturation. In an attempt to discuss the significance of the water quality parameters to fish health, deliberations may begin with temperature. Every fish has an optimum temperature for its growth and maturation. The immune response of a fish also depends on temperature. Thus, warmwater mirror carps, may not produce antibodies when ambient temperature is <12 °C, but coldwater trouts may produce antibodies even at 5 °C. Roberts (1975) and Anderson and Roberts (had noted that both defense mechanisms and susceptibility to disease in a fish depend on temperature. With slightly higher than the optimal temperature, the wound healing of fish is quicker. Solubility of oxygen and other gases also depends on temperature. At higher temperature, fish metabolism is more but the solubility of oxygen is less. With regard to **light**, the growth and maturation rates of fishes in general depend on light or photoperiods. Fish food organisms depend on solar energy for photosynthesis and growth. However, excessive solar radiation may inhibit photosynthesis and cause sunburn to fishes (Roberts, 1978). DO is an important factor that influences the growth and survival of fish in a water body. Though oxygen is plentiful in air (20.95%), it is not so abundant in water (0–14 ppm). Moreover, the solubility of oxygen in water is inversely proportional to both temperature and salinity. A water body may get oxygen by four processes, viz. (a) diffusion, (b) wave action, (c) mechanical agitation, and (d) photosynthesis. On the other hand, oxygen is consumed from water bodies by fishes and other macro- and microorganisms present in water and soil. The oxygen consumption of a fish depends on species, size, activity, temperature, feeding rate, and stress. Oxygen content in a water body ranging between 5.0 and 10.0 mg/L during morning, may be optimum for fish health. Low levels of DO (trace–1 ppm) may be lethal to many species if they continue for long periods. Oxygen content ranging between 1.0 and 5.0 mg/L may have adverse effects on growth, feed conversion, and tolerance to disease (Snieszko, 1973; Plumb et al., 1976). Under culture conditions, CO_2 content is often high. Walters and Plumb (1980) had opined that impacts of environmental stresses could cause bacterial infections in fish. A fish might survive in 0.5 mg/L of DO for some time, but may not be able to survive after several days. According to McKee and Wolf (1962), the DO content of warmwater fish habitats be preferably not <5.0 mg/L during at least 16 h of any 24 h period. Supersaturation, of waters falling over high dams, with atmospheric gases could cause gas bubble disease and mortality in fish living in streams below. Fish in a pond may die, when the DO content reached 300% of saturation; the lethal effect was due to oxygen bubbles surrounding the gills (McKee and Wolf, 1962). However, fish kills due to gas bubble disease may not be so common in fishponds. Nevertheless, DO supersaturation may affect the fry and eggs of fish adversely. With reference to **carbon dioxide (CO_2)**, it may be opined that CO_2 is highly soluble in water, but it is present in atmosphere in very small quantity. <1% of CO_2 in water forms carbonic acid. At higher CO_2 concentration, pH will be less. At a concentration of 30 mg/L, CO_2 would give a pH of 4.8. **CO_2 does not occur above pH 8.3; only carbonates occur above this pH.** CO_2 is not very toxic to fish; most species will survive for several days in water containing up to 60 mg/L of CO_2, provided DO is plentiful (Hart, 1944). However, CO_2 content is generally quite high when DO content is low (Boyd, 1982). Under low DO content, high content of CO_2 hinders oxygen uptake by fish, causing respiratory problem and stress. For healthy growth of fish, 3 mg/L or less of free CO_2 is permissible in pond or hatchery water. CO_2 content, ranging between 12 and 50 mg/L, may have sublethal effects that may include respiratory stress and the development of kidney stones. High quantity (50–60 mg/L) of free CO_2 is lethal to many fish species with prolonged exposure (Das and Das, 1997). Concomitantly, **NH_3** may enter into a water body as fertilizer. However, in ponds where high densities of fish are given supplemental feed, the NH_3 content may increase to undesirably high levels. In water, unionized NH_3 exists in a pH- and temperature-dependent equilibrium with ammonium ions.

$$NH_3 + H_2O: NH^+ + OH^-$$

Unionized ammonia is highly toxic to fish but ammonium ion is relatively nontoxic. The higher the pH and temperature, the higher is the percentage of total unionized ammonia (Boyd, 1982).

Notwithstanding the above, in most natural water bodies and fishponds, nitrite content is generally low. But if the water body is contaminated with high organic pollution and has low oxygen content, the nitrite content may increase to toxic levels. Nitrite content between 1.0 and 10.0 mg/L is lethal for many warmwater fishes. In the range of 0.02–1.0 mg/L, it is sublethal to many fish species. Nitrite is highly toxic to fish. When nitrite is absorbed by fish, it reacts with hemoglobin (Hb) to form methemoglobin. Since, methemoglobin is not effective as an oxygen carrier, continued absorption of nitrite may lead to hypoxia and cyanosis. Addition of calcium and chloride may reduce the toxicity of nitrite to fish (Boyd, 1982). With regard to hydrogen ion concentration **(pH)**, pH may be defined as the negative logarithm of the hydrogen ion activity. pH = $-\log$ (H+). Most natural waters have pH ranging between 6.5 and 9.0. However, if waters are more acidic than pH 6.5 or more alkaline than pH 9–9.5 for long periods, reproduction and growth of fish may diminish. As far as the effect of pH on fish health, fishes may not be at home in pH values of 4.0–6.0 and 9.0–10.0. However, the optimum range of pH for a fish to be healthy in water is in the range of pH 7.4–8.2. **Alkalinity** refers to the concentration of bases in water expressed in mg/L of equivalent calcium carbonate. In most waters, bicarbonate, carbonate, or both are the predominant bases. **As far as the effect of alkalinity** on fish health, there is poor fish growth and stress in fishes in the alkalinity range of 0–20 mg/L; low to medium fish growth in the alkalinity range of 20–50 mg/L; and optimum fish growth and production in the alkalinity range of 80–200 mg/L (Boyd, 1982). Liming of fishponds enhances total alkalinity (TA). Total hardness (TH) refers to the concentration of divalent metal ions in water expressed as mg/L or equivalent calcium carbonate. TH in the majority of FW ponds could sometimes be similar to TA. Liming of fishponds enhances the TH values of water. **Turbidity** of water has a profound impact on aquatic life including fishes. Turbidity due to plankton is generally beneficial for fish growth. Turbidity due to a high content of humic matter is not directly harmful to fishes, but such waters are generally dystrophic due to low nutrient levels, acidity, and poor light penetration for photosynthesis. On the other hand, turbidity due to suspended clay particles is very harmful for fishes. Clay turbidity would reduce light penetration, adversely affecting productivity. Some particles may settle to the bottom and smoother fish eggs, and destroy benthic communities. With regard to the effect of turbidity on fish health, carps, tilapia, and catfishes are generally tolerant up to 10,000 mg/L. However, turbidity values >20,000 ppm could cause stress to a fish. High clay turbidity in a water body may be controlled by alum treatment at 150–250 kg/ha or by application of organic matter (barnyard manure at 2.5 ton/ha).

In continuance of the above, it should be noted that unfavorable values for the above parameters may render the habitat unsuitable for fishes.

6.7 WORKS BY THE AUTHOR OF THE PRESENT TREATISE

In India, EUS had made its first appearance in the province of Assam. A number of different approaches have been adopted by workers here, notably the author of the present treatise, to ascertain causes of the disease. These are:

1. whether the outbreak of EUS is due to any severe organic pollution of water and soil;
2. whether the outbreak of EUS is due to any radioactivity;
3. whether the outbreak of EUS is due to any heavy metals contamination;
4. whether bacterial and fungal flora are the causative factors of EUS; and
5. whether there is any primary viral etiology for the initiation of EUS.

These are analyzed in the paragraphs that follow.

Dr. Devashish Kar (Kar and Dey, 1988a,b) began his study with the ichthyological survey of the lentic water bodies in the Barak Valley region of Assam (24°10′N, −93°15′E) with the identification of the four most widely affected species of fishes (Plate 1) due to EUS during July 1988. These are:

1. *Macrognathus aral*, which portrayed lesions generally on the skin of the abdomen and occasionally in the caudal region;
2. *Channa punctata*, which showed lesions generally on the skin at the base of the caudal peduncle, and sometimes also showed reddish swollen eyes (exophthalmos);
3. *Mystus vittatus*, which depicted amber-colored lesions generally at the base of the rayed dorsal fin and sometimes at the base of the caudal fin; and
4. *Puntius conchonius*, which exhibited amber-colored lesions generally at the base of the dorsal and caudal fins and sometimes on other parts of the body.

Other species affected (Kar and Dey, 1988a,b) by the present epidemic during July 1988, but not widely, include:

Mastacembelus armatus, Macrognathus pancalus, Salmophasia bacaila, Gudusia chapra, Badis badis, G. giuris, Ailia coila, Lepidocephalichthys guntea, C. batrachus, Cirrhinus mrigala, Cirrhinus reba, and *Parambassis*

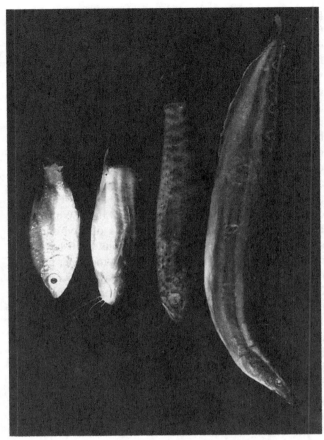

PLATE 1

ranga. W. attu and hilsa species *Tenualosa ilisha* were not found to be widely affected by prevailing epidemics. Among IMCs during July 1988, barring a few *Catla catla* and *C. mrigala*, none were affected by EUS. Nevertheless, *Amblypharyngodon mola, Ctenopharyngodon idella, Rasbora daniconius, Puntius ticto,* and *Channa marulius* were affected in certain areas of the district (Plate 2a–f).

Our continued study revealed that during the period 1992–94, the following species had been found to be severely affected by EUS: *P. ranga, Chanda nama, Nandus nandus,* and *G. giuris* (Kar et al., 1994). Some other species affected by EUS but to a lesser extent during this period included *M. armatus, M. pancalus, Xenentodon cancila,* and *Colisa fasciatus.*

Our recent studies (Kar et al., 1995a,b,c, 1997, 1998a,b,c, 1999a, 2000a; 2001a, 2002d,e,f, 2003a, 2004a) indicated that the following species of fishes have been severely affected by EUS since 1995, particularly during the period November–February, causing large-scale mortality among them (Plate 2a–f):

C. striata, C. punctata, Anabas testudineus, and *C. batrachus.*

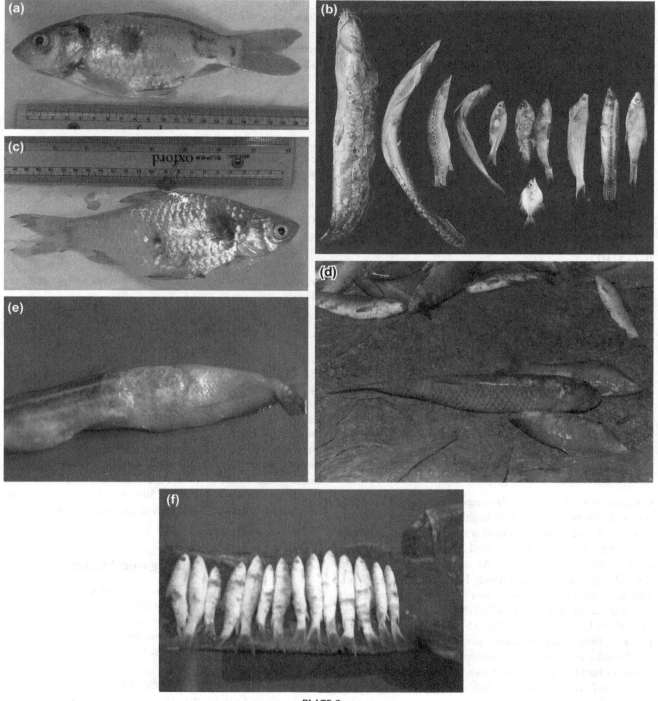

PLATE 2

Thus, there has been a differential pattern of spread of EUS among different fish species during different seasons, as indicated earlier.

Some details of the analysis under different aspects of the study indicated above are given below, preceded by the significance of parameters.

6.7.1 Transparency or Secchi Disk Transparency

The value of Secchi disk transparency (SDT) measured at noon under bright sunlight is a good indicator of fish health. In natural waters, a Secchi disk transparency of 30–80 cm is good for fish health. In intensive fish culture ponds, this value may range from 15 to 40 cm. Fish kills are said to be quite common when the SDT is recorded <12 cm. This low transparency is said to cause stress to fish and outbreak of fish disease had frequently been reported in sewage-fed FW fishing sites.

6.7.2 Turbidity

Appreciable fish mortality had been noted in highly turbid water (>175,000 mg/L).

However, productive waters are said to possess turbidity values around 25 mg/L. Turbid waters having values >25 mg/L are said to have more bacterial density than those of less turbid waters, and consequently may have greater chances of outbreak of diseases.

6.7.3 pH

One of the most important abiotic factors behind the outbreak of fish disease is the unfavorable pH of the water. Fish do not grow well below pH 6.5 and this provides a favorable environment for mycotic diseases. EUS is now supposed to be a fungal disease according to some workers. Fishes, including tilapia, could develop severe ulcers at or < 4.0 pH. Similar disorders may also occur when the pH > 9.5. In fact, fishes may grow comfortably in the pH range of 6.5–8.5. Acidic waters with pH <6.5 and alkaline waters with pH >9.5 retards reproduction and growth of fish (Swingle, 1961; Mount, 1973) and could lead to diseases (Das and Das, 1993, 1994; Kar, 2007a,b,c).

6.7.4 Dissolved Oxygen

Concentration of DO is another important abiotic factor that has a direct bearing on fish health and outbreak of fish disease. DO may not be <5 mg/L for a period of 16 h within a 24 h period. In fact, DO <5 mg/L for a period of 8 h may act as a stress factor on fish. As such, DO, preferably, be not <3 mg/L at any time (Boyd, 1982; McKee and Wolf, 1963). Prolonged exposure to low DO is said to be a precursor to bacterial infection in fish (Snieszko, 1973; Das and Das, 1993; Das, 1997; Kar, 2007a,b,c). Further, Walters and Plumb (1980) had revealed that high dissolved CO_2 (>15 mg/L) and low DO, together are more influential in initiating bacterial infections in fish. Low DO itself (Stewart et al., 1967) retards intake of food and growth. Again, supersaturation of oxygen is believed to cause gas bubble disease and mortality in fish (Nebeker and Brett, 1976).

6.7.5 Carbon Dioxide

CO_2 is said to be accumulated in water due to respiration of aquatic plants and animals. Its concentration <15 mg/L,

is quite conducive for fish health, but when the accumulation of CO_2 is >15 mg/L, it hinders the uptake of oxygen by fish (Boyd, 1982). At higher concentration, water becomes acidic and this favors pathogens to invade fishes and may lead to outbreak of fish diseases (Kar, 2007a,b,c).

6.7.6 Alkalinity

Alkalinities actually mean the total concentration of bases in water expressed in mg/L of calcium carbonate. Waters having alkalinities <20 mg/L have been believed to favor the outbreak of EUS and other parasitic diseases more virulently in fish (Kar and Dey, 1990a; Kar, 2007a,b,c 2013). Waters with low alkalinity show sharp rise or fall of pH even with a small change in the concentration of bases because of its poor buffering capacity. Waters having alkalinities >40 mg/L have been found to be more productive with less possibilities of outbreak of disease.

6.7.7 Total Hardness

Fishes are said to be more susceptible to diseases when the TH of water is <20 mg/L. In fact, severe outbreaks of EUS have been noted more frequently in waters with low hardness (<20 mg/L) (Das and Das, 1995). Productive water could have its hardness >20 mg/L (Ca content >5 mg/L and Mg content >2 mg/L). Very hard water (>300 mg/L) is also not congenial for fish health, because at higher TH, nutrient availability is not optimum. Most suitable value of TH for fish culture is said to be in the range of 75–150 mg/L. Optimum TH is also said to prevent the outbreak of common diseases in fish.

6.7.8 Decomposing Organic Matter

High accumulation of humic substances through the decomposition of profuse organic matter and macro-vegetation is said to allow for the harboring of a higher number of disease-producing organisms than is possible in clean waters, having transparency >20 cm.

6.7.8.1 Whether the Outbreak of Epizootic Ulcerative Syndrome Is Due to Any Severe Organic Pollution of Water and Soil

Field estimation of various water and soil quality and other limnological parameters conducted in various lentic water bodies in the Barak Valley region of Assam during the period July 1988 to October 1988 (i.e., the initial peak period of the EUS outbreak) revealed the following average values (Kar and Dey, 1990a):

Water temperature (°C):	33.4
Turbidity:	61.9
pH:	6.9
Dissolved oxygen:	5.95 mg/L
Free carbon dioxide:	1.25 mg/L
Total alkalinity:	7.5 mg/L
Aquatic macrophytic biomass:	0.406 kg/m^2 (on average)
Total limnoplanktonic count:	144 u/L (on average)

Similar field estimations for the above parameters, conducted in the same water bodies and same region during the predisease outbreak period since January 1979, revealed the following range of values (Dey and Kar, 1987):

Water temperature (°C):	32–34
Turbidity:	50–70
pH:	6.0–7.2
Dissolved oxygen:	3.0–6.0 mg/L
Free carbon dioxide:	0.8–1.9 mg/L
Total alkalinity:	35–65 mg/L
Aquatic macrophytic biomass:	0.308–0.706 kg/m^2 (on average)
Total limnoplanktonic count:	87–173 u/L (on average)

The above data indicated similar parameter values between predisease outbreak and outbreak periods. These results indicated that lentic water bodies of the Barak Valley region of Assam had largely maintained their organically unpolluted nature, and therefore organic pollution of water may not be assigned as a reason for the outbreak of EUS in fishes. However, TA depicted alarmingly lower values during the EUS outbreak period as compared with the pre-EUS outbreak period, and therefore could serve as a predisposing "stress factor" for EUS to set in (Kar and Dey, 1990a,b,c).

6.7.8.2 Whether EUS Is Due to Any Radioactivity

Studies conducted with the help of a gamma-ray spectrometer indicated that the level of radioactive contamination recorded in diseased fishes was too negligible to cause any harm in fish or the human beings who consume fish (radiation level: background, 770, 791; with healthy fish, 741, 716; and background, 400, 460; with EUS-affected fish, 316, 306). As a result, radioactivity may not be regarded as a possible cause of outbreak of EUS (Kar and Dey, 1990a,b,c; Kar et al., 1990a,b).

6.7.8.3 Whether EUS Is Due to Any Heavy Metals Contamination

Quantitative chemical estimation of trace elements (heavy metals) in water, soil, and in EUS-affected fish tissues by atomic absorption spectrophotometry (Vogel, 1978) did not reveal appreciable quantities of mercury, lead, arsenic, and cadmium that could be of lethal proportion (average quantity in mg/kg of Hg, Pb, As, and Cd respectively: (a) in EUS-affected fish tissues: 0.99, 0.60, 0.0001, and 0.0003; and (b) in apparently healthy fish tissues: 0.63, 0.54, 0.00005, and 0.00028) (Kar and Dey, 1990a,b; Kar et al., 1990a, 1993, 2003a). Therefore, with the effect of trace elements being negligible, they may not be regarded as possible causes of outbreaks of EUS.

Notwithstanding the above, results of similar work by others are also appended in the present treatise.

6.7.8.4 Whether Bacterial or Fungal Flora Are the Causative Factors

Bacterial Flora as Causative Factors

Bacterial culture (Cruickshank, 1972) from the surface lesions and gills of each of the four most widely affected fish species revealed isolation of hemolytic *Escherichia coli*, *Pseudomonas aeruginosa*, and *Klebsiella* sp preliminarily (Kar and Dey, 1990a,b,c). All these bacteria have been found to be sensitive to gentamicin, nalidixic acid, nitrofurantoin, ampicillin, chloramphenicol, tetracycline, trimethoprim, and sulfamethoxazole (Kar et al., 1993, 1994, 1995a,b,c) (Plate 3).

Further works have been done on the culture and isolation of bacterial flora from EUS-affected fishes belonging to nine species, viz. *P. conchonius, Notopterus notopterus, C. mrigala, G. giuris, C. punctata, A. testudineus, Labeo gonius, Labeo calbasu*, and *A. mola*, which have been collected from water bodies in the Barak Valley region of

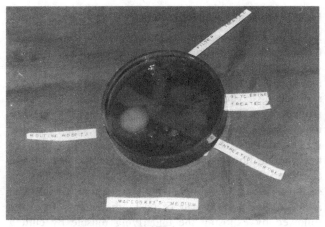

PLATE 3

Assam. Fishes portraying ulcerative lesions have been selected for the isolation of bacteria. Fishes were brought to the laboratory, cleaned properly and the parts showing lesions were sheared for taking swabs. Swabs from lesions and from the different organs, such as gills, liver, viscera, and kidney were streaked on to blood agar (5% sheep blood) and MacConkey lactose agar (MLA). The plates have been incubated at 35 °C aerobically for 24 h. Distinct and isolated colonies were picked-up, purified, and identified on the basis of their morphological, cultural, and biochemical characteristics. All isolates have been tested in vitro against nine antimicrobial agents by disk diffusion method with commercially available discs (HiMedia, India). Results showed that, nine species of fishes showing typical lesions of EUS yielded different types of bacteria. In most cases, lesions and the internal organs yielded similar bacterial isolates. Mixed infection was common, with *A. hydrophila*, *P. aeruginosa*, *Klebsiella* sp. The most common isolate was *A. hydrophila*, followed by *P. aeruginosa*, *Klebsiella* sp., and *Staphylococcus epidermidis*. Strains of *A. hydrophila* were highly susceptible to norfloxacin, tetracycline, and chloramphenicol, while *Pseudomonas* sp. was resistant to most antimicrobial agents (Kar et al., 1999).

Still further works have been done on the characterization of bacterial flora in EUS-affected fishes. Bacterial cultures prepared form body lesions of live EUS-affected *A. testudineus* have been confirmed as *Aeromonas* spp. by biochemical tests including the indole test, oxidase test, TSI test, citrate test, MLA, gas production, DNase test, and slide agglutination with standard antisera AH$_4$. Bacteria were further confirmed to be *A. hydrophila* and *Aeromonas sobria* by polymerase chain reaction (PCR) technique and PCR-RAPD analysis. The bacterial slants, prepared at the same time, were simultaneously tested for the presence of *A. hydrophila*; it was found that slants prepared from EUS-affected fishes contained *A. hydrophila*, whereas slants prepared from corresponding healthy fishes did not contain *A. hydrophila*.

Fungal Flora as Causative Factors

EUS-affected fishes, viz. *C. mrigala* and *P. conchonius*, having pale raised lesions that have not yet completely ulcerated, have been chosen for fungal isolation works. The fish specimen had been killed by decapitation and pinned to a dissecting board with the lesion uppermost. The scales around the periphery of the lesion were removed and the underlying skin was seared with a red-hot spatula to sterilize the surface. The fish with the board was then removed to a laminar flow cabinet with filtered air free from fungal elements. Then, the stratum compactum underlying the seared area was cut through using a sterile scalpel blade and a sterile, fine-pointed pair of rat-tooth forceps. This was followed by exposing the underlying muscle by cutting horizontally and reflecting superficial tissues. Four pieces of affected muscle *c* 2 mm^3 in dimension had been carefully excised up using aseptic technique and they were placed in Petri dish containing Czapek—Dox medium with standard supplements. This experiment led to isolation of *Aphanomyces* sp. with concomitant occurrence of the same fungal genus in histological sections of the same EUS-affected fishes (Kar, 2007a,b,c; 2013) (Plates 4a and b).

It may be mentioned here that the isolated bacterial and fungal flora have been living with fishes in water since time immemorial without generally causing any harm to them. As such, bacteria and fungi may not be considered the primary etiology for outbreak of EUS in fishes. Further, intradermal inoculation of these microbes into a guinea pig (*Cavia* sp) revealed no lesion formation in the latter's body for about two years (Kar, 2007a,b,c; Kar and Dey, 1990a,b,c; Kar et al., 1990a,b).

Viral Particles as Causative Factors

Absence of hemorrhagic septicemia (quite characteristic of *Aeromonas* infection) in almost all of the ulcerated fish (Roberts, 1989); nonoccurrence of these pathogens in the early part of the disease (Lilley et al., 1992); and close

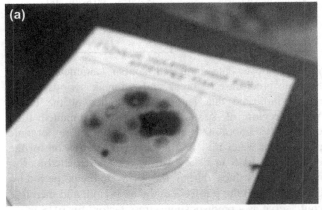

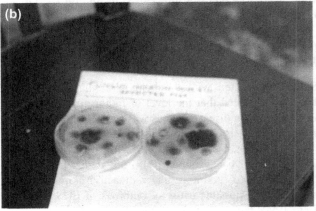

PLATE 4

association of these bacteria with fishes in water (as stated earlier) since time immemorial (Kar and Dey, 1990a,b,c) suggest that *Aeromonas hydrophila* is unlikely to have any primary infective role in the pathogenesis of EUS. Likewise, a primarily viral (?) etiology (filterable biological agent) is always suspected to be associated with the initiation of EUS (FAO, 1986; Kar, 2007a,b,c, 2013; Kar et al., 1990a,b).

As such, in order to confirm the presence of virus, EUS-affected fish tissues have been subjected to virological examination. Ulcerated tissues from EUS-affected *C. batrachus* have been homogenized in PBS, clarified in high-speed centrifugation (10, 000 rpm) for 10 min, and decontaminated through membrane filtration (0.22 μm). Monolayer cultures of BF2 fish cell line have been grown in 96-well microtiter plate at 28 °C in Leibovitz's L-15 Medium supplemented with 10% FCS and vitamin mixture. A 50-μL aliquot of each dilution (up to 10^{-7}) of the filtered supernatant was added to an 80% confluent monolayer of BF2 fish cell line grown in 96-well microtiter plate. After 2 h of adsorption at room temperature, cultures were overlaid with 150 μL of maintenance medium (L-15 supplemented with 2% FCS) and incubated at

28 °C. Progressive cytopathic effect (CPE) appeared between 48 and 72 h postinfection day (PID) in 10^{-1} to 10^{-4} dilution. The infection was passable in the subsequent cultures indicating the isolation of virus (Kar et al., 1993, 1994, 1999a, 2003a, 2004a; Kar, 2007a,b,c, 2013) (Plates 5a and b and 6a and b) Efforts are being made to characterize the virus and correlate the same with viruses isolated elsewhere (Frerichs et al., 1989; Ahne et al., 1988; Wattanavijarn et al., 1985). Electron microscopical (EM) studies (Kar and Dey, S 1988d) of the TCF showed occasional round particles unlike those of Roberts' group in Scotland (Frerichs et al., 1986).

The above study confirms that EUS outbreak in FW fishes is initiated by primary viral infections in the form of very small red spots, which ultimately could be transformed into amber-colored lesions and may terminate into ulcers; followed by secondary bacterial/fungal/parasitic contaminations; and could culminate in the death of the fish (Kar, 2000). Viral presence is said to be only for a very brief period and the havoc is caused mainly by bacteria and fungi (Kar, 2002, 2007a,b,c, 2013). Similar findings were reported by Tripathi and Qureshi (2012), which has been elucidated later in this chapter.

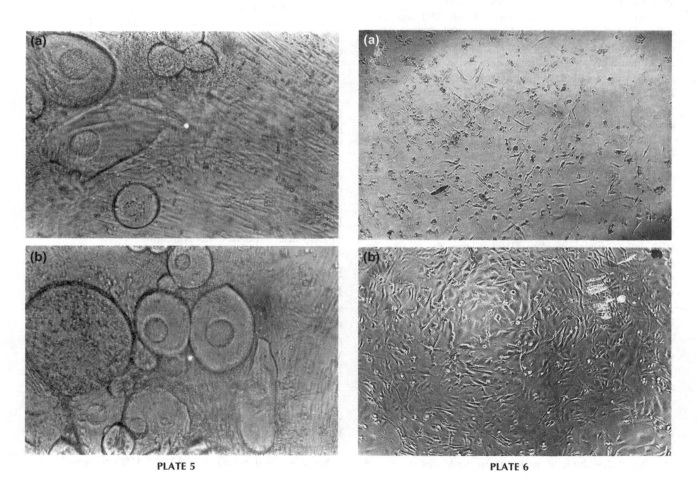

PLATE 5   PLATE 6

Some of our observations on tissue culture works done during 2003–04 are as follows:

Successful explants culture from *C. catla* was grown in L-15 medium and the cells showed good attachment to the tissue culture flask (Plate 5a and b).

The BF2 fish cell lines have shown best results at room temperature (RT) and at 28 °C. When BF2 cells were infected with filtered suspension (10–20%) tissue extract, the cells started showing signs of rounding and detachment from the second day postinfection in 24-well plates. This phenomenon persisted on serial passaging in 25 cm² flasks and CPE was observed from the second day onward. The same was observed when the cells were grown in 75 cm² flasks. The cells showed bulging and rounding, which persisted for 7–8 days.

When another fish cell line, viz. SSN1, was infected with the undiluted TCF collected from BF2 cells, at passage level 4, after two blind passages, the infected SSN1 cells started showing arrested growth with a few distorted cells. This effect persisted on further passage.

The infected TCF collected after three to four passages in BF2 cells was titrated. The CPE was observed up to the dilution of 10^{-6} to 10^{-10} in freshly collected TCF and in stored TCF at −20 °C with no loss of titer.

Test on various cell lines showed that BF2 and SSN1 cells are susceptible to the viral isolates. The CPE consisted of granulation, rounding, and clumping followed by detachment of cells in the case of BF2. In SSN1, cell growth was arrested, while another fish cell line, viz. RTG2, was found to be resistant to EUS infection.

After inoculation of suspected virus/filterable biological agent, best infection result was obtained at 21 ± 1 °C.

In all experiments with tissue culture, display of steady CPE was observed up to the fifth passage level on average.

There has been a gradual decrease in CPE with increase in dilution.

Experiment was conducted to see the effect of UV radiation on susceptibility of cells *vis-à-vis* enhancement of sensitivity of the cells to viruses. Confluent monolayers of BF2 cells were exposed to UV radiation for various periods, viz. 15, 35, 45, and 60 s, and then, the irradiated cells were infected with viral suspension. Nonirradiated cells were used as control. It was found that UV-irradiated BF2 cells showed CPE in the second passage instead of normally at the fourth or fifth passage. It could be because the virus was inhibited by certain cellular mechanism that was unlocked due to UV radiation.

Experiment was conducted to see the effect of heat on activities of the virus (filterable biological agent). Aliquots of viral isolate were heated in a water bath up to 56 °C. After 1, 2, 5, 10, 15, 30, 45, and 60 min of incubation, treated virus was inoculated into confluent monolayer of BF2 cell line. Unheated viral isolate was used as control. It was found that the heat-treated virus could not show CPE beyond 10^{-2}, while viral isolates not treated with heat showed usual CPE generally up to 10^{-6}.

Further effects of temperature on viral virulence were ascertained through experiments.

In temperature sensitivity experiment, cells infected with virus and incubated at 20°–25 °C were found to show prominent CPE. In contrast, cells incubated at 28°–30 °C, after viral inoculation, did not show prominent CPE or show late CPE. This indicates that temperature plays a very important role in the pathogenesis of EUS. This lower temperature range corresponds with that of the ambient water temperature during November–February when natural outbreaks of EUS generally occur.

Parasites as Causative Factors

Parasitism is an association in which one organism, known as the "parasite," lives on or inside the body and at the expense of another organism known as the host. Parasitic mode of life could be a secondary adaptation. The parasites might have arisen independently from free-living ancestors. Nevertheless, the relationship of a parasite to its host is of varying degrees of intimacy. The parasite, however, may be **epizoic** or **ectoparasitic,** which lives on the external surface of the host, though it may wander into the buccal cavity or rectum, particularly some protozoans. Conversely, a parasite may also be **entozoic** or **endoparasitic** (living within the body of the host).

Kar et al. (2011) performed a detailed study on fish diversity and helminth fauna in fishes of some wetlands of Assam and Manipur, India. Twenty-five species of fishes in Dolu Lake, 34 species in Chatla Haor, 40 species in Karbala Lake, 31 species each in Awangsoi Lake and Oinam Lake, 33 species in Loktak Lake, and 32 species in Utra Lake were recorded during the study period. The study revealed that 22.72% of fishes of Dolu Lake, 5.76% of Chatla Haor, 19.16% of Karbala Lake, 23.83% of Awangsoi Lake, 33.86% of Oinam Lake, 49.75% of Loktak Lake, and 48.8% of Utra Lake were parasitized. Incidence of EUS-affected fishes was common.

Study Areas

Dolu Lake (24°55′23.4″ N and 92°47′21.5″ E) is situated along the Silchar–Haflong highway. **In Dolu wetland,** out of 396 fishes examined, 90 (22.72%) fish specimens were parasitized by one or more species of helminth parasites with an overall intensity of 4.32. A total of 389 helminth parasites belonging to four different groups of helminths were found in the analyzed fishes. Of the infected fishes, 35 fishes (38.88%) were infected with only one group of helminth parasite, while 55 (61.11%) were infected with two or three species of helminth parasites. Regarding frequency distribution, it was found that 30 (33.33%) fishes contained nematodes, 12 (13.33%) trematodes, 21

(23.33%) cestodes, and 27 (30.00%) acanthocephalans. *N. notopterus* had the overall highest prevalence (54.76%) and highest intensity of infection (9.91) as well (Singha et al., 2014). EUS-affected fishes are often encountered.

Chatla Haor ($92°46'11.8''$E and $24°42'38''$N) is situated approximately 5 km away from Assam University, Silchar. It is one of the biggest seasonal floodplain wetlands located in the Cachar district of Assam. **In Chatla wetland,** out of 833 fishes examined, 48 (5.76%) fish specimens were infected by helminth parasites with an overall intensity of 4.95. A total number of 238 helminth parasites, belonging to four different groups of helminths have been found in the analyzed fishes. Among fishes infected, 10 fishes (20.83%) contained nematodes, 5 fishes (10.41%) trematodes, 14 fishes (29.16%) cestodes, and 19 fishes (39.58%) carried acanthocephalans. *C. batrachus* had the overall highest prevalence of infection and *C. punctata* had the overall highest intensity of infection (Puinyabati et al., 2013). EUS-affected fishes are often encountered.

Awangsoi Lake ($24°39'48''$N and $93°46'90''$E) is located to the south of Keinou village in Bishnupur district, Manipur, about 22 km away from Imphal city. It contains about 31 species of fishes belonging to 20 genera, 5 orders, and 14 families. **In Awangsoi Lake,** out of 3206 fishes examined, 764 fishes (23.83%) were found to be infected by helminth parasites with an overall intensity of 5.06. A total of 3866 helminth parasites were found to have infected the analyzed fishes. Frequency distribution showed that 434 (56.8%) fishes contained nematodes, 355 (46.46%) trematodes, 74 (9.68%) cestodes, and 178 (23.29%) acanthocephalans. *A. testudineus* depicted the overall highest prevalence of infection (69.4%) while the overall highest intensity was found in *C. fasciatus* (9.2) (Puinyabati et al., 2013). EUS-affected fishes are often encountered.

Karbala wetland (area 0.2345 ha at FSL) lying between $24°$ N and $92°42'$ E and situated 5 km from the Assam University, Silchar is a potential fish habitat for earning their livelihood by the riparian fishers. The wetland has a maximum depth of 2 m at FSL. **In Karbala wetlands,** 11 species of fishes were found to be parasitized with helminth parasites. Out of 663 individual fish of 11 species examined, 127 individual fish were parasitized with 665 helminth parasites. The overall prevalence percentage was 19.16% and intensity 1.0 was recorded. Among fishes examined, *C. punctata* was found to be infected with four groups (trematode, cestode, nematode, and acanthocephalan) of helminth parasites with maximum prevalence percentage (10.8%) and intensity (0.1) of trematode infection. *Monopterus cuchia* was found to be infected with three groups (cestode, acanthocephalan, and nematode) of helminth parasites with maximum percentage (33.3%) and intensity (0.5) of cestode infection. *P. ticto* and *Puntius sophore* showed the infection of both cestode (3.92%, 0.16) and acanthocephalan (5.3%, 0.21) parasites. *C. fasciatus*

was found to be infected with trematode (1.92%, 0.04) and nematode (13.5%, 1.02) parasites. *C. batrachus* was found to be infected with cestode (60.0%, 5.74) and nematode (11.43%, 0.77) parasites. *A. testudineus* (9.09%, 0.09), *L. guntea* (11.11%, 0.15), and *Mystus bleekeri* (4.76%, 0.05) were parasitized with nematode parasite only. Further, *H. fossilis* (21.2%, 0.30) and *C. striata* (16.7%, 0.17) were found to be parasitized with cestode and acanthocephalan respectively (Binky et al., 2014). EUS-affected fishes are generally encountered, particularly during the dry season.

Oinam Lake is a shallow and very old semiterrestrial FW lake having an area of about 0.71 km^2 It is located in the intersection of $24°25'-24°40'$N latitude and $93°45'-93°55'$ longitude and is situated about 783 m MSL. It has a maximum depth of 0.82 m. **In Oinam Lake,** out of 313 fishes examined (belonging to four different orders), 106 specimens were parasitized by one or more species of nematodes. The prevalence of infection is highest in *A. testudineus* (44.44%) and lowest in *N. notopterus* (16.66%), while the maximum intensity of infection was observed in *C. fasciatus* with the intensity of infection to be (6.33%). The lowest infection was observed in *M. bleekeri* with the intensity of infection 1.5% (Sangeeta et al., 2014). EUS-affected fishes are usually encountered during the winter–spring.

Loktak Lake is the largest natural lake in NE India and situated about 38 km away from Imphal city in Manipur state. It has an area of 26,000 ha with a catchments area of more than 98,000 ha, and is situated between $93°46'$E$-93°55'$E and $24°25'$N$-24°42'$N. The lake stretches from downstream of the Iril confluence near Lilong to the Khuga confluence of Ithai. The maximum depth of the lake is 4.58 m, and the average depth is 2.07 m. The unique feature of Loktak Lake is the floating swamps or mats, locally called Phumdi. **In Loktak Lake,** out of 2265 fishes examined, 1127 fishes (49.75%) were found to be infected with four different groups of helminth parasites, viz. Acanthocephala, Trematoda, Cestoda, and Nematoda. Among the four groups, nematode parasites are the most commonly found, with 11 different species from 10 different fish species recorded. This was followed by trematodes, with five different species from eight different fish species. Among cestodes, three species have been identified from three different fish species, and some are yet to be identified. Two different species of Acanthocephala are recorded from four different fish species (Ranibala et al., 2013). EUS-affected fishes are often encountered.

Utra Lake is an old eutrophic wetland lying at a distance of 20 km to the south of Imphal town. It is situated between $93°45'-94°0'$E longitude and $24°30'-25°45'$N latitude and is located about 787 m above MSL. The total surface area of the lake is about 0.366 km^2. The dimensions are found to be maximal in the rainy season. A rich diversity of about 32 species of fishes has been recorded in this lake. **In Utra Lake** of Manipur, out of 684 fishes examined, 334 fishes (48.8%) were found to be infected with various helminth parasites with an overall intensity of 2.6. From this investigation, peak prevalence of infection was found in *Channa orientalis* by the nematode parasites (66.5%) while infection by the acanthocephalan parasite in *C. punctata* (1.1%) was found to be the lowest. The highest intensity of infection was observed in *A. testudineus* by the trematode parasite (9.2) (Geetarani et al., 2011). EUS-affected fishes were common.

Singh (2015) reported about parasite fauna in fishes of **Sone Beel, river Jatinga** in Assam, and in **Pumlen Lake in Manipur. Sone Beel**, situated between 92°24′50″−92°28′25″E and 24°36′40″−24°44′30″N, within the Karimganj district of Assam, 3458.12 ha at FSL, 409.37 ha at DSL, probable origin in Mio-Pliocene, having continuous inlet called river Singla and outlet called river Kachua, falls in a valley geologically called syncline (Kar, 2007a,b,c). It is the biggest wetland in Assam. Concomitantly, Pumlen Lake, lying between 93°50′E to 94°0′E and 24°20′N to 24°35′ N. at an altitude of 768.5 m MSL, located in Thoubal district of Manipur state and is situated at a distance 55 km from Imphal city (Singh et al., 1993, 2015).

Notwithstanding the above, **river Jatinga, another site of study of fish parasites,** is a second-order stream that originates from the North Cachar Hills in Assam and flows down the side of Jatinga village. It is one of the north bank tributaries of river Barak. The reach type is mainly pool-riffle to braided-type, while the valley segment of the river is colluvial to alluvial type. Bedrocks, boulders, cobbles, gravels, and fine sand form the main components of the river substratum. The study spots (selected reaches) of the river were situated within the geographical limits of 24°57′49.4″N and 92°45′41.3″E having an altitude of 19 m MSL on average. **In river Jatinga**, fishes like *Danio aequipinnatus, Botia dario, Barilius vagra, Barilius tileo,* etc., take shelter mostly in the interspace between logs and fallen trees. Further, fishes like *Garra gotyla gotyla, Psilorhynchus balitora, Glyptothorax striata, Crossocheilus latius latius, Labeo pangusia,* and *Neolissochilus hexagonolepis* occur mostly in the rheophilic upstream region where mainly riffle-pool characteristics exist. LGF-like *Bagarius bagarius* occur mostly in the deeper pools in the upstream region of the river (Kar, 2003a, 2005a; Kar et al., 2002a; Kar, 2005a). The critically endangered mahseer fish *Tor progenius* has been recorded in the riffle-pool region of upstream of river Jatinga (Kar et al., 2002b).

In addition to the above, Devi (2015) had portrayed preliminary report about the status EUS-affected fishes in relation to parasitic infestations and parasitization. Her work revealed the occurrence of two species belonging to the groups Trematoda, Cestoda, Acanthocephala, and Nematoda. Out of these, nematodes were collected from

C. striata; cestodes were collected from *H. fossilis;* and acanthocephalans were from *C. striata* and *C. punctata.* However, trematodes were collected from *C. punctata.* Her work further revealed that the percentage of prevalence was found to be highest and lowest in *M. armatus* and *C. striata* respectively, while *M. armatus* and *C. punctata* had depicted the highest and lowest values of intensity of infection respectively. The total percentage of prevalence of infection was found to be 65.12%, while the intensity of infection was 1.3. Limnological parameters such as temperature, pH, DO (mg/L), FCO$_2$ (mg/L), and TA (mg/L) were recorded as **11.67 ± 0.88, 6.63 ± 0.12, 4.33 ± 0.15, 10.29 ± 1.62, and 35 ± 0.58 respectively.** Further, Choudhury and Barbhuiyan (2015) had also reported about incidence of EUS-affected fishes in Assam University hatchery.

Concomitant to the above, Sone Beel wetland (Kar, 1984) distinguished itself, in harboring a bewildering diversity of 70 species of fishes belonging to 49 genera, 24 families, and 11 orders (Kar, 2013). Incidentally, in the Sone Beel wetland, two species of Trematoda, seven species of Cestoda, two species of Acanthocephala, and four species of Nematoda have been recorded. The Pumlen Lake, situated in a different eco setup, had recorded 49 species of fishes belonging to 32 genera, 16 families, and 5 orders. The parasite load in Pumlen Lake revealed three species of trematodes, six species of cestodes, two species of acanthocephalans, and eight species of nematodes. Concomitantly, the river Jatinga, a lotic environment, revealed the occurrence of 61 species of fishes belonging to 36 genera and 16 families. About 14 species of parasites belonging to Nematoda, Acanthocephala, and Trematoda, including 2 new species of trematodes, were recorded from the Jatinga River.

Notwithstanding the above, works done in other prominent habitats by different workers in the region, notably Mandal (2003); Dhar (2004); Mazumdar (2007); Deb (2009); B. Das (2009); Sinha (2010); M.H. Barbhuiya (2010); Barman (2011); Das (2013a); Kar and Kar (2014); Das (2013b); Devi et al. (2014, 2015); and Singha (2015) also often reveal the occurrence of EUS-affected and parasitized fishes. Likewise, other rivers such as the Barak River in Assam−Manipur and its tributaries (Kar, D., 1996, 1998, 2007) the river Kopili in Assam (B. Das, 2009); the river Dhaleshwari in Assam (A.H. Barbhuiya, 2011); the Siang River in Arunachal Pradesh (Das, 2015); the Towkak River at the Assam−Nagaland border (Dutta, 2015); the Tuirial, Kolodyne, and Karnafuli Rivers in Mizoram; and the Singla and Longai Rivers along the Assam - Mizoram - Tripura borders (Das et al., 2015) may also reveal EUS-affected and parasitized fishes.

Riji et al. (2014), as a midterm report of the DBT (government of India)-sponsored Project on Molecular Characterization of Pathogens Associated with Fish

Diseases in Assam, acknowledging DBT, had reported that the collected fishes were processed and analyzed for the presence of viruses by diagnostic PCR for four DNA and three RNA viruses. The samples were also processed for virus isolation using cell-culture systems. The analysis depicted that infected fishes collected from NE India had carried ranavirus infections indicated by amplicons of specific sizes obtained by PCR. The sequence analysis of fragment also had displayed that the infected virus is a ranavirus with 90% homogeneity. In addition, Kar et al. (2015) and Singha et al. (2015) reported about preliminary epidemiological studies and associated limnological parameters in connection with EUS fish disease.

Prevention

There has been no major attempt for the prevention of fish parasites in the wild, although they may pose potential health hazards. The main reason could be nonownership of water bodies in the wild. However, attempts are being made by the present team of workers for finding ways and means of their control.

Hematological, Histochemical, and Enzymological

Preliminary hematological studies revealed higher DLC and ESR but low Hb content in the blood of EUS-affected fishes as compared with corresponding healthy fishes of the same species (Kar et al., 1994).

Preliminary histochemical studies (Kar and Das, 2004a) revealed interruption of glycogen synthesis in the liver and blockade of respiratory pathways in the gills. Further studies in this aspect are in progress.

Preliminary enzymological studies (Kar and Das, 2004b) with EUS-affected fishes revealed high values of alkaline phosphatase, SGOT, SGPT, and LDH. Further works in this direction are in progress.

Electron Microscopical Studies

Preliminary EM works done with EUS-affected tissues of *C. punctata* and *A. testudineus* revealed the occurrence of viruslike particles (Kar and Dey, S 1988d; Kar et al., 1993, 1995a). Detailed transmission electron microscopical (TEM) studies done with ultrathin sections of EUS-affected tissues of *C. striata* and *A. testudineus* revealed membrane bound and granular abnormal structures, hepatocytes with vacuolations, and viruslike inclusion bodies in the nucleus (Kar, 1999). Further, EM works done with EUS-affected tissues of *W. attu* revealed the presence of viruslike particles (Kar, 2004). More works in this direction are in progress (Plates 8a—i).

Histopathology

Detailed HP studies done with tissues of EUS-affected fishes revealed focal areas of increased fibrosis and chronic inflammatory cell-infiltration in muscles; focal areas of fatty degeneration of hepatocytes surrounding the portal triads in liver; hypertrophy of the muscle coat with mild chronic inflammatory cell infiltration in the gut; and dense infiltration with chronic inflammatory cell in the interstitium of the kidney. The liver section further showed occasional infiltration by mononuclear cells in the subcapsular region, while the skin of diseased fish showed epidermal necrosis and desquamation; the stomach portrayed atrophy of the mucosa showing necrotic villi; and the section of gill reflected infiltration by large number of mononuclear cells. Section through heart of EUS-affected fish showed coagulative necrosis of muscle fibres (Plates 7a—h) (Kar and Upadhyaya, 1998; Kar et al., 2003a).

Our View

Although the water quality/limnological parameters were by and large almost the same between the EUS outbreak period and the pre-EUS outbreak period, there had always been a sudden alarming fall in TA values just before the start of EUS outbreaks, indicating that TA could serve as a predisposing "stress factor" leading to decreased immunity levels in fishes, thus enabling viruses to attack fishes "primarily" for a brief period, causing trauma that makes way for bacteria and fungi to infect fishes secondarily and cause havoc in them.

The bacteria that have been isolated, are mostly acidophilic and psychrophilic. Around November—February every year, when there is sudden fall in air and water temperature, these psychrophilic bacteria grow prolifically, producing large amounts of acid in water. High acidity (caused due to large bacterial population) lowers TA and makes the aquatic environment favorable for the "viruses" to be activated. These viruses cause the trauma and disintegrate soon. The bacteria and fungi then attack secondarily and cause havoc in the affected fishes.

With the gradual rise in air and water temperature, with effect from March, the psychrophilic bacteria show decline in their population. Hence, there is less secretion of acid in water. As a result, TA goes high and the aquatic environment is no longer favorable for viruses. Therefore, EUS almost completely disappears during the period March to October every year.

Works on Epizootic Ulcerative Syndrome by the Author of the Present Treatise Are Summarized Below

EUS has been causing large-scale mortality among FW fishes since 1988, initially affecting four species of fishes very widely. Our study revealed fluctuation in the intensity of the disease in relation to species affected. Large hemorrhagic cutaneous ulcers, epidermal degeneration, and necrosis followed by sloughing of scales are the principal symptoms of EUS. Low TA could be a predisposing "stress

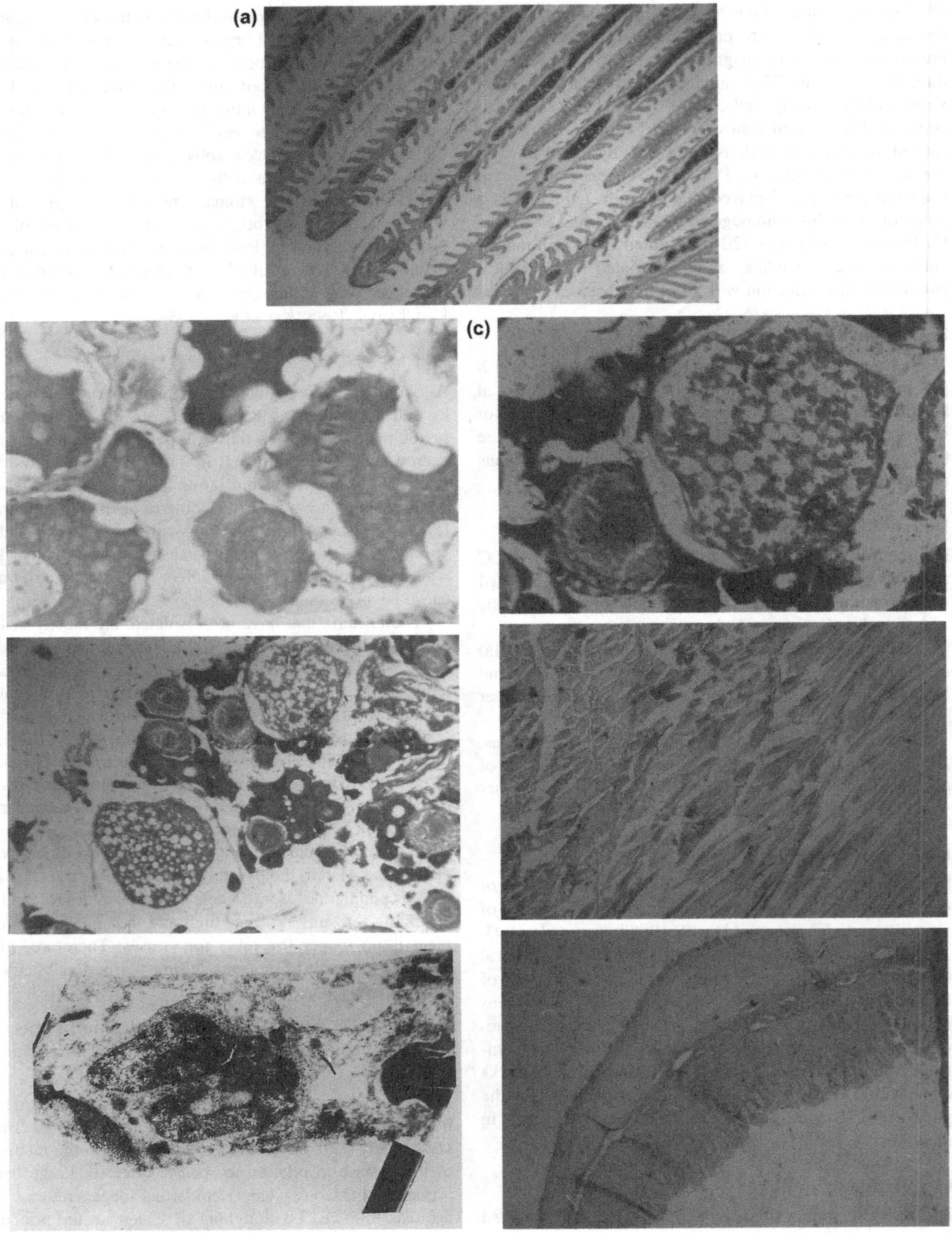

PLATE 7a–c

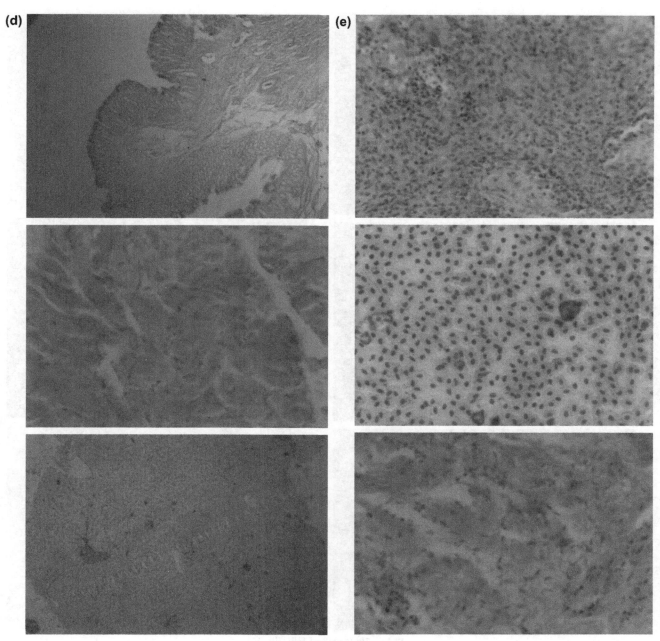

PLATE 7de

factor." Sick fishes show low hemoglobin and polymorphs, but high ESR and lymphocytes. Communicative nature of EUS revealed variation in time gap between fish and infection in different species. Inoculation of microbes in the test animals did not reveal of any sign of ulcerations for two years. Bacterial culture revealed occurrences of hemolytic *E. coli, A. hydrophila, P. aeruginosa, Klebsiella* sp., and *S. epidermidis* in the surface lesions and gut, liver, gills, heart, kidney, and gonads of sick fishes, all of which were found to be sensitive to chloramphenicol, trimethoprim, gentamicin, etc. Fungal isolation revealed the occurrence of

Aphanomyces sp with concomitant occurrence of the same fungal genus in histological sections of EUS-affected fishes. HP studies indicated focal areas of increased fibrosis and chronic inflammatory cell infiltration in muscles, focal areas of fatty degeneration of the hepatocytes surrounding the portal triads in the liver. Preliminary histochemical (HC) studies with regard to interruption in glycogen synthesis and blockade of respiratory pathways are being conducted. Similarly, preliminary enzymological studies are being conducted with regard to amount of alkaline phosphatase, SGOT, SGPT, and LDH. Inoculation

(f) (g)

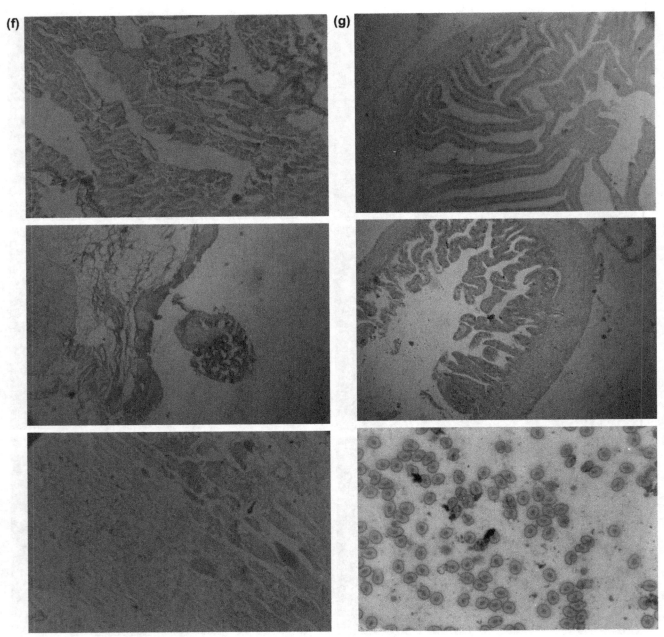

PLATE 7fg (g—HP of Fish sill)

of 10% tissue homogenate of EUS-affected *C. batrachus* into an 80% confluent monolayer form BF2 fish cell line in Leibovitz's L-15 Medium revealed progressive CPE that was passable in subsequent cultures, thus indicating the "isolation" of virus.

EM studies with ultrathin sections of still-occurring EUS in affected fish tissues revealed the presence of viruslike particles (inclusion bodies). Preliminarily, the picobirnavirus has been electron microscopically identified as the primary etiological agent of EUS. Further studies in this regard are being conducted.

On the basis of experiments conducted and observations made consistently, EUS could be defined as "seasonal epizootics of inland fishes with chronic mycotic granulomatous ulcers on the body, initiated with red spots followed by red circular ring due to the SHRV under thermal and chemical stress."

Since its origin, life has developed in an Al rich environment. But this element is usually found in insoluble chemical forms. Under acidic conditions, aluminum compounds become ionic, in a soluble or exchangeable form in soils (Duchaufour).

(h)

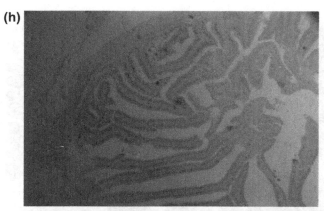

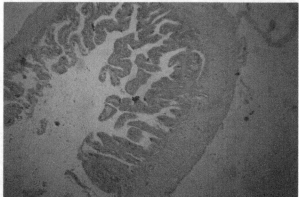

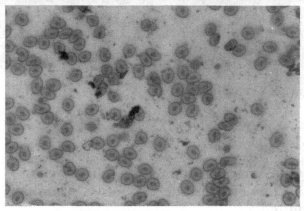

PLATE 7h (HP of Fish gut).

6.8 SIMILAR WORKS DONE ON EPIZOOTIC ULCERATIVE SYNDROME BY OTHER WORKERS

6.8.1 Similar Works Done by Other Workers Are Discussed Below

Work done by Das and Das (1993), which had similarities with work done by Kar (Kar and Dey, 1990a,b,c; Kar et al., 1993, 1994, 1995a,b,c; 2007; Kar, 2007a,b,c, 2013) revealed that fishes were affected by EUS in various water bodies spread across the length and breadth of India. The disease could be encountered during any time of the year. However, the period of occurrence of EUS may vary in different places. In Kerala, which is fed by two monsoons—that is, the southwest in June—August and the northeast in October—November, the disease may be found throughout the year. In Assam, EUS generally occurs toward the end of autumn and beginning of winter, and usually with sudden and rapid falls in the temperature of the environment. However, in West Bengal (WB), EUS outbreak occurs at the time of waning of rainfall and fall in water temperature. EUS first appeared in NE India during July 1988 (Kar and Dey, 1990a,b,c; Kar et al., 1990a,b, 1993, 1994, 1995a,d,c, 2003; Kar, 2007a,b,c, 2013; Das and Das, 1993). It gradually spread to the eastern, central, western, southern, and northern provinces of India. The disease in any particular area had been generally severe during the initial outbreak and gradually had usually diminished during the subsequent years and usually remained confined to localized areas. In addition, the use of adult and juvenile fishes from disease-prone or affected water bodies has often resulted in EUS outbreaks in previously unaffected water bodies. India witnessed the first major outbreak of EUS in July 1988 in the provinces of Assam, Tripura, and West Bengal. It gradually spread until 1992 in the provinces of Orissa, Bihar, Uttar Pradesh, Maharashtra, Andhra Pradesh, Tami I Nadu, Kerala, Karnataka, etc. (Kar, 2007a,b,c, 2013; Das and Das, 1993). A large number of ichthyospecies (numbering >100) living in both fresh and brackish water have been affected by EUS, out of which four are exotic and the rest are indigenous. Study revealed that the following genera of fishes had been found to be highly susceptible to infection by EUS: *Channa, Puntius, Mastacembelus, Mystus, Glossogobius, Anabas, Clarias, and Heteropneustes.* With regard to semiotics of the disease, it may be important to note here that symptoms and other characters of EUS are conspicuously different from the other low level ulcerative conditions reported earlier. It has distinct manifestations. Fishes in both lentic and lotic water bodies were said to exhibit abnormal swimming behavior with head projected out of water. In rivers, abnormal swimming behavior was witnessed with several fishes floating near the bank. In the initial stages of the disease, the infection usually began in the form of multiple inflammatory minute red spots on the body causing localized hemorrhage. In carps, these appear within the scale pockets. In advanced stages of infection, the ulceration covered larger areas with sloughing of scales and degeneration of epidermal tissue. With further advancement of the disease, ulcers became deep hemorrhagic and necrotic often with black melanistic rim. In advanced stages of the disease, large and deep ulcers were very commonly seen in all parts of the fish, especially on

(a)

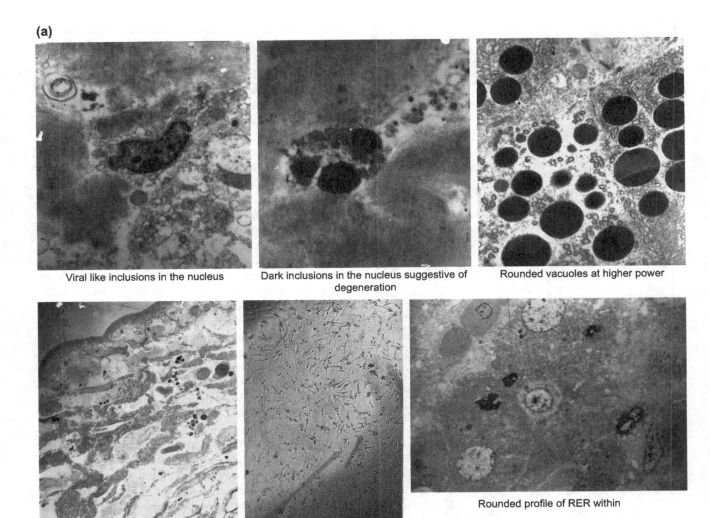

Viral like inclusions in the nucleus

Dark inclusions in the nucleus suggestive of degeneration

Rounded vacuoles at higher power

Gill of diseased C. striata

Rounded profile of RER within

PLATE 8a

the head, abdomen, and caudal peduncle. HP studies conducted on ulcerated fishes showed identical HP manifestations. In heavily ulcerated fishes, there is degeneration of epidermis of skin at the ulcerated areas and granulomatous formations. Basically, fungal granulomata occur in the dermis and hypodermis. A high degree of inflammatory reactions involving infiltrations by macrophage cells and lymphocytes around some of these granuloma formations were found. Liver-affected fishes did not show any significant change except vacuolization in certain cases. However, frequently most of the sinusoidal space and blood vessels were congested (hyperemia) and wandering lymphocytes were plenty in liver parenchyma. No changes were observed in kidney of affected fishes. Hematological parameters of affected fishes showed higher counts of phagocytic cells and reflected initiation of defense phagocytosis in blood circulation.

6.8.2 Investigation on the Causative Factors of Epizootic Ulcerative Syndrome in India

The causative agent of this dreaded disease has baffled scientists. it is widely suspected that a biological infectious agent is the primary cause of EUS and certain abiotic factors are responsible for creating stress to fish. The suspected biological agents are viral, bacterial, fungal, and other animal parasites. The investigations carried out in India until now on the different probable causative agents reveal that **environmental factors** such as the physicochemical parameters of water, and anthropogenic factors such as pesticides, fertilizers, and heavy metals, could play important roles in the outbreak of EUS. As such, data were recorded at specific water bodies in EUS disease-prone areas of West Bengal and in various other provinces, on selected

(b)

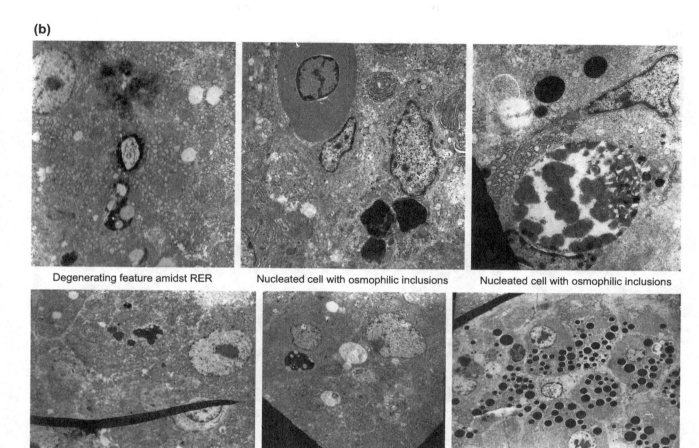

Degenerating feature amidst RER   Nucleated cell with osmophilic inclusions   Nucleated cell with osmophilic inclusions

Electron dense material in the cytoplasm   Osmophilic inclusions within the nucleated cell cytoplasm   Nucleated cell with rounded profile of RER & rounded osmophilic inclusions

PLATE 8b

physicochemical parameters having relevance to the EUS outbreak. It revealed that the affected water areas with severe EUS outbreak showed low alkalinity and hardness. The observation was in agreement with earlier reports from other countries affected by EUS that low alkalinity, hardness, chloride concentration, and fluctuating pH showed a link with EUS outbreak. EUS had also occurred in water areas with high alkalinity and hardness, but with lesser intensity. Investigations carried out at disease-prone areas in Assam, revealed that EUS generally occurs toward the end of autumn and beginning of winter, and usually with sudden and rapid falls in the temperature of the environment. In addition, studies carried out at disease-prone sites in West Bengal had showed that EUS outbreak did not occur during the monsoon period. Sharp falls in the hardness of water from higher summer values due to dilution during the rainy season seems to be another predisposing factor for triggering the outbreak of EUS disease. With regard to heavy metal concentration in water, though in some affected water areas significantly high values of Zn, Cu, and mercury were found, the data collected so far did not suggest any

perceptible role of heavy metal content in creating stress to fishes and subsequently predisposing it to EUS outbreak. With regard to pesticides and other agrochemicals, since the incidence of EUS is high in rice field environments in India, as in case of other countries where EUS had also occurred, pesticides were suspected to be associated with outbreak. Most outbreaks of EUS in India occurred after rainfall. This observation is in agreement with reports from-other countries leading to suspicion that drainages of agricultural chemicals might have an important role as predisposing factor for EUS outbreak. Further analyses of pesticide residue in water, fish, and plankton of specific EUS-affected water areas in India were carried out to assess the relation between pesticide use and EUS outbreak. The studies indicated that although occasionally higher concentrations or organochlorine and organophosphorus pesticides had been found in water and fish samples, no correlation could be made with the presence of pesticide residues and disease outbreak. The study indicated that the extent of pollution may create a stressed condition for aquatic life and may be the predisposing factor for EUS outbreak.

(c)

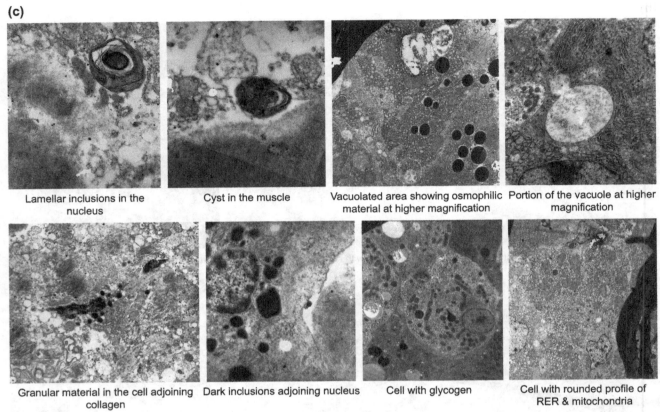

| Lamellar inclusions in the nucleus | Cyst in the muscle | Vacuolated area showing osmophilic material at higher magnification | Portion of the vacuole at higher magnification |

| Granular material in the cell adjoining collagen | Dark inclusions adjoining nucleus | Cell with glycogen | Cell with rounded profile of RER & mitochondria |

PLATE 8c

Tripathi and Quresh (2012) worked on the immunological diagnosis of naturally EUS-infected *Channa gachua*. The pattern of their investigation and the results obtained showed much similarity with results obtained by Kar and Dey (1990a,b,c); Kar et al. (1990a,b, 1993, 1994, 1995a,b,c); and Kar (2007a,b,c, 2013). Tripathi and Quresh (2012) had stated that an immunological approach had been adopted to diagnose the causative agents of EUS. As opined by them, it had been clearly established by the observation of the syndrome under natural condition that it usually develops at three different consecutive stages; that is, the **initial stage** (red spots occur on whole body), followed by an **advanced stage** (a red circular ring, distinguished by the circular hemorrhagic ring), and ultimately terminating in an **ulcer**. **To assess the involvement of various claimed pathogens in EUS,** two serological tests, DIBA (Dot-Blot ELISA) and IEM (immunoelectron microscopy) had been conducted at the three different stages of syndrome. It was clearly revealed by both tests that **only the virus SHRV-19E had been found at the initial stage of EUS,** as was also revealed by kar (2000). At the next stage of involvement, the virus SHRV-19E and the fungus *A. invadans* had been reported; at the last ulcerative stage, meanwhile, only the involvement of the bacterium *A. hydrophila* and the fungus *A. invadans* had been

reported. To test the virulence of the previously mentioned pathogens of the infected *C. gachua*, a lymphocytes transformation test (LTT) and serum profiling had been conducted. The maximum transformation had been reported against the SHRV-19E followed by *A. invadans* while least was against *A. hydrophila*. In serum profiling also, similar results were observed. As opined by Tripathi and Quresh (2012), their **present study had concluded that** the virus is the triggering factor for EUS in presence of thermal and chemical stress, resulting in damage to the physiological and immunological barriers of the host reflected in the form of a "red circular ring" (inflammation) and predisposed fish for the fungal attack. After fungal invasion and their proper vegetative growth, zoospores are liberated and probably make the host a receptor for all opportunistic pathogens and lead to severe granulomatous ulcers of EUS.

Based on experiments conducted and observations made consistently for a period of three years, EUS could be defined as "seasonal epizootics of inland fishes with chronic mycotic granulomatous ulcers on the body, initiated with red spots followed by red circular ring due to the SHRV under thermal and chemical stress."

Tripathi and Quresh (2012) had further stated that the conflicting views of various workers regarding the

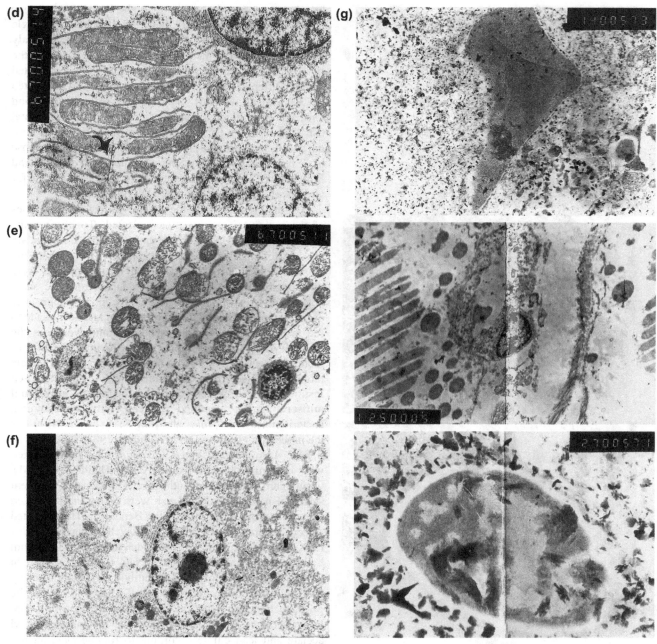

PLATE 8d—g

causative agent of EUS have created confusion rather than providing solution of the problem. It is observed that scientists are working separately on different aspects of EUS and everybody has been trying to prove the aspect of his/her study as the cause of the disease. The complicated nature of the problem thus requires well connected and cooperative efforts. Attempts had been made to develop symptoms of EUS through experimental infection trails since 1988 by various workers, based on which EUS had been considered to be of viral origin by Miller (1994), Kanchanakhan (1996a,b, 1997), Lio-Po et al. (2000), John and George (2012), Kar (2007a,b,c, 2013), Kar and Dey (1990a,b,c); bacterial origin by Jhingran and Das (1990), Pradhan and Pal (1993), Pal and Pradhan (1990), Ali and Tamuli (1991), and Shariff and Subasinghe (1994) and fungal origin by Roberts (1994), Robert et al. (1994a,b), and Saylor et al. (2010). Tripathi (2008) claimed to have developed the exact symptoms of EUS as they happen in nature. The present communication attempted to isolate and identify pathogens at various stages of EUS syndrome. It

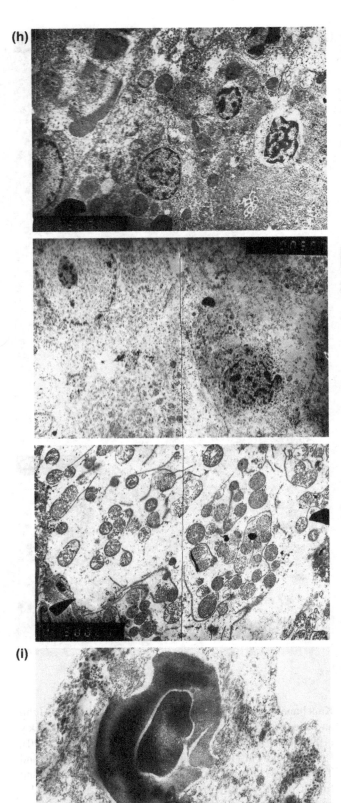

PLATE 8h—i

had been presumed that there might be more than one pathogen involved rather than involvement of single causative agent. To work out the same, a different study plan had been designed. Under this present study, the samples for the recovery of pathogens have been collected at three different stages that were then processed and tested by serological methods. To test the pathogenicity and virulence of individual pathogens against the experimental fish *C. gachua*, LTT and serum profiling had been conducted. Activation and proliferation is a frequent response of T lymphocytes to antigenic stimulation (Pechhold and Kabelitz, 1998), and thus it could be taken-up as test for virulence of pathogens against any given host.

For the present study, the severely infected fish species of the region, *C. gachua*, was selected. To observe the development of EUS syndrome under natural condition, six heavily polluted water reservoirs of the region, where the maximum incidences of EUS had been reported, were selected. Some 25—30 fish were caught from the same pond with help of local fishers and kept in happa after proper tagging. A continuous observation had been made to see the development of the syndrome and work out the proper symptomatology. With regard to immunization of rabbits to get PCA (**polyclonal antisera**), three female rabbits obtained from the Indian Veterinary Research Institute, Bhopal, were immunized to produce **polyclonal antisera** against identified antigens (virus: rhabdovirus; bacterium: *A. hydrophila*; and fungus: *A. invadans*) as per the procedure given by Delves (Plumb, 1983). The PCA of SHRV, *A. hydrophila,* and *A. invadans* were crosschecked with antiserum obtained from AAHRI, Thailand, and IMT Chandigarh, India. The antisera and spleens at different stages of naturally infected fish were taken and processed as per the procedure given by the Bendayan. The dissected spleens were fixed with McDowell—Trump fixative (4% formaldehyde, 1% glutaraldehyde, 80 μM sodium hydrogenphosphate, and 68 μM NaOH) and postfixed in 2% osmium tetrachloride. The antibodies so produced by the immunization were conjugated with gold. Float the grids, section side down on 1% gelatin in PBS for 10 min. Resin sections were stained with 1% methanolic uranyl acetate and lead citrate according to routine electron microscopy procedures. Ultrathin sections were stained with 2% aqueous uranyl acetate, neutral uranyl acetate, and embedded in methylcellulose. Sections were observed at different magnifications at AIIMS, New Delhi. To assess the presence of the three different pathogens indicated above at three different stages of the syndrome—that is, red spot, red circular ring, and finally ulceration—by dot-blot ELISA (DIBA), infected fish were collected from the happa that was kept in natural water bodies. To assess the presence of various pathogens, the spleen, skin, and liver were selected. Fishes with above-mentioned symptoms were brought into laboratory and dissected aseptically to take out the spleen, skin, and liver. The dissected organs were thoroughly washed in PBS and sonicated separately with help of ultrasonicator. This was followed by centrifugation of tissue at 300 × g to remove tissue debris. Supernatants were taken and again filtered by various pore size filter papers to segregate pathogens. For the purpose of DIBA, the procedure given by Yamaura (Cunningham and John, 1976) was followed. Infected fish were brought to the laboratory and sacrificed to dissect out the spleen under aseptic condition. The spleens from various samples were thoroughly washed by the PBS and preserved. For the purpose of IEM, PCA of SHRV, *A. hydrophila*, and *A. invadans* samples at different stages of naturally infected fish and artificially infected fish were taken and processed per the procedure given by Beasley (Thompson et al., 1973). A total of 27 DIBA tests were conducted for this purpose. The first nine tests were conducted with the three antisera raised against three claimed pathogens (rhabdovirus SHRV-19E, bacterium *A. hydrophila*, and fungus *A. invadans*) and the antigen/s obtained from three different samples of skin, spleen, and liver recovered at the initial stage of syndrome. This test was positive only with virus (rhabdovirus SHRV-19E) in all samples of skin, spleen, and liver. The next nine tests were conducted at the circular ring stage of syndrome with the same antigen/s and antisera. In this case, fungus was reported from all three samples, virus was reported from the spleen and liver only, and bacterium could not be recovered from any sample. Similarly, the last nine tests were conducted at the ulcerative stage of syndrome and with the same combination of antigen/s and antisera. In this case, fungus was reported from all samples, bacterium reported from skin and liver while virus reported from the spleen only. With regard to lymphocyte transformation, the degree of transformation was measured as "count per minute" (cpm) of H1 count incorporated into lymphocytes, and results are expressed in terms of the SR ratio that may be defined as "the ratio of cpm in stimulated culture to the cpm in unstimulated culture. The increase in number of transformed cells is recorded in all three experimental antigens. The net cpm is observed against the G-Protein of SHRV and it showed a dose-dependent rise. The serum protein profile of fish naturally affected by EUS was found to be in accordance with the interpretation given by Laurell (Mulcahy, 1971) for the stress conditions. The total protein of naturally infected fishes was found quite low as against the healthy fish. Albumin, a transport protein, was found to be quite reduced in naturally infected fish as compared with healthy fishes that might be due to the conversation of albumin in to protective protein (globulin). The acute phase protein CRP's 2 mobility was found to be too high in naturally infected fish, thereby indicating an extreme stress condition. The increase in globulin was a clear-cut sign of the infectious nature of the disease.

6.9 RELATED WORKS ON EPIZOOTIC ULCERATIVE SYNDROME DONE BY OTHER WORKERS

6.9.1 Some of the Other Studies Done

In addition to the above, works had been done on different parameters in order to reveal an answer to the enigmatic EUS. Hatai et al. (1984) had studied alterations in blood components in natural and experimentally infected ayu. Their results depicted alterations in levels of different blood parameters that, as they opined, are said to be the salient features of early phases of MG. Further, statistically significant differences had been revealed between healthy (control) and experimental (inoculated) fishes with regard to RBC count, Hb content, total cholesterol, and glutamate pyruvate *trans*-aminase (GPT). Moreover, as reported, fishes injected with *A. piscicida* SA7610 culture portrayed reduced RBC, Hb content, alkaline phosphatase, leucine amino-peptidase, and total cholesterol, but with concomitantly higher levels of GPT and blood urea-N.

Notwithstanding the above, Melba et al. (2005) had dealt with **disease and health management in Asian AQC**. Also, **Stephens et al. had worked on ornamental fishes in relation to EUS**.

Also, Rodger (2010) had reported about the *Fish Disease Manual*.

6.10 WORKS DONE BY DIFFERENT WORKERS IN DIFFERENT PLACES

6.10.1 Prevalence in Japan

An epizootic disease, considered the first of its kind in Japan (Egusa and Masuda, 1971), had been revealed to be of mycotic origin. It had affected the cultured FW ayu (*Plecoglossus altivelis*) in Oita Prefecture on the Kyushu Island. *Aeromonas liquefaciens* had been isolated from the ulcerative lesions. Workers were almost certain about the mycotic etiology of the disease. There had been fish mortality, sometimes on a large scale, to the extent of about 500 kg of fish. Localized swellings of the body wall, erosion of skin with lifted scales were some principal symptoms of the aliments. Necrosis in the lesions' centers, loss of scales and overlying skin, and exposure of skeletal muscles are some main symptoms of advanced stages of the disease. Further, expressions of the disease had differed significantly from those of *Vibrio* infection. Furthermore, antibiotics like chloramphenicol, nalidixic acid were also said to be ineffective. On the other hand, branched fungal hyphae were consistently detected under light microscopy from the skeletal muscle of lesion tissues. The hyphae had been observed to be surrounded by epitheloid cells associated with fibrosis in which multinucleate giant cells (30–50 μm) were quite common. The diameters of the hyphae were reported to vary from 5 to 25 μm. However, the fungus was later identified as *Aphanomyces* belonging to the family Saprolegniaceae. Subsequently, further information regarding its mycotic cause along with more detailed pathology had been reported by Miyazaki and Egusa (1972) in cultured goldfish and ayu by Miyazaki and Egusa (1973a). Concomitant to the above, in goldfish the details of morphological changes of granuloma progression, with regard to tissues of gills, skin, skeletal muscles, brain, spinal cord, liver, pancreas, spleen, kidney, intestine, etc., have been known. The multinucleated giant cells at the lesion site varied from species to species and from locations to location, and were of two types. Studies were also made on the behavior of the hyphae at different temperatures. In addition, Miyazaki and Egusa (1973b,c) had worked on the HP aspects of certain other icthyospecies like the bluegill (*Lepomis macrochirus*), wild snakehead (*Channa argus*), grey mullet (*Mugil cephalus*), etc. Dermatitis and myositis are the general pathological features associated with deep-penetrating granulomatous inflammation diagnosed by several layers of epithelioid cells surrounding the fungal hyphae that is the disease condition known as MG. Occasionally, multinucleated giant cells with or without fungal hyphae were also found.

Kumamaru (1973) had done epidemiological studies on EUS fishes in Lake Kasumigaura during 1971–72. He found yellowfin goby (*Acanthogobius flavimanus*) grey mullet, oily gudgeon (*Sarcocheilichthys variegates*), trident goby to be highly susceptible to MG. Conversely, silver carp (*Hypophthalmichthys molitrix*), common carp (*C. carpio*) and certain eels (*Anguilla japonica*) were generally found to be unaffected by MG.

Studies on MG-affected fishes with regard to gross pathology and light microscopy revealed localized swellings on the body surface, protruding scales, hemorrhage, loss of scales and exposure of underlying muscles and formation of ulcers. Spreading of ulcers over the broad area with exposed scarlet granuloma, and spreading of hyphae several millimeters below the skin, was observed in the advanced stages of MG. Under the light microscope, nonseptate branching of fungal hyphae and fungal granuloma were generally observed in MG-affected tissue sections.

Concomitant to the above, HP studies had been done from naturally affected fish (Egusa and Masuda, 1971; Miyazaki and Egusa, 1973a,b,c; Hatai et al., 1994); and from artificially infected fish (Hatai, 1980; Hatai et al., 1994). Lesions were found in those studies. These hyphae were said to be typically associated with granulomatous response and necrosis in tissues. Other things that were consistently observed in older lesions were marked inflammatory infiltration and formation of granulation tissue. In this regard, it may be mentioned here that a granulation tissue, as explained, is a reparative response not to be confused with granulomas. It comprises a fibrous network, regenerating capillaries and muscle fibres, mild hemorrhage and extensive inflammatory infiltration.

6.10.2 Epizootic Ulcerative Syndrome, Aphanomycosis, and Studies in Different Regions of the World: Australia, Asia, the United Kingdom, and the United States

Concomitant to occurrences and analyses of MG in Japan, studies on EUS outside Japan followed in many affected countries. Such studies had included detection of outbreaks followed by analyses of tissues, etc. In fact, works in this direction had continued from the early 1970s, beginning with Australia and followed by SE and South Asia due to sweeping of the region by the EUS epidemic, and probably continuing with its spread to the United States. Incidentally, the pattern of spread within and among the nations was, more or less, consistent with progressive dissemination of a single infectious agent. In fact, there had been a number of reviews on EUS that had been published during the 1990s (Roberts et al., 1993; Chinabut, 1995, 1998; Lilley et al., 1998)...... Earlier studies are said to have confirmed that EUS is caused by *A. invadans (A. piscicida).*

6.10.3 Works Done in the United Kingdom

There possibly had been no large-scale outbreak of EUS in the United Kingdom. Like many other workers, Lilley and Roberts (1997) had opined that *Aphanomyces invadans* is possibly the primary causative agent that could be responsible for the development of the salient pathology of EUS. In order to initiate EUS in possible susceptible host fishes, they had inoculated zoospores from 58 mycotic isolates intramuscularly into *C. striata*. The list of the fungi are as follows: (a) *Aphanomyces* strains isolated from EUS-affected fishes in Asia and Australia; (b) saprophytic *Aphanomyces, Achlya, and Saprolegnia* spp. from infected waters in different places; and (c) other saprolegnious fungi supposedly associated with causing other types of diseases in aquatic animals. It may be noted here that out of the above mentioned fungal genera, only *Aphanomyces* strains isolated from fishes affected by EUS, RSD, and MG could grow invasively through the muscles of fishes and also produce distinctive lesions of EUS. These pathogenic strains from the snakehead fish could further be demarcated from other comparable mycotic forms by their typical growth profile in relation to temperature and their inability to grow on any mycotic media.

In order to have more supportive background, Lilley et al. (1997a, 2003) further had attempted for analysis of r-DNA by using RFLP technique through sequencing of the ITS one region and RAPD in order to ensure that the 20 *A. invadans* isolates, which had been collected from EUS-affected fishes in Asia and Australia, could be representing a single fish-pathogenic species. Moreover, it had been portrayed that *A. invadans* seemed to be clearly distinct from a host of other aquatic animal pathogenic Saprolegniaceae collected from EUS-affected countries. The outcome of the study seemed to have indicated a gross absence of genetic diversity among all *Aphanomyces* isolates. In fact, some authors prefer to consider them not only to be conspecific, but also to represent a single clonal genotype. In view of the above results, some workers opined that *Aphanomyces* might had been able to colonize in Australia, Asia, and subsequently in the United States in relatively quick successional episodes associated with concomitant reports of disease outbreaks.

6.10.4 Research in Australia

The laboratory-based studies on EUS with estuarine fishes in Australia was said to have begun with two key field observations. The first was the prevalence of EUS in the susceptible wild estuarine fish population in the main channel of the Richmond River. In fact, EUS is supposed to be having its maximum prevalence at the sampling site close to the junction with tributaries that contain acidified runoff water from leached acid sulfate soils. The second was the prevalence and sampling of EUS-affected fishes from the wild EUS-susceptible FW population that could be confined to a drainage canal with a water pH <4.0, and that were exhibiting necrotizing dermatitis.

Attempts to initiate EUS symptoms in healthy fishes have been made preliminarily in trial tanks by exposing fishes to zoospores of *A. invadans*. Results did not reveal any expression of EUS symptoms. Further, it was believed that prior damage to skin was perhaps necessary before EUS-specific symptoms could be induced in the experimental fishes. As such, fingerlings of healthy experimental fish, sand whiting (*Sillago ciliata*), were sublethally exposed to acid sulfate runoff (pH 3.1). The subjected fish had developed severe epidermal necrosis. Concomitantly, fishes subjected to somewhat less acidic acid sulfate runoff (pH 5.1) had also developed mild to moderate epidermal necrosis. Further, when these fishes were later exposed to *A. invadans* zoospores, typical EUS lesions consisting of necrotizing granulomatous dermatitis and myositis occurred (with and intensity that was possibly much greater than in controls) associated with invasive nonseptate fungal hyphae. Subsequently, the fungus was also recovered in culture from the newly affected fish. These outcomes portrayed that *A. invadans* is possibly the primary infectious

agent of EUS and disruption of epidermal continuity might be a prerequisite to fungal attachments and lesion initiation. It was further opined that sufficient causes developed mainly artificially included a putative agent, *A. invadans*, and a concomitant abiotic cause in the form of chemically induced epidermal damage. There was possibly no evidence for bacteria, viruses, or other biotic agents as the possible etiology of EUS.

6.10.5 The Indian Scenario

A host of various studies had been done on EUS by the workers in India, as referred to above. Among these, experiments with cohabitation of EUS-affected and corresponding healthy fish of the same species yielded both positive and negative results—that is, in some cases EUS was reproduced, while in others it was not. Studies on communication nature of EUS revealed variation in disease occurrence with fish species and time (Kar, 2007a,b,c, 2013). Further experimental infectivity studies have been done under the IFS-funded project (1999–2001) in order to evaluate the disease susceptibility and sequential inflammatory response among different age groups of IMCs, viz. *Labeo rohita*, *C. mrigala*, and *C. catla*. Comparisons had been made with corresponding age groups of susceptible snakeheads (*Channa* sp.), *Puntius* sp., and the resistant common carp (*C. carpio*). Experimental infectivity studies had been carried out using *A. invadans* (Strain B 99C, provided by J.H. Lilley, as reported). It was found that bath exposure to zoospores often failed to reproduce the disease. Nevertheless, EUS could be initiated in a small population of fish by this method of challenge. Conversely, EUS could be consistently reproduced by injecting spores under the dermis.

Concomitant to the above, there is said to be variations in resistance to *Aphanomyces invadans* infection among the species and age-classes of different ichthyospeces, as reported by various workers. In this regard, it may be pointed out that the fry and fingerlings of IMCs seemed to be more susceptible to *A. invadans* infection, while yearlings of the same group appear to be less susceptible and resistant. Further, there is said to be decrease in susceptibility *vis-à-vis* increase in resistance with age among IMCs, contrary to the status in *Puntius* spp. and snakeheads. It could be mainly because the cellular defense mechanisms against the fungus appear to be well developed among yearlings of IMCs and the common carp, as compared with *Puntius* spp., and snakeheads, whose well-developed monoclonal antibodies (MAbs) against *A. invadans* had been developed of late. The strain B 99C had been given by JH Lilley. The MAbs (using an immunoperoxidase test) are said to had reacted consistently (a) with the fungal hyphae in HP sections of experimentally infected fish; (b) with EUS-affected fish collected from various parts of India during 2002; and (c) with the retrospective EUS tissue samples of 1995 and 1997. Such a result is said to support the view that a single mycotic pathogen has been involved in all EUS-affected cases.

Concomitant to the above, a simple immunodot test on nitrocellulose paper had been devised and tested on a number of suspected EUS-affected samples. Consistent results had been obtained in the immunodot test, if the samples for dotting had been taken from below the ulcer followed by the sides and contralateral part of the ulcer. On the other hand, it had been observed that samples dotted from the ulcer's surface did not yield consistent results. As reported, this could suggest that there had possibly not been any mycotic antigen in this area.

6.10.6 Works Done in Thailand

Quite a host of works had been done on EUS by different workers in Thailand. Kanchanakhan et al. (2002) was reportedly successful in inducing lesions of EUS in 20 out of 20 (20/20) juvenile snakehead fish (*C. striata*) maintained at 20 °C and injected intramuscularly with a rhabdovirus strain T9412 followed by bath challenge with spores of *A. invadans*. Concomitantly, it may be noted here that experimental fish that had received only growth media through injection, followed by spore challenges, also developed EUS, but strikingly fewer in number; that is, only 7 out of 20 were affected. Further, rhabdovirus injection alone could induce only small hemorrhagic lesions at the site of injection, most of which had healed by the end of the experiment. Concomitantly, a similar induction experiment, conducted at 29 °C, could not induce EUS. As reported, this had led the authors to infer that low temperature coupled with an infection with rhabdovirus could possibly be a prerequisite for initiation of EUS infection in murrels. It could subsequently be followed by an *Aphanomyces* challenge as a combination of events for transforming the disease into a virulent form.

6.10.7 Works Done in the United States

Different workers in the United States had worked on EUS. Notably, Kiryu et al. (2002) is reported to have generated lesions of EUS by both injection and bath exposure to zoospores of *A. invadans*. In dose–response studies with menhaden (*Brevoortia tyrannus*), it had been pointed out that *c* 31% of fish specimens inoculated with as few as one zoospore is said to had developed characteristic lesions of EUS within two weeks of inoculation. On the other hand, LD_{50}, by injection had been estimated to be only 10 zoospores per fish (Kiryu et al., 2003). These findings seemed to corroborate that oomycete is a primary pathogen and is highly virulent. These also seemed to support Koch's hypothesis.

6.10.8 Works Done in Pakistan

The financial losses due to the impact of EUS fish disease in some parts of the Punjab province in Pakistan had been estimated and reported (Iqbal, 1992, 1998). It had been reported that fish losses to the tune of Rs 1,25,000 ± (US$ 3125.0) is said to relate to losses incurred in fish farms. In fact, losses spread over *c* 10% of fish farms are included in the five affected districts, notably Kausar, Lahore, Gujranwala, Sialkot, and Narowal in the province of Punjab. The total economic loss amounted to US$ 300,000.0 approx. (AAHRI, NACA, 1989, IoA, and NSW-Fisheries, 1997). Moreover, wild fish catch had been reported to had declined by *c* 25% due to EUS. Further, complete disappearance of murrels, notably *C. punctata* in portions of the rivers Chenab, Ravi, and Sutlej, as well as associated canals, was reported.

Notwithstanding the above, loss of fishes due to EUS also had social and economic impact son fishing communities. There had been a loss of *c* 80% in the daily income of fishers in Maliwal village, on the bank of river Ravi, said to represent a reduction of Rs. 1.92 million (US$ 48,000.0) over one fishing season. In addition, there was also a *c* 20% reduction in government revenue from auction of fishing rights during August 1996, which had been caused, perhaps due to EUS. Further, it had represented an economic loss of revenue of *c* Rs 8.0 million (US$ 200,000.0), which could accumulate to a total of Rs. 40.0 million in successive years (AAHRI, NACA, 1989, IoA, NSW-Fisheries, 1997).

6.11 WORKS DONE ON DIFFERENT PARAMETERS RELATED TO EPIZOOTIC ULCERATIVE SYNDROME

6.11.1 Epizootic Ulcerative Syndrome Study Based on Epidemiology

The spread of EUS had been quite fast across the continents leading to large-scale mortality of fishes. Lilley et al. (1998) had used the following three components in connection with epidemiology of EUS:

6.11.1.1 Component Cause

The various causes that lead to a disease are individually called "component cause."

6.11.1.2 Sufficient Cause

Each combination of the various "component causes" that result in a disease is called "sufficient cause."

6.11.1.3 Necessary Cause

Different combinations of "component causes" may constitute "sufficient cause" for a disease to happen. These "sufficient causes" for a particular disease may have at least one "component cause" in common. This is known as the necessary cause that must always be present for that disease to happen.

In view of the above, believers in "EUS as an aphanomycosis" are said to consider *A. invadans (A. piscicida)* to be the only necessary infectious cause for EUS. To the contrary, proponents of "EUS as a polymicrobial disease" suggest that EUS has three necessary infectious causes, viz. (a) a virus (which in some cases is a rhabdovirus); (b) a fungus (which in some cases is *A. invadans);* and (c) a bacterium (which in some cases is *A. hydrophila* or sometimes *Vibrio* sp.). All three components may act in sequence or in concert to initiate EUS.

It is quite evident from the above that one or more additional component causes are usually involved in the causation of the outbreak. These could include the environmental predisposing "stress factors" that are said to increase the probability of infection by the necessary infectious agents that could infect the host and induce lesions; for example, exposure to acidified water, or water with low alkalinity, trauma-induced epidermal damage or conditions that lead to immunosuppression, which have been depicted as increase of susceptibility of fish to EUS.

EUS was first seen in the Northern Territory (of Australia) during 1986 on mullet from the Mary River. It was subsequently reported from many other river systems in the Top End in 1986, and fish of all ages appeared to be affected. There were reports depicting fewer outbreaks of EUS as the wet season approached. The disease had continued to appear in subsequent years, and has occurred particularly in the early dry season around May and June. Fish less than one year of age appeared to be very badly affected, while only sporadic cases had been reported in mature fish. There had been no confirmed reports of EUS during the wet season, from December to April and the syndrome seemed to be restricted to fish in FW and in water of low salinity in the upper estuarine reaches of rivers.

Koch's Hypothesis with Regard to Disease Reproduction

An infectious pathogen would possibly be able to cause a disease, only if it could come in contact with a susceptible host. Host-susceptibility is said to depend on a number of innate factors and environmental, nutritional, and toxic influences said to influence disease resistance. Henle-Koch's postulates have been in use for >100 years in evaluating the causal relationship of a new infectious agent to a clinical cause (Davis et al., 1980). In fact, Koch's postulates, when applied, tries to distinguish a pathogen from an adventitious microbe (Davis, 1980). The principal criteria of the postulates are as follows:

1. The organism is expected to be found regularly in the lesions of the disease.
2. The organism is expected to isolated in pure culture in artificial media.

3. Inoculation of this culture is expected to produce a similar disease in experimental animals.
4. The organism is expected to be recovered from the lesions of the experimental animals.

Concomitant to the above, there is said to be strong evidence that postulates have been satisfied with regard to *A. invadans* as the causal infectious agent of EUS. HP examination of lesions consistent with EUS from most affected nations has shown beyond reasonable doubt that one or more invasive fungi could be responsible for most of the host responses and tissue destructions. Various works had shown that *A. invadans* could be consistently recovered from progressing lesions, subject to the fact that stringent focus had been laid on obtaining uncontaminated inocula; as well, favorable culture conditions had been used (Lilley et al., 1998; Blazer et al., 2002).

Concomitant to the above, various works are said to have now fulfilled Koch's postulates and have led us to conclude, at least tentatively, that *A. invadans* (=*A. piscicida*) is the primary etiological agent of EUS (Hatai, 1980; Hatai et al., 1977, 1994, Catap and Munday, 1985; Kiryu et al., 2002, 2003. However, it is said that there had been apparent differences between the studies in rates and the severity in lesion-induction (Kiryu et al., 2003). Differences in the susceptibility of exposed fish species coupled with progressive loss of pathogenicity and/or virulence of the pathogen *A. invadans* isolates maintained for different periods in artificial media could serve as the possible reasons for the variations in severity of lesion induction.

Notwithstanding the above, a number of viruses, bacteria, and putatively saprophytic fungi have been inconsistently recovered from lesions of EUS-affected fishes. But attempts to fulfill Koch's postulates and initiate EUS lesions using these agents alone have had been largely not successful. For instance, an *Aphanomyces* sp. (not *Aphanomyces invadans*) had first been isolated from UM lesions in menhaden. This fungus was said to be incapable of initiating EUS-like lesions, and is now considered to be a saprophytic species (Blazer et al., 2002). Likewise, other workers had also failed to induce lesions, which could be said to be histopathologically consistent with the lesions of EUS (Lilley and Roberts, 1997). Other than *Aphanomyces,* the workers had exposed fish to many other fungi, including isolates from EUS-endemic areas. Nevertheless, immunohistochemical or in situ hybridization studies could be required to prove conclusively whether *A. invadans* is the sole fungal element that could be said to be involved in granulomatous lesions of EUS.

Notwithstanding the above, viruses isolated from EUS-affected tissue lesions include birnavirus (Wattamavijarn et al., 1985), a number of rhabdoviruses (Frerichs et al., 1989), reoviruses and a distinct group of type-C retroviruses. Frerichs et al. (1986) had suggested a virus to be the causative etiology of EUS. Further, infectivity experiments done by using a rhabdovirus alone could induce EUS ulcers but not lesions in exposed striped *Channa* sp. In fact, EUS lesions were induced only in fishes that were first exposed to rhabdovirus and then to *A. invadans.*

Notwithstanding the above, a number of bacteria were also isolated from ulcerative lesions in EUS-affected fishes, notably *A. hydrophila, Vibrio* sp, *Pleisomonas* spp. (Iqbal et al., 1998; McGarey et al., 1990). These bacteria had either been proposed as the causative agent of EUS (Rahman et al., 2002), or had been regarded as capable of inducing lesions macroscopically. In fact, these bacteria have been known to cause skin lesions in fishes and may act as secondary etiological agents or as "component causes." It had further been argued that if by definition, EUS lesions include a granulomatous response around invasive fungal hyphae, the lesions caused by bacteria could not be those of EUS.

6.11.2 EUS Study Based on Pathology

Biswas et al. (2006) had reported about **pathomorphological changes in fish affected with EUS.** A total of 67 inland FW bodies including floodplain wetlands (beels), culture ponds, and community tanks of seven development blocks spread over three districts in West Bengal were studied for pathomorphological changes of FW fishes affected with EUS during the period from May 2001 to April 2002. Hemorrhagic and necrotic ulcerative lesions were observed in tissues of EUS-affected FW fishes. In addition, granulomas of central necrotic area with presence of fungal spores, surrounded by inflammatory cells with fibroblast encapsulation were observed in skeletal muscles of *C. striata, C. mrigala,* and *Labeo bata.*

6.11.3 Epizootic Ulcerative Syndrome Study Based on Histopathology

Concomitant to the above, HP studies had been done from naturally affected fish (Egusa and Masuda, 1971; Miyazaki and Egusa, 1973a,b,c; Hatai et al., 1994), and from artificially infected fish (Hatai, 1980; Hatai et al., 1994). Studies portrayed similar pathological alterations. A number of branched aseptate hyphae had been found in the lesions. These hyphae were said to be typically associated with granulomatous response and necrosis in tissues. Other things that were consistently observed in older lesions were marked inflammatory infiltration and formation of granulation tissue. In this regard, it may be mentioned here that a granulation tissue, as explained, is a reparative response not to be confused with granulomas. It comprises a fibrous network, regenerating capillaries and

muscle fibres, mild hemorrhage and extensive inflammatory infiltration. With regard to **HP procedure,** dehydration through ascending alcoholic grades forms part of the processing of fixed tissues. The process also involves clearing in a wax-miscible agent and impregnation with wax. Fish tissue blocks are cut at about 5 µm and are mounted on a glass slide. The section is completely dewaxed before staining. Staining is done with hematoxylin and eosin (H and E). H and E are general fungal stains (e.g., Grocott's stain) that could demonstrate typical granulomas and invasive hyphae. **With regard to diagnosis that could be confirmatory,** it may be noted here that earlier EUS lesions were erythematous dermatitis with perhaps no oomycete involvement. The hyphae of *A. invadans* could be observed to be growing in skeletal muscle as the lesion progressed from a mild chronic active dermatitis to a severe, but locally extensive, necrotizing granulomatous dermatitis with severe floccular degeneration of the muscles. The oomycete elicits a strong inflammatory response, and granulomas are formed around the penetrating hyphae.

Further detailed HP studies done with tissues of EUS-affected fishes revealed focal areas of increased fibrosis and chronic inflammatory cell-infiltration in muscles; focal areas of fatty degeneration of hepatocytes surrounding the portal triads in liver; hypertrophy of the muscle coat with mild chronic inflammatory cell infiltration in the gut; and dense infiltration with chronic inflammatory cell in the interstitium of the kidney. The liver section further showed occasional infiltration by mononuclear cells in the subcapsular region, while skin of diseased fish showed epidermal necrosis and desquamation; the stomach portrayed atrophy of the mucosa showing necrotic villi; and the section of gill reflected infiltration by large number of mononuclear cells. Section through heart of EUS-affected fish showed coagulative necrosis of muscle fibres (Kar and Upadhyaya, 1998; Kar et al., 2003a).

Chavda and Bhatt (2014) had reported about the HP and ultrastructural analysis of microsporidian infection in catla *(C. catla)* and their effects on antioxidant enzymes and protein expression. They had stated that the microsporidian infection had been detected in catla and catla fry. The sporophorous vesicles, necrosis, and liquefaction of muscles, as well as hypertrophy, hyperplasia, and necrosis in gills and liver, were observed. In the ultrathin muscle section, ellipsoidal (1.95 × 2.12 µm) spore with anchoring disc, protuberant area and basal vacuole; oval meronts (3.2 X 3.1 µm), bounded by the membrane stakes with interrupted pores were observed. In the gill epithelia, disrupted organelles, tubular structures, coated vesicles, and round spores (1.43 × 1.47 µm) in direct contact with cytoplasm were found with TEM. The microsporidia appear to belong to *Pleistophora* species. The significant changes in antioxidant enzyme activities in liver in the infected catla, indicates activation of defense mechanism. The water analysis showed high level of ammonia, COD, and alkalinity. The heavy metal contents and the level of indicator microorganisms were found to be low in pond during infection. The protein analysis indicated expression of new proteins in muscle, gill, and liver in the infected catla. The black spot region of catla fry has shown large number of spores; TEM analysis revealed the presence of spores in muscle (0.41 × 0.58 μm) and defense cells (0.28 × 0.39 μm).

Anon (1992) had reported about **HP changes in the skin of *C. batrachus* exposed to acidified water.** They had stated that fisheries and AQC in India are threatened by several fish disease problems often caused by aquatic pollution. The exploitation of water resources for various purposes; for example, abstraction for drinking and for industrial and agricultural use, energy, and transportation; indiscriminate discharge of industrial effluents, sewage disposal, uncontrolled use of pesticides and fertilizers and agricultural activities, etc., could influence the outbreak of EUS.

Chandra et al. (2012) had made some **HP observations of farmed carp fingerlings of Mymensingh area. They had stated that** HP studies were conducted on carp fingerlings of *L. rohita, C. catla,* and *C. mrigala* during October 2003 to May 2004 from fish farms of Mymensingh area. Monthly samplings were done with 10 fish of each species with the help of seine net. Skin, muscle, gill, and liver were collected, processed, and stained in hematoxylin and eosin. Histologically, carp fingerlings of different fish farms were more affected in colder months by protozoan, monogenean, bacterial, and fungal pathogens. Pathological changes like necrosis, fungal granuloma, protozoan cyst, vacuolation, melanomacrophage, hemorrhage, hypertrophy, hyperplasia, and clubbing were recorded in all investigated organs. The fish *C. mrigala* had pathological symptoms like hypertrophy, hyperplasia, necrosis, hemorrhage, inflammation, vacuolation, and clubbing. Similar signs were also recorded in *L. rohita* and *C. catla,* but degrees of severity were less. Among fish, gills were the organs most affected by protozoans, followed by the skin and the liver—the less affected portion was the muscle. In some cases, a large bacterial colony and protozoan cysts were observed in the secondary gill lamellae of *C. mrigala*

Mastan and Qureshi (2003) had reported about observations on the histological alterations in various tissues of EUS-affected fish *C. striata* (Bloch)

6.11.4 Epizootic Ulcerative Syndrome Study Based on Hematology

Pathiratne and Rajapakshe (1998) worked on hematological changes associated with EUS in the Asian cichlid fish *E. suratensis.* As reported, apparently healthy and EUS-positive *E. suratensis* had been collected from Hamilton Canal, Negombo, Sri Lanka. They had been subjected to a hematological examination of total RBC count, total WBC count, hematocrit, Hb content, mean RBC volume, mean RBC-Hb content, mean RBC-Hb concentration, and differential leukocyte count (DLC). Results had indicated that severely affected fish were anemic as shown by significant decrease in total RBC count, hematocrit, and Hb content. In addition, the DLC of severely EUS-affected fish indicated a significant increase in the percentage of neutrophils. However, the anemic condition could be attributed to the hemorrhagic lesions that resulted in the loss of blood in the severely EUS-affected fishes. Nevertheless, increase in neutrophils may be due local inflammation and tissue damage due to severe EUS lesions.

Sharma and Sihag (2013) had worked on the effect of probiotic on hematological parameters of diseased fish *(C. mrigala)*. They had stated that EUS is a dreadful disease of several aquatic animals including mrigal (*C. mrigala* [Ham]). Prevention of the disease with the help of chemical compounds may prove harmful for fish and the end consumer. Nowadays, the use of probiotics has become firmly established due to the beneficial effects on nutritional and therapeutic levels. Therefore, research efforts have been concentrated on optimizing production with ecofriendly alternatives to the therapeutic use of antimicrobials. Commercially available probiotics were used for controlling this disease in mrigal in the present study. The effect of probiotics on four hematological parameters, viz. Hb, hematocrit/packed cell volume, total erythrocyte count (TEC), and total leukocyte count (TLC) were studied over a period of eight weeks. The levels of these parameters decreased significantly in the blood of fish treated with pathogenic bacteria. Fish administered probiotics showed significant increases in hematological parameters, thus contributing to overall fish health.

Innocent et al. (2011) worked on the **hematology of *C. mrigala* fed a vitamin C−supplemented diet and postchallenged by *A. invadans.*** They reported that disease outbreaks in fish culture systems interfere with productivity. Oral immunostimulant containing vitamin C could result in activating the immune system in a nonspecific way, thus providing resistance against pathogens. In the present study, the fish *C. mrigala* was given feed supplemented with vitamin C (100 mg/100 g) for 40 days and post challenged with two different dilutions (102 and 105) of *A. invadans*. The hematological parameters like TEC, TLC, and differential leukocyte counts were analyzed 24 h, 72 h, and 7 days after infection. TEC counts and lymphocyte counts of infected fishes previously fed with control diet decreased significantly (43%, 36%), whereas it was minimal in fishes (21%, 8%) fed a vitamin C−supplemented diet. Postchallenged fishes exhibited an increase in TLC in the vitamin-supplemented diet (20%, 32%) over control diet–fed fishes. In differential leukocyte counts, lymphocytes decreased in both experimental groups (9%, 11%), whereas neutrophils increased in both. However, the increase was higher in the control diet–fed group (34%). Basophils and eosinophils increased significantly in vitamin C–supplemented diets only (10%, 60%). Thus, supplementation of feed with vitamin C improved fishes' tolerance to infection.

6.11.5 Epizootic Ulcerative Syndrome Study Based on Immunology

Bhatt et al. (2014) had worked on the immunological role of C4 CC chemokine-1 from snakehead murrel *C. striata*. They had reported that a cDNA sequence of C4 CC chemokine identified from snakehead murrel (also known as striped murrel) *C. striata* (named CsCC-Chem-1) normalized cDNA library constructed by genome sequencing FLX™ technology. CsCC-Chem-1 is 641 base pairs (bp) long and contain 438 bp open reading frame (ORF). The ORF encodes a polypeptide of 146 amino acids with a molecular mass of 15 kDa. The polypeptide contains a small cytokine domain at 30–88. The domain carries the CC motif at Cys(33)-Cys(34). In addition, CsCC-Chem-1 consists of another two cysteine residues at C(59) and C(73), which together with C(33) and C(34) make CsCC-Chem-1 as a C4-CC chemokine. CsCC-Chem-1 also contains a "TCCT" motif at 32–35 as CC signature motif; this new motif may represent new characteristic features, which may lead to some unknown function that needs to be further focused on. Phylogenetically, CsCC-Chem-1 clustered together with CC-Chem-1 from rock bream (*Oplegnathus fasciatus)* and European sea bass (*Dicentrarchus labrax)*. Significantly, ($p < 0.05$) highest gene expression was noticed in the spleen and is upregulated upon fungus (*A. invadans*), bacterium (*A. hydrophila)*, and virus (poly I:C) infections at various time points. The gene expression results indicate the influence of CsCC-Chem-1 in the immune system of murrel. Overall, the gene expression study showed that the CsCC-Chem-1 is a capable gene to increase the cellular response against various microbial infections. Further, we cloned the coding sequence of CsCC-Chem-1 in pMAL vector and purified the recombinant protein to study the functional properties. The cell proliferation activity of recombinant CsCC-Chem-1 protein showed a significant metabolic activity in a concentration-dependent manner. Moreover, the chemotaxis assay showed the capability of recombinant CsCC-Chem-1 protein that can induce the migration of spleen leukocytes in *C. striata*. However, this remains to be verified further at molecular and proteomic level.

Kurata et al. (2000) were reported to be some workers on fish immunology related to EUS, MG etc. Some earlier reports of a hemagglutinin in fish pathogenic peronosporomycete had portrayed a galactose-binding protein (GBP) from *A. piscicida* activated carp WBC. This could be responsible for the inflammatory response, said to be unique to MG. Such a revelation could also throw some light on the mechanism of host immune system (Valsa Judit Anto and Balasubramanian, 2014).

Miles et al. (2003) worked on the immunofluorescence of the EUS pathogen *A. invadans,* using MAb. They had reported that an MAb designated 3gJC9 had been raised against a protein antigen (Ag) of *A. invadans,* the oomycetous pathogen that causes EUS. The Ag was expressed on the surface of hyphae and secreted extracellularly. MAb 3gJC9 did not cross-react with other oomycetous or fungal pathogens of fish, although it did react to the crayfish plague pathogen *A. astaci*. The MAb had been used for immunofluorescent staining on HP sections of fish infected with EUS, and was found to be more sensitive than conventional staining methods for detecting *A. invadans*. It may therefore have some usefulness in framing the case definition of EUS. It also portrayed very small filamentous structures, but the significance of these was not clear. However, they could represent an early stage of infection. This could help in earlier detection of the disease. It may be noted here that they are not detected using conventional staining methods.

Sudhakaran et al. (2006) had worked on the immunostimulatory effect of *Tinospora cordifolia* Miers leaf extract in *O. mossambicus*. As reported, they investigated the immunostimulatory effect of the leaf extract of *T. cordifolia* on: (a) specific immunity (antibody response); (b) nonspecific immunity (neutrophil activity); and (c) disease resistance against *A. hydrophila* in *O. mossambicus*. Ethanol and petroleum ether extracts of leaves were used. Both ethanol and petroleum ether extracts administered at doses of 0.8, 8.0, and 80.0 mg/kg body weight, prolonged the peak primary antibody titers by one to three weeks. Further, ethanol extract at a dose of 8 mg/kg body weight and petroleum ether extract at doses of 0.8 or 8.0 mg/kg body weight enhanced secondary antibody response. Moreover, it had been observed that all doses of ethanol extract had significantly enhanced neutrophil activity. Fishes injected with petroleum ether or ethanol extract at a dose of 8 mg/kg body weight were protected against experimental infection from virulent *A. hydrophila*. The results thus portrayed the potential of *T. cordifolia* leaf extracts as an immunoprophylactic to prevent diseases in finfish AQC.

Rattan and Parulekar (1998) had worked on diseases and parasites of laboratory reared and wild population of banded pearl spot *Etroplus suratensis* (Cichlidae) in Goa. They had reported that the banded pearl spot (*E. suratensis)* is an inhabitant of coastal backwaters and lagoons. Significantly, it is one inhabitant of coastal backwaters and lagoons and is identified as a species for backwater fish farming. They had described the common diseases and parasites from the wild population of Goa and from the laboratory reared fish. Their study had revealed that the finrot disease and *Ancyrocephalus* infestations could lead to heavy losses in the lentic water systems.

Mayo (2012) reported about the immune response of common carp (*C. carpio*) under stress. It had been stated that the effect of simulated stress induced by cortisol implantation in common carp (*C. carpio*) had been studied. Humoral-mediated responses to injected antigens following cortisol administration were assessed by using the passive hemolytic plaque and indirect hemagglutination (HA) assays. A single cortisol implant and subsequent challenge to *A. hydrophila* elicited modulation in the fish immune system and HP changes in the thymus, kidney, and spleen. There had been significant decreases in hemolytic plaque-forming cells and HA antibody (Ab) titer in cortisol-administered carps, which indicated the Ab-mediated responses.

Valsa and Balasubramanian (2014) had reported about immunological studies of disease-induced common carp *C. carpio* fed with neem extract added to the feed. They had stated that low concentrations of plant extracts can act as an immunostimulant. It is biodegradable and environmental friendly. *Azadirachta indica* (neem) extract was used to prepare artificial feed at concentrations of 250, 500, 750 mg/kg of dry diet. The prepared diets were fed to experimental fishes common carp (*C. carpio*) for 30 days and then injected with 0.1 mL of 105 CFU/mL of *A. hydrophila* and at seven-day intervals, the following immunological aspects were studied, including antibody titer, phagocytic activity, and hepatosomatic index. In the plant extract treated groups, no mortality was seen. Low dose of plant extract (250 mg/kg) showed maximum antibody titer, phagocytic activity, and hepatosomatic index compared with control and other experimental groups. A dose of 250 mg/kg feed was found more effective than control and other fish groups.

Miles et al. (2003) reported on the immunofluorescence of the EUS pathogen *A. invadans* using a monoclonal antibody. They had stated that a monoclonal antibody (MAb), designated 3gJC9, was raised against a protein antigen of *A. invadans*, the oomycetous pathogen that causes EUS. The antigen was expressed on the surface of hyphae and secreted extracellularly. MAb 3gJC9 did not cross-react with other oomycetous or fungal pathogens of fish, although it did react to the crayfish plague pathogen *A. astaci*. The MAb was used for immunofluorescent staining on histological sections of fish infected with EUS, and was found to be more sensitive than conventional staining methods for detecting *A. invadans*. It thus has utility in confirming the case definition of EUS. It also revealed very small filamentous structures, the significance of which was unclear—they may represent an early stage of infection, thus allowing earlier detection of the disease because they are not detected using conventional staining methods.

Kales (2006) had reported about a **functional study of major histocompatibility expression and immune function in rainbow trout (*Oncorhynchus mykiss*).** He had stated that the major histocompatibility complex (MHC) receptors serve a critical role in self/nonself recognition through the presentation of peptide antigen to circulating T lymphocytes and are believed to play a role in mate selection. Through the development of antibodies to MHC homologues in trout, this report demonstrates the presence of MHC expression in germ cells and a soluble form in seminal fluid.

6.11.5.1 Immunodiagnostic Methods for Microbial Fish Diseases

Bhatia (1998) **reported about immunodiagnostic methods for microbial fish diseases. He had stated that** the term immunodiagnosis refers to any diagnostic technique based on detection of either antigen or antibody inside or outside the host system (e.g., fish). The antigen, defined as a macromolecule (usually a protein of viral or bacterial origin) that can evoke specific immunoresponse inside the body of the host, and the antibody, a globular protein that can neutralize a specific antigen, are the two most important variables of an immune process. Nearly all immunodiagnostic techniques are based on either the detection of an antigen with antibody of known specificity, or the detection of an antibody with a known antigen. When an antibody complements a known antigen, antigen and antibody molecules complement each other when mixed, the reaction being manifested by precipitation, agglutination, etc. On the basis of the reaction that results from the interaction of antigen and antibody, one possible immunodiagnostic method is an agglutination test, which is based on the visible clumping (agglutination) of a particulate antigen with an antibody when the two test reagents are mixed together on a glass slide. Bhatia concluded that immunodiagnostics have great prospects in fish disease diagnosis since they can specifically detect different microbial diseases with great sensitivity. The advantage of these methods over other conventional tests is that they can differentiate between closely related strains of same species of bacteria or virus. This is an important aspect in case of fish pathogen, since many of the bacteria are normal inhabitant of aquatic bodies with only some strains being pathogenic to fish.

6.11.6 Epizootic Ulcerative Syndrome Study Based on Tissue Culture

6.11.6.1 Tissue Culture

Much Work on Epizootic Ulcerative Syndrome Has Been Done Using the Tissue Culture Technique

Ahmed et al. (2011) had worked on the **health condition of juvenile exotic carp *C. carpio* from various fish farms of**

the Mymensingh area, Bangladesh. In this work, clinical and HP studies were done with juvenile common carp (*C. carpio*) from a government fish farm and an NGO fish farm in Mymensingh area for nine months from April to December 2005. Clinically, red spots, loss of scales, weak body, hemorrhage, and subcutaneous lesions were noticed, especially during November and December. A moderate pathological change in the investigated organs of all fishes was recorded in summer season (April—May), whereas in rainy season (June—July) pathological symptoms were significantly reduced, which again increased to some extent in the autumn season (August—September).

6.11.7 Epizootic Ulcerative Syndrome Study Based on Electron Microscopy

6.11.7.1 Viruses Observed in Electron Microscopy

The presence of virus particles in the tissues of EUS-infected fishes was first demonstrated by electron microscopy. The initial observations of viral association with EUS were noted by TEM studies of tissues of diseased striped snakehead fish and walking catfish specimens obtained from the epizootic in Thailand during 1983 and 84. Viruslike particles were reported in hepatocytes, the spleen, the kidney, blood cells, the capillary endothelium, and muscle myofibrils taken from the site of ulceration in diseased specimens. These particles observed in the cytoplasm were approximately 70 nm in diameter but could not be precisely identified, as they were pleomorphic with various shapes including round, oval, elongate, and kidney. Particles of rhabdovirus-like morphology were noticed in skin lesions of infected barramundi *(Lates calcarifer)*. Presence of rhabdovirus and infectious pancreatic necrosis virus like particles in cell cultures inoculated with tissue extracts from EUS-infected snakeheads was noted by Wattanavijarn et al. Rhabdoviruses were also observed in cell cultures inoculated with tissue extracts from a variety of infected fishes including that of a barramundi, which showed extensive CPE.

Further preliminary EM works done with EUS-affected tissues of *C. punctata* and *A. testudineus* revealed the occurrence of viruslike particles (Kar et al., 1993, 1995a). Detailed TEM studies done with ultrathin sections of EUS-affected tissues of *C. striata* and *A. testudineus* revealed membrane bound and granular abnormal structures, hepatocytes with vacuolations, and viruslike inclusion bodies in the nucleus (Kar, 1999). Further EM works done with EUS-affected tissues of *W. attu* revealed the presence of viruslike particles (Kar, 2004). More works, in this direction, are in progress.

Irina et al. (2011) had reported about SEM investigations regarding bacterial infections in *C. carpio* and *Salmo trutta fario*. They had stated that *Aeromonas* is among the most common bacteria in FW habitats throughout the world, and these bacteria frequently cause disease among cultured fishes It is the predominant bacteria isolated from the hemorrhagic lesions and necrotic ulcers in fish skin. Bacterial infection of fish constitutes a huge menace for AQC farming, leading to disastrous economic loss and health risks for the consumer. *Aeromonas* are often referred to as a complex of disease organisms associated with bacterial hemorrhagic septicemias and other ulcerative conditions in fishes. These bacteria can cause the following diseases: furunculosis of salmonids, goldfish ulcer disease, carp erythrodermatitis, and trout ulcer disease. Regarding problem formulation, they had stated that ulcerative lesions in fish are usually a consequence of bacterial, viral, or fungal pathogens, as well as parasites. They had also opined that fish skin is an important defense system against pathogens because they are in contact with numerousness microorganisms in the aquatic habitat. Fish skin poses a highly developed antimicrobial defense mechanism. The antibacterial role of fish skin mucus has been known for many years. It contains antimicrobial bioactive substances including lysozyme, complement, C-reactive protein, hemolysins and lectins, and the epidermal migration of inflammatory cells and their secretions, which are of interest with regard to fish immunity. The mucus plays a role in the prevention of colonization by parasites, bacteria, and fungi. Fish bacteriosis was investigated by numerous authors having histological, microbiological, and taxonomical points of view. Since the 1980s, molecular methods have provided useful tools for diagnostic microbiology and for taxonomic and epidemiological studies on pathogenic bacteria. With regard to taxonomical point of view, they had observed that since the 1980s, molecular methods have provided useful tools for diagnostic microbiology and for taxonomic and epidemiological studies on pathogenic bacteria. They had further stated that increased preoccupations in fish culture have increased the interest regarding bacteria that affect fish health, growth, and survival. The majority of fish bacteriosis may only be identified by the use of microbiological and molecular techniques. In this study, the scanning electron microscope (SEM) was used for the first time in characterization of the bacterial lesions produced by *Aeromonas* sp. on *C. carpio* skin.

Chakrabarty et al. (1996) had reported about EM characteristics of actinomycetic agents having etiological association with human leprosy and EUS of fish.

Rattan and Parulckar (1998) had reported on EM characteristics of actinomycetic agents having etiological association with human leprosy and EUS of fish.

6.11.8 Epizootic Ulcerative Syndrome Study Based on Nanoparticles

Kuan et al. (2013) had dealt with gold-nanoparticle based electrochemical DNA sensor for the detection of fish pathogen *A. invadans*.

Sarkar et al. (2012) had worked on the inhibitory role of silver nanoparticles against the important fish pathogen *A. hydrophila*. They had stated that *A. hydrophila* is a heterotrophic, Gram-negative bacterium that is a primary or secondary cause of ulcers, fin rot, tail rot, and hemorrhagic septicemia in fish. The treatments for this infection are only restricted to some antibiotics. Therefore, novel materials are being sought to combat bacterial infections and their consequences Biplab et al. In the present study, silver nanoparticles were synthesized chemically & biologically and characterized by UV−visible spectrophotometer and TEM. The antimicrobial activities of silver nanoparticles, synthesized by both the methods were checked by both well diffusion and turbidimetric (spectrophotometric) method. Both biologically and chemically synthesized nanoparticles exhibited their antimicrobial efficacy in both the standard inhibitory assays; however chemically synthesized particles that were also smaller in size, exhibited more inhibitory effect than biologically synthesized nanoparticles. These results thus provide a scope for further research on the application of silver nanoparticles as disinfectant and/or antibiotic in the fishery industry.

6.11.9 Epizootic Ulcerative Syndrome Study Based on PCR

Works related to EUS revelation were also done using sophisticated know-how. As reported, PCR-based analyses had been done using ITS. Primers had been sequenced from *A. piscicida* NJN 0204 isolated from *Channa* sp. and registered under the accession number AY283640 in GenBank. The study had involved the use of a number of isolates of pathogenic *Aphanomyces* spp., that had been collected from different countries, notably 20 isolates of *Aphanomyces* recovered from EUS lesions from different fish species from different regions across the globe, like grey mullet from Australia; dwarf gourami from Singapore; eel and snakeheads from Thailand and the Philippines; ayu and golden gourami from Japan; menhaden from the United States; and so on. Results preliminarily revealed that 20 *Aphanomyces* spp., pathogenic to fish, had tested positive for PCR. In addition, nonpathogenic *Aphanomyces* isolates, and other fungi used in the study, tested negative to PCR.

6.11.10 Epizootic Ulcerative Syndrome Study Based on Mycology

6.11.10.1 In Vitro Screening of Novel Treatments for Aphanomyces invadans

Campbell et al. (2001) had worked on the in vitro **screening of novel treatments for** *A. **invadans**.* In their study,

49 compounds were tested in vitro for fungicidal activity (Volf and Steinhauer, 1997; Yuasa and Hatai, 1995) against hyphae of the fish-pathogenic Oomycetes fungus *A. invadans*. These were: (a) chemicals with previous reported activity against Oomycetes fungi; (b) chemicals in use in Asia to treat ulcerative disease outbreaks; (c) commercial biocides and fungicides; (d) natural products with potential antimicrobial potency; and (e) surfactants tested separately and in combination with some of the above-mentioned treatments. No compounds tested had proved as effective as malachite green. However, some low-toxicity natural compounds and particular surfactants showed potential for further studies. Some compounds currently in use in Asian AQC were shown to have no effect on *A. invadans* hyphae at recommended treatment rates. A range of compounds that demonstrated activity against the mycelium was selected for further testing in a zoospore motility assay. Lower treatment concentrations were required to inhibit zoospore motility than were required to inhibit hyphal growth.

Recent works seemed to have confirmed that a single species of fungus, viz. *Aphanomyces*, is said to be a necessary cause of EUS; that is, it occurs in all outbreaks. In some outbreaks (e.g., in Australian estuaries), it may be the only biological factor required for the disease to occur.

The Pathogenic Fungal Genus *Aphanomyces*

Fungi have been known to be involved in the etiology of EUS in SE Asia (Min et al., 1994; Miyazaki, 1994) since the initial outbreaks in Thailand. Limsuwan and Chinabut (1983) described a "severe chronic granulomatous mycosis" in histological sections of affected fish. However, the dominance of saprophytic fungal contaminants on the surface of EUS lesions led to the identification of *Achlya* and *Saprolegnia* spp. from affected fish. These were soon recognized as secondary agents (Tonguthai, 1985), but it was also assumed that this may be the case for all mycotic involvement in EUS (Tomlinson and Faithfull, 1979).

As discussed, before the first appearance of EUS in SE Asia, the pathogenic *A. piscicida* had been isolated from MG-affected fish in Japan (Hatai et al., 1977), but MG had not yet been recognized as synonymous with EUS. An *Aphanomyces* fungus was subsequently obtained from RSD outbreaks in Australia in 1989 (Fraser et al., 1992), and independently from EUS outbreaks in Thailand in 1991−1992.

The genus *Aphanomyces* is contained within the family Saprolegniaceae and class Oomycetes, and it should be noted here that Oomycetes are no longer regarded as true fungi, but rather fungal-like protists. They are now often classed alongside diatoms, brown algae and xanthophytes within the phylum Heterokonta as part of the third botanical kingdom, the Chromista. They are sometimes called pseudofungi as either a general term or a formal taxon. They are still commonly referred to as fungi, however, and that term will be used in this treatise.

These isolates were shown to be capable of reproducing typical EUS lesions when injected below the dermis of susceptible fish. All these pathogens of MG, RSD, and EUS isolates were shown to be slow growing and thermolabile in culture (Roberts et al., 1993). Similar isolates were obtained from the Philippines, Indonesia, Bangladesh (Lilley and Roberts, 1997), and India. Further pathogenic *Aphanomyces* cultures from most of these countries were compared directly and shown by means of protein banding profiles (Callinan et al., 1995b; Lilley et al., 1997b), growth characteristics (Lilley and Roberts, 1997), and chemical susceptibility (Lilley and Inglis, 1997) that all three belong to the same species. Genetic fingerprinting techniques were also used to show that the various isolates were genetically quite similar (Lilley et al., 1997a). This probably had served as a proof that isolates are not long-term residents in each locality, as could be expected of an opportunistic fungi. Rather, they are said to be part of one fungal strain that has colonized much of Asia and Australia, probably for decades, and resulted in the spread of EUS. Notwithstanding this, the pathogenic *Aphanomyces* had been named variously as *A. piscicida* (Hatai, 1980), *A. invaderis* (Willoughby et al., 1995), and ERA (EUS-related *Aphanomyces* sp.: Lumanlan-Mayo, 2011). This could be mainly because isolates in each case had been shown to be conspecific. However, it is believed that *Aphanomyces invadans* is probably the only valid taxon, according to the International Code of Botanical Nomenclature, and this name was adopted.

A. invadans is known to grow fastest in culture at a temperatures between 26 and 30 °C (Fraser et al., 1992; Lilley and Roberts, 1997), and has been shown to grow in snakehead muscle tissue between 19° and 31 °Chinabut et al., 1995). However, further investigations revealed that snakeheads are able to recover from *A. invadans* infection at higher temperatures (26–30 °C), but are unable to prevent fungal invasion and eventually succumb to the disease at lower temperatures (<19 °C) (Chinabut et al., 1995). The humoral and cellular immune responses of fish are suppressed at low temperatures (Bly et al., 1996), which may explain why mortalities from EUS occur when water temperatures are low. Naturally and artificially infected snakeheads have been shown to produce an antibody response against *A. invadans*, and the cellular macrophage response is considered important in enabling fish to resist infection.

A summary of various published descriptions of the characteristics of *Aphanomyces invadans* from the tissues of fishes affected by EUS, MG, and RSD had been dealt with. Moreover, techniques for isolating *A. invadans* from fish and water, and confirmed identification of the candidate fungus as belonging to the genus *Aphanomyces*, were also dealt with. As with other saprolegnious fungi,

A. invadans is aseptate and produces two zoospore forms, viz. the secondary form being free-swimming and laterally biflagellate. No sexual reproductive structures had been observed in any isolates from EUS, MG, or RSD outbreaks. In fact, it may be noted here that the lack of sexual structures is considered a particularly common phenomenon among the more pathogenic members of Saprolegniaceae (Alderman and Polglase, 1988). The Fisheries Research and Development Corporation plans, invests in, and manages fisheries research and development throughout Australia. It is a federal statutory authority jointly funded by the Australian government and the fishing industry. They had reported that disease outbreak in fish culture system interferes with productivity. Oral immunostimulant containing vitamin C may result in activating the immune system in a nonspecific way, thus providing resistance against pathogens. In the present study, the fish *C. mrigala* was given feed supplemented with vitamin C (100 mg/100 g) for 40 days and postchallenged with two different dilutions (102 and 105) of *A. invadans*. The hematological parameters like TEC, TLC, and differential leukocyte counts were analyzed 24 h, 72 h, and 7 days after infection. Fishes are considered one of the important food sources for human beings because their flesh contains a high percentage of proteins, calcium, and phosphorus. Therefore, increased attention is given to fish farms and their diseases. Pathogens such as viruses, bacteria, and fungi are all items of concern to aquaculturists (RantamaEki et al., 1992). In fish farms many of these pathogens cause widespread mortality. The use of antibiotics to treat bacterial infections has resulted in a global increase of resistance. For this reason, studies are conducted on immunostimulants that represent a modern and promising tool in AQC as they enhance the resistance of cultured fish to disease and stress. It comprises a group of biological or synthetic compounds that enhance the hormonal and cellular response both in specific and nonspecific way. Immunostimulants also stimulate the natural killer cells, complement, and lysozyme antibody responses of fish (Sakai, 1999). Investigations regarding the health status of a fish may be assessed by hematological parameters (Hickey, 1976). By analyzing blood cells, characteristic clues for diagnosis and prognosis of diseases may be found (Anderson, 2003). Ascorbic acid is an indispensable and multifunctional micronutrient. It plays important roles in improving immune function, improving growth (Boonyaratpalin et al., 2001), providing good health, converting food, survival (Journal of Applied Pharmaceutical Science 01 (09); 2011: 141–144), resisting stress, and oxidation. Most fish species are not capable of vitamin C biosynthesis due to the absence of the enzyme L-gulonolactone oxidase necessary for ascorbic acid synthesis. Vitamin C also plays an important role in animal health as antioxidants by inactivating damaging free radicals produced through normal cellular activity and from various stress. It has been suggested that the antioxidant function of these micronutrients could enhance immunity by preserving the functional and structural integrity of immune cells. In this respect, the need for specific nutrients may be increased during infection that could require the feeding on diets formulated for optimal immune competence rather than growth and survival. Supplementation of vitamin C—enhanced antibody production against *E. ictaluri* in channel cat fish (Li and Lovell, 1985).

Adil et al. (2013) had reported about the development and standardization of a monoclonal antibody-based rapid flow-through immunoassay for the detection of *A. invadans* in the field.

Adil et al. (2013) developed a monoclonal antibody (Mab)-based flow-through immunoassay (FTA) using a nitrocellulose membrane placed on the top of adsorbent pads enclosed in a plastic cassette with a test zone at the center. The FTA could be completed within 10 min. Clear purple dots against a white background indicated the presence of *A. invadans*. The FTA limit of detection was 7 µg/mL for *A. invadans* compared with 56 µg/mL for the immunodot. FTA and PCR could detect *A. invadans* in fish tissue homogenates at 10^{-11} dilution, compared with 10^{-8} dilution by immunodot. FTA and PCR could detect *A. invadans* in 100% of the samples compared with 89.04% detected by immunodot in fishes suffering from natural cases of EUS collected from Mangalore, India. Moreover, FTA reagents are said to be stable and produced expected results for four months when stored at 4—8 °C. This rapid test could serve as a simple and cost-effective on-site screening tool to detect *A. invadans* in fish affected by EUS outbreaks in required areas, and in ports, during the shipment of live or frozen fish. The authors opined that EUS, caused by *Aphanomyces (A.) invadans*, is one of the most destructive diseases affecting several species of fresh and brackish-water fishes in both farms and the wild. The salient diagnostic features of EUS are the presence of the fungus and MG in the tissues of the affected fishes. The fungus is said to be largely responsible for extensive damage to the tissues. They are also said to cause ulcers on the fish body. Ulcers and fish mortality are considered to be due to the invasive activity of the fungus. However, not all ulcers in a fish are related to EUS, because lesions do not occur in epizootic proportions and are not seasonal in nature. In addition, they had further opined that incidence of the disease coupled with large-scale mortality due to EUS are said to be accelerated by several environmental situations, such as low temperature along with rapid change in salinity and dissolved oxygen. Recently, EUS has started recurring in endemic regions like the NE India (from 1988 to present), Bangladesh (2001—2003, even today), Nepal (2005),

Kerala (1999), Karnataka (2005), and so on. New occurrences of EUS are still being reported in previously unaffected areas and farming systems. **In 2009, snakehead fingerlings with characteristic lesions collected from the wild in southern Florida (United States) were confirmed to be affected by EUS.** More recently, the first outbreaks of EUS were reported in Canada, and could be said to have recurred in Australia during 2010. Traditionally, EUS is diagnosed by selectively growing the fungus from fish tissue in glucose peptone (GP) broth. This method is not ideal for routine diagnosis. Histological examination, the most common technique used to diagnose EUS, is time-consuming. Furthermore, in situ diagnostic methods such as in situ DNA hybridization, immunohistochemistry, and immunofluorescence assays are specific but time-consuming and require sophisticated equipment. Hence, the present method has been developed. The development of field-friendly, rapid, sensitive, and low-cost tests could be, according to the authors, highly useful for managing diseases such as EUS. The goal of the present study was to develop a simple and rapid monoclonal antibody-based flow-through immunoassay (FTA) for detecting *A. invadans* in fish.

Huchzermeyer and van der Waal (2012) had reported about EUS, an exotic fish disease that threatens Africa's aquatic ecosystems.

Boys et al. (2012) had reported about emergence of EUS in native fish of the Murray-Darling river system, Australia: hosts, distribution, and possible vectors.

Saikia and Kamilya (2012) had worked on immune responses and protection in catla (*C. catla*) vaccinated against EUS. They had stated that three different antigenic preparations from the EUS pathogen *A. invadans* were evaluated as vaccine candidate in catla (*C. catla*). Anti-catla enzyme immunoconjugate was prepared after isolating catla immunoglobulin and raising hyperimmune sera against it, in rabbit. Three antigens, namely fungal extract (FE), fungal extract mixed with Freund's incomplete adjuvant (FIA) in a 1:1 (v/v) ratio (FE + A) and extracellular product were prepared, and three groups of catla were vaccinated intramuscularly with all these antigens (200 µg/fish). Different cellular and humoral immune responses were measured for the all vaccinated and control groups at 0, 5, 15, and 25 days postvaccination. Thirty days after vaccination, fish were challenged with an *A. invadans* zoospore dose of $1 \times 10(5)$ mL(-1), and mortality and relative percent of survival were recorded. Study of cellular immunological parameters including antigen-specific leukocyte proliferation, antigen-specific nitric oxide production, and superoxide anion production showed significantly higher values ($p < 0.05$) in general at 5 and 15 days postvaccination, compared with day zero.

Oidtmann (2012) had reviewed the biological factors relevant to import risk assessments for EUS.

Songe et al. (2011) had reported about their field observations of fish species susceptible to EUS in the Zambezi River basin in Sesheke District, Zambia.

Saylor et al. (2010) had reported about EUS caused by *A. invadans* in captive bullseye snakehead *C. marulius* collected from south Florida in the United States.

Choongo et al. (2009) had worked on the environmental and climatic factors associated with EUS in fish from the Zambezi floodplains in Zambia (Africa).

Oidtmann et al. (2008) had reported about experimental infection and detection of *A. invadans* in European catfish, rainbow trout and European eel.

Callinan et al. (2005) had reported about dermatitis, bronchitis, and mortality in empire gudgeon, *Hypseleotris compressa*, exposed naturally to runoff from acid sulfate soils.

Lio-Po et al. (2003) had reported about horizontal transmission of EUS-associated virus in the snakehead *Ophicephalus striata* under simulated natural conditions.

Lilley et al. (2003) had dealt with the molecular characterization of the fish-pathogenic fungus *A. invadans*.

Lilley et al. (2003) had worked on the immunofluorescence of the EUS pathogen *A. invadans* using a monoclonal antibody (MAb)

Lilley et al. (2001) had worked on the characterization of *A. invadans* isolates using pyrolysis mass spectrometry (PyMS).

Anon (2012) **had reviewed the biological factors relevant to import risk assessments for EUS in relation to *A. invadans*.** They had opined that EUS is a disease affecting both wild and farmed fish in FW and estuarine environments. After it was first described in Japan in 1971, the disease had spread widely across Asia and to some regions of Australia, North America, and Africa. In Asia and Africa, the spread of the disease had substantially affected livelihoods of fish farmers and fishers. Reports had not yet been published showing the presence of the disease in Europe or South America. Given its epizootic nature and its broad susceptible fish species range, it would appear that the disease has the potential for further spread. This study provides a review of the scientific literature on several biological factors of the pathogen *A. invadans* associated with the disease EUS and aspects of the disease that are relevant to undertaking import risk assessments (IRA) covering (i) life cycle and routes of transmission; (ii) minimum infectious dose; (iii) tissue localization and pathogen load; (iv) predisposing factors for infection and factors influencing expression of the disease; (v) carrier state in fish; (vi) diagnostic methods; (vii) survival in the environment; (viii) permissive temperature range; (ix) stability of the agent in aquatic animal products; (x) prevalence of infection; and (xi) affected life stages.

Much of the biological information presented is relevant to a broad range of risk questions. Areas where data are lacking were identified, and the information provided is put into context with other aspects that need to be addressed in an IRA.

Miles et al. (2003) had reported on immunofluorescence of the EUS pathogen *A. invadans* using a monoclonal antibody.

Nuryati et al. (2005) had studied the antimycotic potency of *Terminalia catappa*, and *Andrographis paniculata* on the growth of *Aphanomyces* sp. in vitro as an effort to prevent its infection in fish. Use of such a natural material is said to be easy and economic, as they are not much difficult to obtain and said to be safe for the environment. There have been possibly some works on the antimycotic potency and efficacy of scalded-leaf extract of *T. catappa*, *Psidium guajava*, and *A. paniculata* on the prevention of growth of *Aphanomyces* sp. in vitro in GYA medium. The results of the study indicated that *T. catappa* in a dosage of 40 g/L had the most potent prevention activity.

Nuryati et al. (2005) had studied on **antifungal potency of *T. catappa*, Piper betle, *P. guajava*, and *A. paniculata* on the growth of *Aphanomyces* sp.** in vitro. **They had stated that** an effort to prevent aquatic fungi *Aphanomyces* sp. infection on fish using natural material could be an economic way with easy-to-find materials that are easy to apply and safe for the environment. The antifungal potency and efficacy of scalded-leaf extract of *T. catappa*, Piper betle, *P. guajava*, and *A. paniculata* on the prevention of *Aphanomyces* sp. growth in vitro in GYA medium. Scalding was performed in the water at 50 °C. Concentration of leaf extracts tested was 0, 10, 20, 40 and 80 g/L. The results of the study showed that *T. catappa* in a dosage of 40 g/L had the best prevention activity, followed by Piper betle in the same dosage. *P. guajava* and *A. paniculata* had no prevention activity on growth of *Aphanomyces* sp.

Gozlan et al. (2014) had reported about the current ecological understanding of fungal-like pathogens of fish. He had queried about what lies beneath.

Concomitant to the above, there had been **extensive fungal studies on MG since the early 1970s**. The ambit of studies had ranged from evolving an artificial culture medium for MG fungus coupled with studies related to isolation of the fungus (Hatai, 1980); pathogenicity studies (Hatai et al., 1977, 1994); the effect of chemicals on mycelial growth of the fungus; and cultural, biological, and biochemical characterization of the fungus (Hatai et al., 1994). Incidentally, Hatai (1980) had described the fungus and named it *A. piscicida*. In addition, fungal isolates were also recovered from ayu (Hatai, 1980), dwarf gourami, and ornamental fish (Hatai et al., 1994), and were said to be morphologically and culturally consistent with *A. piscicida*.

In addition to the above, there had been comparative studies with the Japanese isolates done by scientists from different parts of the world (Lilley and Roberts, 1997; Lilley et al., 1997b; Blazer et al., 2002).

Artificial Initiation of the Infection

There had been a number of artificial infection experiments to reproduce the disease. Hatai et al. (1977) had been considered the pioneer in generating the same granuloma formation as seen in the naturally affected fish. They had injected the cultured hyphae into the muscles of ayu after successful isolation of the suspected fungi. Subsequently, Hatai (1980) is said to have reproduced the same disease by inoculating the hyphae of *A. piscicida* from ayu to a number of fish species; for example, rosy bitterling carp (*Rhodeus ocellatus ocellatus)*, crucian carp, bluegill, goldfish (*Carassius auratus auratus*), rainbow trout (*O. mykiss),* eel, and loach. It was found that of the above-mentioned fishes, rosy bitterling was highly susceptible to EUS, but rainbow trout was less susceptible. Incidentally, loach was not susceptible.

Notwithstanding the above, different workers had used various animals as **test models** for conducting tests of pathogenicity. In this regard, Hatai et al. (1994) used goldfish (*C. auratus auratus*) and ayu to ascertain the pathogenicity of a strain of *Aphanomyces* that had been isolated from an outbreak of a disease supposedly caused by *Aphanomyces* sp. in dwarf gouramies, notably *Colisa lalia,* The consignment had possibly arrived from Singapore and the experiment had been done through zoospore injection. Likewise, in another study had attempted to generate the disease by artificially infecting ayu and carp with *A. piscicida* using zoospore injection of *A. piscicida* 1989 isolates (NJM8997) from ayu. By following the methods of Wada et al. (1996), Bondad-Reantaso et al. (1999b) had initiated the disease in goldfish by Artificial injection of *Aphanomyces* sp. isolated from EUS-affected snakeheads in the Philippines (1998); from MG-affected ayu in Japan (1998); and from EUS-affected *C. mrigala* in Bangladesh (1999). The results depicted not only successful reproduction of the disease, but also the finding that HP lesions caused by the reisolated fungus were quite indistinguishable from those found in naturally occurring cases.

Characterization of the fungus in connection with EUS-affected tissues is usually done in HP sections, which alone may not always be conclusive. Further, there are variations in the rate of isolation of *A. invadans* from EUS-affected tissues; the presence of dead mycelia in the lesions is sometimes considered the possible reason for low rates of isolation of *A. invadans* from EUS-affected tissues (Roberts et al., 1993). By following the method of Willoughby and Roberts et al. (1993), some workers had tried

to isolate *A. invadans* from fish showing clinical signs consistent with EUS. However, the fungus had been recovered only from *c* 5% of these fishes.

Concomitant to the above, there had been efforts toward **initiation of EUS lesion**s **experimentally** in various species of fishes, by (a) intramuscular injection of fish with a large number of viable zoospores; (b) the same thing done with small pieces of fungal hyphae; or (c) exposure of sacrificed fish to spores (Roberts et al., 1993; Hatai et al., 1994). However, there had not been any portrayal of a high rate of lesion-induction in the experimental fishes without a breach of the physical barriers of the skin of the test fish. Nevertheless, it had been depicted by Kiryu et al. (2002) about the formation of lesions by both injection and bath exposure.

In view of the above, the definition of EUS as an epizootic, with the involvement of *A. invadans* as a pathogen, warrants further study on the involvement of other microbes in the epizootic syndrome. Incidentally, as reported, there had been an outbreak of ulcerative disease among fishes in Pakistan, in two locations fed by the same river. However, experts had identified two different and distinct diseases, although microscopical examinations showed similar lesions. One symptom had been identified as that of EUS because of the presence of *A. invadans*. The other symptom could not lead to any identification of disease because of the absence of *A. invadans*, and had been left as unidentified. The fish—fungus relationship could be less specific if we consider other fungal forms such as *Achlya, Saprolegnia,*etc.

Notwithstanding the above, workers had compared the RAPD patterns of two isolates from India and two isolates form Thailand (Karunasagar, 1995), in order to study the genetic diversity of *A. invadans* isolated from EUS-affected fish. Results had indicated possible genetic diversity among the few isolates that had been studied using four RAPD primers.

6.11.10.2 Further Inputs to Fungal Study

Several workers have successfully and consistently **isolated** *A. invadans* from EUS-affected fishes, although isolation of the putative causal fungus from EUS-affected fishes was one of the significant controversies about regarding *Aphanomyces* sp. as one of the etiological causes of EUS. Ever since the pioneering successful isolation of the fungus in Japan (Hatai et al., 1977, 1980), there had been a series of successful isolations from naturally EUS-affected fishes in several countries, notably Australia, the Philippines, Thailand, Bangladesh, India, the United States, and so on (Fraser et al., 1992; Roberts et al., 1993; Paclibare et al., 1994; Roberts, 1994; Blazer et al., 2002; Bashar et al., 2009). Subsequently, stringent application of advanced methods of isolation (Fraser et al., 1992; Roberts, 1994) coupled with standardization of methods for growing and sporulating the fungus (Lilley et al., 1998) are said to have accelerated the rate of successful recovery of the fungus. This is said to have helped the workers in the field in comprehending the etiology of EUS.

EUS-affected fishes, viz. *C. mrigala* and *P. conchonius,* having pale raised lesions that have not yet completely ulcerated, have been chosen for fungal isolation works. The fish specimen had been killed by decapitation and pinned to a dissecting board with the lesion uppermost. The scales around the periphery of the lesion were removed and the underlying skin was seared with a red hot spatula so as to sterilize the surface. The fish with the board was then removed to a laminar flow cabinet with filtered air free from fungal elements. Then, the stratum compactum underlying the seared area was cut through using a sterile scalpel blade and a sterile, fine-pointed pair of rat-tooth forceps. This was followed by exposing the underlying muscle by cutting horizontally and reflecting superficial tissues. Four pieces of affected muscles *c* 2 mm^3 in dimension had been carefully excised up using aseptic technique and they were placed in a Petri dish containing Czapek−Dox medium with standard supplements. This experiment led to isolation of *Aphanomyces* sp. with concomitant occurrence of the same fungal genus in histological sections of the same EUS-affected fishes (Kar, 2007a,b,c, 2013; Kar, 2007a,b).

It may be mentioned here that the isolated bacterial and fungal flora have been living with fishes in water since time immemorial without generally causing any harm to them. As such, bacteria and fungi may not be considered the primary etiology for outbreak of EUS in fishes. Further, intradermal inoculation (Dey and Dey, 1980) of these microbes into a guinea pig (*Cavia* sp) revealed no lesion formation in the latter's body for about two years (Kar, 2007a,b,c; Kar and Dey, 1990, a,b,c; Kar et al., 1990a,b).

Lilley et al. (2001) had worked on fungal pathogens of EUS. They had opined that *A. invadans* is the fungal pathogen associated with the Asian fish disease EUS. A total of 21 isolates of the fungus were compared with other Oomycetes fungi in terms of their pyrolysis mass spectrometry (PvMS) profiles. Canonical variate analysis (CVA) of pyrolysis mass spectra distinguished *Aphanomyces* species from a wide scatter of *Achlya* and *Saprolegnia* isolates. Further CVA and hierarchical cluster analysis (HCA) separated *Aphanomyces* species into two main groups. The first group clustered *A. invadans* isolates from EUS outbreaks in Thailand, Bangladesh, Indonesia, Philippines, Australia, and Japan together. However, HCA also included the crayfish plague fungus *A. astaci* within this group. Nonpathogenic *Aphanomyces* strains isolated from UM-affected fish were shown to be distinct from *A. invadans,* instead, clustered with saprophytic

Aphanomyces strains to form the second group. Of late, an invasive *Aphanomyces* pathogen has been isolated from UNI-affected fish, but that was not tested here. This is the first report using PyMS in the study of Oomycetes systematics. The technique was not sensitive enough to show any intraspecific differences, but it was considered a useful technique for the discrimination of species where taxonomic relationships are uncertain.

The microbial flora of the liver, gills, intestine, and muscles of EUS-infected murrel *C. striata* in Tirunelveli, Tamilnadu was estimated quantitatively and qualitatively by Muthukrishnan Dhanraj et al. (2008). The isolates were identified generically and specifically. Total viable microbial count was measured as highest range in gills (5.9 ± 0.5 × 10^7 CFU (colony forming units) g^{-1}) and lowest range in intestine (8.7 ± 1.8 × 10^4 CFU g^{-1}). In total, 17 bacterial and mycotic species were isolated and identified. Among them, *A. hydrophila, Enterobacter* sp., *Vibrio* sp., *Pseudomonas* sp., *E. coli, A. invadans,* and *Aspergillus* sp. were the most predominant. in all samples. The purpose of this study is to isolate and identify the microflora of EUS-infected *C. striata.*

Murrels, commonly called snakeheads, belonging to the family Channidae, constitute the most common and dominant group of air-breathing FW fishes and are highly regarded as food fish in the South and SE Asian countries. It had long been commercially cultured in Thailand, Taiwan, and the Philippines. There are several species of murrels belonging to the genus *Channa,* but only one species, namely *C. striata,* also called the striped murrel, enjoys a good deal of popularity as a food fish in many parts of India. Besides the high quality of their flesh in terms of taste and texture, they also have good market value due to the low fat, fewer intramuscular spines, and medicinal qualities. Over the past two decades, EUS has had a serious impact on tropical fisheries, resulting in heavy economic losses. It is one of the most destructive diseases among FW and brackish-water fish in the Asia−Pacific region and it is very common in both northern and southern India and has spread though rivers, reservoirs, and paddy fields to neighboring provinces and countries, causing considerable losses to fish farmers. In the initial stages, the disease is marked by little red spots on the skin surface, which progress in size until eventually a circular to oval deep hemorrhagic ulcer exposing the skeletal musculature is visible. A diverse group of biotic agents such as viruses, bacteria, and cutaneous ectoparasites may initiate skin lesions that are subsequently colonized by *A. invadans* and ultimately lead to EUS. Different pathogenic organisms, including bacteria, fungi, and viruses, have been reported to be isolated from naturally infected fish. Roberts et al. had reported that in natural outbreaks, >100 fish species, especially air-breathing fishes, had been affected by EUS.

Theirs is said to be the first study undertaken to assess quantitatively and qualitatively the bacterial and mycotic species from EUS-infected murrel *C. striata*, collected in and around Tirunelveli region during 2006. Fourteen bacterial species and three mycotic flora were isolated and identified. Among the 17 isolates, dominant bacterial and fungal species were *A. hydrophila*, *Flavobacterium* sp. *Aspergillus flavus*, and *A. invadans*. The following were the most frequently isolated bacterial and mycotic flora with a prevalence of >10%: *A. hydrophila*, *Aeromonas* sp., *Flavobacterium* sp., *Vibrio vulnificus*, *Staphylococcus* sp., *Yersinia enterocolitica*, *Shigella* sp., *A. salmonicida*, *A. flavus*, *Aspergillus* sp., and *A. invadans*. Further, 11 colonies of *A. hydrophila* were identified predominantly from 25 microbial colonies in gill samples.

Notwithstanding the above, in Bangladesh, *C. punctata* and *L. rohita* were affected by various ulcer-type diseases, such as EUS, bacterial hemorrhagic septicemia, tail rot, fin rot, gill rot, and fungal and parasitic diseases. Conversely, in India, different species of fishes have been affected by EUS; for example, in Tirunelveli district in Tamil Nadu (TN), *C. striata* was said to be largely affected by EUS. Concomitantly, *A. hydrophila* is said to be one of the most significant FW fish pathogens isolated from EUS-affected fish. Survey of bacterial flora in EUS-affected fishes from different parts of the world in general and Asia in particular have largely showed that *Aeromonas* sp. was generally one of the most predominant microorganisms isolated from different sites of the fish body, viz. the skin, gills, intestine, and so on. Sufficient information is also said to be available regarding the pathogenic mechanism and virulence of *A. hydrophila*, said to have caused mass mortality in cultured Japanese catfish larvae. Further, *A. hydrophila* was considered the main cause of >75% mortality among IMCs that resulted from outbreak of the disease. *A. hydrophila* was also isolated from channids. Furthermore, Lio-Po et al. had reported that a number bacterial and fungal flora are said to be associated with EUS-affected snakehead, *C. striata*, and *c* 89% of the total isolates were *A. hydrophila*. In addition, 25 bacterial and mycotic species were identified from a number of FW carps in Himachal Pradesh in India. Interestingly, 15 isolates of bacteria have been reported from hybrid tilapia in **Saudi Arabia**. Notably, 17 bacterial and mycotic forms had been isolated from EUS fishes in Tirunelveli district in TN, with most isolates coming from muscle and gills. Incidentally, there had been gill diseases among the sheath fishes and silver carps in **Hungary** during winter and spring. It has been opined by many that *A. invadans*, a highly invasive, specific, slow-growing fungus, causes EUS in fishes.

Concomitant to the above, mycobiotic agents *A. flavus* and *Aspergillus fumigates* have been considered the principal fungal flora isolated from the **Nigerian FW fish** in cultured systems. Even in the Indian scenario, mycotic species such as *A. invadans*, *A. flavus*, and *Aspergillus* sp. had been isolated from the affected fish samples.

In fine, it has been suggested by many that further molecular level research on this subject is required to identify the causative agents, disease mechanisms and methods of, prophylaxis and treatment of EUS.

Harikrishnan et al. (2010) had reported on their work on a supplementation diet containing probiotics, herbs, and azadirachtin and the resulting hematological and biochemical changes in *C. mrigala* against *A. invadans*. In their study, *C. mrigala* (63 ± 2 g) were injected intramuscularly (i.m.) with 2.3×10^7 CFU mL/1 of *A. invadans* (isolate B99C). The hematological and biochemical parameters were studied in the experimental and control groups for 30 days. The WBC count depicted a significant increase (WBC: 10^{-6} mm^{-3}) ($p < 0.05$) in the infected untreated group (I), as compared with control. But there was no change in groups fed with probiotics, herbal, and azadirachtin supplementation diets. A similar trend was found in the hemoglobin (Hb: g/dL) and hematocrit (Hct: %) levels. It may be noted here that infected fish fed with probiotics, herbal, and azadirachtin supplementation diets, did not differ ($p > 0.05$) from the control. A similar trend had been found to prevail in the percentage proportion of lymphocytes, monocytes, eosinophils, and neutrophils. The total protein (g dL^{-1}), glucose (mg dL^{-1}), calcium (mmol L^{-1}), and cholesterol (mmol L^{-1}) levels were affected ($p < 0.05$) in the affected group. It had been reported that the administration of probiotics, herbal, and azadirachtin-supplemented diets (for 30 days) could protect the hematological and biochemical parameters in *C. mrigala* from *A. invadans*.

Gayathri et al. (2008) had worked on the "Determination of Specific Sampling Site on EUS Affected Fish for Diagnosis of *Aphanomyces invadans* by Immunodot Test Using Monoclonal Antibodies." This paper is repeated here (vide p. 88 in this file). They had stated that the current study was the first in India to utilize a MAb based immunodot test duly developed for the early, speedy, and specific diagnosis of EUS pathogen *A. invadans* in fish. It aimed at evaluating a specific site for sampling on fish to detect the pathogen. The test was standardized to obtain the result within 3 h with sensitivity range of 45–99 ng of the fungal protein using MAb C1, C13, and C14. Serotyping was also done to deduce the class of antibodies and their consistency in the result on immunodot test. Furthermore, it had been demonstrated by present immunodot test that *A. invadans* antigen was consistently found in deeper portion of the ulcer and very poorly at the surface of the ulcer. Therefore, it was determined that the ideal site for early diagnosis of *A. invadans* on fish would be the deeper portion of the ulcer. This technology would certainly help fish farmers to prevent further economic losses in cases of EUS incidence.

Further, more emphasis currently is given to expand FW AQC in all government sectors utilizing vast inland aquatic resources available worldwide. This increased attention to nonfarming sector such as AQC will help to substantially alleviate the pressure on agriculture sector. Intensive inland AQC system essentially demands a strict implementation of good management practices (GMPs). In such a system of AQC, new problems will inadvertently emerge when GMPs are not followed diligently. Anticipating such a crisis, advanced disease diagnostic techniques, and innovative preventive and curative methods need to be continuously developed to effectively combat the same. Moreover, EUS is one of the most destructive diseases affecting fresh and brackish water fish (cultured and wild) in the Asia—Pacific region, resulting in decreased production and high economic losses, although incidence has come down in FW fish in the recent past. Routinely, EUS is diagnosed by histopathology of multiple sampling sites demonstrating mycotic granuloma and by isolation of the fungus for study of hyphal characters and sporulation, which is laborious. Rabbit antisera based detection and PCR have been developed for EUS, but have their own disadvantages for cost-effective, rapid, specific, and simple detection. Therefore, specific, rapid, and cost-effective immunodot test using monoclonal antibody (MAb) was developed for the first time in India. However, it was necessary to detect the pathogen at an early stage of infection so as to avoid the spread of disease and further losses. Therefore, a study was conducted to determine a specific sampling site to detect the pathogen at a most possible early stage of infection in fish. Technique developed in the present study has immense applications in aquatic animal health management.

Chauhan (2012) had reported on a study of certain fungal diseases in culturable and nonculturable species of fishes of Upper Lake, Bhopal. It was opined that fungal diseases of FW fishes are known to be problematic diseases. The aim of this study was to investigate the increasing aquatic fungal flora of the Upper Lake and the percentage of infection of various species in fishes. For this, investigation was carried out on fungi-infected fishes from November 2011 to April 2012. A total number of 2066 fishes were examined. Out of them, about 287 fishes were found to be infected. The most affected species of fish were *C. batrachus* (24.6%). A total of 13 species of fishes were found to be infected, from which 27 isolates had been found out, which belonged to 8 genera and 14 species. Cultures were made on hempseeds, sesame seeds, and mustard seeds. Pure cultures were made on Sabouraud dextrose agar, glucose yeast agar, and potato dextrose agar. Identification of various species was made based on their vegetative and reproductive characteristics and through consultation of standard literature. During investigation period, among all eight genera, *Achlya* and

Saprolegnia were found to be most virulent. Maximum percentage of infected fishes were found in December (32.3%), followed by November (24.6%) and January (22.8%). *Aspergillus niger* and *Fuzarium* sp. were isolated first from fungi-infected fishes of this region. To find the pathogenicity of these species, there is a need for further study.

Anon (2008) reported on experimental infection and detection of *A. invadans* in European catfish, rainbow trout, and European eel. They had stated that European catfish (*Silurus glanis*), European eel (*Anguilla Anguilla*), and rainbow trout (*O. mykiss*) were challenged by intramuscular injection of zoospores of *A. invadans*, the oomycete associated with EUS. The tropical three-spot gourami *Trichogaster trichopterus* is known to be highly susceptible and was used as a positive control. European catfish were highly susceptible and rainbow trout had moderate to low susceptibility, whereas eels appeared largely unaffected. Inflammatory host response in European catfish deviated from effects seen in most other susceptible fish species and was characterized by a more loosely arranged accumulation of macrophages, small numbers of lymphocytes and multinucleated giant cells without occurrence of EUS-characteristic mycotic granulomas. Semi-nested and single-round PCR assays were developed for this study to detect *A. invadans* DNA in clinical samples of experimentally infected fish. The detection limit of the assays equals one genomic unit. Specificity was examined by testing the DNA of various oomycetes, other relevant pathogens and commensals and host DNA. The single round assay used was fully specific, whereas cross-reaction with the closely related *Aphanomyces frigidophilus* was observed using the semi-nested assay. Analysis of samples by PCR allowed detection prior to detectable HP lesions. Two other published PCR protocols were compared with the PCR protocols presented here.

Mohan et al. (2001) had opined about the possibility of specific fungus as the primary pathogen of EUS in fishes (Mohan et al. (1999) had posed the question, "Is EUS-specific fungus of fishes a primary pathogen?" They had stated that earlier findings on EUS, and the present observation of the authors on the transmission of EUS to snakehead (*Channa* sp.) without skin damage, provide evidence to suggest that the invasive fungus associated with EUS is a primary pathogen.

Campbell et al. (2001) reported on the in vitro **screening of novel treatments for *A. invadans*. They had stated that 49** compounds were tested in vitro for fungicidal activity against hyphae of the fish-pathogenic Oomycetes fungus *A. invadans*. These comprised: (a) chemicals with previous reported activity against Oomycetes fungi; (b) chemicals in use in Asia to treat ulcerative disease outbreaks; (c) commercial biocides and fungicides; (d) natural products with potential antimicrobial activity; and (e) surfactants tested separately and in combination with some of the above treatments. No compounds tested proved as effective as malachite green, but some low-toxicity natural compounds and particular surfactants showed potential for further studies. Some compounds currently in use in Asian AQC were shown to have no effect on *A. invadans* hyphae at recommended treatment rates. A range of compounds that demonstrated activity against the mycelium was selected for further testing in a zoospore motility assay. Lower treatment concentrations were required to inhibit zoospore motility than were required to inhibit hyphal growth. Zoospore activity ceased within 1 h of exposure to 2.5 ppm coconut diethanolamide; 1.25 ppm propolis +0.5 ppm 13/6.5; 5 ppm neem (*Azadirachta siamensis*) seed extract +0.01 ppm, OP10; 20 ppm tea tree (*Melaleuca alternifolia*) oil; and 25 ppm D-limonene + 0.05 ppm E-Z-MulseTM. The treated spores were shown to be nonviable in culture medium. Selected compounds were further tested for ability to inhibit zoospore production by *A. invadans* mycelium over a 72-h period. In toxicity trials, silver barb [*Barbonymus gonionotus* (Bleeker)] exposed to 2.5 ppm coconut diethanolamide; 2.5 ppm OP10; 0.5 ppm E-Z-MulseTM; 20 ppm neem seed extract + 0.5 ppm OP10; and 5 ppm D-limonene + 0.5 ppm E-Z-MulseTM suffered no mortalities and no obvious behavioral changes. Similarly, rainbow trout *O. mykiss* (Walbaum), exposed to 25 ppm propolis + 1 ppm OP10; 10 ppm neem seed extract + 0.01 ppm 13/6.5; and 10 ppm D-limonene + 0.01 ppm OP10 suffered no mortalities and no obvious behavioral changes.

Callinan et al. (1995a,b) had done preliminary comparative studies on *Aphanomyces* species associated with EUS in the Philippines and RSD in Australia. They had reported that fungi morphologically consistent with the class Oomycetes had been recovered in primary culture from 20 out of 22 ulcers in 21 fish species affected by EUS and collected from five sites in the Philippines. Eleven primary isolates and the unifungal cultures derived from them were identified as *Aphanomyces* spp. The *Aphanomyces* isolates seemed to be morphologically and culturally indistinguishable from those reported from RSD in Australia. Comparison of four representative *Aphanomyces* isolates from Australian fish with RSD and three representative *Aphanomyces* isolates from the Philippine fish with EUS, using SDS-PAGE, portrayed similar peptide banding profiles. This could be an indication of a single *Aphanomyces* species. These findings, coupled with epizootiological and pathological similarities between EUS and RSD, seemed to suggest the two syndromes to be identical; and that a single *Aphanomyces* sp. could be the primary infectious etiology.

Severe periodic outbreaks of EUS have affected wild and cultured FW fishes, and wild estuarine fishes, in many countries of South and SE Asia since the late 1970s (Lilley

et al., 1992). Fishes, affected typically by EUS, are characterized in having one or more large dermal ulcers with varying degrees of destruction of underlying tissues. Further, mortality rates are often high. Periodic outbreaks of EUS have occurred in the Philippines since 1985 affecting a number of fish genera, including *Mugil, Anabas, and so* on.

On the other hand, periodic outbreaks of a similar syndrome, known colloquially as "red spot disease" in eastern and northern Australia, have occurred since 1972 in wild FW and estuarine fish (McKenzie and Hall, 1976). Affected fish genera included *Mugil* sp., among others.

The patterns of spread of both EUS and RSD suggested the involvement of one or more primary infectious agents (Rodgers and Burke, 1981, Roberts et al., 1992). No etiological agent had been conclusively identified for either condition to date, although a number of viruses or bacteria have been recovered inconsistently from fish affected by EUS (Frerichs et al., 1986, Roberts et al., 1992) and RSD (Burke and Rodgers, 1981). Further, a number of invasive fungal hyphae, morphologically consistent with the classes Oomycetes or Zygomycetes among fungi, have been recorded in histological sections of advanced ulcers from EUS-affected fish (Tonguthai, 1985; Roberts et al., 1993) and in early and advanced ulcers in RSD-affected fishes (McKenzie and Hall, 1976). Moreover, severe necrotizing granulomatous dermatitis and myositis are associated with the invasive hyphae. Similar fungal morphology and host tissue responses in numerous ichthyospecies with EUS and RSD had suggested that fungi may be the primary infectious agents in both conditions, although there have been conflicting opinions regarding this. While it is possible that these fungi are opportunists that have invaded ulcers initiated by other agents. In that case, the other so-called agents are "viruses" (considered "primary etiological agents" by many schools of thought); the opportunistic agents could also be "bacteria" (in addition to "fungi") that are "secondary opportunistic agents" (Kar, 2013). There have been several studies to identify them. In addition, Fraser et al. (1992) had isolated an apparently single *Aphanomyces* sp. from 27 out of 28 RSD lesions in three species of estuarine fish from widely separated river systems in eastern Australia. They had used methods, however, that are supposed to minimize contamination of inocula. These isolates were morphologically and culturally very similar.

The authors had thus suggested that the putative species was the cause of the typical RSD granulomas. To the contrary, *Achlya* spp. had been recorded from EUS-affected fishes in Thailand and Sri Lanka. *Achlya* spp. occurred commonly in the freshwaters of SE Asia. It had been experienced that the isolation methods used in these studies were likely to result in a high rate of recovery of contaminant fungi. Later, Roberts et al. (1993)

had recovered three putative *Aphanomyces* pathogens from EUS-affected striped snakeheads (*C. striata*) in Thailand. These isolates were said to be morphologically and culturally consistent with those recovered by Fraser et al. (1992), except that the RSD isolates typically had one or two lateral evacuation tubes on each sporangium, rather than the four reported as typical for the Thai EUS isolates.

Identification of *Aphanomyces* spp. was based on differential morphology of oogonia and antheridia. None of these structures were observed in the study of Callinan et al. (1995a,b). Also, none were said to be observed in RSD isolates by Fraser et al. (1992), and none were described for the Thai EUS isolates by Roberts et al. (1993).

In addition to the above, total protein electrophoresis is said to provide useful but not definitive information about relationships within and between species of Oomycetes fungi. In order to determine possible relationships between isolates, peptide extracts from four representative RSD *Aphanomyces* isolates, three representative EUS *Aphanomyces* isolates, as well as an *Aphanomyces* laevis type isolate, an *Aphanomyces* cochloides type isolate, and an *Aphanomyces* euteiches type isolate, had been compared electrophoretically in a study. Peptides resolved by SDS-PAGE were visualized by silver staining. Band similarities or differences between isolates were assessed based on band clusters over specific molecular weight class ranges. On this basis, the banding profiles of isolates from striped snakehead (Bautista), striped snakehead (Laguna), sea mullet (Queensland), and yellowfin bream (Clarence) were similar over the 14–94 kDa range, indicating a close degree of relatedness between these Philippine and Australian strains. The remaining EUS and RSD isolates had similar banding profiles to the above group and to each other. However, they showed various band cluster differences across this molecular weight range, suggesting they may be less closely related to the above isolates.

Manna (1998) had reported about diagnosis and therapy of fungal and bacterial diseases of fish. He had stated that fish husbandry had been practiced for at least 3000 years in China and a little less so in other parts of the world. But AQC had grown into a significant industry in many countries only during the last few decades through gradual transition from extensive to semi-intensive and intensive farming systems. However, outbreak of infections diseases, caused by fungi, bacteria, viruses, and parasites, are becoming more common today mainly due to increased farming intensity and high population density in culture systems; they pose serious social and economic threats to successful fish and prawn husbandry (Siddique et al., 2009). Concomitantly, extensive works have been done on fish diseases and fish pathogens in developed countries that practice large-scale fish farming activities. However, such studies, in India, is scarce and had just picked-up in the recent years. Diseases caused by fungi have long been recognized in fish, but far less is known about the mycotic diseases than those of bacterial or viral origin. Fungi associated with fish diseases belong to a wide range of taxa, the most frequent being the so-called water molds known as oomycetes. The principal members are *Saprolegnia, Achlya,* and *Aphanomyces.* Two important pathogens, viz. *Branchiomyces* and *Ichthyophonus* are still of uncertain taxonomic affinity. The majority of these fungi are at best facultative parasites and attack their hosts when they are injured or as secondary infectants. Only *A. astaci, Ichthyophonus,* and *Trichomaris invadans* are obligate parasites and may cause severe mortalities. Mycotic infections of fish by FW oomycetes can develop at all stages of a fish's life cycle and are of considerable economic significance. Unless treated, such infections are usually lethal to fish and extensive zoospore production ensures that infections spread rapidly through a population. Nearly every important fish species in India has been affected by *Saprolegnia* infection. These wool-like lesions are normally white in color, but may be discolored by the accumulation of debris between the fungal hyphae or as a result of simultaneous bacterial infection. Generally fungal growth is restricted to epidermis and dermis and occasionally growth extends to or beyond muscles. Histologically, there are focal areas of cellular necrosis, spongiosis or intercellular oedema, and ultimately epidermis sloughs-off. Inflammatory responses are normally absent or weakly developed. Fishes die from massive osmoregulatory problems caused by destruction of the superficial tissues. The most common pathogen is the *Saprolegnia parasitica.* Other species of interest are *Saprolegnia diclina* and *Saprolegnia ferax.* Among the **other fungal infections,** infection of fishes by *Achlya, Pythium, and Aphanomyces* have been recorded by many workers. Notwithstanding the above, EUS is the most dreaded fish disease causing severe and chronic skin ulceration and heavy mortality. Nearly all fish species, both cultured and wild, had been affected. In India, the disease was first reported in 1988 in northeastern provinces (Kar and Dey, 1988; Kar and Dey, 1990a,b,c; Kar et al., 1990a,b, 1993, 1994, 1995a,b,c, Kar, 2007a,b,c, 2013), which gradually spread to almost the whole of India. The disease gradually diminished in subsequent years and is restricted mainly to some areas now. The disease is most commonly observed during postmonsoon periods and falls in water temperature. The incidence is more in acidic low calcium soil areas. Pesticides, fertilizers, and heavy metals are suspected to have some role in EUS outbreak. Clinically disease commences with inflammatory red spots that gradually ulcerate. Histologically, the changes are mycotic granuloma with cellular infiltrations.

Etiology of EUS is still confusing. Some rhabdovirus-like particles could be detected from the affected fishes. However, the pathogenicity of the virus particles could not be proved. A number of bacterial floras, such as *A. hydrophila, Pseudomonas* spp., *Staphylococcus* sp., members of the Enterobacteriaceae group, and chemoautotrophic nocardioform bacteria have been isolated from EUS-affected diseased fishes. Initially *Achlya* and *Saprolegnia* spp. were identified from EUS-affected fishes, but these were later dismissed as secondary agents. However, *A. invadans,* has been believed to be one cause of EUS, according to some workers. Classically, diagnosis of infectious diseases is done from natural history, clinical findings, gross pathology and histopathology, and perhaps most importantly, from isolation and identification of the causative organism. Often, particularly for sporadic outbreaks, natural history is not available. Clinical signs exhibited by affected fishes are a few and often overlapping and confusing. Environmental impact on fish health makes the problem more complex. Perhaps. the best way of diagnosis is to isolate and identify the causative agent through Koch's postulates. Sometimes, the concentration of bacteria used to induce disease may be critical or particular environmental situation may be required for clinical manifestation of the disease. For example, pathogenic strain of *A. hydrophila* may not induce disease unless fishes are exposed to environmental stress of low DO and high stocking density and acute injury. However, *Saprolegnia* is said to affect fishes only when there are surface abrasions. Notwithstanding the above, it may be noted here that the bacteria are traditionally isolated on different bacteriological media and are identified by their morphology, cultural and biochemical characters, genetic relatedness, and the production of virulent factors such as toxins and enzymes. On the other hand, fungi are identified by their morphology and characteristics of reproduction. Serodiagnosis of bacterial fish diseases is becoming common with availability of diagnostic antisera.

6.11.11 Epizootic Ulcerative Syndrome Study Based on Bacteriology

Das et al. (2009) reported about Antimicrobial resistance and in vitro gene transfer in bacteria isolated from ulcers of EUS-affected fish in India. They had stated that disk diffusion assay showed that all strains were resistant to ampicillin and sensitive to streptomycin. Of the 15 isolates examined, 93.3% isolates were resistant to erythromycin, sulfadiazine, and novobiocin, while 66% were resistant to rifampin and 20% to chloramphenicol. All isolates harbored plasmids with sizes ranging from 64 to 23 kbp with a 23-kbp plasmid in common. Plasmids from

11 *Aeromonas* strains were transferred to *E. coli* DH5 alpha recipient strain along with the transfer of ampicillin, erythromycin, and chloramphenicol resistance determinants with frequencies ranging from 7.0 × 10(−6) to 1.8 × 10(−5) transconjugants per recipient cell. The resistance to ampicillin, erythromycin, sulfadiazine, novobiocin, and chloramphenicol is prevalent among the bacteria isolated from EUS-affected fish, and resistant determinants of some of these antibiotics have been transferred to the bacteria of other origin. The emergence of antibiotic resistance bacteria and gene transfer in vitro suggests that antibiotics should be used more cautiously to treat *Aeromonas* infections in AQC.

Miles et al. (2001) had reported about the effect of macrophages and serum of fish susceptible or resistant to EUS on the EUS pathogen *A. invadans.*

Mastan and Qureshi (2001) had reported about the role of bacteria in EUS of fishes.

Rahman et al. (2002) had worked on the identification and characterization of pathogenic *Aeromonas veronii* biovar sobria associated with EUS in fish in Bangladesh.

Hassan et al. (2010a,b) had reported on studies of the effects of bacterial diseases on the skin and gill structure of *Clarias gariepinus* in the Dakahlia Governorate, Egypt (Hassan et al., 2010a,b).

The sharp tooth catfish (*C. gariepinus*) is one of the most important FW fishes in the river Nile in Egypt. A total of 94 samples of the live mature catfish *C. gariepinus* were collected from tributaries of the river Nile, Egypt during summer and winter seasons of 2008−2009. Both healthy and diseased fish were examined macroscopically for skin and gill disease. The skin and gills of both healthy and diseased fish were subjected for microbiological identification of pathogenic bacteria. SDS-PAGE and DNA fragmentation analysis and light, scanning electron, and electron microscopical investigations were done. The infected fishes possessed ulcerative and hemorrhagic skin patches, which were more intensified in winter than in summer. The isolated bacteria from both gills and skin of diseased fish were *Acinetobacter lwoffii, Enterobacter amnigenus, E. coli, Citrobacter amalonaticus, Serratia odorifera* and *Aeromonas jandaei* (Gram negative), *S. epidermidis* (Gram positive). SDS-PAGE revealed no variations between healthy and diseased skin and gills. However, DNA fragmentation was moderately detected in diseased skin, which was more than that of the gill. Examination by SEM revealed necrotic patches in both skin and gills. Light microscopical study showed massive atrophy, thinning, and degenerative changes of both skin and gill. TEM study exhibited abundant distribution of cytoplasmic vacuoles and alterations of cytoplasmic organelles including mitochondria, rough endoplasmic reticulum, and lysosomes of club cells and epithelial cells. The nuclear envelope of both club and epithelial cells became convoluted. The mucous cells showed widespread of vacuoles. The macrophage cells were apparently degenerated. The epidermal—dermal junction lacked irregular pattern and the dermis exhibited internal hemorrhage and hyaline necrosis. Finally, *C. gariepinus* was found to be highly susceptible to bacterial infection inducing pathological alterations in skin and gills of fishes and may lead to mortality of fish.

Satyalatha and Vardhani (2014) had reported about **SDS-PAGE analysis of protein in muscle and head kidney of *L. rohita* in response to aeromoniasis.**

Aeromonas infections are regarded as the most common bacterial infections in FW fish. Fish suffering due to aeromoniasis show external symptoms like hemorrhages at the base of fins or on the skin, distended abdomen, and protruding eyes. *L. rohita*, one of the most important FW IMCs, is selected to understand the pathogenic effects of *A. liquefaciens*. Tissues of muscle and head kidney were separated and analyzed for protein profile by SDS-PAGE analysis. The protein bands in muscle and head kidney showed remarkable changes in the experimental groups of fish (groups A, B, C, D, and E treated with *A. liquefaciens*) on day 1—4 of experimental period due to the sensitization of bacterial antigens.

Pramanika and Santrab (2012) had reported about **fish pathogenic *A. hydrophila*)-induced HP attributes in spleens and lymph nodes of Swiss albino mice.** They stated that the effect of the fish pathogenic bacteria *A. hydrophila* on the spleen and lymph nodes of male Swiss albino mice has been investigated and examined histopathologically. Morphological and histological changes were noticed 7, 14, 21, and 30 days postinfection. Time exposure showed no significant difference in body weight between control and treated mice but the value showed a tendency toward increment. Spleen weight had increased significantly ($p < 0.001$) at 14 and 21 days for posttreated mice. The treatment groups showed hepatocellular necrosis of the spleen and other immunoresponsive tissue like lymph-node of mice also. The damaging nature of the immunoresponsive tissues of mice to the pathogenesis of bacteria *A. hydrophila* is provoked in the results obtained.

Sharma and Sihag (2013) had reported about Pathogenicity test of bacterial and fungal fish pathogens in *C. mrigala* infected with EUS disease. The study was conducted for eight weeks to determine pathogenic effect of different fish pathogens in *C. mrigala* that had been affected by EUS disease. The pathogenic organisms (bacteria and fungi) were collected from the infected part of the diseased fish. Isolation and screening of microbes were carried out with the help of a number of biochemical tests. A total of eight bacterial isolates were obtained from the diseased fish. Out of eight, six bacteria viz. *Streptococcus* grp Q1, *A. hydrophila*, *Shigella* spp., *Streptococcus*

faecalis, *Cellobiosococcus sciuri*, *Micrococcus luteus* were found to be pathogenic. The fungus *A. invadans* was also isolated from the diseased fishes. The pathogenicity of disease causing organisms was tested through in vitro and in vivo experiments in different treatments. The result of experiment was found to be significant at level of $p \leq 0.05$.

Harikrishnan et al. (2009) had reported about the **use of an herbal concoction in the therapy of goldfish (*Carassius auratus*) infected with *A. hydrophila*.** The aim of the study was testing the efficacy of a herbal concoction in goldfish (*C. auratus*) experimentally infected with *A. hydrophila*. After an intramuscular injection of the pathogen, scales sloughed off on the site of the injection, with an appearance of a muscular hemorrhagic protuberance, which progressed into an extensive ulcerative dermatitis associated with focal hemorrhage, oedema, and dermal necrosis exposing the underlying muscle (Stanghellini 1997). The progression of the disease affected organs in the following order: muscle, gills, liver, and finally the heart. The dip treatment with the 1% herbal concoction for 5 min daily restored the macro- and microscopic structure of the altered primary gill lamellae, liver, heart, and muscles. The changes after recovery were described.

Saha and Pal (2002) had worked on in vitro antibiotic susceptibility of bacteria isolated from EUS-affected fishes in India. They had evaluated 12 antibiotics for in vitro sensitivity against 16 bacterial strains isolated from surface lesions of fishes affected by EUS, in connection with their studies on in vitro antibiotic susceptibility of bacteria isolated from EUS-affected fishes in India. Disk diffusion assay in Mueller—Hinton agar, in their study, showed that pseudomonads and aeromonads were mainly resistant to penicillin, ampicillin, and erythromycin. However, additionally, some were resistant to gentamycin and amoxycillin. But resistance toward antibiotics, which were previously recommended for EUS treatment, such as oxytetracycline and chloramphenicol, was not observed. It was further revealed in their study that four aeromonads and two pseudomonads were found to induce ulcers when injected intramuscularly in healthy *A. testudineus*. Nevertheless, all six pathogenic isolates were sensitive toward oxytetracycline, chloramphenicol, and nalidixic acid. Oxytetracycline seems to be an effective antibiotic. Further investigations to determine the mode of treatment and dose appear to be worthwhile.

Roy et al. (2013) studied isolates of multiple antibiotic resistant *Aeromonas* sp. from the skin, gill, and gut of FW loach *L. guntea* collected from four different sites of river Lotchka in the Darjeeling district of West Bengal in India. Isolated bacteria were identified by different biochemical procedures. PCR was performed using genus-specific 16S rDNA primers for confirming the identification of *Aeromonas* sp. Antibiotic susceptibility test of bacterial isolates was done by disk diffusion method. A total of 18 *Aeromonas* spp. were isolated from skin, gill, and gut of fish. Maximum resistance was exhibited for penicillin-g, ampicillin, and cephalothin (100%, 92%, and 84% resistant in the gut of fish). Resistance was almost nil in ciprofloxacin and tetracycline. Among all isolates, one particular strain showed resistance against eight antibiotics. The bacterial strain was subjected to PCR analysis with *Aeromonas*-specific 16S rRNA primers and the PCR product was sequenced. The sequence was performed by comparative analysis with GenBank, and maximum similarity was found with *A. veroni*. Presence of *A. veronii* resistance to multiple antibiotics indicated that loaches were contaminated with multiantibiotic resistant enteric pathogenic bacteria. Their study concluded with the view that the presence of *A. veronii* and its resistance to multiple antibiotics indicated that fish were contaminated with multiantibiotic-resistant enteric pathogenic bacteria. This study evidently provides valuable information for making policy decisions aimed at reducing microbial contamination of fish and the indiscriminate use of antibiotics. Further research on antibiotic susceptibility in the aquatic environments of northern region of West Bengal (India) may be needed to throw more light in the matter.

Das et al. (2009) worked on the antimicrobial resistance and in vitro gene transfer in bacteria isolated from ulcers of EUS-affected fish in India with the aim to reveal the occurrence of drug resistance and plasmid-mediated transferability. Disk diffusion assay portrayed that all strains were resistant to ampicillin and sensitive to streptomycin. Further, *c* 93.3% of isolates were resistant to erythromycin, sulfadiazine, and novobiocin, out of the 15 isolates examined; 66% were resistant to rifampin and 20% to chloramphenicol. Nevertheless, all isolates harbored plasmids, with sizes ranging from 64 to 23 kbp with a 23-kbp plasmid in common. Further, plasmids from 11 *Aeromonas* strains were transferred to *E. coli* DH5 alpha recipient strain along with the transfer of ampicillin, erythromycin, and chloramphenicol resistant determinants with frequencies ranging from $7.0 \times 10^{(-6)}$ to $1.8 \times 10^{(-5)}$ transconjugants per recipient cell.

The resistance to ampicillin, erythromycin, sulfadiazine, novobiocin, and chloramphenicol are prevalent among the bacteria isolated from EUS-affected fish; resistant determinants of some of these antibiotics have been transferred to the bacteria of other origin.

The main significance and impact of the study could be the fact that the emergence of antibiotic resistant bacteria and gene transfer in vitro seem to suggest that antibiotics be used more cautiously to treat *Aeromonas* infections in AQC.

It is common sight that murrels are often affected by the dreadful disease EUS. The primary causative agent of EUS,

according to some workers, is said to be the fungus *A. invadans* and the opportunistic bacterial pathogen is said to be *A. hydrophila*. performed in vitro and in vivo antimicrobial effects of *Wrightia tinctoria* (Roxb.) R. Br. against EUS in *C. striata*.

The antibacterial and antimycotic activity of *Wrightia tinctoria* was studied against selected pathogens. Also, in vivo antibiotic effect in EUS-affected striped murrel *C. striata* was examined. The leaves of the medicinal herb, viz. *W. tinctoria* (Roxb.) R. Br. was evaluated against five fish pathogens. Phytochemical screening of leaves revealed the presence of steroids, reducing sugars, alkaloids, phenolic compounds, flavonoids, saponins, and tannins. The TLC chromatogram of extracts revealed several colored bands. Antibacterial activities of extracts were determined by the agar disk diffusion method. Chlortetracycline (5mcg/disc) was used as positive control for comparison of the inhibition zones. The extract depicted efficient antibacterial activity against *A. hydrophila* with maximum zone of inhibition. The minimum inhibitory concentration of the extract against pathogens was determined. Growth of *A. invadans* was found to be inhibited by 85%—90%. The herbal paste applied on lesions showed a positive effect by controlling and curing lesions. Thus, the active extract of *W. tinctoria* could be taken to the next step of bioassay guided purification to characterize the novel antimicrobial agents. Almost similar work was done by Rungratchatchawal (1999).

Twenty-five bacterial strains were isolated by Hossain et al. (2011) from various locations (culture and capture sources) in different geographical areas of Bangladesh. Six pathogenic bacterial isolates had been chosen from among the 25 isolates of *Aeromonas* and *Pseudomonas* spp. in order to characterize them and analyze their potential virulence status. The bacterial pathogens were put in four types of water, viz. pond water, physiological saline, distilled water, and tap water. The isolates were injected into silver barb (*B. gonionotus*) with low to high doses. Out of the six isolates, *Pseudomonas fluorescens Pf-16* was detected as highly virulent. This pathogen had resulted in 40% lesions and 30% mortality among the experimental fish. On the other hand, *A. hydrophila, Ah-25*, was detected as medium virulent. They reported that this pathogen could successfully produce 30% lesions with 20% mortality among the experimental fishes. Concomitantly, *A. veronii* biovar sobria A 44 was detected as low virulent. It could produce 50% lesions and 40% mortality among the experimental fishes. Bacterial diseases are a potential threat to both wild and AQC fishes in Bangladesh. They have been isolated from food and from different aquatic systems such as lentic bodies, lotic bodies, and drinking water, and a variety of foods. Various diseases, like EUS, bacterial hemorrhagic septicemia, tail and fin rot, bacterial gill rot, dropsy, and different fungal and parasitic ailments are said to be significant limiting factors for sustained fish production.

Datta Ray and Homechaudhuri (2014) have worked on the morphological and functional characterization of hepatic cells in IMC, *C. mrigala* against *A. hydrophila* infection. As reported, the IMC mrigala (*C. mrigala*) was exposed to sublethal doses of *A. hydrophila* in order to study the innate immune responses. Hepatic morphology, characterized by light microscopy and flow cytometry, revealed production of oxygen radicals, hepatic catalase and significant ($p < 0.05$) increase of super oxide dismutase activity in liver of treated fish. Condition factor and hepatosomatic index (HSI) were calculated as indices of energy status. Results portrayed a significant ($p < 0.05$) increase in HSI; the condition factor had significantly ($p < 0.05$) decreased in the treated fishes. HP studies showed changes including focal necrosis, hemorrhage, hyperplasia, and early stages of granulomatous formation in hepatocytes of infected fishes. Cytolysis due to bacterial toxins was also noted. Flow cytometry study had confirmed the presence of apoptotic cells among the treated fish. As such, distinctive immune responses had been noted in *C. mrigala* due to bacterial infection. The work thus is said to present an insight into the innate immune response in conjunction with the identification of potential fish biomarkers, which could be screened for timely application of proper preventive measures.

Kurhalyuk and Tkachenko (2011) worked on the induction of oxidative stress and antioxidant defenses in livers of sea trout, *S. trutta* L. with ulcerative dermal necrosis. As reported, the aim of this study was to investigate the lipid peroxidation (thiobarbituric acid reactive substance content) and oxidative modified protein levels (stable 2.4-dinitrophenyl hydrazine derivatives of the carbonyl groups), and antioxidant defense system (superoxide dismutase, catalase, glutathione peroxidise, glutathione reductase activity, and total antioxidative activity) in livers of males and females from a control group (healthy specimens) and a group of sea trout, *S. trutta* L. affected by ulcerative dermal necrosis (UDN). UDN tends to induce protein oxidative destruction and negatively affects livers of infected trout by initiating oxidative stress. UDN-affected males are characterized by higher expressed oxidative-modified protein levels in their livers compared with those of healthy specimens. Our results seem to indicate a link between the level of oxidative stress and antioxidant defenses in livers of trout with UDN. Increased levels of protein oxidation and modified activities of antioxidant defenses had caused declines in total antioxidant activity in livers of UDN-affected females. The glutathione-mediated antioxidant defense system and endogenous catalase play a critical role in the intracellular antioxidant defenses under UDN-induced oxidative stress.

Rahman et al. (2004) had worked on the distribution of *Aeromonas* spp. emphasizing on a newly identified species *Aeromonas* sp. T8 isolated from fish and aquatic animals in SE Asia. As reported, the distribution pattern of a collection of 106 *Aeromonas* strains, isolated from both healthy and EUS-affected fish as well as septicemic disease-affected frogs and turtles in Bangladesh, Japan, Malaysia, the Philippines, and Thailand had been investigated. The study had been done through the physiobiochemical characterization of the strains and subsequent confirmation by analysis of the 16S rDNA sequences of some randomly chosen representative strains from all identified phenotype. Special emphasis had been laid on confirmation of a group of strains, which possibly belonged to a newly identified species *Aeromonas* sp. T8 bDNA−DNA hybridization method; this was particularly found in different species EUS-affected fish in the Philippines and Thailand. *A. hydrophila* subsp. *hydrophila* and *A. hydrophila* subsp. *ranae* were found only in Thailand. *A. veronii* biotype *sobria* and *A. veronii* biotype *veronii* had been found to dispersed mostly in EUS-affected fish in various countries. *A. jandaei* was obtained from EUS-positive fish in Bangladesh and Malaysia, but *Aeromonas media* came from healthy fish in Bangladesh.

Ramkumar et al. (2013) had reported on studies on ulcerative disease caused by *Providencia stuartii* bacteria in the IMC *L. rohita* (Ham).

Rohu (*L. rohita*) is a species of fish of the carp family Cyprinidae. A survey was conducted from September 2012 to August 2013 to study the prevalence of bacterial ulcerative disease in *L. rohita*, cultivated in a FW farm located at the Moongilthuraipattu village in the Villupuram district of Tamil Nadu, India. To mimic the infection in healthy fish, organs such as liver, gill, and ulcerative skin were used to prepare the inoculum in PBS buffer. The inoculum was injected into healthy fishes through intramuscular injection, immersion challenge, and oral route. The tissue supernatant from infected rohu fish was screened for isolation and screening of the causative organism responsible for ulcerative lesion in *L. rohita*. Based on the biochemical, morphological, and molecular features, the bacterial culture isolated from infected rohu was tentatively identified as *Providencia* species like bacterium that was further confirmed by analysis of its 16S rRNA gene using PCR and DNA sequencer and it has been identified as *P. stuartii*. Further, its partial sequence was deposited in GenBank (KF155520.1).

Notwithstanding the above, with regard to *Signal transduction pathway of A. hydrophila*, their results suggest that there is a signal transduction pathway participating in the interaction between the pathogen *A. hydrophila* and EPC cells. It is likely that *A. hydrophila* adheres to the host cell surface before internalization. The adhesins may be the 0-antigen of lipopolysaccharide or the outer-membrane protein isolated by Lee et al. (1997) that mediate the adhesion and internalization of the bacterium. At present it is not clear whether *A. hydrophila* also produces invasins to aid in internalization. After adhesion, *A. hydrophila* probably initiates a signaling cascade that involves tyrosine kinase. Hence this kinase appears to function as part of an uptake signal for this fish pathogen. At the end of the signaling pathway, *A. hydrophila* causes the rearrangement of microfilament (F-actin) in EPC cells to form actin clouds. These actin clouds might be associated with the internalization of bacteria. Subsequently, the internalized bacteria replicate intracellularly and cause morphological changes in the EPC monolayer.

6.11.12 Epizootic Ulcerative Syndrome Study Based on Viruses

Rodger (1994) had reported about the **NRED-ODA Aquaculture Research Program. The executive summary revealed that** in order to investigate the significance of viruses isolated from fish affected by EUS, a series of viral characterization and pathogenicity studies were undertaken (Juan, 1994). Further investigation of EUS and non-EUS diseased fish were undertaken with regard to viral involvement. A rhabdovirus (T9412) was demonstrated as a primary pathogen for striped snakehead fish and a combination of this virus and *Aphanomyces* spp. fungi appeared to produce more severe EUS disease than a single infection with the fungus. Improved virus isolation methodology resulted in higher viral recovery rates from EUS outbreaks. The genome of the snakehead cell line-isolated retrovirus was fully sequenced and cloned and demonstrated to be distinct from other gene sequenced retroviruses. The EUS-associated reovirus (T9231) was shown to be a member of the genus *Aquareovirus*, family Reoviridae, but was a new serotype. Experimental studies with this virus in snakehead fish demonstrated nonpathogenicity. The BSN cell line-associated virus was shown to be a new member of the family Birnaviridae and was also shown to be nonpathogenic for snakehead fish. Enzyme-linked immunosorbent assays (ELISA) were developed for both the aquareovirus and the BSN virus and new cell lines developed from Asiatic snakehead fish. A regime to remove *mycoplasma* contamination from cell lines was developed and demonstrated as effective. The piscine neuropathy nodavirus was isolated from diseased spotted brown grouper from Thailand. As background information, EUS disease is **widely recognized** as the most significant limiting factor in the development of food fish culture, as dramatically illustrated by the annually recurring outbreaks of EUS that have now become established throughout the Indo-Pacific region. The rational implementation of methods of treatment and control of infectious fish disease requires a thorough understanding of the etiology,

epidemiology, and pathogenesis of such conditions and the availability of rapid and accurate means of diagnosis of the clinical and latent states of infection.

The additional factors that emerged were:

1. Higher virus isolation rates were obtained when tissues were processed and inoculated into cell cultures within 24 h of sampling without storage or transport.
2. Virus could only be isolated from fish during the early stages of a disease outbreak and not from the same population two weeks later.

Similar results were also obtained by Kar and Dey (1990a,b,c) and Kar (2007a,b,c, 2013).

3. Virus could be isolated directly from skin lesion material, demonstrating local association of the virus with the lesion.
4. Virus could be isolated from more than one fish species (snakehead and gourami) in the same environment, at the same time.

b) Reexamination of 100 selected fish tissue extracts from SE Asian FW species and mullets from southern India, affected with ulcerative disease conditions, did not result in any further virus isolations. There was no apparent survival of EUS-associated rhabdoviruses in original tissue extracts after prolonged storage at low temperatures.

It is important to note the isolation of the piscine neuropathy nodavirus, a very significant breakthrough in fish disease research and opens up a vista of opportunities for the diagnosis, control, and prevention of this devastating disease of marine fish, both farmed and wild. The specific cell line utilized for the isolation of the virus is available to other fish disease centers through the ECACC.

Concomitant to the above, with regard to **epizootic hematopoietic necrosis** is an infection caused by EHNV, a member of the genus *Ranavirus* in the family Iridoviridae with the type species frog virus. It causes systemic clinical or subclinical infection in rainbow trout and redfin perch. The virus is currently restricted to Australia. Further, ***Bonamia exitiosa*** is a protistan parasite of the phylum Haplosporidia. ***Perkinsus marinus*** is a pathogenic dinoflagellate of oysters. ***Mikrocytos mackini*** is an eukaryotic protistan of uncertain taxonomy. **Taura syndrome** is due to a small icosahedral virus not enveloped. The agent is an RNA virus classified in the Discistroviridae family. Taxonomy of the **yellowhead virus, which causes yellowhead disease**, had been the subject of some controversy.

John et al. (2001) had worked on the characteristics of a new reovirus isolated from EUS-infected snakehead fish.

John and George (2012) had reported about an update of viruses associated with EUS in fishes.

Frerichs (1995) had reported about viruses associated with EUS) of fish in SE Asia.

Workers had to take recourse to virological study, when studies with bacteria and fungi were not fully able to answer to the riddles of EUS. However, virological studies could be possible, only where there had been tissue culture laboratory with electron microscope. This could be the possible reason for paucity of works on the virological aspects of EUS. Nevertheless, a range of viruses has been recovered from whatever virological works have been done to date (Frerichs et al., 1986; Frerichs, 1995; Kar et al., 1993; Kar, 2007a,b,c, 2013). It is believed that *A. invadans* could invade more readily and induce lesions in the affected fish, but it appears to take place only if the skin of the affected fish is damaged rather than intact (Kiryu et al., 2002). Further works done by in Thailand had portrayed that rhabdoviruses could be isolated only from samples collected during the early stages of outbreaks in snakeheads. It has been suggested that rhabdoviruses usually initiate a very preliminary and primary lesion on the skin, said to provide a portal of entry for the bacteria (notably *A. hydrophila)* and the fungi (notably *A. invadans)* to cause EUS.

John and George (2012), while reporting about "Viruses Associated with Epizootic Ulcerative Syndrome: An Update," had stated that since 1983, several viruses had been observed and isolated from fishes infected with EUS. These viral isolates were principally rhabdoviruses and birnaviruses. A single isolation of a reovirus-like agent was also made during the 1992 Thailand epizootic. A ranavirus isolation was also reported recently from an ulcerated fish from China.

With regard to **viruses observed in electron microscopy**, it was reported that the presence of virus particles in tissues of EUS-infected fishes was first demonstrated by electron microscopy. The initial observations of viral association with EUS were noted by TEM studies of tissues of diseased striped snakehead fish and walking catfish specimens obtained from the epizootic in Thailand during 1983 and 84. Virus-like particles were reported in hepatocytes, the spleen, the kidney, blood cells, the capillary endothelium, and muscle myofibrils taken from the site of ulceration in diseased specimens. These particles observed in the cytoplasm were of approximately 70 nm diameter but could not be precisely identified as they were pleomorphic with shapes varying from round, oval, elongate to kidney shaped. Particles of rhabdovirus-like morphology were noticed in skin lesions of infected barramundi *(L. calcarifer)*. Presence of rhabdovirus and infectious pancreatic necrosis viruslike particles in cell cultures inoculated with tissue extracts from EUS-infected snakeheads was noted by Wattanavijarn et al. Rhabdoviruses were also observed in cell cultures inoculated with tissue extracts from a variety of infected fishes including that of a barramundi, which showed extensive CPE. In addition, with regard to viruses isolated in cell culture, birnaviruses belonging to coldwater infectious pancreatic necrosis virus (IPNV) serotypes and different from existing reference strains (Sp, Ab, VR299 and TV-1) were isolated from EUS-infected fishes. The first isolation of a virus associated with EUS was from an ulcerated sand goby (*Oxyeleotris marmoratus*) in Thailand. The agent named sand goby virus (SGV) was found to belong to the family Birnaviridae although the biological, serological, and biochemical characteristics of the virus differed from the existing reference strains of IPNV VR299, Sp, and Ab. Although SGV was found to replicate well at temperatures ranging from 5 to 30 °C, its role in inducing EUS was not determined. SGV was later reported to be a mixture of two strains Sp and Ab of IPNV based on a comprehensive reciprocal cross-neutralization study. Second isolation of another birnavirus was from EUS-infected snakehead fish (*C. striata*) in Thailand in 1986. The snakehead fish virus was found to be similar to IPNV based on morphometric characteristics and biophysical properties such as resistance to chloroform and heat treatment, but was not serologically compared with the existing strains of IPNV. Third birnavirus isolation was made from pooled organ extracts of EUS-infected snakehead fishes from Myanmar and Thailand and eye spot barb (*Hampala dispar*) from Lao PDR in SE Asia in 1988. The isolate was serologically compared and identified as Sp serotype of IPNV. A fourth isolation of an aquatic birnavirus was reported in 1993 from the spleen of an ulcerated giant snakehead (*Channa micropeltes*) in Singapore. The isolate had differed from the existing strains of IPNV in the relative mobility patterns of nucleic acid and structural polypeptides. Further comparative studies of the isolate, however, were not reported. However, none of the above viruses have been reported to have any causal role in the pathogenesis of EUS. Rhabdoviruses are quite well known with regard to EUS. Since its first isolation from striped snakeheads and a FW eel (*Fluta alba*) in 1986 in Northern Thailand and in Myanmar, rhabdoviruses were further isolated from infected snakehead fish. The five ulcerative rhabdo-virus isolates (UDRV) that had been obtained, grew in AS, BF-2, snake-head (*C striata and C. micropeltes*) and climbing perch (*A. testudineus*) cell lines. CHSE-214, RTG-2, EPC, nile tilapia (*Oreochromis niloticus*) and grass carp (*C. idella*) cell lines were all refractory to infection. Characterization of these isolates by cross-neutralization showed that they were serologically distinct from the common pathogenic fish rhabdoviruses VHSV, IHNV, EVA, EVX, SVCV, PFRV, and perch Rhabdovirus. Rhabdovirus isolates were subsequently isolated from infected snakehead fish (SHRV) in central Thailand in 1986. UDRV and SHRV were distinguishable by serology and the relative mobility pattern of their structural polypeptides. Both viruses were serologically unrelated to the other known fish rhabdoviruses and phylogenetic analysis of the SHRV G protein indicated that the SHRV belonged to the genus *Novirhabdovirus*. The rhabdovirus isolates obtained from EUS-infected fishes have been found to grow well in the cell lines established from the susceptible fishes. The UDRV failed to induce any clinical lesions in snakeheads subjected to intraperitoneal inoculation or bath exposure. EUS lesions were however induced in snakeheads (*C. striata*) held at 20 °C injected intramuscularly with a rhabdovirus (strain T9412) followed by bath challenge with *Aphanomyces invadans* spores. Rhabdovirus injection alone induced only small hemorrhagic lesions at the injection site most of which had healed by the end of the experiment. They did not get any EUS induction in a similar experiment conducted at 29 °C indicating the combinatorial conditions that lead to the development of EUS lesions. Cell culture grown EUS-associated rhabdovirus was experimentally shown to induce slight to moderate skin lesions in all tested naive snakeheads at rearing water temperatures of 20—22.5 °C, but not at 28—32 °C. LioPo et al. transmitted rhabdovirus isolated from the Philippines horizontally from virus infected fish to naive snakehead (*C. striata*) by cohabitation and lesions appeared in 6—14 days on the naive fish. Naive fish exposed to lake water only developed lesions in 16—21 days. Rhabdovirus was isolated from fish in both treatments suggesting transmission via water from infected fish. Fish in aquifer water only without contact with EUS-infected fish did not develop the disease. **Further isolation of rhabdoviruses** was reported during the later incidences of EUS in Thailand, Sri Lanka, and the Philippines, making this group of viruses the most isolated from EUS-infected fishes. Though the rhabdoviruses isolated were heterogenic, their frequent isolation, range of areas from which isolates have been obtained and ability to induce dermal lesions in experimental infectivity studies make rhabdoviruses one of the likely potential primary etiological agents that could be triggering the onset of EUS paving way for the invasive fungus *A. invadans* to develop characteristic dermal lesions. While reporting about **reoviruses,** a reovirus had been isolated from EUS-infected snakehead fish in 1992 following a cohabitation experiment. EUS was induced in six healthy snakehead fish by cohabitation with six characteristically EUS-infected snakehead fish. All six EUS-infected fish died within five days of cohabitation trial. Tail ulcerations developed in the healthy snakehead fish 12 days post cohabitation. They were sampled for virological examination by inoculating the tissue homogenate onto SSN-1 and BF-2 cells. The CPE was developed in SSN-1 cells incubated at 25 °C after eight days. Virus was further passaged onto fresh SSN-1 cells and CPE observed in two days. This isolate (snakehead reovirus SKRV) was characterized in comparison with other European and American aquareoviruses and observed to be different due to a 10-segmented dsRNA genome, lack of syncytium formation, and the inability to multiply in coldwater fish

cell line CHSE-214. The snakehead reovirus (SKRV) was found to belong to the *Orthoreovirus* genus and is the first orthoreovirus to be isolated from a fish host. SKRV was found to grow in SSN1 and SSN3 cell lines at 25—30 °C. Serologically, the virus was distinctly different from golden shiner virus, catfish reovirus, tench reovirus, and chub reovirus. Pathogenicity studies indicated that the virus was not capable of producing EUS-associated lesions in snakehead juveniles. While discussing about **ranaviruses,** it was reported that a virus belonging to ranavirus of Iridoviridae family was isolated from cultured largemouth bass (*Micropterus salmoides*) experiencing extensive mortality in the Guangdong province, China in 2008. Though not categorized into EUS, affected fish had ulcerations on the skin and muscle. The virus grew well in EPC cell cultures and TEM study of the ulcerative muscle tissue and the infected EPC cells revealed cytoplasmic, icosahedral virions of 145 nm in diameter. The virus resembled doctor fish virus and was closely related to largemouth bass virus upon sequence analysis of viral major capsid protein and DNA methyltransferase. Intramuscular injection of the virus resulted in clinical signs of the disease and caused 100% mortality of healthy largemouth bass.

With regard to viruses associated with cell culture, it had been stated that some cell cultures developed from tropical warm-water fishes were found to harbor persistent retrovirus and birnavirus-like infections. Spontaneous production of C-type retrovirus particles was noticed in cell lines developed from healthy striped snakehead, snakeskin gourami (*Trichogaster pectoralis*), and climbing perch (*A. testudineus*), which are known to be susceptible to EUS. Similarly, a cell line developed from blotched snakehead (*Channa lucius*) was found infected with a birnavirus that was found to be a new serotype (serotype C) of the IPNV group. Pathogenicity studies indicated that both the snakehead retrovirus and BSNV were unable to induce the disease condition in snakeheads. The role of these persistent viral infections in acting as potential reservoirs of viruses and their effect on the immune system of the carrier fish had not been fully investigated.

Absence of hemorrhagic septicemia, (quite characteristic of *Aeromonas* infection) in all but most of the ulcerated fish (Roberts, 1989); nonoccurrence of these pathogens in the early part of the disease (Lilley et al., 1992); and close association of these bacteria with fishes in water (as stated earlier) since time immemorial (Kar and Dey, 1990a,b,c); suggest that *Aeromonas hydrophila* is unlikely to have any primary infective role in the pathogenesis of EUS. Likewise, a primarily viral (?) etiology (filterable biological agent) is always suspected to be associated with the initiation of EUS. (FAO, 1986; Kar, 2007a,b,c, 2013.

Therefore, in order to confirm the presence of virus, EUS-affected fish tissues have been subjected to virological examination. Ulcerated tissues from EUS-affected *C. batrachus* have been homogenized in PBS, clarified by high-speed centrifugation (10,000 rpm) for 10 min, and decontaminated through membrane filtration (0.22 µm). Monolayer cultures of BF2 fish cell line have been grown in 96-well microtiter plate at 28 °C in Leibovitz L-15 medium supplemented with 10% FCS and vitamin mixture. A 50-µL aliquot of each dilution (up to 10^{-7}) of the filtered supernatant was added to 80% confluent monolayer of BF2 fish cell line grown in 96-well microtiter plate. After 2 h of adsorption at room temperature, cultures were overlaid with 150 µL of maintenance medium (L-15 supplemented with 2% FCS) and incubated at 28 °C. Progressive CPE had appeared between 48 and 72 h postinfection day (PID) in 10^{-1} to 10^{-4} dilution. The infection was passable in the subsequent cultures indicating the isolation of virus (Kar et al., 1993, 1994, 1999a, 2003a, 2004a; Kar, 2007a,b,c, 2013). Efforts are being made to characterize the virus and correlate the same with viruses isolated elsewhere (Frerichs et al., 1989; Ahne et al., 1988; Wattanavijarn et al., 1985). EM studies of the TCF showed occasional round particles unlike those of Roberts' group in Scotland (Frerichs et al., 1986).

The above study confirms that EUS outbreak in FW fishes is initiated by primary viral infection followed by secondary bacterial and fungal contaminations. Viral presence is said to be only for a very brief period, and the havoc is caused mainly by bacteria and fungi.

Gilda et al. (2000) is said to have published a first report of the isolation and characterization of a virus from the Philippines that supposedly causes EUS in fishes. As reported, the virus was isolated using snakehead spleen cells (SHS) from EUS-affected snakehead fishes (*C. striata*) with severe lesions collected from Laguna de Bay during January 1991. The virus was said to induce CPE in SHS cells and yielded titer of 3.02×10^6 TCID$_{50}$/mL at 25 °C within two to three days. As reported, other susceptible cell lines included bluegill fry (BF-2) and catfish spleen cells. However, it was reported that the replication in Chinook salmon embryo cells (CHSE-214) was minimal. On the other hand, *Epithelioma papulosum cyprinid* cells (EPC) and rainbow trout gonad cells (RTG-2) were refractory. Incidentally, there was no replication of the virus at 37 °C. Further, the virus could be stored at −10 °C for 30 days without much loss of infectivity. Exposure of the virus to chloroform or heating to 56 °C for 30 min inactivated the virus. Further, there was a loss of >100-fold viral infectivity after exposure to a medium with pH 3.0 for 30 min. It was revealed to be an RNA genome, as virus replication remained unaffected in 5-iododeoxyuridine (IUdR). Neutralization tests using the Philippine virus, the ulcerative disease rhabdovirus (UDRV), and infectious hematopoietic necrosis virus (IHNV) poly-valent antisera, depicted slight cross-reaction between the Philippine virus antiserum and the UDRV. It, however, did not reveal any serological

relationship with SHRV and IHNV. TEM studies with SHS cells infected with the virus showed virus particles with typical bullet morphology with an estimated size of 65 × 175 nm. The Philippine virus was therefore regarded as a rhabdovirus with no possible role in the epizootiology of EUS.

Lio-Po et al. (2000) had reported about the characterization of a virus obtained from snakeheads *O. striata* with EUS in the Philippines.

Zhang et al. (2000) had reported the isolation of a lethal rhabdovirus from the cultured Chinese sucker *Myxocyprinus asiaticus*. It had been stated that a rhabdovirus was found to be associated with a lethal hemorrhagic disease in the cultured Chinese sucker *M. asiaticus* (Bleeker). The rhabdovirus was amplified and isolated from infected grass carp ovary cells. The virus particles (150−200 nm in length and 50−60 nm in diameter) from the ultrathin sections of the liver cells had been reported to exhibit the characteristic baciliform morphology with two rounded ends rather than a typical flat base, budded through vesicle membranes of the infected cells. Most of the other characteristics, like size, virus infectivity to fish cell lines, cytopathogenic effects, stability at high temperatures, vesicle formation in infected cells, electrophoretic patterns of structural proteins and the presence of an RNA genome, seem to reflect close resemblance to other fish rhabdoviruses. However, it is not very well known, whether it is a novel virus species or it is an isolate of a known fish rhabdovirus. The authors opined to name it tentatively as Chinese sucker rhabdovirus until its confirmed identification is done.

John et al. (2002) had reported on VP2 protein carries neutralization epitopes in blotched snakehead virus (BSNV). They had stated that new aquabirnavirus BSNV had been isolated from a persistently infected warmwater fish cell line (BSN) developed from blotched snakehead (*C. lucius*). Four polypeptides of molecular mass 112 (polyprotein) (VP1), 44 (VP2), and 37 (VP3) kDa had been identified as BSNV. Reciprocal cross-neutralization analysis involving BSNV and four classical strains of infectious pancreatic necrosis virus (IPNV) strains WB, Sp., Ab, and TV-1 revealed that BSNV antiserum was capable of neutralizing only homologous virus, while no cross-reaction had been evident against the other viruses. Western blot analysis of separated viral proteins had been done using antisera against BSNV and IPNV Sp in order to study and compare the antigenic properties of the virion structural proteins of BSNV and IPNV strain Sp. The analysis indicated that VP2 p2 protein (44 kDa) of BSNV alone was recognized by homologous antiserum while antiserum against IPNV Sp and SGV using BSNV antiserum had showed that VP2 is probably the only protein expressed on the virion surface. This protein carries neutralization epitopes in BSNV.

6.11.12.1 Viruses Observed in Electron Microscopy

Presence of virus particles in tissues of EUS-infected fishes was first demonstrated by electron microscopy. The initial observations of viral association with EUS were noted by TEM studies of tissues of diseased striped snakehead fish and walking catfish specimens obtained from the epizootic in Thailand during 1983 and 84. Presence of viruslike particles was reported in hepatocytes, the spleen, the kidney, blood cells, the capillary endothelium, and muscle myofibrils taken from the site of ulceration in diseased specimens. The particles observed in the cytoplasm were approximately 70 nm in diameter, but could not be precisely identified as they were pleomorphic and varied among round, oval, elongate, and kidney-shaped. Particles of rhabdovirus-like morphology were noticed in skin lesions of infected barramundi (*L. calcarifer*). Presence of rhabdovirus and infectious pancreatic necrosis viruslike particles in cell cultures inoculated with tissue extracts from EUS-infected snakeheads was noted by Wattanavijarn et al. Rhabdoviruses were also observed in cell cultures inoculated with tissue extracts from a variety of infected fishes including that of a barramundi, which showed extensive CPE.

6.11.13 Molecular Diagnosis

6.11.13.1 Aphanomyces invadans: Its Identification

As reported, *A. invadans* does not produce any sexual structure. It thus may not be diagnosed by morphological criteria alone. Nevertheless, the oomycete could be identified to the genus level by inducing sporogenesis, demonstrating typical asexual characteristics of *Aphanomyces*. In fact, *A. invadans* is characteristically slow growing in cultures and fails to grow at 37 °C on GPY agar. However, the two procedures that confirm *A. invadans* are (a) bioassay and (b) PCR amplification of the rDNA of *A. invadans*.

6.12 WORKS DONE ON EUS IN RELATION TO SOCIOECONOMICS

EUS fish disease has been having a devastating effect on the socioeconomic conditions of fishers and fish farmers. In connection with discussing aquatic animal health in rural and small-scale AQC, Subasinghe et al. (2001) had reported that the incurred losses due to EUS fish disease affected the livelihoods of people involved in AQC and the communities in which they live. The effect of the disease is reflected through the reduced availability and loss of food, income, and employment; and other social consequences, notably shortfalls, shortages, and destabilization of markets.

While addressing disease problems in small-scale AQC, it had been opined that rural, small-scale farmers are generally resource-poor with usually limited knowledge of disease and health management among fishes. Thus, it is perhaps essential to understand how the rural small-scale AQC sector is being managed by farmers and their associates engaged in various sectoral activities. This could, concomitantly, help us in framing appropriate interventions through better health management practices. Some significant health management practices could be as follows: (a) framing of effective quarantine measures at the national level; (b) access of farmers to the basic aquatic animal health services; (c) basic health management measures are to be incorporated into fisheries enhancements and rural livelihood programs; and (d) strengthening the extension services and acceleration of communication exchange could bring about a let-up in the situation through fast responses to disease situations. In this connection, primary health care may be defined as that kind of care that could be considered practical, community-based, scientifically sound, socially acceptable, and last, but not the least, appropriate to the needs of the small-scale farmers. In others words, "primary health care" generally refers to an appropriate approach for small-scale AQC, which reduces the risk of disease outbreaks and concomitant loss of production. Primary health care also stresses that the emphasis should be on "people" and not to be narrowed down to pathogens and technology only. The main objective of any such study could be the unraveling of the intensity and scale of aquatic animal disease problems and their impacts in order to gain better insights into their risks to rural livelihood.

Concomitantly, Chris (2002) had highlighted the health management issues in the rural livestock sector and the useful lessons for consideration when formulating programs on health management for rural, small-scale AQC. He had further stated that there could be a number of similarities between a smallholder of livestock and a big AQC system. In both systems, similar outcomes may result after the application of same principles of health management. Thus, it may appear appropriate that lessons learned from the livestock sector could be applied to AQC sector and vice versa. However, improvement in the situation of disease prevention and control may happen only when there is a clear understanding of the etiology of various disease syndromes and their epidemiology. In this regard, effective measures could be adopted only when the main reasons could be found through different means including epidemiological survey programs, adopted at the village level, may gradually be extended to national level. These disease-control strategies could form strong database for present and future works on fish health management. However, such databases for those countries should invariably be based on competent diagnosis, well-designed disease reporting systems, and specific field surveys. However, collection of high-quality representative basic data from the farms and villages forms a prerequisite need for building a sound database. Differences exist in capabilities of governments at different levels for improving the health of animals in the small-holder sector. The responsibilities of livestock and aquatic animal health are said to be entrusted with the same governmental department in countries like Lao PDR and the Philippines. Thus, marvelous opportunities exist for sharing experiences, expertise, and resources. Further, the livestock and aquatic health surveillance and control systems for the small-holders are said to be quite well-developed in some Asian nations. This could be expanded to other nations through network collaborations.

Concomitant to the above, governments of different nations now seem to be under increasing pressure to supply better-quality information about the health status of livestock and aquatic animals. This could become more increasingly important, over the coming decades, for those countries that export aquatic animal products, particularly the live animals and their products. The *International Aquatic Animal Health Code* of The Office International des Epizooties has guidelines about the modus operandi of a surveillance to be conducted and concomitant reporting of diseases, if any. Improvements in the reporting system could be an ongoing process, with inputs in the form of tables, figures, maps, plates, etc., so that targeted goals could be achieved. Nevertheless, it may be appreciated that the modus operandi of disease surveillance for livestock, fisheries, and AQC may not be the same, and thus may call for different strategies. It is often appreciated that there are similarities between smallholder livestock and AQC systems with regard to application of the similar principles of health management and their outcomes. Similar problems often occur with increased stocking densities coupled with commercial pressures for continual production-efficiency gains. This often may lead to enhanced chances of infectious diseases.

Concomitant to the above, global production from AQC is said to be increasing at approximately 13%/y, and thus production is expected double every six years if this rate of production is sustained, along with almost uniform rate of production from different countries. Conversely, fast growth may also help potential pathogens to find additional susceptible hosts. Also, rapid unchecked growth might boost unhealthy competition among the neighboring farms; this might thereby lead to the spread of infectious diseases in epidemic dimension, leading to devastation among not only fishes, but also both large-scale and small-scale aquafarms. In this communication, certain issues from the livestock and AQC sectors have been outlined with some discussion regarding how issues of the livestock sector might suitably be applied to the rural small-scale AQC sector. **With regard to the statement, how important is**

the small-scale livestock sector, it had been reported that the percentage of the population involved in agriculture ranges from 50% to 90% in the Asian region (FAO, 1999). Approximately 10–20% of GDP is derived from livestock in these countries. Further, small-scale producers are said to raise 60–90% of the livestock. **With regard to organization of the small-scale livestock and AQC sectors,** the target of a service provider such as an elected government is the "individual farmer," particularly a small-scale farmer in a farming system who may not always be competent enough to handle and tackle situations such as floods and droughts. On the other hand, apex bodies such as governments operate on a collective basis like a "group of farmers" rather than as "individuals." In fact, the small-scale livestock and AQC sectors usually operate in hierarchical manner, beginning with small-scale farmers and followed by villages, provinces, and countries.

On a larger scale, international trade at the global level has enhanced the risks of disease emergence to almost an exponential scale. The situation seems to have been aggravated with the raising of GMOs, cross-breedings, and other modern applications, sometimes without considering the consequences. As such, there have been emergence of global bodies like the WTO, with the responsibility for administration of a number of multilateral agreements, notably the GATT, Agreement on Agriculture, Agreement on Sanitary and Phytosanitary measures (SPS Agreement), and so on, to take care of frameworks to facilitate trade, along with steps to evolve mechanisms to reduce risks associated with free trade. The SPS agreement provides the core principles, while the *International Animal Health Code of the OIE* (OIE, 2000) provides standards for management of these risks. Concomitantly, the Animal Health Service of the FAO, after appreciating and recognizing the international spread of diseases as one of the potentially greatest downsides in emerging changes to world trade, consolidated its activities under the EMPRES (Emergency Prevention System for Animal and Plant Pests and Diseases) to control and contain serious and severe epidemics among the livestock and AQC sectors, such as *Transboundary animal diseases* (TADs), notably rinderpest, FMD, Newcastle disease, EUS, and so on. **In keeping an eye on the interests of consumers**, awareness among consumers is said to be on the rise. Concomitantly, it had been noted that **government involvement had been reduced**. In view of the global economic crisis, there have been an increasing trend in the reduction of government subsidies in different countries, notably Australia, New Zealand, Zimbabwe, and so on. As such, **services have become more focused**. In view of curtailing and controlling costs, there have been attempts to adopt a more businesslike approach in some countries, like the outsourcing of works to private laboratories rather than doing the same works in government laboratories.

In this connection, it may be emphatically noted here that the **welfare of animals has** of late been a major issue in livestock and fisheries management systems.

A brief account about the *Trans*-boundary animal diseases (TADs) is given below.

It is a common notion that diseases know no boundaries and generally spread in an epidemic form like wild fire. Nevertheless, they have enough potential to affect the human population adversely (Baldock, 2002). Concomitantly, TADs may be defined as "those epidemic diseases that are severely contagious and have the potential to spread very fast irrespective of national boundaries or borders, causing serious socioeconomic and public health consequences." Such high risks of morbidity and mortality among susceptible animals pose a serious threat to the livelihood of livestock and AQC farmers. With regard to the aspect that TADs to be combated, an effective national animal quarantine system could always be the first line of defense against the entry and establishment of TADs. There could be a second line of defense as well. As such, in order to deal appropriately with TADs, there is a need for the application of appropriate technology. Some points in this regard could be as follows: (a) disease surveillance and animal health information systems; (b) To device various alternative methods for studying epidemiology of disease outbreaks; (c) disease diagnosis and characterization of the etiological agents; and (d) production of vaccines for disease control and eradication. It has been detailed later.

In fine, it could be inferred **that** there is the possibility for more disease occurrences, rather than less, in AQC systems as they expand. However, the underlying endemic diseases may be managed at the farm level. Moreover, the rapidly sweeping infectious diseases are to be managed at different levels. Producers, consumers, stakeholders, and almost the whole nation could be affected by impacts of certain virulent diseases.

Solutions are supposed to lie at all levels of production, marketing, and policy framing. Treatment with chemicals and antibiotics, proper hygienic measures, adequate quarantine steps, appropriate policies, etc., could go a long way in the proper healthy management of farms.

Further, Mazid and Banu (2002) had reported about an overview of the Social and Economic Impact and Management of fish and Shrimp disease in Bangladesh, with an emphasis on small-scale AQC. They had stated that fisheries have been a lucrative sector today in Bangladesh, in view of burgeoning population pressure with increasing demands for finfish and shellfish. In fact, pisciculture in the country seemed to have been progressing toward semi-intensive AQC. Conversely, shrimp culture seems to have been moving toward an improved traditional system. In fact, uncalculated, unrestricted, and indiscriminate use of fertilizers and feed has caused severe stress on fishes due to changes in water quality parameters in the aquatic domain, and thereby has rendered fishes susceptible to pathogens in foul water. At the same time, not much rich database are seem to be available on the disease and health problems of finfish and shellfish, although mortality and economic losses are the inevitable consequences of disease problems. Mass mortality of carp juveniles due to protozoan and metazoan parasites seems to be often reported. Nevertheless, EUS has been one of the most virulent, enigmatic, hitherto unknown diseases among fishes and has swept through FW fishes semiglobally since about 1972. Concomitantly, the white spot disease (WSD), said to be caused by white spot syndrome virus (WSSV, reported as a systematic epidermal and meso-dermal baculovirus or SEMBV) is believed to have caused the loss of 44.4% of shrimp production alone during the mid-1990s in Asia in general and the Indian subcontinent in particular. It is believed that the shrimp culture industry has been suffering a setback since then. About 350,000 people associated with the shrimp culture were thrown out of employment. Consequently, it had appeared that diseases among finfishes and shellfishes had severe economic impact on the life of the LIG. **However,** production through AQC is said to be continuing leaps and bounds in the developing countries in Asia in order to meet the needs of nutrition and avocation. Burgeoning population pressure vis-a-vis a shortage of alternative employment opportunities in these countries had increased the attraction of fisheries and AQC to people, mainly for avocation. But these countries, notably Bangladesh, India, etc., had faced many difficulties in the appropriate management of AQC, mainly due to lack of up-to-date knowledge of management practices, and lack of awareness among growers (producers). Information pertaining to appropriate management of emerging problems is to be disseminated among fisheries and AQC sectors.

With regard to the FW (freshwater) fisheries sector, there seems to be widespread appreciation that the fisheries and AQC sectors play significant roles in meeting national goals in social, economic, and nutritional arenas. As already mentioned, indiscriminate use of fertilizers and feed had rendered fishes susceptible to diseases due to foul water, and consequently the effects of diseases on improved systems of AQC are said to be significant. At the same time, outbreaks of disease had been insignificantly reported and poorly documented. While dealing with landless farmers and their seasonal occupations, c 80% of the population in Bangladesh is poor and landless, their livelihood is uncertain, and they live mostly hand to mouth. As such, they have to be always on the move in the lookout for improved living conditions. They often may migrate to the urban areas for so-called economic opportunities. Some of such villagers may settle in the "char" areas, which may be found to be uninhabited, and earn their livelihood mainly through fishing. Further, they operate traps in rivers during fish breeding season (February—July) for catching juvenile carps (notably *C. catla, C. mrigala,,* etc.) that are usually sold to pond owners on the mainland. In addition, landless people also fish in canals and open waters during the rainy season. Similar situation is also said to exist in India. **With regard to communities that depend on fish and fisheries,** fish and fishery-dependent people generally dwell in boats, floating houses, or on the banks of rivers and shores of wetlands. They are said to depend exclusively on fishing or fish-related avocations such as fabrication and construction of fishing gears (fish catching devices) and crafts (mainly boats), growing and processing of fishes, fish trade, etc. However, the people belonging to these communities are usually socioeconomically poor and downtrodden. **In dealing with the development in AQC and problems of diseases, it is usually opined that** there has been enhancement in AQC in countries of the region, mainly through increased stocking densities. This may rather invite disease and health problems among intensely stocked fishes (mainly due to overcrowding), and may lead to large-scale mortalities and consequently poor production rather than so-called enhanced production. **While deliberating upon diseases among FW fishes, the occurrence of** fish diseases among carp-rearing farms in Asia, notably Bangladesh, had been reported in *c* 31% of the extensive farms and in *c* 24% of the semi-intensive farms. The most common and virulent disease problems among FW fishes is EUS, which has caused large-scale mortalities among FW fishes since the early 1970s and rendered many ichthyospecies endangered.

Concomitant to the above, there have been risk factors and socioeconomic impacts associated with EUS in Bangladesh. While dealing with this, Khan and Lilley (2002) had stated that an interview-based questionnaire survey of a fish farmer and a fisher, who had been randomly selected from each of the 64 districts of Bangladesh, was conducted in order to study the risk factors associated with outbreaks of EUS. The survey was conducted during the EUS outbreak season between December 1998 to April 1999. Data had revealed a significantly higher relative risk of EUS among the farmed fishes when wild fishes had also occurred in the same pond at the same time. EUS might have occurred in the same pond during the previous season when pond embankments not high enough to prevent incoming floodwater from entering the pond. However, ponds are usually connected to natural water bodies; at the same time, they are not usually dried up and limed before or even after stocking. Further, fish catching devices like nets are generally not properly washed, dried, or disinfected. All these things aggravate the risk of initiation of fish diseases, notably EUS, etc. In addition, it could be further revealed that fishes sampled from the "haor" (seasonal floodplain wetlands) depicted a greater risk of being affected by EUS, as

compared with their counterparts in rivers and other types of lentic water bodies.

Concomitant to the above, of 64 districts in Bangladesh, fish farms in 32 (50%) of them recorded fishes with lesions; fishes in 30 (47%) of the 64 districts were confirmed as EUS-positive. Concomitantly, wild fishes from 52 (81%) of the 64 districts depicted lesions; wild fishes with lesions in 49 (77%) of the 64 districts were confirmed as EUS-positive. Incidentally, in total 6433 wild fishes and 6401 farmed fishes were examined for lesions, and average prevalence was estimated as 16.0% and 15.5% respectively. Moreover, out of the total ichthyospecies found in water bodies, 40 species were recorded as possessing lesions; of those 40 species, 31 were confirmed as EUS-positive.

Notwithstanding the above, *c* 88% of fish farmers interviewed had one to four ponds at their disposal. These small-scale marginal farmers are, usually, at risk of suffering from serious financial constraints and significantly depleted production levels due to abrupt and large-scale outbreaks of diseases. On the other hand, losses in wild fisheries could deprive poorer sections of society of relatively cheap sources of animal protein.

Notwithstanding the above, EUS, is, still perhaps the most damaging disease among the FW fishes (according to fishers and fish farmers), and probably has significant effects on fish production. However, no authentic information about mortalities of fishes resulting from EUS could be obtained as yet, although *c* 86% of farmers and *c* 89% of fishers interviewed had regarded EUS to be a significantly major problem. Nevertheless, the total loss of fishes due to EUS during 1998—99 had been approximately quantified to be 39,797 metric tonnes (mt) in Bangladesh, which had been estimated to be approximately equivalent to US$ 3.97 million, calculated based on the prevailing data from the study.

While analyzing aquatic animal health management issues in rural AQC development in Lao PDR, Funge-Smith and Dubeau (2002) had stated that their communication tried to highlight the role of small-scale AQC in subsistence farming systems in rural Lao PDR. In fact, small AQC is said to be a popular component of subsistence farming systems in Lao PDR. Moreover, rice-cultivation happens to be one of the principal activities during the rainy season. Concomitantly, collection of aquatic products from the rice fields is also quite common. Results from a production-cum-consumption survey, among the subsistence farmers of rural Lao PDR had revealed that many (*c* 84%) of them had been engaged in fish culture. However, consumption of fish and other aquatic products had been estimated to be 13—48 kg/*capita*/y, thereby, accounting for *c* 22%—55% of the total consumption of animal products. Further, fish production and livestock farming are said to be the principal sources of income generation, and the value of fish production/household was US$ 81 on average. In addition, the overall income of families had ranged between US$ 372−594/household/y.

One of the principal strategies in subsistence farming is to minimize the "risk factors." This is possibly reflected through the occasional low input vis-a-vis low productivity from rural Lao AQC. Size of Lao ponds are said to range from 550 to 1520 m² with a water depth of c 50 cm on average. Productivity is sometimes said to be low, ranging from 417 to 708 kg/ha/y, mainly due to low stocking densities (generally one to four fish/m²), and limited artificial feeding. It may be important to note here that low-input AQC systems are generally not much disease-prone, but may become so during the dry season, or with increased inputs.

Notwithstanding the above, Lilley et al. (2002) had reported about the social, economic, and biodiversity impacts of EUS as part of the major treatise on primary aquatic animal health care in rural, small-scale AQC development. According to them, not many surveys had been conducted to accurately assess the impact of EUS on fish populations and associated fishing communities. A review of previous information on various aspects of EUS is said to have failed to prominently focus the **social, economic, and biodiversity impacts of EUS** in view of the advocated fact that HP-based diagnosis of EUS had not been much applied in most studies on EUS before 1994. Information from the Bangladesh Food Action Plan 17 (FAP 17) Project from October 1992 to March 1994 had portrayed that c 26% of the 34,451 FW fishes examined had lesions of some kind. Further, a recent cross-sectional survey in Bangladesh displayed that c 80% of the 471 fishes with lesions sampled from 84 sites had been diagnosed as displayed-positive. This points out that studies on EUS in Bangladesh, which had just examined fishes with lesions, had not been grossly overestimating the prevalence of EUS disease.

Concomitant to the above, outbreaks of EUS had been subsiding in many regions. However, new occurrences are being reported from previously unaffected areas as well as in newly developed fisheries and farming management systems. It has been recently revealed that EUS may not always occur as seasonal outbreaks, and may not always cause high mortalities, but may prevail at low levels throughout the year. It may thus portray some effect on productivity that may not be measured in terms of mortalities alone. Nevertheless, as reported, communities that depend on local fisheries had been affected much by outbreaks of EUS. On the other hand, societal impacts may sometimes extend beyond persons who are directly affected by the loss of fishes. Notwithstanding the above, there have been reports of direct economic losses due to EUS-caused deaths in various affected nations. In addition, further estimates are being prepared based on recent survey data. But losses due to lower productivity could be of greater

significance. Naturally, during periods of severe EUS outbreaks, reduced AQC and fish production could be demonstrated. But that the decline in fish production was positively due to EUS, could not be positively ascertained. However, during periods of severe EUS outbreaks, it may sometimes be difficult to locate the susceptible species as revealed from anecdotal evidence. There had not been much information available, however, on the long-term effects of EUS on aquatic ecology, as reported.

Concomitant to the above, Little et al. (2002) had surveyed **health management issues in FW fish hatcheries, nurseries, and fry distribution, with emphasis on experiences in Vietnam and Bangladesh. They had opined that**, an understanding of the status of fish seed production and marketing in various countries in Asia calls for an analysis of the health management issues. There had been rapid development of the sector in dealing with fish seed production by small and marginal entrepreneurs during the last few decades. However, the concentration of institutions of this sector at certain favorable locations might have either positive or negative impacts on the health status of fish seed. Attempts had been made to identify the key features for hatcheries, nurseries, and traders. Decentralized fish seed production approaches had been found to have the greatest potential for minimizing health management problems.

In addition to the above, Hasan and Ahmed (2002) had dealt with issues **in carp hatcheries and nurseries in Bangladesh, with special reference to health management. Their study** presents results of a case study that had been carried out in 180 hatcheries and nurseries in the NE and SW regions of Bangladesh over a 30-day period during August−September 1999. Study of various aspects of management issues in small-scale carp hatcheries and nurseries, with particular reference to their health management was the principal objective of the survey. Three IMCs (rohu, catla, and mrigal) and three exotic carps (common, silver, and grass) were the ichthyospecies generally cultured in most hatcheries and nurseries in Bangladesh. In this context, the average production of spawn in hatcheries was 844 kg/ha, while in nurseries, the production was found to depend on the size of the fry. Nevertheless, the average production of fry in nurseries had been found to be 1.722 million/ha, 1.339 million/ha, and 0.837 million/ha respectively for early fry, fry, and fingerlings respectively. Moreover, the average survival of spawn, fry, and fingerlings in hatcheries and nurseries had been reported to be reasonably high, varying between 74% and 82%. The study further portrayed that the major source of spawn for nurseries was hatcheries, while the hatchery broodstock had been reported to be collected mostly from farmers' grow-out ponds. In general, hatcheries had been reported to be more profitable than nurseries. Further, the profitability of nursery operations appeared to fluctuate mainly because of high variability in market prices of fry and fingerlings. Furthermore, hatcheries and nurseries had been reported to provide full-time employment to farmers and labors. It is significant to note here that the average contribution of AQC to the household income of hatchery and nursery owners had been reported to vary between 79.3% (nursery owners) and 95.1% (hatchery owners).

Concomitant to the above, occurrence of diseases, and incidences of droughts and floods happened to be the major management problems faced by hatcheries and nurseries. However, reports had revealed that diseases are less prevalent among the former than the latter. However, major ailments reported from nurseries include the following: white spot, tail and fin rot, gill rot, sudden spawn mortality, dropsy, malnutrition, EUS, etc. Conversely, the major diseases reported from hatcheries are fish lice and sudden spawn mortalities. As reported, the economic loss due to disease had been about 7.6% of the profit in Bangladesh. Nevertheless, gill rot probably caused the greatest financial losses to affected farms, followed by EUS, sudden spawn mortality, fish lice, and malnutrition. The case studies in Bangladesh indicated that on average, disease is a significant issue in hatcheries and nurseries. The issue of disease and health management must be taken care of with due emphasis. In spite of adversities, hatchery and nursery operations have remained profitable enterprises in Bangladesh.

Notwithstanding the above, Phan Thi van et al. (2002) had reported about the **impacts of RSD on small-scale AQC in Northern Vietnam. They had stated that** small-scale and marginal AQC is said to play a significant role in the livelihoods of rural Vietnam. The principal culturable species are, *L. rohita, C. mrigala H. molitrix, C. idella, Aristichthys nobilis, C. carpio, Tilapia* spp., etc. A hitherto unknown, yet-to-be identified RSD, opined by some quarters as similar to EUS in its gross symptoms, has been causing significant losses to the economy of Vietnam and had been a potential constraint in the development of AQC in Vietnam.

The objectives of such studies are generally aimed at providing an overview of RSD in Vietnam and impacts the disease might have on the small-scale AQC and on the socioeconomic aspects of the marginal farmers in Vietnam. As reported, a total of 145 fish farmers from Thai Nguyen and Bac Ninh provinces, representing highland and lowland AQC systems, had been interviewed using questionnaires provided by the Network of Aquaculture Centres in Asia—Pacific (NACA, 1989). The **Epi-Info software** program (it is a specialized program for epidemiological analyses, and is distributed free of cost by the Centers for Disease Control, Atlanta, Georgia, United States. A copy (obtainable at the CDC's website at http://www.cdc.gov) was used for analyzing the data. Results of the study had reported grass carp to be the most susceptible species to RSD. This was reflected by the fact that out of the *c* 81.4%

of farms that had disease problems, *c* 73.1% had depicted RSD problem. Consequently, the disease had portrayed a strong negative impact on the socioeconomics of both lowland and highland AQC systems in the north of Vietnam. However, it did depict a seasonal pattern, registering its occurrences mainly during March—April and during October—November. The results had also reported about a close liaison between the AQC farms and extension officers of the government in tackling the disease. However, as reported, farmers may have less knowledge and experience about various aspects of RSD, including its prevention and control. Even government extension officers may be ignorant about modern know-how for tackling the disease. As suggested, trainings could be arranged for (a) fish farmers in basic aspects of disease recognition and fish health management, and (b) government extension officers in advanced modus operandi for disease and health management. In fact, research is needed to (i) identify the etiology of RSD; (ii) investigate its epidemiology; and (iii) find methods for its prevention and control. As was opined, knowledge generated and experience gained through these exercises could go a long way toward serving as a useful model on which effective aquatic health management strategies could be built up for success in small-scale AQC systems in Vietnam. **With regard to fisheries and AQC in Bangladesh,** the fisheries sector is said to contribute *c* 5.3% to the gross domestic product (GDP) of Bangladesh. Also, >7% of the country's population is supported by fisheries and their allied activities. Further, *c* 6% of export-related earnings is contributed by the fisheries sector, thus paving the way to earn the third rank after jute products and leather. Furthermore, fish accounts for *c* 6% of per capita protein intake and contributes about 60% of the animal protein consumed throughout Bangladesh. Moreover, the fisheries sector is said to provide income to *c* 1.5 million full-time and 11.0 million part-time fishers. A similar situation is said to exist in India.

Notwithstanding the above, *c* 1.31 million mt of fish were reportedly produced during 1996—97. Out of this, *c* 1.0 million mt had been obtained from the inland waters contributing to *c* 79% of the total fish production. Further, open water capture fishery was said to contribute *c* 58% of total inland production, with the remaining *c* 42% contributed by closed-water culture fishery. However, the quantum of production from culture fishery has been increasing since 1983—84. Its yearly growth rate over previous years had varied between 4% and 20%. Further, the contribution of culture fisheries to inland fish production had been *c* 16% during 1983—84; this increased to *c* 33% by 1996—97. However, the average rate of depletion from capture fisheries was *c* 1.7% per year during the period 1983—84 to 1989—90. Moreover, a trend of increasing fish production had been registered since 1991—92, perhaps due to large-scale open-water stocking programs initiated by the government of Bangladesh. Concomitant to the above, the projected target of fish production during 2001—02 was 2.08 million mt. It was anticipated that this goal could be achieved only through intensification of AQC, and possibly through enhanced fisheries from floodplains. However, sustained supply of large-sized quality fish fry and fingerlings could be a prerequisite for both the intensification of AQC and enhancement of fisheries through open-water stockings. A similar situation is also said to exist in India.

Concomitantly, Ahmed et al. (2011) had reported on the health condition of juvenile exotic carp, *C. carpio*, from various fish farms of Mymensingh area in Bangladesh. They had reported that clinical and HP studies were conducted with juvenile common carp (*C. carpio*) from a government fish farm and an NGO fish farm in Mymensingh area for nine months from April to December 2005. Clinically, red spots, the loss of scales, a weak body, hemorrhage, and subcutaneous lesions were noticed, especially during November and December. Moderate pathological changes in the investigated organs of all fishes were recorded in the summer season (April—May), whereas in the rainy season (June—July), pathological symptoms were significantly reduced, before again increasing to some extent in the autumn season (August—September). However, marked pathological changes such as necrosis, protozoan cyst, bacterial colony, vacuum, melanomacrophage, hemorrhage, hypertrophy, hyperplasia, and clubbing were recorded in all investigated organs like skin, muscle, gill, liver, and kidney in the months of October, November, and December. Among the affected organs, gills were more affected with the presence of numerous protozoan cysts followed by skin, liver, and the less affected portion was the muscle. Clinically and histopathologically, juvenile common carp of government fish farms were more affected than NGO fish farms, especially in colder months of the year.

In addition to the above, Faruk et al. (2004) had reported on the status of fish disease and health management Practices in rural FW AQC of Bangladesh. They had stated that a questionnaire survey and participatory rural appraisal (PRA) tools were used in five districts of Bangladesh, viz. Mymensingh, Comilla, Jessore, Natore, and Dinajpur, in order to examine the current status of fish disease and health management practices in rural FW AQC. A total of 500 farmers had been interviewed and 25 PRA sessions had been conducted. The majority of farmers (*c* 87.8%) interviewed had experienced disease problems in their fishponds. Prevalence of fish disease had varied with districts and the size of farms. However, average prevalence of fish disease had been reported to highest in Jessore district (*c* 18.2%), followed by Comilla (*c* 13.4%), Mymensingh (11.4%), and Dinajpur district (*c* 10.4%). On the other hand, minimum prevalence (*c* 5.5%) had been recorded from Natore district. Further, small farms (*c* 0.4 ha) had

suffered from high disease prevalence (*c* 13.8%) followed by medium (*c* 0.2—0.4 ha) and large (>0.4 ha) farms with *c* 12.4% and 9.3% of diseases, respectively. In fact, the most prevalent disease was tail and fin rot (*c* 20.5%), followed by EUS (*c* 18.9%), nutritional diseases (*c* 15.3%), red spot (*c* 13%), and gill rot (*c* 12.3%). Further, other conditions of ailment like argulosis, dropsy, and white spot were also reported by farmers, but with lower incidence. Notably, diseases had occurred mainly during the winter season. Incidentally, the overall knowledge of farmers was found to be very poor. However, most farmers (*c* 83.8%) had used treatment measures in response to particular fish disease problem. Also, *c* 46% of farmers had used combination of lime and potassium permanganate; *c* 26% of farmers had used only lime and *c* 10% of farmers had used lime and salt together. In addition, some fish health management-related problems had been identified in rural AQC, such as lack of assistance, poor technical knowledge and lack of suitable therapeutics and their proper uses.

It is a well-known fact that diseases affect health, survival, and recruitment of any individual susceptible for diseases. As a consequence of disease, harvests from natural resources, and in particular those from AQC dwindle quite severely. While an appreciable volume of information on the variety of mycotic diseases in marine organisms is available on a global scale, studies from Indian waters are few. There have been attempts to bring together a set of information deemed useful for stimulating marine mycopathological investigations in our waters.

Further, Bhaumik (1998) had reported about the **impact assessment of fish disease** epizootics. India is endowed with a vast expanse of inland waters. Maximum sustained yield of fish from natural waters and the assurance of a recurring bountiful harvest of fish without the depletion of resources and wastage of fishing effort, are necessary for the nation. Although AQC is an age-old practice, it has acquired the shape of an industry during the last two decades due its rapid development and high profit turnover. During this period, there had been fast expansion of fish and prawn farming and products in the country had increased many-fold. But the growth and development of fisheries had been hampered mainly by outbreaks of various fish diseases of bacterial and viral origin, poor management, and improper planning that had led to drastic reductions in production and colossal economic losses. Generally, during the outbreak of diseases, society is afflicted on three tiers—producers, fish traders, and consumers. Of course, producers become the worst victims, especially when disease takes shape in epidemic form. Thus, it becomes imperative to know the impact of diseases in society. Keeping this in view, it is highly essential to know:

1. the socioeconomic conditions of fish farmers and extent of suffering caused due to disease;

2. the impact of the disease on fish traders and consumers; and

3. the role of communications media toward the creation of mass awareness.

It is also imperative to harmonize the AQC system for sustainable development, so that fish farmers do not suffer much. It is important to involve all groups of people including scientists, extension functionaries and fish farmers in the process of formulating and implementing the package of management practices toward the development process. Effective communication is warranted to motivate fish producers toward adoption of proper management measures for sustainable production for larger benefit. A case study on the impact of EUS, a dreadful fish disease reported in the country during July 1988, became a matter of grave concern for fishery scientists, administrators, and policy makers. The disease had a striding progress to several provinces, and had caused havoc in the provinces of Assam, Tripura, Meghalaya, West Bengal, Bihar, and Uttar Pradesh resulting the poor fish farmers to become poorer. Importantly, fish serves as a significant item of diet of c 70% population of the country. However, consumers were baffled by various rumours related to the diseased fish and finally, fish markets were affected. People stopped consuming even the unaffected and healthy fishes. Moreover, some provincial governments had banned consumption of fish during the outbreak period. In this sequel, fish farming community and fish traders started suffering adversely. Thus, the outbreak of the disease established a setback in biological and socioeconomic conditions of society.

EUS had been a *hither to* unknown virulent, enigmatic disease among the FW fishes ever since its inception in SE Asia. **EUS had also spread to Pakistan.** It had inflicted colossal losses to fish farmers mainly because of its sweeping the rice fields and fish farms. As usual, there had been a widespread fear psychosis of EUS disease transmission into the human consumers, which has been found to be totally unfounded and baseless. But this consequently led to a drastic decrease in the market demand for fish including marine species that were not, in fact, affected by EUS.

Notwithstanding the above, there had been obvious nutritional deficiency among fish consumers mainly due to the exclusion of fish, because Pakistan happened to be one of the countries where "fish" happens to be a main source of animal protein available to accompany a rice meal. It has been estimated that c 250 million families in SE Asia cultivate paddy as a principal crop and eat rice as a staple food item. As such, a big portion of the incidental fish harvest from paddy fields forms an important part of the family diet (Macintosh, 1986). In fact, it is quite difficult to estimate the economic value of fish losses due to virulent fish diseases like EUS. It is mainly because in rural areas, fish is mostly consumed at a local level and only a small portion is marketed. Tonguthai (1985), however, had estimated direct losses to the culture fishery in Central Thailand to be approximately US$ 8.0 million.

Concomitantly, the average daily income of c US$ 4.0 for Filipino fishers is said to had declined to US$ 1.50 during the EUS disease outbreak period in Laguna de Bay mainly due to the rejection of EUS-affected fishes by consumers. Further, it had been opined that the repeated yearly outbreak of EUS, especially in Laguna de Bay, could lead to a significant decline in natural fish population and could, eventually, result in dwindled stock of the affected ichthyospecies.

In addition to the above, there have been reports of economic losses due to fish mortalities caused by EUS in countries like, India, Bangladesh, Nepal, Pakistan, Sri Lanka, Myanmar, Thailand, Lao PDR, Kampuchea, Vietnam, The Philippines, Malaysia, Indonesia, and in the Australian continent, etc.

As in other countries, the outbreak of EUS in India created panic in the affected areas, with a sizable loss of valuable edible fish. This unprecedented appearance of the disease caused grave bioecological and socioeconomic consequences. Large lentic water bodies and occasionally rivers were much affected at the initial stages, with heavy mortality of valuable stock of fish. As a result, there could be destruction of fisheries and depletion of fish yields, with consequent impacts on fishers' socioeconomic development.

Concomitant to the above, investigations carried out in five districts of West Bengal revealed that c 73% AQC operation units were adversely affected by EUS. The outbreak of the disease depressed fish consumption rate by 28.7%, 23.3%, and 20.5% in urban, suburban, and rural sectors respectively. Consequently, the fish trade was also affected seriously. Owing to consumer resistance, traders did not accept such fish for selling. In rural markets, diseased fishes were sold at a very low price. About 42.19% of aquaculturists suffered a 31−40% loss of fish in their culture ponds. The pecuniary loss faced by 50% aquaculturists was in the range of Rs. 1001 to Rs. 5000+/−, while 19.73% culturists suffered a greater loss ranging from Rs. 5001 to 10,000. A section of farmers had to look for alternate jobs. 88.9% fish traders also suffered losses to some extent during the affected period. Another study undertaken in five districts of Kerala revealed that the spread of EUS completely paralyzed the inland fish market and threw fishers out of their occupation. Further, the women fish vendors were particularly subject to severe hardship. They had to seek alternative employment as agricultural laborers, head-load and quarry workers, etc., but without much success.

6.13 TAXONOMIC ANALYSIS

Changes in nomenclature of organisms often happen in the field of taxonomy in response to different contemporary needs. As such, there had been deliberations among taxonomists with regard to *Aphanomyces* sp. Hatai (1980) had described a new species by the name *A. piscicida* based on salient features of asexual reproductive stages of isolates. It may be mentioned here that the species *A. piscicida* was differentiated from *Aphanomyces* spp. and other aquatic biota, notably *A. laevis* and *A. astaci*. Other pertinent research in the field of mycotic systematics includes that of, Yuasa and Hatai (1996), and so on. The former had ascertained the importance of using ubiquinone systems as probable new taxonomic feature for the fungal biota belonging to Oomycetes. On the other hand, the latter had contributed to identification of some biochemical features that could be used to demarcate among the mycotic genera *Achlya, Aphanomyces, and Saprolegnia.*

Notwithstanding the above, Japanese works had yielded powerful evidence that *A. piscicida* is the primary etiology of EUS. This had been achieved by fulfilling Koch's postulates based on inoculation of the putative agent. Consequently, the new name "epizootic granulomatous aphanomycosis" (EGA) was proposed. As such, under the said case definition, occurrences of epizootic fungal infection, secondary ulcer formation, nonseptate fungi, and granulomas nearly every time in the lesion are salient features of EGA. However, PCR-based work is done for confirmed diagnosis of EGA.

6.14 ETIOLOGICAL AGENTS OF EUS

According to OIE Manual (1997), diagnosis of EUS is based on clinical signs and histopathology. No specific diagnostic tests are seemed to be available. The fungus could be isolated and cultured without much difficulty, provided that measures are taken to exclude bacterial and other fungal contaminants. Besides their typical vulnerability to temperatures >30 °C, *A. invadans* and *A. piscicida* are very similar to nonpathogenic opportunistic *Aphanomyces* spp. that readily contaminate the surface of affected fish and often interfere with isolation attempts. In the laboratory, the fungus was also shown to be pathogenic to a wide range of fish, inducing similar pathology and mortality under various predisposing experimental conditions.

ODA (1994) had defined EUS as a "seasonal epizootic condition of FW and estuarine warm water fishes of complex infectious aetiology characterized by the presence of invasive *Aphanomyces* infection and necrotizing ulcerative lesions typically leading to a granulomatous response." There have been wide ranging interactions at different times on various aspects of EUS since its inception during the 1970s. As such, a large amount of scientific information has been generated during this long period. Consequently,

different views and definitions have cropped-up to unravel the etiology of this hitherto unknown enigmatic fish disease. However, two groups of inconclusive case definitions have been briefly stated below to take care, as far as possible, of different situations and views:

1. Case definitions for screening programs, surveys, etc.:
   a. A fish with focal to locally extensive cutaneous ulceration.
   b. A fish with focal to locally extensive cutaneous erythema or ulceration.
2. Case definitions for definitive diagnosis
   a. A fish with necrotizing granulomatous dermatitis and myositis associated with *A. invadans* hyphae.
   b. A fish with necrotizing granulomatous dermatitis and/or myositis and/or granulomas in internal organs, associated with the presence of *A. invadans* (=*A. piscicida*) found within the lesion.

6.14.1 New Common Names for Epizootic Ulcerative Syndrome

Concomitantly, two new common names have been suggested for EUS. These are:

1. epizootic granulomatous aphanomycosis (EGA) and
2. ulcerative aphanomycosis.

6.14.2 Synonymy of *Aphanomyces* sp.

Synonymy is a matter of opinion, often with formal nomenclatural requirements. As such, it is sometimes proposed that in cases other than purely taxonomic contexts, the *Aphanomyces* causing EUS, MG, RSD, and UM could be initially referred to as *A. invadans* (=*A. piscicida*) followed by *A. invadans*.

From the foregoing discussion, it could be tentatively concluded that etiology of EUS is based on two schools of thoughts, viz. EUS as (a) an "aphanomycosis" and (b) a "polymicrobial infection."

6.14.2.1 Definition of Epizootic Ulcerative Syndrome

A seasonal epizootic condition of FW and estuarine warm-water fish of complex infectious etiology characterized by the presence of invasive *Aphanomyces* infection and necrotizing ulcerative lesions typically leading to a granulomatous response.

6.14.2.2 Similar Diseases

Pathological and epizootiological evidence had indicated that the condition known as RSD in Australia is indistinguishable from EUS. Similarly, all available evidence seemed to suggest that the condition known in Japan as MG is indistinguishable from EUS.

6.14.2.3 Extension of Range

EUS is endemic in many countries and is still extending its geographical range even into subtropical, subtemperate, and temperate climates. Experimental evidence indicates that the *Aphanomyces* involved is capable of causing disease in temperate species.

6.14.3 Human Significance

All available evidence suggests that consumption of EUS-infected fish poses no proven specific health problems to human beings, provided that they are properly prepared in sanitary conditions.

6.14.4 Present State of Knowledge on Epizootic Ulcerative Syndrome

Till now, investigators throughout the world have put in a great deal of effort toward ascertaining the etiology of EUS in fishes, but to date no firm conclusions have been reached regarding the cause(s) of the disease. During January 1994, in the regional seminar on Epizootic Ulcerative Syndrome (EUS) organized by Overseas Development Administration (ODA) and Aquatic Animal Health Research Institute (AAHRI) in Bangkok, scientists from affected countries had presented up-to-date findings on EUS and its relationship to Red Spot Disease (RSD) in Australia, menhaden disease in the United States, and Piscida disease in Japan. The conclusions and recommendations that emanated from the deliberations revealed the latest state of knowledge on EUS.

6.15 RECOMMENDATIONS FOR FUTURE WORK

In view of the importance and continuing extension of this serious disease, there had been emphatic recommendations for conducting researches on EUS on a priority basis.

6.15.1 Investigation of Early Stage

It was clear from the information presented at the seminar that there is a distinct and critical early premycotic stage of the pathogenesis of the disease, and it is essential that detailed multidisciplinary research is carried out at this stage.

6.15.2 Virology

Studies have shown the presence of a wide range of viral agents in fish affected with EUS. It had been recommended that further extended works be undertaken to determine

more accurately the incidence and distribution of tropical food fish viruses throughout the region, in addition to specific EUS related-investigations.

6.15.3 Epidemiology

Proposed studies are likely to produce a large amount of complex data relating to EUS outbreaks in fish populations. It is, however, recommended that epidemiological expertise be developed within the region to enable these data to be effectively utilized.

6.15.4 Environment

The evidence presented at the seminar had strongly pointed to a relation between the initiation of EUS and environmental factors. It was recommended that further studies on environmental conditions, including physical, chemical, and biological factors, be carried out to better understand their role in outbreaks of EUS. The seminar expressed concern over the limited understanding of the relationship between fish health and environmental conditions in general, and recommended expansion of research efforts in this important subject area.

6.15.5 Speciation of Fungus

Aphanomyces sp. is a significant fungus related to EUS. Detailed mycological and molecular biological works be carried out on *Aphanomyces*. The isolates are fully characterized and their relationship defined.

6.15.6 Diagnosis

Currently diagnosis is of necessity based upon a number of clinical and pathological features of the disease. It is important that a rapid, specific, accurate, low cost diagnostic test capable of being used under field conditions is developed.

6.15.7 Development of Resistance

Evidence from some countries suggests that after the initial outbreak, an element of resistance to the disease might develop in fish. This resistance may be ecological, genetic, or associated with some acquired immunity. It is important that the mechanism for this be now investigated.

6.15.8 Mode of Transmission

The evidence suggests that the spread of the disease in Asian countries was largely via the lotic systems and natural waterways. Nevertheless, there was said to be evidence of distinct transfer over marine barriers. It is important that there could be an understanding as to the mechanism of transfer between various water bodies. Further, it is essential that attention be paid to the development of quarantine measures to prevent the transfer of these via transportation or live fish or infected materials.

6.16 CONCLUDING REMARKS

The welfare of an animal is its state as regard its attempts to cope with its environment Ulcerative Fish Disease Committee (1983). This means that welfare is a characteristic of an individual animal that can vary from good to very poor and can be assessed scientifically. Fish welfare or animal right law is related to farmed fish. It means that fish have a right to live a life as good as possible and express their natural behavior as much as possible while being free of negative experiences. Fish are in intimate contact with their environment through the large surface of their gills and skin, and of necessity they defecate into the medium in which they live, so water quality (in terms of DO, FCO_2 ammonia, and pH) and the presence of contaminants (organic and inorganic pollutants) are probably the most critical aspects of the environment for fish welfare and also the best defined. Optimal conditions vary markedly between species; for example, catfish do poorly in clear water, whereas salmon do poorly in cloudy water and cyprinid fish are very tolerant of low dissolved oxygen levels, whereas salmonid fish are not (Kramer, 1987). The flow characteristics of the fish's natural habitat are also of importance. A degree of environmental complexity may be important, depending on the species concerned. Conditions that produce unacceptable levels of anxiety, fear, distress, boredom, sickness, pain, thirst, hunger, and so on should be minimized in fish as in other vertebrates.

6.17 SUMMARY

1. A living body is prone to suffer from a disease being attacked by pathogens or parasites. This is true also with a fish body. Often, a disease may be so virulent that it could sweep unabated in an epidemic dimension.
2. The temporal pattern of an outbreak is described in terms of its epidemic curve. The epidemic curve is a graph showing the onset of cases of the disease either as a bar graph or frequency polygon. In general, an epidemic curve has four and sometimes five segments: (i) the endemic level; (ii) an ascending branch; (iii) a peak or plateau; (iv) a descending branch; and (v) a secondary peak.
3. If transmission is rapid and the incubation period is short, then the ascending branch will be steeper than if transmission is slow or the incubation period is long. The length of the plateau and slope of the descending branch are related to the availability of susceptible animals, which in turn depends on many factors, such as stocking densities, the changing importance of different mechanisms of transmission and the proportion of resistant or immune fish in the population at risk.
4. Snieszko (1974) had stated that an overt infectious disease could occur when a susceptible host is exposed to a virulent pathogen under stress.
5. In order to understand the influence of environmental factors on fish disease, one must realize that there is much greater chemical and physical variability in aquatic environments than in terrestrial ones.
6. Epizootic ulcerative syndrome or EUS is a hitherto unknown, dreadful, virulent, and enigmatic disease among FW fishes that has swept through water bodies in an epidemic dimension almost semiglobally causing large-scale mortality among FW fishes, thus rendering many of them endangered and throwing the lives of fishers out of gear though the loss of avocation and causing difficulties to fish consumers through scarcity of fish flesh as a protein-rich source of nutrition. This dreaded fish disease has thus been a major concern in several countries of the world, particularly the Asia—Pacific region.
7. The epidemiology of EUS seemed not to have been studied in much detail. However, a number of factors had been hypothesized as "risk factors," predisposing and precipitating "stress factors," or "determinants" of EUS outbreaks. These factors are based on observations of the mode of disease transmission, species, habitats, systems of culture, and so on. However, identification of the true "risk factors" of EUS could allow rational control measures to be developed and adopted.
8. There seems to be widespread appreciation that the fisheries and AQC sectors play significant roles in meeting the national goals of social, economic, and nutritional arena. As already mentioned, indiscriminate use of fertilizers and feed had rendered fishes susceptible to diseases due to foul water, and consequently the effects of diseases on improved systems of AQC are said to be significant. At the same time, outbreaks of disease had been insignificantly reported and poorly documented.
9. Small-scale fish farming has become increasingly popular among unemployed youths as a potential source of income, in view of the burgeoning unemployment problem. Some of the small-scale farmers own ponds, but instead of culturing, they give their ponds on lease to others leading to a kind of ownership—tenant business. In view of this, diseases like EUS, in the event of its outbreak, causes colossal losses to AQC, mainly because the tenant farmer may not always be competent enough to tackle and manage grievous situations posed by the outbreak of EUS.

10. Concomitant to the above, governments of different nations now seem to be under increasing pressure for supplying with better quality information for the health status of livestock and aquatic animals. This could become more increasingly important, over the coming decades, for those countries that export aquatic animal products, particularly the live animals and their products. The *International Aquatic Animal Health Code* of The Office International des Epizooties has guidelines about the modus operandi of a surveillance to be conducted and concomitant reporting of diseases, if any.

11. Literature on EUS and other possible related diseases in fishes was reviewed by different workers long ago.

12. **Stress inducing parameters, and the underlying concept of stress:** Mortality of fish or decline in a fish population in a water body is, at present, the sole indicator that the effects of environmental stress factors are exceeding the acclimation tolerance limit of fish.

13. The term "stress," "stressor," or "stress factor" could be defined as a force or challenge in response to which there is a compensatory physiological change in fish. Thus, an environmental or biological stress is of significance, if it requires a compensating response by a fish population or ecosystem. Stresses could be due to many parameters, notably **(a) stress due to metal toxicity; (b) conceptual framework of stress response; (c) secondary stress response; and (d) tertiary stress response**.

14. **GAS**: The various physiological changes that occur as a fish responds to stressful stimuli are compensatory, or in other words adaptive in nature, and are required for acclimation. Collectively, these phenomena have been termed GAS.

15. There are environmental parameters that are of significance to fish health

16. **The author of the present treatise has been working on EUS fish disease pertaining to different aspects, notably** whether (a) the outbreak of EUS is due to any severe organic pollution of water and soil; (b) the outbreak of EUS is due to any radioactivity; (c) the outbreak of EUS is due to any heavy metallic contaminations; (d) bacterial and fungal flora are the causative factors of EUS; and (e) there is any primary viral etiology for the initiation of EUS.

    EUS has been causing large-scale mortality among FW fishes since 1988, initially affecting four species of fishes very widely. Our study revealed fluctuation in the intensity of the disease in relation to species affected. Large hemorrhagic cutaneous ulcers, epidermal degeneration, and necrosis followed by sloughing of scales are the principal symptoms of EUS. Low TA could be predisposing "stress factor."

Sick fishes show low hemoglobin and polymorphs, but high ESR and lymphocytes. The communicative nature of EUS revealed variation in time gap between fish and infection in different species. Inoculation of microbes in the test animals did not reveal of any sign of ulcerations for two years. Bacterial culture revealed occurrences of hemolytic *E. coli, A. hydrophila, P. aeruginosa, Klebsiella* sp., and *S. epidermidis* in surface lesions and the gut, liver, gills, heart, kidney, and gonads of sick fishes, all of which have been found to be sensitive to chloramphenicol, trimethoprim, gentamicin, etc. Fungal isolation revealed the occurrence of *Aphanomyces* sp with concomitant occurrence of the same fungal genus in histological sections of EUS-affected fishes. HP studies focal areas of increased fibrosis and chronic inflammatory cell infiltration in muscles, focal areas of fatty of degeneration of hepatocytes surrounding the portal triads in the liver. Preliminary histochemical (HC) studies with regard to interruption in glycogen synthesis and blockade of respiratory pathways are being conducted. Similarly, preliminary enzymological studies are being conducted with regard to the amount of alkaline phosphatate, SGOT, SGPT, and LDH. Inoculation of 10% tissue homogenate of EUS-affected *C. batrachus* into an 80% confluence monolayer form BF2 fish cell line in Leibovitz's L-15 Medium revealed progressive CPE that was passable in subsequent cultures, thus indicating the "isolation" of virus. EM studies with the ultrathin sections of EUS-affected fish tissues revealed the presence of viruslike particles (inclusion bodies); preliminarily, the picobirnavirus has been electron microscopically identified as the primary etiological agent of EUS. Further studies in this regard are being conducted.

17. **Various parasites are associated with EUS fish disease.** Kar et al. (2011), while doing a detailed study on fish diversity and helminth fauna in fishes of Assam and Manipur, India had preliminarily displayed a survey of fish diversity and helminth fauna in fishes of some wetlands of Assam and Manipur. 25 species of fishes in Dolu Lake, 34 species in Chatla Haor, 40 species in Karbala Lake, 31 species each in Awangsoi Lake and Oinam Lake, 33 species in Loktak Lake and 32 species in Utra Lake have been recorded during the study period. Study revealed that 22.72% of fishes of Dolu Lake, 5.76% of Chatla Haor, 19.16% of Karbala Lake, 23.83% of Awangsoi Lake, 33.86% of Oinam Lake, 49.75% of Loktak Lake and 48.8% of Utra Lake were found to be parasitized. Incidence of EUS-affected fishes was not very uncommon. Singh (2015) reported about parasite fauna in fishes of **Sone Beel, river Jatinga** in Assam, and in **Pumlen Lake in Manipur**. Umi (2015) had portrayed preliminary

report about the status EUS-affected fishes in relation to parasitic infestations and parasitization. **Further** works had been done in other prominent habitats by different workers.

18. Riji et al. (2014), as a midterm report of the DBT (government of India)-sponsored Project on Molecular Characterization of Pathogens Associated with Fish Diseases in Assam, acknowledging DBT, had reported that the collected fishes were processed and analyzed for the presence of viruses by diagnostic PCR for four DNA and three RNA viruses.

19. **Prevention:** There has been no major attempt at the prevention of fish parasites in the wild, although they may pose potential health hazards. The main reason could be nonownership of water bodies in the wild. However, attempts are being made by the present team of workers for finding ways and means of their control.

20. **Preliminary hematological, histochemical, and enzymological works had been with EUS-affected fishes.**

21. **EM studies:** Preliminary EM works done with EUS-affected tissues of *C. punctata* and *A. testudineus* revealed the **Australia, Asia, the United Kingdom, and the United States** occurrence of viruslike particles. Detailed TEM studies done with ultrathin sections of EUS-affected tissues of *C. striata* and *A. testudineus* revealed membrane bound and granular abnormal structures, hepatocytes with vacuolations, and viruslike inclusion bodies in the nucleus. Further EM works done with EUS-affected tissues of *W. attu* revealed the presence of viruslike particles

22. **Histopathology:** Detailed HP studies done with tissues of EUS-affected fishes revealed focal areas of increased fibrosis and chronic inflammatory cell-infiltration in muscles; focal areas of fatty degeneration of hepatocytes surrounding portal triads in liver; and hypertrophy of the muscle coat with mild chronic inflammatory cell infiltration.

23. **Similar works were done on EUS by other workers.** Works done by Das and Das (1993), Tripathi and Quresh (2012), and had similarities in their works and results. Works on EUS were also done by different workers in different countries, notably Japan, **Australia, Asia, the United Kingdom, the United States, India, Thailand, Pakistan, and so on.**

24. **There has been much work done along different parameters related to EUS, and based on different aspects, notably study based on (a) epidemiology; (b) pathology; (c) histopathology (hp); (d) hematology; (e) immunology; (f) tissue culture; (g) electron microscopy; (h) nanoparticles; (i) PCR; (j) mycology; (k) bacteriology; (l) viruses; and (m) molecular diagnosis; as well as (n) works done on EUS in relation to socioeconomics.**

25. Works have been done to ascertain the etiological agents of EUS.

REFERENCES

APHA, 1995. Standard Methods for Examination of Water and Waste Water. Washington, DC, 1995, pp.1−1268.

Adil, B., Shankar, K.M., Kumar, B.T., Patil, R., Ballyaya, A., Ramesh, K.S., Poojary, S.R., Byadgi, O.V., Siriyappagouder, P., 2013. Development and standardization of a monoclonal antibody-based rapid flow-through immunoassay for the detection of *Aphanomyces invadans* in the field. J. Vet. Sci. 14 (4), 413−419. Epub 2013 June 30.

Ahmed, G.U., Akter, M.N., Islam, M.R., Rahman1, K.M.M., 2011. Health condition of juvenile exotic carp *Cyprinus carpio* from various fish farms of Mymensingh area, Bangladesh. Int. J. Nat. Sci. 1 (4), 77−81. pISSN: 2221-1012 eISSN: 2221-1020.

Ahne, W., Jørgensen, P.E.V., Olesen, N.J., Wattanavijarn, W., 1988. Serological examination of a rhabdovirus isolated from snakehead fish (*Ophicephalus striatus*) in Thailand with ulcerative syndrome. J. Appl. Ichthyol. 4, 194−196.

Ali, A., Tamuli, K., 1991. Isolation of an aetiological agent of epizootic ulcerative syndrome. Fish. Chimess 11, 43−46.

Anderson, D.P., 2003. Disease of Fishes. Narendra Publishing House, Delhi, pp. 22−73.

Anon, February 2012. Review of biological factors relevant to import risk assessments for Epizootic Ulcerative Syndrome in relation to *Aphanomyces invadans*. Transbound. Emerg. Dis. 59 (1), 26−39. http://dx.doi.org/10.1111/j.1865-1682.2011.01241.x. Epub 2011 July 7.

Anon, December 22, 2008. Experimental infection and detection of *Aphanomyces invadans* in European catfish, rainbow trout and European eel. Dis. Aquat. Organ 82 (3), 195−207. http://dx.doi.org/10.3354/dao01973.

Australian Fisheries Management Authority, 2003. Australian Fisheries Management Authority 2002−2003 Annual Report and Report of Operations, Canberra.

Barbhuiya, M.H., 2010. Studies on Limnology and Fishery of Salchapra Wetland in Cachar District of Assam (PhD thesis). University of Gauhati, Assam.

Barbhuiya, A.H., 2011. Habitat Mapping of Mahseer Fisheries and Development of Spatial Database for Rivers Barak, Jatinga and Dhaleswari in North-East India (PhD thesis). Assam University, Silchar.

Barua, G., 1994. The status of epizootic ulcerative syndrome of fish of Bangladesh. In: Proceedings of the ODA. Regional Seminar on EUS. Aquatic Animal Health Research Institute, Bangkok, Thailand, 1994, pp. 13−20.

Bashar, M.A., Kibria, M., Hussian, M., Siddiqui, M., 2009. Fungal disease of fresh water fishes of Natore District of Bangladesh. J. Bangladesh Agrl. Univ. 7 (1), 157−162.

Barman, R.C., 2011. Impact of Development and Management of Selected Floodplain Wetlands (Beels) in Assam through Community-based Fisheries Management (PhD thesis). Assam University, Silchar.

Binky, K., Shomorendra, M., Kar, D., 2014. Incidence and intensity of trematode metacercariae infection in *Channa punctata* of Karbhala wetland, Cachar district, Assam. In: Kar, D. (Ed.), Research Frontiers in Wetlands, Fishery and Aquaculture. Dominant Publishers and Distributors Pvt. Ltd, New Delhi, pp. 201−206, xi + 320. ISBN: 978-93-82007-39-5, ISBN 10: 93-82007-39-3.

Binky, K., Singh, M.S., Kar, D., 2015. A Treatise on Helminthology Fauna of Fishes, pp. 268. Asiatic Publishing House, New Delhi.

Blazer, V.S., Lilley, J.H., Schill, W.B., Kiryu, Y., Densmore, C.L., Panyawachira, V., Chinabut, S., 2002. *Aphanomyces invadans* in Atlantic Menhaden along the East Coast of the United State. J. Aquat. Anim. Health 14, 1−10.

Baldock, C., 2002. Health management issues in the rural livestock sector: useful lesions for consideration when formulating programmes on health management in rural, small-scale AQC for livelihood, pp. 7-19. In: Arthur, J.R., Phillips, M.J., Subasinghe, R.P., Reantaso, M.B., MacRae, I.H. (Eds.), Primary Aquatic Animal Health Care in Rural, Small-scale, AQC Development, pp. 54, FAO Fisheries Tech. Pap No. 406.

Boys, C.A., Rowland, S.J., Gabor, M., Gabor, L., Marsh, I.B., Hum, S., Callinan, R.B., 2012. Emergence of epizootic ulcerative syndrome in native fish of the Murray-Darling river system, Australia: hosts, distribution and possible vectors. PLoS One 7 (4), e35568. http://dx.doi.org/10.1371/journal.pone.0035568. Epub 2012 April 25.

Barua, G., 1990. Bangladesh report. In: Phillips, M.J., Keddie, H.G. (Eds.), Regional Research Programme on Relationships between Epizootic Ulcerative Syndrome in Fish and the Environment. A Report on the Second Technical Workshop, August 13−26, 1990. Network of Aquaculture Centres in Asia-Pacific, Bangkok.

Baldock, F.C., Blazer, V., Callinan, R., Hatai, K., Karunasagar, I., 2005. Outcome of a short expert consultation on epizootic ulcerative syndrome (EUS): re-examination of causal factors, case definition and nomenclature. In: Walker, P., Laster, R., Bondad-Reantaso, M.G. (Eds.), Disease in Asian Aquaculture V. Manila, Phillippines. Fish Health Section. Asian Fisheries Society, 555−585.

Bhaumik, U., 1998. Impact Assessment of Fish Disease Epizootics- a Case Study on EUS. Central Inland Capture Fisheries Research Institute, Barrackpore, West Bengal, pp. 106−115.

Biswas, et al., June, 2006. Pathomorphological changes in fish affected with epizootic ulcerative syndrome. Indian J. Animal Health 45 (1).

Bhatt, et al., February, 2014a. Immunological role of C4 CC chemokine-1 from snakehead murrel *Channa striatus*. Mol. Immunol. 57 (2), 292−301. http://dx.doi.org/10.1016/j.molimm.2013.10.012. Epub 2013 November 12.

Bhatt, P., et al., 2014b. Immunological role of C4 CC chemokine-1 from snakehead murrel, Channa Striatus. Mol. Immunol. 57 (2), 292−301.

Bly, J.E., Quiniou, M.A., Lawson, L.A., Clem, L.W., 1996. Therapeutic and prophylactic measures for winter saprolegniosis in channel catfish. Dis. Aquatic Org. 24, 25−33.

Biplab, S., Mahanty, A., Netam, S.P., Mishra, S., Pradhan, N. and Samanta, M. Inhibitory role of silver nanoparticles against important fish pathogen, *Aeromonas hydrophila*. Int. J. Nanomater. Biostructures. ISSN 2277−3851.

Bondad-Reantaso, M.G., Lumanlan-Mayo, S.C., Natividad, J.M., Phillips, M.J., 1990. Environmental monitoring of the epizootic ulcerative syndrome (EUS) in fish from Munoz, Nurva Ecija, the Philippines. In: Symposium on Diseases in Asian Aquculture, November 26−29, 1990, Bali, Indonesia. Asian Fisheries Soc., Manila, Philippines.

Bondad-Reantaso, M.G., Paclibare, J.O., Lumanlan-Mayo, S.C., Catap, E.S., 1994. EUS outbreak in the Philippines: a country report. In: Roberts, R.J., Campbell, B., Mac Rae, I.H. (Eds.), In: Proceedings of the ODA Regional Seminar on Epizootic Ulcerative Syndrome, January 25−27, 1994. Aquatic Animal Health Research Institute, Bangkok, pp. 61−87.

Boonyaratpalin, S., 1985. Fish Disease Outbreak in Burma. FAO Special Report, TCP/BUR/4402, p. 7.

Boonyaratpalin, S., Supamataya, K., Duangsawad, M., Reungprach, H., Tayapachara, N., 1983. Relationship between Pesticide Residues and Disease: Practical Report of the Ulcerative Fish Disease Committee, 1982−83. Bangkok, pp. 226−252.

Brett, J.R., 1958. Implications and Assessments of Environmental Stress in the Investigation of Fish Power Problems. H.R. MacMillan Lectures on Fisheries, University of British Columbia.

Callinan, R.B., Paclibare, J.O., Bondad-Reantaso, M.G., Chin, J.C., Gogolewski, R.P., 1995a. *Aphanomyces* species associated with epizootic ulcerative syndrome (EUS) in the Philippines and red spot disease (RSD) in Australia: preliminary comparative studies. Dis. Aquatic Org. 21, 233−238.

Callinan, R.B., Paclibare, J.O., Reantaso, M.B., Lumanlan-Mayo, S.C., Fraser, G.C., Sammut, J., 1995b. EUS outbreaks in estuarine fish in Australia and the Philippines: associations with acid sulphate soils, rainfall and *Aphanomyces*. In: Shariff, M., Arthur, J.R., Subasinghe, R.P. (Eds.), Diseases in Asian Aquaculture II. Fish Health Section. Asian Fisheries Society, Manila, pp. 291−298.

Callinan, R.B., Sammut, J., Fraser, G.C., 1996. Epizootic ulcerative syndrome (red spot disease) in estuarine fish - confirmation that exposure to acid sulfate soil runoff and an invasive aquatic fungus, *Aphanomyces* sp., are causative factors. In: Proceedings of the Second National Conference on Acid Sulfate Soils. Roberts J Smith and Associates and ASSMAC, Australia.

Callinan, R.B., Sammut, J., Fraser, G.C., February 28, 2005. Dermatitis, branchitis and mortality in empire gudgeon, *Hypseleotris compressa* exposed naturally to runoff from acid sulfate soils. Dis. Aquat. Organ 63 (2−3), 247−253.

Cartwright, G.A., Chen, D., Henna, P.J., Gudkovs, N., Tajima, K., 1994. Immunodiagnosis of virulent strains of *Aeromonas hydrophila* associated with epizootic ulcerative syndrome (EUS) using monoclonal antibody. J. Fish. Dis. 17, 123−133.

Chakrabarty, et al., 1996. Electron microscopic characteristics of actinomycetic agents having aetiological association with human leprosy and epizootic ulcerative syndrome of fish. PMID: 8979491; [PubMed - indexed for MEDLINE] Int. J. Nanomater. Biostructures. ISSN 2277−3851.

Chakraborty, A., Mukharjee, M., Chakraborty, A.N., Dastidar, S.G., Basak, P., Saha, B., 1996. Electron microscopic characteristics of actinomycetic agents having aetiological association with human leprosy and epizootic ulcerative syndrome of fish. Indian J. Exp. Biol. 34, 810−812.

Chattopadhyay, U.K., Pal, D., Das, M.S., Das, S., Pal, R.N., 1990. Microbiological investigations into epizootic ulcerative syndrome (EUS) in fishes. In: The National Workshop on Ulcerative Disease Syndrome in Fish, March 6−7, 1990, Calcutta.

Chavda, D., Bhatt, S., 2014. The histopathological and ultrastructural analysis of microsporidian infection in catla *(catla catla)* and their effects on anti-oxidant enzymes and protein expression. Int. J. Adv. Res. 2 (1), 608−624.

Chauhan, R., 2012. Study on certain fungal diseases in culturable and non-culturable species of fishes of Upper Lake, Bhopal. J. Chem. Biological Phys. Sci. 2 (4), 1810−1815.

Chinabut, S., 1994. Environmental factors in relation to epizootiology of EUS in Thailand. In: Proceedings of ODA Regional Seminar on Epizootic Ulcerative Syndrome. Aquatic Animal Health Research Institute, Bangkok, Thailand, pp. 25–27.

Choongo, K., Hang'ombe, B., Samui, K.L., Syachaba, M., Phiri, H., Maguswi C., Muyangaali, K., Bwalya, G., Mataa, L. June 30, 2009. Environmental and climatic factors associated with epizootic elcerative eyndrome (EUS) in fish from the Zambezi floodplains in Zambia (Africa). PMID: 19565173([PubMed - indexed for MEDLINE]).

Choudhury, M.I., Barbhuiyan, R., 2015. Induced Breeding of *Cyprinus carpio* and Occurrence of EUS in Fishes of Assam University Pond at Silchar, Assam.

Chauhan, R., Qureshi, T.A., 1994. Host range studies of *Saprolegnia ferax* and *Saprolegnia hypogyana*. J. Inland Fish. Soc. India 26 (2), 99–106.

Chinabut, S., 1995. Histopathology of snakehead, *Channa striatus* (Bloch), experimentally infected with the specific Aphanomyces fungus associated with epizootic ulcerative syndrome (EUS) at different temperatures. J. Fish. Dis. 18, 41–47.

Chinabut, S., 1998. Epizootic ulcerative syndrome: information up to 1997. Fish. Pathol. 33, 321–326.

Cunningham, R.S., John, A., 1976. Hansen and Bo Dupont: lymphocytes transformation in vitro in response to mitogens and antigens. In: Bagh, F.H., Good, R.A. (Eds.), Clinical Immunology, vol. 3, pp. 151–194.

Campbell, R.E., Lilley, J.H., Taukhid, Panyawachira, V., Kanchanakhan, S., 2001. *In vitro* screening of novel treatments for *Aphanomyces invadans*. Aquac. Res. 32, 223–233.

Das, A., Saha, D., Pal, J., October 2009. Antimicrobial resistance and *in vitro* gene transfer in bacteria isolated from the ulcers of EUS-affected fish in India. Lett. Appl. Microbiol. 49 (4), 497–502. http://dx.doi.org/10.1111/j.1472-765X.2009.02700.x. Epub 2009 July 21.

Das, M.K., Das, R.K., 1993. A review of the fish disease epizootic ulcerative syndrome in India. Etiviron. Ecol. 11 (1), 134–148.

Das, M.K., Das, R.K., 1994. Outbreak of the fish disease epizootic ulcerative syndrome in India - an overview. In: Roberts, R.J., Campbell, Biand Mac Rae, I.H. (Eds.), Proc. Regional Sent. Epizootic Ulcerative Syndrome, pp. 21–36.

Datta Ray, S., Homechaudhuri, S., 2014. Morphological and functional characterisation of hepatic cells in Indian major carp, *Cirrhinus mrigala* against *Aeromonas hydrophila* infection. J. Environ. Biol. 35, 253–258.

Das, M.K., 1997. Epizootic Ulcerative Syndrome in Fishes - Its Present Status in India. Bull. 69. CIFRI, Barrackpore, pp. 22.

Das, U., Kar, D., 2013a. A comparative study on qualitative and quantitative analysis of zooplankton in relationship with physico-chemical properties of water between Karbala Lake and Baram Baba Pond of Cachar district, Assam. Int. J. Curr. Res. 5, 3038–3041.

Das, P., Kar, D., 2013b. Studies on zooplankton diversity and physico-chemical parameters of Ramnagar Anua, Assam. Int. J. Curr. Res. 5, 3058–3062.

Das, B.K., 2015. Studies on the Fish Diversity and Habitat Mapping of River Siang in Arunachal Pradesh (Ph.D. thesis). University of Assam, Silchar.

Dutta, B., 2015. Studies of Fish Diversity and Habitat Mapping in the Towkak River in Assam and Nagaland (Ph.D. thesis). University of Assam, Silchar.

Devi, B.M., Singh, O.S., Kar, D., 2014. Studies on indigenous fishing techniques in Loktak Lake, Manipur. Research frontiers in wetlands,

fishery and aquaculture, pp. 163–174. In: Kar, D. (Ed.), Research Frontiers in Wetlands, Fishery and Aquaculture. Dominant Publishers and Distributors Pvt. Ltd, New Delhi. ISBN: 978-93-82007-39-5; ISBN: 10: 93–82007-39-3, pp. XI + 320.

Devi, U.N., 2015. Studies on Fishes Affected by Epizootic Ulcerative Syndrome (EUS) with Special Emphasis on Parasitic Infestations (M.Sc. dissertation). Assam University, Silchar.

Das, S., Das, T., Das, B.K., Kar, D., 2015. Length-weight relationship and condition factor in *Channa punctata* of River Manu in Tripura. Int. J. Fish. Aquatic Stud. (IJFAS). 2 (3), 56–57.

Dhar, N., 2004. Limnological Studies in Baskandi Anua in Cachar District of Assam (Ph.D. thesis). University of Gauhati, Assam.

Deb, S., 2009. Integrated Fish Farming System Around Silchar City in Assam (M.Phil dissertation). Assam University, Silchar.

Das, B., 2009. Study of Habitat and Fish Biodiversity of River Kopili in Assam (M.Phil dissertation). Assam University, Silchar.

Das, S., 2009. Composite Fish Farming System Around Bhagabazar in Assam (M.Phil dissertation). Assam University, Silchar.

Das, M.K., 1994. Outbreak of the fish disease epizootic ulcerative syndrome in India - an overview. In: Roberts, R.J., Campbell, Biand Mac Rae, I.H. (Eds.), Proc. Regional Sent. Epizootic Ulcerative Syndrome, pp. 21–36.

Devi, U.N., Singh, N.R., Kar, D. Preliminary studies on fish affected with epizootic ulcerative syndrome with special emphasis on parasitic infestations. Biol. Forum Int. J., in press.

Dykstra, M.J., Noga, E.J., Levine, J.F., Moye, D.W., Hawkins, J.H., 1986. Characterization of the *Aphanomyces* species involved with ulcerative mycosis (UM) in menhaden. Mycologia 78 (4), 664–672.

Egusa, S., Masuda, N., 1971. A new fungal disease of *Plecoglossus altivelis*. Fish. Pathol. 6, 41–46.

FAO, 1986. Report of the Expert Consultation on Ulcerative Fish Diseases in the Asia-Pacific Region. (TCP/RAS/4508). Bangkok, August 1986. FAO, Regional Office for Asia and the Pacific, Bangkok.

Fraser, G.C., Callinan, R.B., Calder, L.M., 1992. *Aphanomyces* species associated with red spot disease: an ulcerative disease of estuarine fish of Eastern Australia. J. Fish Dis. 15, 173–181.

Faruk, M.A.R., Alam, M.J., Sarker, M.M.R., Kabir, M.B., 2004. Status of fish disease and health management practices in rural freshwater aquaculture of Bangladesh. Pak. J. Biological Sci. 7 (12), 2092–2098.

Frerichs, G.N., 1995. Viruses associated with the epizootic ulcerative syndrome (EUS) of Fish in SE Asia. Vet. Res. 26 (5–6), 449–454.

Frerichs, G.N., Hill, B.J., Way, K., 1989. Ulcerative disease rhabdovirus: cell-line susceptibility and serological comparison with other fish rhabdoviruses. J. Fish. Dis. 12, 51–56.

Frerichs, G.N., Millar, S.D., Roberts, R.J., 1986. Ulcerative rhabdovirus in fish in South-East Asia. Nature 322, 216.

Funge, S.S., Dubeau, P., 2002. Aquatic Animal Health Management Issues in Rural AQC Development in Lao PDR, pp. 41–54.

Gayathri, D., Shankar, K.M., Mohan, C.V., Devaraja, T.N., 2008. Determination of specific sampling site on EUS affected fish for diagnosis of *Aphanomyces invadans* by immunodot test using monoclonal antibodies. Res. J. Agric. Biological Sci. 4 (6), 757–760.

Gozlan, R.E., Marshall, W., Lilje, O., Jessop, C., Gleason, F.H., Andreou, D., 2014. Current ecological understanding of fungal-like pathogens of fish: what lies beneath? Front. Microbiol. 5, 62.

Geetarani, B., Shomorendra, M., Kar, D., 2011. Occurrence of trematodes in some fishes of utra Lake in Manipur. Environ. Ecol. 29 (2A), 864–865.

Hassan, I., El-Sayyad, et al., 2010a. Haematological changes associated with EUS in Asian Cichlid Fish, *Etroplus suratensis*. Ann. Biological Res. 1 (4), 106–118.

Hassan, I.El-S., Viola, H.Z., Abdallah, M.El S., Dina, A.El-B., 2010b. Studies on the effects of bacterial diseases on the skin and gill structure of *Clarias gariepinus* in Dakahlia Provinence, Egypt. Scholars Res. Libra Ann. Biological Res. 1 (4), 106–118.

Hatai, K., 1980. Studies on Pathogenic Agents of Saprolegniasis in Fresh Water Fishes. Special Report of Nagasaki Prefectural Institute of Fisheries No. 8, Matsugae-cho, Nagasaki, Japan (in Japanese).

Hatai, K., Egusa, S., Takahashi, S., Ooe, K., 1977. Study on the pathogenic fungus of mycotic granulomatosis—I. Isolation and pathogenicity of the fungus from cultured ayu infected with the disease. Fish. Pathol. 11 (2), 129–133.

Hatai, K., Nakamura, K., Rha, S.A., Yuasa, K., Wada, S., 1994. Aphanomyces infection in dwarf gourami (*Colisa lalia*). Fish. Pathol. 29 (2), 95–99.

Hossain, M.M.M., Richards, M.A., Mondal, S Sadat, Mondal, A.S.M., Chwdhury, M.B.R., September, 2011. Isolation of some emergent bacterial pathogens recovered from capture and culture fisheries in Bangladesh. Bangladesh Res. Publ. 6 (1), 77–90 (Sep, 2011).

Hasan, M.R., Ahmed, G.U., 2002. Issues in Carp Hatcheries and Nurseries in Bangladesh, with special reference to Health Management. In: Arthur, J.R., Phillips, M.J., Subasinghe, R.P., Reantaso, M.B., MacRae, I.H. (Eds.), Primary Aquatic Animal Health Care in Rural, Small-scale Aquaculture Development. FAO Fish. Tech, pp. 147–164. Pap. No. 406.

Harikrishnan, R., Balasundaram, C., Heo, M.S., Moon1, Y.-G., Kim, M.-C., Kim, J.-S., 2009. Use of herbal concoction in the therapy of goldfish (*Carassius auratus*) infected with *Aeromonas hydrophila*. Bull. Vet. Inst. Pulawy 53, 27–36.

Harikrishnan, R., Balasundaram, C., Heo, M.S., 2010. Supplementation diet containing probiotics, herbal and azadirachtin on hematological and biochemical changes in *Cirrhinus mrigala* against *Aphanomyces invadans*. Fish. Aquac. J. 4, 32–38.

Hickey, C.R., 1976. Fish haematology its uses and significance. N.Y. Fish Game J. 23, 170–175.

Huchzermeyer, van der Waal, September 25, 2012. Epizootic ulcerative syndrome, an exotic fish disease which threatens Africa's aquatic ecosystems. J. S Afr. Vet. Assoc. 83 (1), 204.

Iqbal, J.E., 1992. In: Manson, M.M. (Ed.), Immunogold Probe in Electron Microscopy: Methods in Molecular Biology Immunochemical Protocols, 10, pp. 177–185.

Iqbal, M.M., Tajima, K., Ezura, Y., 1998. Phenotypic identification of motile *Aeromonas* isolated from fishes with epizootic ulcerative syndrome in Southeast Asian countries. Bull. Fac. Fish. Hokkaido Univ. 49, 131–141.

Innocent, et al., 2011. Haematology of *Cirrhinus mrigala* fed with vitamin C supplemented diet and post challenged by *Aphanomyces invadens*. J. Appl. Pharm. Sci. 01 (09), 141–144.

Irina, N.G., Neagu, A.N., Vulpe, V., 2011. SEM investigations regarding skin micromorphology and modification induced by bacterial infections in *Cyprinus carpio* and *Salmo trutta fario*. Int. J. Energ. Env. 5, 274–281.

Jhingran, A.G., Das, M.K., 1990. Epizootic Ulcerative Syndrome in Fishes. Bulletin of the Central Inland Capture Fisheries Research Institute (No. 65). CURL, Barrackpore, India, 14.

Jobling, S., Sheahan, D., Osborne, J.A., Matthiessen, P., Sumpter, J.P., 1996. Inhibition of testicular growth in rainbow trout (*Oncorhynchus mykiss*) exposed to estrogenic alkylphenolic chemicals. Environ. Toxicol. Chem. 15, 194–202.

John, K.R., George, M.R., Richards, R.H., Frerichs, G.N., 2001. Characteristics of a new reovirus isolated from epizootic ulcerative syndrome infected snakehead fish. Dis. Aquat. Organ 46 (2), 83–92, 12.

John, R.K., George, R.M., 2012. Viruses associated with epizootic ulcerative syndrome: an update. Indian J. Virology 23 (2), 106–113.

Juan, D.A., 1994. Virus-Screening in Cultured Carp and Tilapia Species and the Virus Sensitivity of Cell Cultures from Indigenous Fish Species (M.Sc. dissertation). University of Pertanian, Malaysia, pp. 23.

Kar, D., 1984. Limnology and Fisheries of Lake Sone in the Cachar District of Assam (India), VIII + 201 (PhD thesis). University of Gauhati, Assam.

Kar, D., 1990. Limnology and Fisheries of Lake Sone in the Cachar District of Assam (India). Matsya 15-16, 209–213.

Kar, D., 1991. Fish disease syndrome affects the people of Barak valley. Fish. Chimes 10 (11), 63.

Kar, D., 1996. Biodiversity conservation prioritisation project (BCPP) in India. In: Proc. International Project Formulation Workshop of BCPP, World Wide Fund (WWF) for Nature-India, 1, New Delhi.

Kar, D., 1998. Biodiversity conservation prioritisation project (BCPP) in India. In: Proc. International Project Finalisation Symposium of BCPP, World Wide Fund (WWF) for Nature-India, 1, New Delhi.

Kar, D., 1999a. Preliminary study of Limnology and aquatic biota of Rudra Sagar wetland in South Tripura. In: Proc. Regional Seminar on Biodiversity, Guwahati, Assam, 1.

Kar, D., 1999b. Microbiological and environmental studies in relation to fish disease in India. In: Gordon Research Conference, Connecticut, USA.

Kar, D., 2000a. Present status of fish biodiversity in South Assam and Tripura. In: Ponniah, A.G., Sarkar, U.K. (Eds.), Fish Biodiversity of North-East India. NBFGR-NATP Publication, Lucknow, pp. 80–82. No. 2: pp. 228.

Kar, D., 2000b. Socio-economic development of the fisherwomen through aquaculture with emphasis on integrated farming in the villages around Chatla Haor Wetland in Silchar, Assam. In: Seminar Presented on DBT-sponsored Awareness Workshop on Biotechnology-based Programmes for Women and Rural Development, NEHU, Shillong, pp. 13–14, 1.

Kar, D., 2001. Species composition and distribution of Fishes in the rivers in Barak valley region of Assam and the principal rivers in Mizoram and Tripura in relation to their habitat parameters. In: Proc. National Workshop, NATP-ICAR Project Mid-term Review. Central Marine Fisheries Research Institute, Cochin, p. 25, 1.

Kar, D., 2002. Present status of biodiversity off fishes of Barak valley region of Assam with a note on their management and conservation. In: Bhattacharjee, M.K., Dattachoudhury, M., Mazumder, P.B. (Eds.), Proc. UGC-Sponsored State-level Seminar on Biodiversity of Assam, Session Chairman's Lecture, 30 Jan 2000. Karimganj College, Assam University, Assam, pp. 3–10.

Kar, D., 2003a. Fishes of barak drainage, Mizoram and Tripura. pp. 203–211. In: Kumar, A., Bohra, C., Singh, L.K. (Eds.), Environment, Pollution and Management. APH Publishing Corporation, New Delhi, pp. xii + 604.

Kar, D., 2003b. An Account of the Fish Biodiversity in South Assam, Mizoram and Tripura along with a Brief Account of Epizootic Ulcerative Fish Disease Syndrome in Freshwater Fishes. UGC-sponsored Invited Lecture in Dept. of Environmental Engg., Guru Jambheshwar University, Hisar, Haryana.

Kar, D., 2003c. Peoples' Perspective on fish conservation in the water bodies of south Assam, Mizoram and Tripura. pp. 325—328. In: Mahanta, P.C., Tyagi, L.K. (Eds.), Participatory Approach for Fish Biodiversity Conservation in North-East India. National Bureau of Fish Genetic Resources (ICAR), Lucknow, v + 412.

Kar, D., 2004a. Inventorying of fish biodiversity in North-East India with a note on their Conservation. In: Proc. National Conference on Fish and Their Environment, 9—11 Feb 2004. B.A. Marathwada University, Aurangabad, 1.

Kar, D., 2004b. Conservation and ichthyodiversity in the river systems of barak drainage, Mizoram and Tripura. In: Proc. National Symp. Management of Aquatic Resources for Biodiversity Maintenance and Conservation, 2—4 Feb 2004. J.N. Vyas University, Jodhpur, 1.

Kar, D., 2004c. A glimpse into the fish bioresources of North-East India with a note on their management, conservation and biotechnological potential. In: Proc. (Invited Lecture) DBT-sponsored National Symposium on Biodiversity Conservation and Sustainable Utilisation of Environmental Resources, Tripura University, 10-11 Jan., 2004, 9 p.

Kar, D., 2005a. Fish genetic resources and habitat diversity of the barak drainage. pp. 68—76. In: Ramachandra, T.V., Ahalya, N., Rajsekara Murthy, C. (Eds.), Aquatic Ecosystems, Conservation, Restoration and Management. Capital Publishing Company, Bangalore, pp. xiii + 396.

Kar, D., 2005b. Fish fauna of river Barak, of Mizoram and of Tripura with a note on Conservation. J. Freshw. Biol. 16.

Kar, D., 2005c. Fish diversity in the major rivers in southern Assam, Mizoram and Tripura. In: Nishida, T., Kailola, P.J., Hollingworth, C.E. (Eds.), Proc. 2nd International Symposium on GIS and Spatial Analyses in Fisheries and Aquatic Sciences, 2-6 Sep 2002, University of Sussex, Brighton (UK), vol. 2. Fisheries and Aquatic GIS Research Group, Kawagoe, Saitama, Japan, pp. 679—691.

Kar, D., 2007a. Fundamentals of Limnology and Aquaculture Biotechnology. Daya Publishing House, New Delhi, pp. xiv + 609.

Kar, D., 2007b. Sustainability issues of inland fish biodiversity and fisheries in barak drainage (Assam). pp. 555—560. In: Kurup, Madhusoodana, B., Ravindran, K. (Eds.), Mizoram and Tripura, Sustain Fish, Cochin University of Science and Technology (CUSAT): Proceedings of International Symposium on 'Improved sustainability of Fish Production Systems and Appropriate Technologies for Utilisation', 16-18 March, 2005 (Cochin). School of Industrial Fisheries, pp. xii + 863.

Kar, D., 2007c. Lentic fishery: fishery of a tropical wetland (Beel) in Assam. Sci. Soc. 5 (1), 53—72.

Kar, D., 2010a. Biodiversity Conservation Prioritisation. Swastik Publications, Delhi pp. X + 180.

Kar, D., 2010b. Fish diversity in rivers of Mizoram. Fish. Chimes 30 (7), 18—22.

Kar, D., 2013. Wetlands and Lakes of the World. Springer, London pp. xxx + 687.

Kar, D., Das, M., 2004a. Preliminary histochemical studies with diseased fishes suffering from epizootic ulcerative syndrome in Assam. In: Proc. National Symposium for Management of Aquatic Resources for Biodiversity Maintenance and Conservation, vol. 1, p. 27.

Kar, D., Das, M., 2004b. Preliminary enzymological studies with diseased fishes suffering from epizootic ulcerative syndrome in fishes. In: Proc. National Conference on Fish and their Environment, vol. 1, p. 25.

Kar, D., Dey, S.C., 1982a. Hilsa ilisha (Hamilton) from Lake Sone in Cachar, Assam. Proc. Indian Sci. Congr. 69, 77.

Kar, D., Dey, S.C., 1982b. An account of Hilsa ilisha (Hamilton) of sone beel (Cachar district, Assam, India). Proc. All-India Sem. Ichthyol. 2, 3.

Kar, D., Dey, S.C., 1986. An account of ichthyospecies of Lake Sone in Barak valley of Assam. Proc. All-India Sem. Ichthyol. 2, 3.

Kar, D., Dey, S.C., 1987. An account of the fish and fisheries of Lake Sone in the Barak valley of Assam (India). Proc. Workshop Dev. Beel Fish. Assam 1, 13.

Kar, D., Dey, S.C., 1988a. A critical account of the recent fish disease in the Barak valley of Assam. Proc. Regional Symp. Recent outbreak Fish Dis. North-East India 1, 8.

Kar, D., Dey, S.C., 1988b. Impact of recent fish epidemics on the fishing communities of cachar district of Assam. Proc. Regional Symp. Recent outbreak Fish Dis. North-East India 1, 8.

Kar, D., Dey, S.C., 1988c. An account of fish yield from Lake Sone in the Barak valley of Assam. Proc. Indian Sci. Congr. 75, 49.

Kar, D., Dey, S., 1988d. Preliminary electron microscopic studies on diseased fish tissues from Barak valley of Assam. Proc. Annu. Conf. Electron Microsc. Soc. India 18, 88.

Kar, D., Dey, S.C., 1990a. Fish disease syndrome: a preliminary study from Assam. Bangladesh J. Zool. 18, 115—118.

Kar, D., Dey, S.C., 1990b. A preliminary study of diseased fishes from Cachar district of Assam. Matsya 15-16, 155—161.

Kar, D., Dey, S.C., 1990c. Epizootic ulcerative syndrome in fishes of Assam. J. Assam. Sci. Soc. 32 (2), 29—31.

Kar, D., Bhattacharjee, S., Kar, S., Dey, S.C., 1990. An account of 'EUS' in fishes of Cachar district of Assam with special emphasis on Microbiological studies. In: Souv. Annual Conference of Indian Association of Pathologists and Microbiologists of India (NE Chapter), vol. 8, pp. 9—11.

Kar, D., Dey, S.C., Kar, S., Bhattacharjee, N., Roy, A., 1993. Virus-like particles in epizootic ulcerative syndrome of fish. In: Proc. International Symp. On Virus-Cell Interaction: Cellular and Molecular Responses, 1, p. 34.

Kar, D., Dey, S.C., Kar, S., Roy, A., Michael, R.G., Bhattacharjee, S., Changkija, S., 1994. A candidate virus in epizootic ulcerative syndrome of fish. In: Proc. National Symp. Of the Indian Virological Society, 1, p. 27.

Kar, D., Roy, A., Dey, S.C., Menon, A.G.K., Kar, S., 1995a. Epizootic ulcerative syndrome in fishes of India. World Congress of in vitro biology. Vitro 31 (3), 7.

Kar, D., Kar, S., Roy, A., Dey, S.C., 1995b. Viral disease syndrome in fishes of North-East India. In: Proc. International Symp. Of International Centre for Genetic Engg, 1. And Biotechnology (ICGEB) and the Univ. of California at Irvine, p. 14.

Kar, D., Dey, S.C., Kar, S., 1995c. A viral disease among the fishes of North-East India. In: Annual Congress on Man and Environment, 10. National Environment Sci. Acad. And National Institute of Oceanography, p. 62.

Kar, D., Dey, S.C., Kar, S., Michael, R.G., Gadgil, M., 1996. Ichthyoecology, management and conservation fish resources of Lake sone in Assam (India). Tiger Pap. (FAO, UN) XXIII (3), 27—32.

Kar, D., Dey, S.C., Purkayastha, M., Kar, S., 1996a. An overview of the impediments in conservation of biodiversity of Lake Sone in Assam. In: Proc. Seminar on Conservation of Biodiversity in Indian Aquatic Ecosystems. JawaharLal Nehru University, New Delhi, 1.

Kar, D., Dey, S.C., Roy, A., Kar, S., 1996b. Viral disease syndrome in fishes of India. Proc. Int. Congr. Virology 10.

Kar, D., Purkayastha, M., Kar, S., 1996c. Biodiversity conservation prioritisation project: a case study from sone beel in Assam. In: Proc. National Workshop on Biodiversity Conservation Prioritisation Project (BCPP). World Wide Fund (WWF) for Nature-India and Centre for Ecological Sciences, Indian Institute of Science, Bangalore.

Kar, D., Saha, D., Laskar, R., Barbhuiya, M.H., 1997. Biodiversity conservation prioritisation project (BCPP) in barak valley region of Assam. In: Proc. National Project Evaluation Workshop on BCPP, Betla Tiger Reserve and National Park, 1 (Palamu).

Kar, D., Saha, D., Dey, S.C., 1998a. Epizootic ulcerative syndrome in barak valley of Assam: 2—4. In: Project Report Submitted and Presented at the National Symposium of Biodiversity Conservation Prioritisation Project (BCPP) Held at WWF-India, 18—19 Jan 1998 (New Delhi).

Kar, D., Dey, S.C., Roy, A., 1998b. Present status of epizootic ulcerative syndrome (EUS) in southern Assam. In: Proc. Regional Project Initiation Workshop for NATP-ICAR-NBFGR Project, 1, p. 9.

Kar, D., Dey, S.C., Kar, S., Roy, A., 1998c. An account of epizootic ulcerative syndrome in Assam. In: Final Project Report of Biodiversity Conservation Prioritisation Project (BCPP). Submitted and Presented at the International Project Finalisation Workshop of BCPP, 1, 22—28 April 1998, New Delhi, pp. 1—3.

Kar, D., Dey, S.C., Roy, A., 1998d. Fish disease diagnosis and fish quarantine problems in India and south-east Asia with particular emphasis on epizootic ulcerative syndrome (EUS) fish disease problems. In: Proc. International Workshop on Fish Disease Diagnosis and Quarantine, IndiaN Council of Agricultural Research (ICAR)-Ministry of Agriculture (MoA), Govt. Of India (GoI)-network of Aquaculture Centres in Asia (NACA)-food and Agricultural Organisation (FAO) of the United Nations (UN)-oie, France: Held at the Central Institute of Freshwater Aquaculture (CIFA), 1.

Kar, D., Upadhyaya, T., 1998. Histopathological studies of fish tissues affected by epizootic ulcerative syndrome in Assam. In: Technical Bulletin, XIII Convention and National Symposium of Indian Association of Veterinary Anatomists (IAAM), 11—13 Dec 1998. College of Veterinary Sciences, Assam Agricultural University, Guwahati.

Kar, D., Dey, S.C., Kar, S., Ramachandra, T.V., 1999a. Trawls of Lake Sone in Assam. J. Appl. Zool. Res. 10 (2), 170—172.

Kar, D., Rahaman, H., Barnman, N.N., Kar, S., Dey, S.C., Ramachandra, T.V., 1999b. Bacterial pathogens associated with epizootic ulcerative syndrome in freshwater fishes of India. Environ. Ecol. 17 (4), 1025—1027.

Kar, D., Mandal, M., Bhattacharjee, S., 1999c. Fungal pathogens associated with epizootic ulcerative syndrome in fishes of barak valley region of Assam. In: Proc. 1st National Conference on Fisheries Biotechnology, CIFE, 1, p. 34.

Kar, D., Dey, S.C., Roy, A., 2000a. Fish genetic resources in the principal rivers and wetlands in North-East India with special emphasis on Barak valley (Assam), in Mizoram and in Tripura, with a note on epizootic ulcerative syndrome fish disease. In: Proc. National Project Initiation Workshop of the NATP-ICAR World Bank-aided Project on 'Germplasm Inventory, Evaluation and Gene Banking of Freshwater Fishes', National Bureau of Fish Genetic Resources (NBFGR), Lucknow, 1, p. 12.

Kar, D., Mandal, M., Laskar, B.A., Dhar, N., Barbhuiya, M.H., 2000b. Ichthyofauna of some of the Oxbow lake in Barak valley region of Assam. In: Proceedings of the National Symposium on Wetlands and Fisheries Research in the New Millennium, 1, p. 16.

Kar, D., Dey, S.C., Roy, A., Mandal, M., 2000c. Epizootic ulcerative syndrome fish disease in Barak valley region of Assam, India. In: Proc. Nat. Symp. Current Trends in Wetlands and Fisheries Research in the New Millennium, 1, p. 2.

Kar, D., Dey, S.C., Mandal, M., Lalsiamliana, 2000d. Epizootic ulcerative syndrome among the fishes of Assam. In: Proc. National Workshop of NATP-ICAR-NEC North-East Programme, Shillong, 1, p. 24.

Kar, D., Mandal, M., Lalsiamliana, 2001a. Species composition and distribution of fishes in the rivers in Barak valley region of Assam and in the principal rivers of Mizoram and Tripura in relation to their habitat parameters. In: Proc. National Workshop, NATP-icar Project, Mid-term Review, 1: Central Marine Fisheries Research Institute, Cochin, 25 pp.

Kar, D., Laskar, B.A., Lalsiamliana, 2001b. Further studies on the ichthyospecies composition and distribution of freshwater fishes in Barak drainage and in the principal rivers in Mizoram and in Tripura with a note on their feeding and breeding biology. In: Proc. National Project Monitoring Workshop of NATP-ICAR Project, 1. National Bureau of Fish Genetic Resources, Lucknow, p. 22.

Kar, D., Laskar, B.A., Mandal, M., Lalsiamliana, Nath, D., 2002a. Fish genetic diversity and habitat parameters in barak drainage, Mizoram and Tripura. Indian J. Environ. Ecoplanning 6 (3), 473—480.

Kar, D., Laskar, B.A., Nath, D., Mandal, M., Lalsiamliana, 2002b. *Tor progenius* (McClelland) under threat in river jatinga, Assam. Sci. Cult. 68 (7—8), 211.

Kar, D., Laskar, B.A., Nath, D., 2002c. *Tor* sp. (Mahseer fish) in river Mat in Mizoram. Aquaculture 3 (2), 229—234.

Kar, D., Dey, S.C., Roy, A., 2002d. Prevalence of epizootic ulcerative syndrome among fishes of Assam. In: Proc. Regional Symp. On Biodiversity, S.S. College, Assam (Central) University, Hailakandi (Assam).

Kar, D., Dey, S.C., Roy, A., 2002e. On the diseased fishes suffering from hitherto unknown epizootic ulcerative syndrome in fishes in India. In: Proc. All-India Congress of Zoology, Dec., 2002 (Bangalore).

Kar, D., Dey, S.C., Roy, A., 2002f. Prevalence of epizootic ulcerative syndrome (EUS) among fishes of Mizoram. In: Proc. Regional Symp. On Aquaculture, 1.

Kar, D., Laskar, B.A., Lalsiamliana, 2002g. Germplasm Inventory, evaluation and gene banking of freshwater fishes. In: Third Annual Technical Report: pp. 57. National Project Evaluation Workshop, 3: NATP-icar World Bank-aided Project: National Bureau of Fish Genetic Resources, Lucknow.

Kar, D., Dey, S.C., Mandal, M., Laskar, B.A., Lalsiamliana, 2002h. Preliminary survey of the fish genetic resources of the rivers in barak drainage, Mizoram and Tripura. pp. 73—81. In: Ramachandra, T.R., Rajasekahra, Murthy, C., Ahalya, N. (Eds.), Restoration of Lakes and Wetlands. Allied Publishers (P) Ltd, Bangalore, pp. xxii + 400.

Kar, D., Dey, S.C., Roy, A., Mandal, M., 2002i. Epizootic ulcerative syndrome in fishes of barak valley of Assam, India. pp. 303—307. In: Ramachandra, T.V., Murthy Rajasekhara, S., Ahalya, N. (Eds.), Restoration of Lakes and Wetlands. Allied Publishers (P) Ltd, Bangalore, pp. xii + 400.

Kar, D., Dey, S.C., Datta, N.C., 2003a. Welfare Biology in the New Millennium. pp. xx + 97. Allied Publishers Pvt. Ltd, Bangalore.

Kar, D., Dey, S.C., Roy, A., 2003b. Capture fishery in lentic systems with some light on Sone Beel in Assam with special reference to prevalence to EUS in the Barak valley region of Assam. In: Proc. Lecture Series in UGC-sponsored Vocational Course in Industrial Fish and Fisheries, Cachar College, Assam University, Silchar, Abstracts, 1.

Kar, D., Roy, A., Dey, S.C., 2003c. Epizootic ulcerative disease syndrome in freshwater fishes with a note on its management for sustainable fisheries. In: Proc. International Conference on Disease Management for Sustainable Fisheries, 26–29 May, 2003, Department of Aquatic Biology and Fisheries, University of Kerala, Trivandrum.

Kar, D., Lohar, M., Ngassepam, Z., Sinom, S., Tiwary, B., Bawari, M., 2003d. Fish Bioresources in certain rivers of Assam and Manipur with a note on their assessment, management and conservation. In: Proc. Nat. Symp. Assessment and Management of Bioresources, North Bengal University and the Zoological Society, Calcutta, 28-30 May, 2003, 56 pp.

Kar, D., Roy, A., Dey, S.C., 2004a. An overview of fish Genetic diversity of North-East India. In: Garg, S.K., Jain, K.L. (Eds.), Proc. National Workshop on Rational Use of Water Resources for Aquaculture, vol. 1. CCS Haryana Agricultural University, pp. 164–171. March 18–19.

Kar, D., Roy, A., Mazumder, J., Patil, P., 2004b. Biotechnological approach for defining epizootic ulcerative syndrome disease in freshwater fishes. In: Proc. Mid-Term Review: DBT-Sponsored Project, New Delhi, 1.

Kar, D., Dey, S.C., Roy, A., Mazumder, J., Patil, P., Kohlapure, R.M., 2004c. Fish disease prevalence in Assam with particular reference to epizootic ulcerative syndrome and its pathogens. In: Proc. International Conference 'Bioconvergence', November 18–20, 2004, vol. 1. Thapar Institute of Engineering and Technology, Patiala, 220 pp.

Kar, D., Mazumder, J., Devi, P., Devi, B.R., Devi, V., 2006. Isolation of *Aeromonas hydrophila* from fishes affected by epizootic ulcerative syndrome as well as from corresponding healthy fish species and from their habitat. J. Curr. Sci. 9 (1), 323–327.

Kar, D., Mazumder, J., Halder, I., Dey, M., 25 January, 2007a. Dynamics of initiation of disease in fishes through interaction of microbes and the environment. Curr. Sci. (Bangalore) 92 (2), 177–179.

Kar, D., Mazumder, J., Barbhuiya, M.A., 2007b. Isolation of Mycotic flora from fishes affected by epizootic ulcerative syndrome in Assam, India. Asian J. Microbiol. Biotechnol. 9 (1), 37–39.

Kar, D., Laskar, B.A., Nath, D., 2006. An account of Fecundity of *Eutropiichthys vacha* (Hamilton-Buchanan): a commercially important fish in Assam. Environ. Ecol. 24 S (3), 726–727.

Kar, D., Shomorendra, M., Singha, R., Puinyabati, H., Geetarani, B., Binky, K., Sangeeta, O., Ranibala, T., 2011. Fish diversity and helminth fauna in the fishes of Assam and Manipur. India: Fish. Chimes 55–65.

Kar, S., Kar, D., 2014. Ecology of four wetlands in South Assam: a comparison. Research frontiers in wetlands, fishery and aquaculture. In: Kar, D. (Ed.), Research Frontiers in Wetlands, Fishery and Aquaculture. Dominant Publishers and Distributors Pvt. Ltd, New Delhi, ISBN 978-93-82007-39-5, pp. 239–246 xi + 320; ISBN 10: 93-82007-39-3.

Kar, D., Singha, R., Riji, J.K., George, R.M., Mansoor, M., Selva, M., 2015. Preliminary epidemiological studies on epizootic ulcerative syndrome in freshwater fishes of Assam. Environ. Ecol. (in press).

Kar, D., 2012. Essential of Fish Biology Dominant Publishers and Distributors Pvt. Ltd. (New Delhi). ISBN: 978-93-82007-39-5. ISBN 10: 93-82007-39-3.

Kanchanakhan, S., 1996a. Epizootic ulcerative syndrome (EUS): a new look at the old story. AAHRI Newslett. 5 (1), 2–5.

Kanchanakhan, S., 1996b. Field and Laboratory Studies on Rhabdoviruses Associated with Epizootic Ulcerative Syndrome (EUS) of Fishes (Ph.D. thesis). University of Stirling, Scotland, pp. 278.

Kanchanakhan, S., 1997. Variability in isolation of viruses from fish affected by EUS. AAHRI Newslett. 6 (1), 1–5.

Khan, M.H., Lilley, J.H., 2002. Risk Factors and Socio-economic Impacts Associated with EUS in Bangladesh, pp. 27–39.

Kiryu, Y., Shields, J.D., Vogelbein, W.K., Zwerner, D.E., Kator, H., Blazer, V.S., 2002. Induction of skin ulcers in Atlantic menhaden by injection and aqueous exposure to the zoospores of *Aphanomyces invadans*. J. Aquatic Animal Health 14, 11–24.

Kiryu, Y., Shields, J.D., Vogelbein, W.K., Kator, H., Blazer, V.S., 2003. Infectivity and pathogenicity of the oomycete *Aphanomyces invadans* in Atlantic menhaden *Brevoortia tyrannus*. Dis. Aquatic Org. 54, 135–146.

Kurata, O., Kanai, H., Hatai, K., 2000. Hemagglutinating and hemolytic capacities of *Aphanomyces piscicida*. Fish. Pathol. 35 (1), 29–33.

Kurhalyuk, N., Tkachenko, H., 2011. Induction of oxidative stress and antioxidant defenses in the livers of sea trout, *Salmo trutta* L. with ulcerative dermal necrosis. Arch. Pol. Fish. 19, 229–240.

Kales, S.C., 2006. A Functional Study of Major Histocompatibility Expression and Immune Function in Rainbow Trout (Oncorhynchus mykiss): To Be Printed (PhD thesis), pp. 156: Not much required by me, April 25, 2015.

Kuan, G.C., Sheng, L.P., Rijiravanich, P., Marimuthu, K., Ravichandran, M., Yin, L.S., Lertanantawong, B., Surareungchai, W., December 15, 2013. Gold-nanoparticle based electrochemical DNA sensor for the detection of fish pathogen *Aphanomyces invadans*. Talanta. Int. J. Nanomater. Biostructures 117, 312–317. http://dx.doi.org/10.1016/j.talanta.2013.09.016. Epub 2013 Sep 19. ISSN 2277–3851.

Khulbe, R.D., 1991. An ecological study of watermolds of some rivers of Kumaun Himalayas. Indian Trop. Ecol. 32, 127–135.

Khulbe, R.D., 2001. A Manual of Aquatic Fungi (Chytridiomycetes and Oomycetes). Daya Publishing Housing House, Delhi, p. 255.

Khulbe, R.D., Sati, S.C., 1981. Studies on parasitic watermolds of Kumaun Himalaya, host range of Achlya Americana Humphrey on certain temperate fish. Mykosen 24, 177–180.

Leclerc, H., Mossel, D.A.A., Edberg, S.C., Struijk, C.B., 2001. Advances in the bacteriology of the coliform group: their suitability as markers of microbial water safety. Annu. Rev. Microbiol. 55, 201–234.

Li, Y., Lovell, R.T., 1985. Elevated levels of dietary ascorbic acid increase immune responses in channel catfish. J. Nutr. 115, 123–311.

Lilly, S.J., Scade, D., Ellis, R., Le Fevre, O., Brinchmann, J., Tresse, L., Abaham, R., Hammer, F., Crampton, D., Colless, M., Glazebrook, K., Mallen-Ornelas, G., Broadhurst, T., 1998. Appl. Phys. J. 500, 75.

Lilley, J.H., Beakes, G.W., Hetherington, C.S., 2001. Characterization of *Aphanomyces invadans* isolates using pyrolysis mass spectrometry (PyMS). Mycoses 44 (9–10), 383–389.

Lilley, J.H., Callinan, R.B., Khan, M.H., 2002. Social, economic and biodiversity impacts of epizootic ulcerative syndrome (EUS). In: Arthur, J.R., Phillips, M.J., Subasinghe, R.P., Reantaso, M.B., MacRae, I.H. (Eds.), Primary Aquatic Animal Health Care in

Rural, Small-scale Aquaculture Development. FAO Fish. Tech, pp. 127–129. Pap. No. 406.

Lilley, J.H., Hart, D., Panyawachira, V., Kanchanakhan, S., Chinabut, S., Söderhäll, K., Cerenius, L., May 2003. Molecular characterization of the fish-pathogenic fungus Aphanomyces invadans. J. Fish. Dis. 26 (5), 263–275.

Little, et al., 2002. Health management issues in freshwater fish hatcheries, nurseries and fry distribution, with emphasis on experiences in Viet Nam and Bangladesh. In: Arthur, J.R., Phillips, M.J., Subasinghe, R.P., Reantaso, M.B., MacRae, I.H. (Eds.), Primary Aquatic Animal Health Care in Rural, Small-scale Aquaculture Development. FAO Fish. Tech, pp. 141–146. Pap. No. 406.

Leobrera, A., Gaeutan, R.Q., 1987. Aeromonas hydrophila associated with ulcerative disease epizootic in Laguna de Bay, the Philippines. Aquaculture 67, 273–278.

Lilley, J.H., Inglis, V., 1997. Comparative effects of various antibiotics, fungicides and disinfectants on Aphanomyces invaderis and other saprolegniceous fungi. Aquac. Res. 28, 461–469.

Lilley, J.H., Roberts, R.J., 1997. Pathogenicity and culture studies comparing Aphanomyces involved in epizootic ulcerative syndrome (EUS) with other similar fungi. J. Fish Dis. 20, 101–110.

Lilley, J.H., Phillips, M.J., Tonguthai, K., 1992. A Review of Epizootic Ulcerative Syndrome (EUS) in Asia. Aquatic Animal Health Research Institute and Network of Aquaculture Centres in Asia-Pacific, Bangkok, pp. 73.

Lilley, J.H., Hart, D., Richards, R.H., Roberts, R.J., Cerenius, L., Söderhä ll, K., 1997a. Pan-Asian spread of single fungal clone results in large-scale fish-kills. Veterinary Rec. 140, 653–654.

Lilley, J.H., Thompson, K.D., Adams, A., 1997b. Characterization of Aphanomyces invadans by electrophoretic and Western blot analysis. Dis. Aquatic Org. 30, 187–197.

Lio-Po, G.D., Traxler, G.S., Lawrence, J.A., Leano, E.M., 2000. Characterization of a virus isolated from the snakehead Ophiocephalus striatus with EUS in Philippines. Dis. Aquatic Org. 43, 191–198.

Lumanlan-Mayo, S.C., 2011. Immune response of common carp, Cyprinus carpio, under stress. Int. J. Energy Environ. 5 (2).

Manna, S.K., 1998. Fungal and Bacterial Diseases of Fish - Diagnosis and Therapy. Barrackpore-743 101. Central Inland Capture Fisheries Research Institute, West Bengal, pp. 16–23.

Mandal, M., 2003. Fishery and Management Aspect of an Oxbow Lake in Assam (India) with Particular Reference to Feeding Habits of Two Fishes (M. Phil. dissertation). Assam University, Silchar.

Mastan, Qureshi, 2001. Role of bacteria in the epizootic ulcerative syndrome (EUS) of Fishes. J. Environ. Biol. 22 (3), 187–192.

Mazid, M.A., 1995. Matshya o chingri chash unnayaner janna gabesanalavda lagshai projuctir utvaban, prayash ebang prosar. In: Fisheries Fortnight '95 Bulletin, Karbo Mora Macher Chash Thakbo Sukhe Baro Mas. Department of Fisheries, Dhaka (in Bengali).

Mazumdar, Y., 2007. Molecular Characterisation of Causative Agents of Atrophic Rhinitis with Special Emphasis on Borditella Bronchiseptica (PhD thesis). Assam University, Silchar.

McGarey, D.J., Foley, D., Lim, D.V., 1990. Rapid identification of motile aeromonads by bacteriological and serological methods. Abstr. Ann. Meeting ASM, p. 351.

Min, H.-K., Hatai, K., Bai, S., 1994. Some inhibitory effects of chitosan on fish-pathogenic Oomycete, Saprolegnia parasitica. Fish. Pathol. 29, 73–77.

Mohan, C.V., Shankar, K.M., 1994. Epidemiological analysis of epizootic ulcerative syndrome of fresh and brackishwater fishes of Karnataka, India. Curr. Sci. 66 (9), 656–658.

Mohan, C.V., Shankar, K.M., Ramesh, K.S., 1999. Is epizootic ulcerative syndrome (EUS) specific fungus of fishes a primary pathogen? Naga, ICLARM Q. 22 (1), 15–18.

Mohan, C.V., Shankar, K.M., Ramesh, K.S., 2001. Is epizootic ulcerative syndrome (EUS) specific fungus of fishes a primary pathogen?-an opinion. Aquac. Res. 32, 223–233.

Mulcahy, M.F., 1971. Serum protein change in ulcerative dermal necrosis (UDN) of trout (S. trutta). J. Fish. Biol. 3199–3203.

Mazid, M.A., Banu, A.N.H., 2002. An Overview of the Social and Economic Impact and Management Fish and Shrimp Disease in Bangladesh, with an Emphasis on Small-scale AQC, pp. 21–25.

Miyazaki, T., Egusa, S., 1972. Studies on mycotic granulomatosis in freshwater fish I. Mycotic granulomatosis in goldfish. Fish. Pathol. 7, 15–25 (In Japanese).

Miyazaki, T., Egusa, S., 1973a. Studies on mycotic granulomatosis in freshwater fish II. Mycotic granulomatosis prevailed in goldfish. Fish. Pathol. 7, 125–133 (In Japanese).

Miyazaki, T., Egusa, S., 1973b. Studies on mycotic granulomatosis in freshwater fish III. Bluegill. Mycotic granulomatosis in bluegill. Fish. Pathol. 8, 41–43 (In Japanese).

Miyazaki, T., Egusa, S., 1973c. Studies on mycotic granulomatosis in freshwater fish IV. Mycotic granulomatosis in some wild fishes. Fish. Pathol. 8, 44–47 (In Japanese).

Mckenzie, R.A., Hall, W.T.K., 1976. Dermal ulceration of mullet (Mugil cephalus). Aust. Veterinary J. 52, 230–231.

Miles, et al., 2001. Effect of macrophages and serum of fish susceptible or resistant to epizootic ulcerative syndrome (EUS) on the EUS pathogen Aphanomyces invadans. Fish. Shellfish Immunol. 7, 569–584.

Miles, D.J.C., Thompson, K.D., Lilley, J.H., Adams, A., 2003. Immunofluorescence of the epizootic ulcerative syndrome pathogen, Aphanomyces invadans, using a monoclonal antibody. Diseases of Aquatic Organisms Dis. Aquat. Org. 55, 77–84.

Munday, B.L., 1985. Ulcer disease in cod (Pseudophycis bachus) from the Tamar River. Tasman. Fish. Res. 27, 15–18.

McGarey, D.J., Beatty, T.K., Alberts, V.A., Te Strake, D., Lim, D.V., 1990. Investigations of potential microbial pathogens associated with ulcerative disease syndrome (UDS) of Florida fish. Pathology Mar. Sci.

Miyazaki, T., 1994. Comparison among mycotic granulomatosis, saprolegniasis and anaaki-byo in fishes: a Japanese experience. In: Roberts, R.J., Campbell, B., MacRae, I.H. (Eds.), Proceedings of the ODA Regional Seminar on Epizootic Ulcerative Syndrome, Aquatic Animal Health Research Institute, Bangkok, 25-27 January 1994. AAHRI, Bangkok, pp. 253–270.

Millar, S.D., 1994. Joint viral/fungal induction of EUS in snakeheads. ODA Regional seminar Epizootic Ulcerative Syndrome 272–278.

Melba, G., Bondad-Reantaso, Rohana, P., Subasinghe, J., Richard, A., Kazuo, O., Supranee, C., Robert, A., Zilong, T., Mohamed, S., 2005. Disease and health management in Asian aquaculture. Veterinary Parasitol. 5 (2005), 172–188.

Manual, O.I.E., 1997. Epizootic Ulcerative Syndrome.

Nuryati, S., et al., 2005. Study on anti-fungal potency of Terminalia cattapa, Piper betle, Psidium guajava, and Andrographis peniculata on the growth of Aphanomyces sp. vitro Fish. Aquac. J. 2010. FAJ 4.

NACA, 1989. Technical Manual 7. A World Food Day 1989 Publication of the Network of Aquaculture Centres in Asia and the Pacific, Bangkok, Thailand.

Noga, E.J., Dykstra, M.J., 1986. Oomycete fungi associated with ulcerative mycosis in Menhaden, *Brevoortia tyrannus* (Latrobe). J. Fish Dis. 9, 47–53.

Oidtmann, B., Steinbauer, P., Geiger, S., Hoffmann, R.W., 2008. Experimental infection and detection of *Aphanomyces invadans* in European catfish, rainbow trout and European eel. Dis. Aquat. Organ 82 (3), 195–207.

Oidtmann, 2012. Review of biological factors relevant to import risk assessments for epizootic ulcerative syndrome. Transbound. Emerg. Dis. 59 (1), 26–39.

Pradhan, K., Pal, J., 1993. Drug sensitivity testing of bacteria isolated from ulcerative diseases of air breathing fishes. J. Parasitol. Appl. Animal Biol. 2, 15–18.

Plumb, J.A., 1983. Viral Fish Diseases in "Microbial Fish Disease: Laboratory Manual". Brown Printing Company, Montgomery, Alabama, pp. 26–41.

Pathiratne, A., Rajapkshe, W., 1998. Haematological changes associated with EUS in Asian Cichlid Fish, Etroplus suratensis. Asian Fish. Sci. 11, 203–211 (Asian Fisheries Soc, Manila, The Philippines).

Pechhold, K., Kabelitz, D., 1998. Measurement of cellular proliferation. In: Kaufman, S.H.E., Kabelitz, D. (Eds.), Methods in Microbiology, Vol. 25. Academic Press, London, pp. 59–75.

Pal, J., Pradhan, K., 1990. Bacterial involvement in ulcerative condition in air breathing fish from India. J. Fish. Biol. 37, 833–839 (1990).

Phillips, M.J., March 20–24, 1989. A Report on the NACA Workshop on the Regional Research Programme on Ulcerative Syndrome in Fish and the Environment. Network of Aquaculture Centres in Asia-Pacific, Bangkok.

Phillips, M.J., Keddie, H.G., 1990. Regional research programme on relationships between epizootic ulcerative syndrome in fish and the environment. In: A Report on the Second Technical Workshop, August 13–26, 1990. Network of Aquaculture Centres in Asia-Pacific, Bangkok, p. 133.

Poonsuk, K., Saitanu, K., Navephap, O., Wongsawang, S., June 23–24, 1983. Antimicrobial susceptibility testing of *Aeromonas hydrophila*: strain isolated from fresh water fish infection. In: The Symposium on Fresh Water Fishes Epidemic: 1982–1983. Chulalongkorn University, Bangkok, pp. 436–438 (in Thai, English abstract).

Prasad, P.S., Sinha, J.P., 1990. Status paper on the occurrence of ulcerative disease syndrome in fishes of Bihar. In: National Workshop on Ulcerative Disease Syndrome in Fish, March 6–7, 1990, Calcutta.

Pramanika, A.K., Santrab, K.B., 2012. Fish pathogenic bacteria (*Aeromonas hydrophilia*)-Induced histopathological attributes in the spleen and lymph-nodes of swiss albino mice. Bangladesh J. Sci. Ind. Res. 47 (1), 43–46.

Priyadi, A., 1990. Indonesia report. In: Regional Research Programme on Relationships between Epizootic Ulcerative Syndrome in Fish and the Environment, August 13–26, 1990. NACA, Bangkok, pp. 15–17.

Puinyabati, B., Shomorendra, M, Kar, D., 2013. Basic Parasitology. Asiatic Publishing House, New Delhi, pp. xxi + 201, ISBN 978-81-923026-9-0 (2013).

Purkait, M.C., 1990. Case studies on the epizootic ulcerative syndrome of fishes in Chanditala: I of Hooghly district. In: National Workshop on Ulcerative Disease Syndrome in Fish, March 6–7, 1990, Calcutta.

Quddus, M.M.A., 1982. Two types of *Hilsa ilisha* and their population biology from Bangladesh waters (Doctoral thesis). University of Tokyo, p. 180.

Rahman, et al., 2002. Identification and characterization of pathogenic *Aeromonas veronii* biovar *sobria* associated with epizootic ulcerative syndrome in fish in Bangladesh. App. Env. Microbiol. 68 (2), 650–655.

Ramkumar, R., Anandhi, M., Rajthilak, C., Natarajan, T., Perumal, P., October 2013. Studies on ulcerative disease caused by *Providencia stuartii*, bacteria in the Indian major carp, Labeo rohita (Ham.). Int. J. Innovative Res. Sci. Eng. Technol. 2 (10).

Rahman, M., Colque-Navarro, P., Kühn, I., Huys, G., Swings, J., Möllby, R., 2002. Identification and characterization of pathogenic *Aeromonas veronii* biovar *sobria* associated with epizootic ulcerative syndrome in fish in Bangladesh. Appl. Environ. Microbiol 68 (2), 650–655.

Rahman, T., Choudhury, B., Barman, N., 1988. Role of UDS affected fish in the health of ducks in Assam: an experimental study. In: Proceedings of the Symposium on Recent Outbreak of Fish Diseases in North Eastern India, December 30, 1988, Guwahati, Assam, India p. 10.

Rahman, M.M., Somsiri, T., Tajima, K., Ezura, Y., 2004. Distribution of *Aeromonas* spp. emphasising on a newly-identified species *Aeromonas* sp. T8 isolated from fish and aquatic animals in Southeast Asia. Pak. J. Biological Sci. 7 (2), 258–268.

Raina, H.S., Sunder, S., Vass, K.K., Langer, R.K., 1986. Artificial breeding and nursery rearing of *S. esocinus* Heckel in Kashmir. Proc. Nat. Acad. Sci. India 56 (B), 335–338.

Ranibala, Th, Shomorendra, M., Kar, D., 2013. Seasonal variation of the nematode *Camallanus anabantis* in the fish *Anabas testudineus* in Loktak Lake, Manipur, India. J. Appl. Nat. Sci. 5 (2), 397–399.

Ranibala, T., Singh, M.S., Kar, D., 2015. Essentials of Fish Helminthology. Asiatic Publishing House, New Delhi, pp. 185.

Reantaso, M.B., 1991. EUS in brackish waters of the Philippines. Fish. Health Sect. Newsl. 2 (1), 8–9. Asian Fisheries Society (Manila).

Reichenbach-Klinke, H., Elkan, E., 1965. The Principal Diseases of Lower Vertebrates: Book I: Diseases of Fishes. Academic Press, New York.

Reinprayoon, S., Lertpo-kasombat, K., Chantaratchada, S., 1983. The incidence of *Aeromonas hydrophila* in asymptomatic people at Bang-Toi. In: The Symposium on Freshwater Fishes Epidemic, June 23, 1983. Chulalongkornkorn University, Bangkok, pp. 534–546.

Riji, J.K., George, R.M., Mansoor, M., Selva, M., Singh, R., Kar, D., 2014. Molecular Characterisation of Pathogens Associated with Fish Diseases in Assam: Mid-term Report: June, 2014. Department of Biotechnology, Government of India, New Delhi.

Rattan, P., Parulekar, A.H., 1998. Diseases and parasites of laboratory reared and wild population of banded pearl spot, *Etroplus suratensis* (Cichlidae) in Goa. Indian J. Mar. Sci. 27, 407–410. Research Journal of Recent Sciences 3 (ISC-2013) 124-126. ISSN 2277–2502.

RantamaÈki, J., Cerenius, L., SoÈderha, È.K., 1992. Prevention of transmission of the crayfish plague fungus (*Aphanomyces astaci*) to the freshwater crayfish *Astacus astacus* by treatment with MgCl2. Aquaculture 104, 11–18.

Richards, R.H., Pickering, A.D., 1978. Frequency and distribution patterns of *Saprolegnia* infection in wild and hatchery-reared brown trout *Salmo trutta* L. and char *Salvelinus alpinus* (L.). J. Fish. Dis. 1, 69–82.

Roberts, R.J., 1994. Pathogenesity Study on Fungi Isolated from EUS; ODA Regional Seminar Report on EUS. AAHRI, Bangkok, Thailand.

Roberts, R.J., 1975. Melanin-containing cells of teleost fish and their relation to disease. In: Ribelin, W., Migaki, G. (Eds.), The Pathology of Fishes. University of Wisconsin Press, Madison, WI, pp. 399–428.

Roberts, R.J., Willoughby, L.G., Chinabut, S., 1993. Mycotic aspects of epizootic ulcerative syndrome (EUS) of Asian fishes. J. Fish Dis. 16, 169–183.

Roberts, R.J., Frerichs, G.N., Miller, S.D., 1992. Epizootic ulcerative syndrome, the current position in disease in Asian Aquaculture. In: Shariff, M., Subasingh, R.P., Arther, J.R. (Eds.), Fish Health Section. Asian Fisheries Society, Manila, pp. 431–436.

Roberts, R.J., Campbell, B., MacRae, I.H., 1994a. Proceedings of the Regional Seminar on Epizootic Ulcerative Syndrome, January 25–27, 1994. The Aquatic Animal Health Research Institute, Bangkok.

Roberts, R.J., Frerichs, G.N., Tonguthai, K., Chinabut, S., 1994b. Epizootic ulcerative syndrome of farmed and wild fishes. In: Muir, J.F., Roberts, R.J. (Eds.), Recent Advances in Aquaculture V. Blackwell Science, pp. 207–239 (Chapter 4).

Roberts, R.J., Macintosh, D.J., Tonguthai, K., Boonyaratpalin, S., Tayaputch, N., Phillips, M.J., Millar, S.D., 1986. Field and laboratory investigations into ulcerative fish diseases in the Asia-Pacific region. In: Technical Report of FAO Project TCP/RAS/4508. Bangkok, p. 214.

Roberts, R.J., Wootten, R., MacRae, I., Millar, S., Struthers, W., 1989. Ulcerative Disease Survey, Bangladesh. Final Report to the Government of Bangladesh and the Overseas Development Administration. Institute of Aquaculture, Stirling.

Rodgers, L.J., Burke, J.B., 1981. Seasonal variation in the prevalence of red spotdisease in estuarine fish with particular reference to the sea mullet, Mugil cephalus L. J. Fish. Dis. 4, 297–307.

Rungratchatchawal, N., 1999 Acute Toxicity of Neem Extract on Some Freshwater Fauna (MSc thesis). Kasetsart University, Bangkok.

Rodger, H.D., 1994. NRED aquaculture research programme studies on virus infections of food fish in the Indo-Pacific region. ODA Res. Proj. R5430.

Saha, Pal, 2002. Had Worked on Antibiotic Susceptibility of Bacteria Isolated from EUS-affected Fishes in India.

Saitanu, K., Poonsuk, K., 1983. Prophylactic and therapeutic studies of antimicrobial agents to the Aeromonas hydrophila infection of snakehead fish (Ophicephalus stiratus). In: Proc. Symp. on Freshwater Fishes Epidemic, June 23–24, 1983. Chulalongkorn University, Bangkok, pp. 461–514.

Saitanu, K., Wongsawang, S., Sunyascotcharee, B., Sahaphong, S., 1986. Snakehead fish virus isolation and pathogenicity studies. In: Maclean, J.L., Dizon, L.B., Hosillos, L.V. (Eds.), The First Asian Fisheries Forum. Asian Fisheries Society, Manila, pp. 327–330.

Sakai, M., 1999. Current research status of fish immuno stimulants. Aquaculture 172, 63–92.

Saikia, Kamilya, February 2012. Immune responses and protection in catla (Catla catla) vaccinated against epizootic ulcerative syndrome. Fish. Shellfish Immunol. 32 (2), 353–359. http://dx.doi.org/10.1016/j.fsi.2011.11.030. Epub 2011 Dec 6.

Singha, R., Kar, D., Riji, J.K., George, R.M., Mansoor, M., Selva, M., 2015. An account of ichthyospecies affected and limnological parameters of water bodies in connection with epizootic ulcerative syndrome in South Assam. Natl. J. Life Sci. (in press).

Singh, N.R., 2015. Studies on Helminth Parasites of the Fishes of Sone Beel, and River Jatinga in Assam and Pumlen Lake in Manipur (PhD thesis). University of Assam, Silchar.

Singha, R., Singh, M.S., Kar, D., 2014. Parasite Fauna in the Wetland Fishes of India, pp. 209. Today and Tomorrow's Printers and Publishers, New Delhi. ISBN: 81-7019-184-9 (India), ISBN: 1-55528-341-1 (USA).

Sangeeta, O., Shomorendra, M., Kar, D., 2014. Nematode parasites in relation to the length of two freshwater fishes of Oinam Lake, Manipur, India. Research frontiers in wetlands, fishery and aquaculture. pp. 221–228. In: Research Frontiers in Wetlands, Fishery and Aquaculture. xi + 320.

Sharma, P., Sihag, R.C., 2013. Pathogenicity test of bacterial and fungal fish pathogens in Cirrihinus mrigala infected with EUS Disease. Pak J. Biol. Sci. 16 (20), 1204–1207, 15.

Sinha, M.. (2010). Anti-microbial Screening of Some Medicinal Plants in Barak Valley Region of Assam (PhD thesis). Assam University, Silchar.

Singha, M., 2015. Epidemiology of Anti-microbial Resistant and Virulence Gene Positive Escherichia coli in Livestock and Poultry (PhD thesis). Assam University, Silchar.

Singh, G., Upadhyay, R.K., Narayanan, C.S., Padmkumari, K.P., Rao, G.P., 1993. Chemical and fungitoxic investigations on the essential oil of Citrus sinensis (L.) Pers. Z. Fuer Anzenkrankh. Panzenschutz 100, 69–74.

Singh, N.R., Das, B.K., Kar, D., 2015. Length-weight relationship (LWR) of Glossogobius giuris (Hamilton Buchanan, 1822) of Pumlen lake-Thoubal, Manipur, India. Int. J. Environ. Nat. Sci. 1, 1–15.

Stanghellini, M., 1997. Inert Components: are they really so? Phytoparasitica 25 (Suppl.), 81S–86S.

Shariff, M., Subasinghe, 1994. Experimental Induction of EUS; ODA Regional Seminar Report on EUS. AAHRI, Bangkok Thailand.

Shariff, M., Saidin, T.H., 1994. Status of epizootic ulcerative syndrome in Malaysia since 1986. In: Roberts, R.J., Campbell, B., MacRae, I.H. (Eds.), Proceedings of the ODA Regional Seminar on Epizootic Ulcerative Syndrome, 25-27 Jan 1994. Aquatic Animal Health Research Institute, Bangkok, p. 48.

Sammut, J., White, I., Melville, M.D., 1996. Acidification of an estuarine tributary in eastern Australia due to drainage of acid sulfate soils. Mar. Freshwater Res. 47 (5), 669–684.

Saylor, R.K., Miller, D.L., Vandersea, M.W., Bevelhimer, M.S., Schofield, P.J., Bennett, W.A., 2010. Epizootic ulcerative syndrome caused by Aphanomyces invadans in captive bullseye snakehead (Channa marulius) collected from south Florida, USA. Dis. Aquat. Org. 88, 169–175.

Siddique, M.A.B., Bashar, M.A., Hussain, M.A., Kibria, A.S.M., 2009. Fungal disease of freshwater fishes in Natore district of Bangladesh. J. Bangladesh Agril. Univ. 7 (1), 157–162.

Satyalatha, B.D.J., Vardhani, V., 2014. V2. SDS-PAGE analysis of protein in muscle and head kidney of Labeo rohita in response to aeromoniasis. Biolife 2 (2), 550–556. ISSN (online): 2320-4257. www.biolifejournal.com. Biolife, Vol 2, Issue 2.

Shariff, M., Law, A.T., 1980. An incidence of fish mortality in Bekok River, Johore, Malaysia. In: Proceedings of the International Symposium on Conservation Input from Life Sciences, October 27–30, 1980. Universiti Kebangsaan, Bangi, Selangor, Malaysia, pp. 153–162.

Shrestha, G.B., 1994. Status of epizootic ulcerative syndrome (EUS) and its effects on aquaculture in Nepal. In: Roberts, R.J., Campbell, B., MacRae, I.H. (Eds.), Proceedings of the ODA Regional Seminar on Epizootic Ulcerative Syndrome, January 25—27, 1994. Aquatic Animal Health Research Institute, Bangkok, pp. 49—57.

Shrestha, T.K., 1990. Resource Ecology of the Himalayan Waters. Curriculum Development Center, Tribhuvan University, Nepal, 61—109.

Snieszko, S.F., 1974. The effect of environmental stress on outbreaks of infectious diseases of fishes. J. Fish. Biol. 6 (2), 197—208.

Soe, U.M., 1990. Myanmar report. In: Phillips, M.J., Keddie, H.G. (Eds.), Regional Research Programme on Relationships between Epizootic Ulcerative Syndrome in Fish and the Environment. A Report on the Second Technical Workshop, August 13—26, 1990. Network of Aquaculture Centres in Asia-Pacific, Bangkok, pp. 35—38.

Subasinghe, G.K.N.S., 1983. Modeling of a Pinched Sluice Concentrator (Ph.D. thesis). University of Auckland, New Zealand.

Songe, M.M., Hang'ombe, M.B., Phiri, H., Mwase, M., Choongo, K., Van der Waal, B., Kanchanakhan, S., Reantaso, M.B., Subasinghe, R.P., 2011. Field observations of fish species susceptible to epizootic ulcerative syndrome in the Zambezi River basin in Sesheke District of Zambia. Trop. Anim. Health Prod. 44 (1), 179—183, 2012 Jan.

Sudhakaran, D.S., Srirekha, P., Devasree, L.D., Premsingh, S., Michael, R.D., 2006. Immunostimulatory effect of Tinospora cordifolia Miers leaf extract in Oreochromis mossambicus. Indian J. Exp. Biol. 44, 726—732.

Tachushong, A., Saitanu, K., 1983. The inactivation of Aeromonas hydrophila by Potassium permanganate. In: Proc. Symp. on Freshwater Fishes Epidemic, 1982—83, June 23—24, 1983. Chulalongkurn University, Bangkok, pp. 394—402.

Tamuli, K.K., Dutta, O.K., Dutta, A., 1995. Probability of use of Musa balbisiana Colla ash in fish farming. Environ. Ecol. 13, 652—654.

Tangtrongpiros, J., Charoennetisart, P., Wongsatayanont, B., Tavatsin, A., Chaisiri, N., 1985. Haematological change in snakehead fish during 1984 outbreak. In: Proceedings of the Technical Conference on Living Aquatic Resources, March 7—8, 1985. Chulalongkorn University, Bangkok (in Thai, English abstract).

Temeyavanich, S., 1983. Acute toxicity of iodine, chlorine and malachite green on three freshwater fishes and its treatment on snakehead fish pond. In: Proc. Symp on Freshwater Fishes Epidemic, 1982—83, June 23—24, 1983. Chulalongkorn University, Bangkok, pp. 415—427.

Tesprateed, T., Rattanaphani, R., Wongsawang, S., Wattanawijarn, W., Ousawaplangehai, L., Tangtrongpiros, J., Saitanu, K., 1983. Pathology of the diseased fish with reference to the recent epidemic in Thailand in 1982. In: Proc. Symp. on Freshwater Fishes Epidemic, 1982—83, June 23—24, 1983. Chulalongkorn University, Bangkok, pp. 255—276.

Tewary, P., 2004. Use of Immunostimulants in Aquaculture. In: Advances in Biochemistry and Biotechnology, vol. I. Daya Publishing House, New Delhi, pp. 183—194.

Thi van, P., Van Khoa, Le, Thi Lua, D., Van van, K., Ha, N.T., 2002. The impacts of red spot disease on small-scale aquaculture in northern vietnam. In: Arthur, J.R., Phillips, M.J., Subasinghe, R.P., Reantaso, M.B., MacRae, I.H. (Eds.), Primary Aquatic Animal Health Care in Rural, Small-scale Aquaculture Development. FAO Fish. Tech, pp. 165—176. Pap. No 406.

Tiffney, W.N., 1939. The host range of Saprolegnia parasitica. Mycologia 31, 310—321.

Tonguthai, K., 1985. A Preliminary Account of Ulcerative Fish Diseases in the Indo-Pacific Region (A Comprehensive Study Based on Thai Experiences). National Inland Fisheries Institute, Bangkok, p. 39.

Tripathi, A., 2008. Elicitation of EUS under laboratory conditions through experimental infection trails. J. Aquat. Biol. 24 (1), 220—233.

Tripathi, A., Qureshi, T.A., 2012. Immunological diagnosis of naturally EUS infected Channa gachu. Int. J. Sci. Res. (IJSR) 2319—7064.

Tomlinson, J.A., Faithfull, E.M., 1979. Effects of fungicides and surfactants on the zoospores of Olpidium brassicae. Ann. Appl. Biol. 93, 13—19.

Thompson, E., Lewis, C.M., Pegrum, G.D., 1973. Changes in antigenic nature of lymphocytes caused by common viruses. Br. Med. J. 4, 709—711.

Ulcerative Fish Disease Committee, 1983. Practical Report of the Ulcerative Fish Disease Committee 1982-1983 (Bangkok).

Valsa Judit Anto, A., Balasubramanian, V., 2014. Immunological studies of disease induced common carp Cyprinus carpio fed with neem extract added. Feed. Res. J. Rec. Sci. 3, 124—126.

Vishwanath, T.S., Mohan, C.V., Shankar, K.M., 1997. Mycotic granulomatosis and seasonality are the consistent features of epizootic ulcerative syndrome of fresh and brackishwater fishes of Karnataka, India. Asian Fish. Sci. 10, 155—160.

Vishwanath, T.S., Mohan, C.V., Shankar, K.M., 1997a. Mycotic granulomatosis and seasonality are the consistent features of epizootic ulcerative syndrome of fresh and brackishwater fishes of Karnataka, India. Asian Fish. Sci. 10, 155—160.

Viswanath, T.S., Mohan, C.V., Shankar, K.M., 1997b. Clinical and histopathological characterization of different types of lesions associated with epizootic ulcerative syndrome (EUS). J. Aquac. Trop. 12 (1), 35—42.

Vishwanath, T.S., Mohan, C.V., Shankar, K.M., 1998. Epizootic ulcerative syndrome associated with a fungal pathogen in Indian fishes-histopathology: a cause for invasiveness. Aquaculture 165, 1—9.

Volf, O., Steinhauer, B., 1997. Fungicidal activity of neemleaf extracts. Meded. Fac. Landbouwkd. En. Toegepaste Biol. Wet. Univ. Gent 62, 1027—1033.

Wattanavijarn, W., 1983. Virus-like particles in the skeletal muscle, capillaries and spleen of sick snakehead fish (Ophicephalus striatus) during a disease epidemic. Thai J. Vet. Med. 13 (2), 122—130.

Wattanavijarn, W., Tesaprateep, T., Wattanodorn, S., Sukolapong, V., Vetchangarun, S., Eongpakornkeaw, A., 1985. Virus-like particles in the liver snakehead fish. Thai J. Vet. Med. 13 (1), 51—57.

Widagdo, D., 1990. Indonesia report. In: Phillips, M.J., Keddie, H.G. (Eds.), Regional Research Programme on Relationships between Epizootic Ulcerative Syndrome in Fish and the Environment. A Report on the Second Technical Workshop, August 13—26, 1990. Network of Aquaculture Centres in Asia-Pacific, Bangkok.

Wikramanayake, E.D., Moylw, P.B., 1989. Ecological structure of tropical fish assemblages in wet-zone streams of Sri Lanka. J. Zool. Lond. 281, 503—526.

Williams, I.V., Amend, D.F., 1976. A natural epizootic of infectious haematopoietic necrosis in fry of sockeye salmon (Oncorhynchus nerka) at Chilko Lake, British Columbia. J. Fish. Res. Bd. Can. 33, 1564—1567.

Willoughby, L.G., 1968. Atlantic salmon disease fungus. Nature (London) 217, 872—873.

Willoughby, L.G., 1970. Mycological aspects of a disease of young perch in Windermere. J. Fish. Biol. 2, 113—116.

Willoughby, L.G., Roberts, R.J., Chinabut, S., 1995. *Aphanomyces invaderis* sp. nov., the fungal pathogen of freshwater tropical fishes affected by epizootic ulcerative syndrome (EUS). J. Fish Dis. 18, 273–275.

Wilson, H.C., 1909. Fischdale Farm, Nilgiris. Government Press, Madras, p. 10.

Wilson, K.D., Lo, K.S., 1992. Fish Disease in Hong Kong and the potential role of the Veterinarians. In: 23rd Annual Conference of the International Association for Aquatic Animal Medicine (IAAAM), May 18–22, 1992, Hong Kong.

Wolf, K., Darlington, R.W., 1971. Channel cat fish virus: a new herpes virus of ictalurid fish. J. Virol. 8, 525–533.

Wolf, K., Herman, R.L., Carlson, C.P., 1972. Fish viruses: histopathologic changes associated with experimental channel cat fish virus disease. J. Fish. Res. Bd. Can. 29, 149–150.

Wolf, K., Quimby, M.C., Pyle, E.A., Dexter, R.P., 1960. Preparations of monolayer cell cultures from tissues of some lower vertebrates. Science (N.Y.) 132, 1890–1891.

Wood, J.W., 1968. Diseases of Pacific Salmon: Their Prevention and Treatment. Department of Fisheries, Hatchery Division, State of Washington (USA), Olympia, Washington.

Yuasa, K., Hatai, K., 1995. Drug sensitivity of some pathogenic water moulds isolated from freshwater fishes. J. Antibact. Antifung. Agents 23, 213–219 (In Japanese, English abstract).

Yuasa, K., Hatai, K., 1996. Investigation of effective chemicals for treatment of saprolegniosis caused by *Saprolegnia parasitica*. J. Antibact. Antifung. Agents 24, 27–31 (In Japanese, English abstract).

Chapter 7

Histopathological, Hematological, Histochemical, and Enzymological Studies of Epizootic Ulcerative Syndrome

7.1 DETAILS OF CLINICAL SYMPTOMS AND HISTOPATHOLOGY

7.1.1 Histopathology (HP)

7.1.1.1 Introduction

Histology is that branch of biology that deals with the study of microanatomy or the internal organization of tissues. The knowledge of the discipline has been successfully employed in biological, medical, and veterinary sciences as a diagnostic tool since the days when the first cellular investigations were carried out during the mid-nineteenth century (Virchow, 1858). Considerable developments have taken place since then in all aspects of cellular biology, with the result that today a number of novel and sophisticated histological techniques are available to piscian histologists.

Notwithstanding the above, stringent attention must be paid to preparation before any satisfactory histological sections can be produced. In this sense, it should be noted that the very fast rate of autolysis of piscian tissues compared with those of homotherms could mean that they are processed rapidly in order to prevent degenerative changes within the specimen, which otherwise could make final diagnosis either unreliable or impossible.

Concomitantly, it should be emphasized that only freshly killed or moribund fishes may be considered for HP studies. However, careful postcapture treatment is essential for most of the external lesions, mainly because of the ease with which the epidermis of the teleosts are generally abraded.

For optimal preparation, the fish may be taken out of water by means of hooks or fine-meshed nets and be quickly transferred to a container of suitable size and containing a general anesthesia, or be decapitated. Subsequently, the fish specimen is handled with a pair of forceps, or if too large, by the tail or fins, assuming that these areas are not under investigation. The lesions can be excised, or if preservation of the whole specimen is planned, a number of deep cuts be given along the length of the body wall from the nostrils to the tail, made in parallel to allow immediate access of fixative. A little caution at this stage will be well rewarded in the quality of information ultimately gained.

On the other hand, in the event of lesions that are internal, or if the whole fish is to be preserved, it is important that the whole length of the body cavity is opened, usually by slitting along the midventral line. The viscera and swim bladder are to be carefully displaced and each organ incised at least once in order to allow maximum penetration of the fixative. Ideally, the organ of the lesion under investigation is carefully dissected from the body, cut into blocks of <1.0 cm^3, and placed in a volume of fixative that is at least 20 times the volume of the tissue (Roberts, 1978).

A bewildering array of tissue fixatives, each with its own special advantages and disadvantages, confronts the ichthyologist. In this regard, it should be noted that proper fixation is a prerequisite to satisfactory histological preparation, and hence its importance may not be overemphasized. If fixation is not satisfactory, the end product could be a direct reflection of this. The primary objective of fixation is to preserve the morphology of the tissue in a condition that is as near as possible to that existing during life. This could presuppose inhibition of both autolysis (the self-destruction of tissues by the intracellular enzymes that are said to be released from their normal membrane-bound site after death and putrefaction) and putrefaction (the effects of bacterial degradation of the tissue).

HP study is an indispensable tool in unraveling the etiology of the hitherto unknown, virulent, and enigmatic epizootic ulcerative syndrome (EUS) fish disease. EUS was first reported in March 1972 from central Queensland, Australia, where several species of estuarine fish had developed large, shallow, and circular or irregular skin lesions. Initially named "Bundaberg fish disease," it displayed a pronounced seasonality and was soon associated with prolonged periods of rain which were thought to alter the quality of water and make it prone to infection by bacteria. As it spread to several species of freshwater fishes in the river systems of Papua New Guinea and Western Australia, the disease soon came to be called "red spot."

Epizootic Ulcerative Fish Disease Syndrome. http://dx.doi.org/10.1016/B978-0-12-802504-8.00007-9

In 1980, a similar hemorrhagic condition was seen among the fishes, including the rice-field fishes in Java, Indonesia. But pathological and epidemiological differences seemed to set this apart from the Australian condition. However, subsequent outbreaks of EUS in brackish-water fish in the Philippines and typically ulcerated snakeheads and catfish in other states of Indonesia had confirmed the link with the Australian red spot.

In 1986, as indicated earlier, a Consultation of Experts on Ulcerative Fish Diseases, organized by the Food and Agriculture Organization (FAO), adopted the name "epizootic ulcerative syndrome" specifically to refer to the Asian condition. The disease is characterized by large cutaneous ulcerative lesions known to cause death in many species of wild and cultured freshwater fishes periodically. In fact, cutaneous ulcerative diseases are common among wild and cultured fish, for the last two decades. Different regions in Australia and Asia—Pacific have been witness to a group of epizootic syndromes, all believed to be involving severe UM.

The first report of classic EUS came from peninsular SE Asia during 1979—1980 from the Bekok River system of Malaysia, and the next year from its northern rice-growing states, where freshwater rice-field species of fishes succumbed to serious ulcerations. In the course of the decades since then, the disease has spread to almost every part of South and SE Asia, specifically Thailand, Lao PDR, Myanmar, Vietnam, Cambodia, Bangladesh, India, and Sri Lanka. In its westward spread from Australia, the occurrence of EUS has been reported from Kerala, Gujarat, and Rajasthan in India. Also, in many of Asia's paddy field systems, EUS has been occurring toward the end of the paddy cultivation period when the water level is usually low, decomposition of organic matter like grass and waterweeds is common, and certain types of fertilizers are believed to accumulate.

7.1.2 Clinical Symptoms

EUS is characterized by the occurrence of large hemorrhagic or necrotic ulcerative lesions on the bases of the fins and other parts of the body that becomes larger inflamed areas with acute degeneration of epidermal tissues (Kar and Dey, 1988, 1990a,b,c).

7.1.3 Background Histopathology

1. In early stages, small hemorrhagic lesions on the skin surface (epidermis, dermis, and hypodermis) may begin, but may not affect the underlying muscle.
2. Early lesions continue to show mild epithelial necrosis, surrounding edema, hemorrhage of underlying dermis, and inflammatory cell infiltration, but accompanied with severe necrotizing myopathy, although only a

few fungal hyphae enclosed in epithelioid capsules are apparent. There is possibly no disruption of the internal organs.
3. Advanced lesions show large bacterial ulcerations and massive necrotizing granulomatous mycosis of underlying muscle fibers involving a distinctive branching aseptate oomycete coated with epithelioid cells. Hyphae may invade abdominal viscera and penetrate renal tubules and glomeruli, causing death. Usually, there may be only mild generalized HP changes in other organs.

Concomitant to the above, routinely, EUS is diagnosed by HP of multiple sampling sites demonstrating mycotic granuloma and by isolation of the fungus for study of hyphal characters and sporulation. But it is laborious. As such, Rabbit antisera-based detection and polymerase chain reaction (PCR) techniques have been developed for EUS. But these techniques may also have their own disadvantages, although they are usually cost-effective, rapid, and specific. In view of this, specific, rapid, and cost-effective immunodot test, using monoclonal antibody (MAb), was developed. However, it was necessary that the pathogen be detected at an early stage of infection, so as to avoid the spread of disease and further loss. Hence, a study was conducted to determine a specific sampling site in order to detect the pathogen at a most possible early stage of infection in fish. The technique, thus developed, is believed to have immense applications in aquatic animal health management.

In addition to the above, MAbs against *Aphanomyces invadans* have been developed of late. The strain B 99C had been given by JH Lilley. The MAbs (using an immunoperoxidase test) are said to have reacted consistently (1) with the fungal hyphae in HP sections of experimentally infected fish; (2) with the EUS-affected fish collected from various parts of India during 2002; and (3) with the retrospective EUS tissue samples of 1995 and 1997. Such a result is said to support the view that a single mycotic pathogen had perhaps been involved in all the EUS-affected cases.

Moreover, observations on the behavior of the hyphae at different temperatures had also been made. At 20 °C, the hyphae tended to invade deeper into the fish tissues, while at temperatures between 25 and 30 °C, the hyphae had grown actively outside the body surface around the lesions.

7.1.4 HP Works on EUS by Different Workers

EUS is one of the most destructive diseases affecting fresh and brackish-water fishes (cultured and wild) in the Asia—Pacific region, resulting in dwindled fish production and colossal economic loss. EUS is believed to be caused by the fungal species, *A. invadans*. The organism is believed to require a specific combination of factors in order to germinate within the dermis of the fish. The disease causes lesions in both the skin and the visceral organs. EUS generally occurs during the periods of **low temperature and heavy rainfall in tropical and subtropical waters**. These conditions are said to favor sporulation, and low temperatures are believed to **delay the inflammatory response** of the fish to infection.

Subsequently and concomitantly, more emphasis had been laid on the expansion of freshwater aquaculture (AQC) in many sectors by different governments through utilization of the vast inland aquatic resources available worldwide. This possibly drew more attention to specific farming sectors such as AQC. This could certainly help to substantially alleviate the pressure on agriculture sector. However, an intensive inland AQC system essentially demands the strict implementation of good management practices (GMPs). In such systems of AQC, however, new problems could inadvertently emerge when GMPs are not followed diligently. Anticipating such crises, advanced disease diagnostic techniques that encompass innovative preventive and curative methods need to be continuously developed to effectively combat the same.

According to Anon (1997), histological occurrences include necrotizing, granulomatous dermatitis, and myositis associated with invasive, nonseptate fungal hyphae, 10−20 μm in diameter. The fungus may penetrate visceral organs, such as the kidney and liver, after it has penetrated the musculature. HP studies had also portrayed swelling of gill filaments and edema in the muscles of the fish. In addition, Anon (1997) brought out excerpts concerning EUS from the **OIE Manual**.

Notwithstanding the above, Vogelbein et al. (2001) dealt with skin ulcers in estuarine fishes and made a comparative pathological evaluation of wild and laboratory-exposed fish (LEF). Field-collected menhaden and moribund LEF tilapia were killed by overdose with tricaine methanesulfonate. Grossly visible skin lesions on wild menhaden, and the skin, gills, liver, spleen, pancreas, gut, kidney, and brain and other neural tissues from LEF tilapia were processed by routine methods for paraffin histology (20). Tissues were fixed in 10% neutral buffered formalin for 48 h (hour), washed overnight in running water, dehydrated, infiltrated and embedded in paraffin, sectioned on a rotary microtome at 5 μm, mounted on slides and stained with hematoxylin and eosin (H&E) or with Grocott's methenamine silver stain for fungal hyphae (20).

Concomitant to the above, the toxic dinoflagellate, *Pfiesteria piscicida* (Steidinger and Burkholder) had subsequently been implicated as the etiologic agent of acute mass mortalities and skin ulcers in menhaden, *Brevoortia tyrannus*, and other fishes from mid-Atlantic US estuaries. Evidence for this association, however, was perhaps largely circumstantial and controversial. The tilapia (*Oreochromis*

spp.) had been exposed to *Pfiesteria shumwayae* (Glasgow and Burkholder) and the resulting pathology had been compared to the so-called *Pfiesteria*-specific lesions occurring in wild menhaden. The tilapia that were challenged with high concentrations (2000−12,000 cells/ml) of *P. shumwayae* exhibited the loss of mucus coat and scales plus mild petechial hemorrhage, but no deeply penetrating chronic ulcers had been found like those in wild menhaden.

Histologically, the fish had exhibited epidermal erosion with bacterial colonization but minimal associated inflammation. In moribund fish, loss of epidermis was widespread over large portions of the body. Similar erosion had occurred in the mucosa lining of the oral and branchial cavities. The gills exhibited epithelial lifting, loss of secondary lamellar structure, and infiltration by lymphoid cells. Epithelial lining of the lateral line canal (LLC) and olfactory organs had exhibited severe necrosis. Visceral organs, kidney, and neural tissues (brain, spinal cord, ganglia, and peripheral nerves) were, however, said to be histologically normal.

Further, in connection with agent detection and identification methods by HP, A. invadans is said to elicit a strong inflammatory response, as well as granulomas are said to be formed around the penetrating hyphae. **This could be followed by** isolation of A. invadans from internal tissues, and identification of pure A. invadans isolates by PCR.

In addition to the above works, Anon (1992) held "Consultation on Epizootic Ulcerative Syndrome vis-à-vis the Environment and the People" under the aegis of the International Collective in Support of Fishworkers. As an invited speaker at the Symposium, Kamonporn Tonguthai of the Aquatic Animal Health Research Institute, Kasetsart University, Bangkok, had opined that in the Asia−Pacific region, there had been several reports of ulcerative disease conditions among wild and cultured fish. While EUS refers specifically to the Asian condition, there are great similarities with other fish disease conditions. However, Tonguthai cautioned, only further research could confirm whether these are indeed the same disease, or not. Yet, as Tonguthai had pointed out, ultimately no definite conclusions about the cause of the disease could be drawn as yet, since outbreaks are considered to be a complication of several factors.

In addition to the above, Anon (1998) had studied the prevalence and geographical distribution of EUS fish disease in the susceptible fish populations in and around the Rivers Indus, Chenab, Jhelum, Ravi, and Sutlej in the Punjab province of Pakistan. It was believed that the sample size could permit the prevalence of EUS to be calculated with 95% confidence limit (CL). Further, fishes had been sampled during the harvesting time from the ponds as well as from the catchment areas, which are adjacent to 10 strategically selected dams and barrages on the rivers in

Punjab in order to have wider coverage in epidemiological and HP studies. Tissue samples had been collected from around the lesions immediately after the selection of the specimen, and the tissues had been kept in 10% formalin for HP studies. Further, as indicated above, Gayathri et al. (2008) worked on determining a specific sampling site on EUS-affected fish for the diagnosis of *A. invadans* by immunodot test using monoclonal antibodies.

Notwithstanding the above, Miyazaki and Egusa (1973b,c) had further reported about **HP** observations in other species such as bluegill (*Lepomis macrochirus*) and wild fish such as snakefish (*Channa argus*) from Chiba Prefecture; gray mullet *(Mugil cephalus)* from Kojima Bay, Okayama Prefecture, crucian carp (*Carassius auratus*), and trident goby (*Tridentiger obscurus*) from Lake Kasumigaura, Ibaragi Prefecture. Further, granulomatous inflammation was observed in all the five species, consistent with earlier observations on goldfish and ayu. The general pathological features observed were dermatitis and myositis associated with deep penetrating granulomatous inflammation characterized by several layers of epithelioid cells surrounding the fungal hyphae. In some cases, multinucleate giant cells with or without fungal hyphae were detected, depending on species and geographic location. The disease was named "mycotic granulomatosis" (MG) (Miyazaki and Egusa, 1972).

Concomitant to the above, HP observations made from both naturally infected fish (Egusa and Masuda, 1971; Miyazaki and Egusa, 1973a,b,c; Wada et al., 1994; Hatai et al., 1994; Hanjanvanit et al., 1997) and artificially infected fish (Hatai et al., 1977, 1994; Hatai, 1980; Rha et al., 1996; Wada et al., 1996; Bondad-Reantaso et al., 1999) showed similar pathological changes. Many branched aseptate hyphae were observed in the lesions. These hyphae were typically associated with a granulomatous response and extensive tissue necrosis. Marked inflammatory infiltration and formation of granulation tissue had also been consistently observed in older lesions. Granulation tissue (a reparative response not to be confused with granulomas) composed of a fibrous network, regenerating capillaries and muscle fibers, mild hemorrhage, and extensive inflammatory infiltration had developed to replace necrotic areas.

Further, Baidya and Prasad (2013) had reported about the prevalence of EUS among the carps in Nepal. As with many workers, Baidya and Prasad (2013) had also opined that EUS disease could be confirmed by histological diagnosis. The early skin lesions of some samples had been observed and found to be principally in the areas of epithelial necrosis with surrounding edema, hemorrhaging of the underlying dermis, and some inflammatory cell infiltration. The epidermis at the margins of the ulcer itself was found to be degenerated and thickened due to the enclosing of a very small number of fungal hyphae within an epithelioid capsule. In advanced lesions, there is massive necrotizing granulomatous mycosis of the underlying muscle fibers.

Concomitantly, FAO (2009) had dealt with HP of EUS-infected dashtail barb (Botswana) and had found typical mycotic granulomas surrounding the invasive fungal hyphae in the skin layer (by the use of Grocott's silver stain). Typical severe mycotic granulomas from muscle section of EUS fish (barb from Namibia) had also been observed (by the use of H&E stains).

In a continuation of the ongoing documentation of works on EUS, Lilley et al. (1998) had given a general description of the typical HP developments that had occurred in EUS-diseased fish. The early skin lesions of some samples had been observed and found to be principally in the areas of epithelial necrosis with surrounding edema, hemorrhaging of the underlying dermis, and some inflammatory cell infiltration. It had not been possible to confirm fungal involvement in most of these early samples, but a few, however, had harbored a small number of hyphae. The presence of fungal hyphae was demonstrated in the epidermis of some early stages of infected fish from India (Vishwanath et al., 1997). Similarly, Roberts et al. (1989) were able to study early lesions during an EUS outbreak in a captive population of IMCs. They observed that an acute necrotizing myopathy, more severe than usually seen in wild fishes, had spread over a wide area below the active skin lesion. The epidermis at the margins of the ulcer itself was degenerated and thickened, and contained only a very small number of fungal hyphae enclosed within an epithelioid capsule. The blood vessels of the dermis were very hyperemic and some had a collar of lymphoid or myeloid cells that might have been associated with virus infection although no viral inclusion bodies were detected.

Subsequent pathological developments in all infected fish species had involved significant degenerative changes in skin and muscle tissues with minimal disruption of internal organs. In advanced lesions, there was massive necrotizing granulomatous mycosis of the underlying muscle fibers, involving the distinctive branching aseptate, invasive fungal mycelia. Large number of bacteria might have been present on the surface of some advanced lesions. With advancing age of the lesion, fungal cells had been reported to have become progressively enveloped by thick sheaths of host epithelioid cells, and some areas might have shown evidence of myophagia and healing. In some advanced lesions, fungal hyphae could be seen invading the abdominal viscera, which could almost certainly be the ultimate cause of death. A large number of mycotic granulomas had been demonstrated in the kidney, liver, and gut of several fishes including spiny eels, *Cirrhinus mrigala, Colisa lalia, Channa* sp., *Puntius* sp., *Esomus* sp., *Mugil* sp., *Valamugil* sp., *Therapon* sp., *Glossogobius* sp., and *Sillago* sp. (Chinabut, 1990; Ahmed and Hoque, 1998;

Wada et al., 1994; Vishwanath et al., 1998). Wada et al. (1994) had also found mycotic granulomas in the abdominal adipose tissue, pancreas, gonad, spleen, central nervous system, and heart of dwarf gourami, and Vishwanath et al. (1998) had further demonstrated fungi penetrating the esophagus and spinal cord of mullet and intermuscular bones of *Puntius*.

Concomitant to the above, the internal organs of diseased fish, other than those invaded by fungal hyphae, had depicted only mild HP changes that Roberts et al. (1986) had pointed out that it might sometimes be the result of background pathology. Palisoc (1990) had observed minimal tissue disruption in the kidney of striped snake-heads in terms of an increased number of melanomacro-phage centers, hemosiderin pigments, and few mitotic figures. Spleen sections had shown a marked increase of white pulp production and the heart, liver, and gills underwent mild HP changes, but none were observed in the stomach and intestine. Other kidney samples had shown tubular, vacuolar degeneration with granular occlusion and hematopoietic tissue degeneration or focal proliferation. Chinabut (1990) had demonstrated that these features were consistently more severe in armed spiny eels. Further, pancreas samples had occasionally shown acinar necrosis (Callinan et al., 1989) with eosinophil and inflammatory cell infiltration. In the liver, mild focal hepatic cellular degen-eration might sometimes also occur in advance of bacterial/fungal involvement. The only consistent hematological change in diseased fish was a significantly lower level of hemoglobin (Hb) as a result of extra- and intravascular destruction of red blood cells (Tangtrongpiros et al., 1990).

Concomitant to the above, histology of early lesions had revealed **acute spongiosis and epithelial cell loss**. Degenerative changes progress through the **dermis with hyperemia, hemorrhages, and inflammatory infiltra-tion**. In advanced stages, **sarcolysis** was also obvious. Fungal hyphae had been enclosed by a **well-defined epithelioid cell layer** and mycotic granulation had spread through the infected muscles and internal organs. **Muscle fibers eventually disappeared altogether and were replaced by fibrosis, inflammatory cells, and new blood vessels**. These distinct features of EUS ulcers had been known to make histological analysis enough for a definitive diagnosis.

In addition, squash preparations of skeletal muscle from beneath an ulcer could also demonstrate septate fungal hyphae for a rapid provisional diagnosis.

Concomitant to the above, Callinan et al. (1989) had done preliminary comparative studies on the *Aphanomyces* species associated with EUS in the Philippines and RSD in Australia. Callinan et al. (1989) had, further, revealed that HP examination of ulcer sections from 21 fish specimens had shown lesions typical of EUS and indistinguishable from those of RSD. In this connection, it may be pointed

out here that identification of *Aphanomyces* spp. was based on differential morphology of oogonia and antheridia (Scott, 1961). None of these structures were observed in their study. Also, none had been observed in RSD isolates by Fraser et al. (1992). Even, none had been described for Thai EUS isolates by Roberts et al. (1993). In addition, total protein electrophoresis had provided useful, but not definitive, information concerning relationships within and between species of oomycete fungi (Bielenin et al., 1988; Chen et al., 1991). In order to examine possible relation-ships between the isolates, peptide extracts from four representative RSD *Aphanomyces* isolates; three represen-tative EUS *Aphanomyces* isolates, as well as an *Aphano-myces laevis*-type isolate, an *Aphanomyces cochloides*-type isolate, and an *Aphanomyces euteiches*-type isolate had been compared electrophoretically. Fungal mats were derived from GY broth (Dykstra et al., 1986). Cultures had been incubated at 30 °C in the dark for 15 days. Mats were then rinsed in distilled water, dried, and stored at −80 °C. Prior to extraction, frozen fungal mats were placed in liquid nitrogen and ground to a powder with mortar and pestle. 100 mg aliquots of fungi were then extracted by boiling for 5 min in 500 ml of reducing mixture (2% sodium dodecyl sulfate, 5% 2-mercapto- ethanol, 10% glycerol in 62.5 mM Tris—HCl at pH 6.8). The mixture was clarified at 5000 g and a known volume of the supernatant was loaded into 12% acrylamide and electrophoresed under reducing conditions according to Laemmli. Peptides resolved by SDS-PAGE were visualized by silver staining.

Band similarities or differences between isolates were assessed based on band clusters over specific molecular weight class ranges. On this basis, the banding profiles of the isolates from striped snakehead (Bautista), striped snakehead (Laguna), sea mullet (Queensland) and yellow-fin bream (Clarence) were very similar over the 14—94 kDa range, indicating a close degree of relatedness between these Philippine types.

Notwithstanding the above, Anon (2011) had dealt with EUS (which is also known as the RSD, MG, and UM), while preparing **Identification Field Guide** for **Aq Animal Diseases significant to Australia**. It had been reported that the microscopic pathological signs were: (1) erythematous dermatitis; (2) hyphae (threads) associated with granulo-matous tissue, sometimes extending into visceral organs; and (3) liquefactive necrosis of muscular tissue.

Baruah et al. (2012) had worked on interspecific transmission of the EUS pathogen, *A. invadans*, and asso-ciated physiological responses. They had worked on different aspects of EUS.

Vogelbein et al. (2001), as briefed earlier, had dealt with the skin ulcers in estuarine fishes and had made a comparative pathological evaluation of wild and LEF. **In an attempt to reiterate their works in a detailed way**, the toxic dinoflagellate, *P. piscicida* (Steidinger and

Burkholder) had of late been implicated as the etiologic agent of acute mass mortalities and skin ulcers in menhaden, *B. tyrannus*, and other fishes from mid-Atlantic US estuaries. However, evidence for this association had been largely circumstantial and controversial. In fact, tilapia (*Oreochromis* spp.) had been exposed to *P. shumwayae* (Glasgow and Burkholder) and the resulting pathology was compared to the so-called *Pfiesteria*-specific lesions occurring in wild menhaden. However, the identification had been based mainly on SEM studies and molecular analyses. The tilapia challenged by high concentrations (2000–12,000 cells/ml) of *P. shumwayae* had depicted loss of mucus coat and scales plus mild petechial hemorrhage, but there had been no deeply penetrating chronic ulcers like those in wild menhaden.

Histologically, the fish had exhibited epidermal erosion with bacterial colonization but minimal associated inflammation. In moribund fish, loss of epidermis was widespread over large portions of the body. Similar erosion had occurred in the mucosa lining the oral and branchial cavities. Gills had exhibited epithelial lifting, loss of secondary lamellar structure, and infiltration by lymphoid cells. Moreover, the epithelial lining of the LLC and olfactory organs had exhibited severe necrosis. Visceral organs, kidneys, and neural tissues (brain, spinal cord, ganglia, and peripheral nerves) have been said to be histologically normal. A so-called significant result was the innumerable *P. shumwayae* cells adhering to damaged skin, skin folds, scale pockets, LLC, and olfactory tissues. In contrast, histological evaluation of skin ulcers in >200 wild menhaden from Virginia and Maryland portions of the Chesapeake Bay and the Pamlico Estuary, North Carolina, had revealed that all the ulcers had harbored a deeply invasive, highly pathogenic fungus now known as *A. invadans*. In menhaden, the infection had always elicited severe myonecrosis and intense granulomatous myositis. The consistent occurrence of this fungus and the nature and severity of the resulting inflammatory response, had indicated that these ulcers were chronic (age >1 week) and of an infectious etiology. They might not be the direct result of an acute toxicosis initiated by *Pfiesteria* toxins. The disease was therefore best called ulcerative mycosis (UM). This study had indicated that the pathology of *Pfiesteria* laboratory exposure was considered by some as basically different from that of UM in menhaden. However, one may not rule out *Pfiesteria* as one of many possible early initiators, predisposing wild fishes to fungal infection under certain circumstances.

Later, the fungi associated with the menhaden ulcers in North Carolina estuaries were hypothesized to be secondary, opportunistic invaders with the lesions caused, in some instances, by exotoxins of *P. piscicida* (Steidinger and Burkholder). This hypothesis was based on the observed co-occurrence of the ulcerous lesions in wild menhaden and

P. piscicida in some acute fish kill events in North Carolina estuaries. A laboratory exposure study had provided some support for these field observations. Striped bass (*Morone saxatilis*) and tilapia (*Oreochromis* spp.) exposed to sublethal concentrations of *P. Piscicida* had exhibited initial widespread loss of epidermis, followed by ulcer development in a few fish. This had suggested a possible early initiating role for *P. piscicida* in menhaden ulcer development. Similar cutaneous ulcerative syndromes have been reported in >100 species of wild and cultured freshwater and estuarine fishes in the Indo-Pacific region since the 1970s. Outbreaks had been referred to, respectively, as EUS in Asia, RSD in Australia, and MG in Japan. This disease had been devastating to several major commercial fisheries, both wild and farmed, and is now collectively referred to as EUS. *Aphanomyces invadans*, an oomycete fungus, was recently implicated as the etiologic agent of EUS. Grossly and histologically, EUS in Indo-Pacific fishes and UM in menhaden from mid-Atlantic US estuaries are essentially identical. Further, an investigation indicated that the fungal agent observed in menhaden UM was also *A. invadans*. Earlier works had indicated that this fungal agent might have been a primary pathogen in menhaden, which was said to be capable of inducing formation of skin ulcers, with other environmental or biological factors, including *Pfiesteria* spp., playing possibly no role whatsoever. The roles supposed to have been played by *Pfiesteria* spp. in the development of skin ulcers in wild menhaden and other fishes had, therefore, possibly remained inconclusive. In spite of this, the menhaden lesions are said to be used in conjunction with presumptive counts of *Pfiesteria*-like cells in water samples and toxic fish bioassays as primary criteria for decisions about human health impacts and river closures in Maryland, Virginia, and North Carolina. As an attempt to better understand the association between *Pfiesteria* spp. and fish lesions, field-collected menhaden and moribund laboratory-exposed tilapia were killed by overdose with tricaine methanesulfonate. Grossly visible skin lesions on wild menhaden and the skin, gills, liver, spleen, pancreas, gut, and kidney, as well as the brain and other neural tissues from laboratory-exposed tilapia were processed by routine methods for paraffin histology as mentioned earlier. It should be noted here that Tilapia exposed to *P. shumwayae* have consistently exhibited widespread partial to complete loss of the epidermal layer of the skin. In moribund fish, the epidermis was completely eroded over much of the body surface. Bacterial colonization of the dermis was common, and *Pfiesteria* cells were frequently seen adhering to the surface. However, inflammation associated with these alterations was said to be minimal. Further, epidermal erosion was often associated with the presence of *Pfiesteria* cells in the skin folds. Similar pathological effects were observed in the oral cavity, which had been normally lined by a multilayered mucosa composed of squamous epithelium and goblet cells. Exposed fish had often exhibited complete mucosal erosion. *Pfiesteria* cells and bacterial contaminants had often been seen adhering to exposed connective tissues within the oral cavity. Commonly observed changes in the gill tissues had included separation and lifting of the respiratory epithelium, accumulation of lymphocytes in secondary lamellae, and gathering of *Pfiesteria* cells within the branchial chamber. Degenerative alterations had also occurred within the LLC system of the head. The epithelial lining of the LLC was eroded and often completely lost, and had been colonized with bacterial contaminants and *Pfiesteria* cells. Similar degenerative changes had also been observed within the nares and olfactory organs. Lesions attributable to *Pfiesteria* exposure were not observed in internal organs or tissues including liver, spleen, kidney, pancreas, gut, brain, spinal cord, and peripheral nervous tissues.

Pradhan et al. (2014) reported about the emergence of EUS and the large-scale mortalities of cultured and wild fish species in Uttar Pradesh, India. HP analysis was carried out according to the method described by Chinabut and Roberts (1999). The tissue sections were stained with hematoxylin and eosin (H&E) for general pathological investigation, and with Grocott's methenamine silver nitrate for the presence of *A. invadans* hyphae. **Gross pathology** of the affected fishes had varied from red hemorrhagic spots and scale losses to severe ulceration and exposure of underlying tissues. In histological examination, mycotic granulomas were demonstrated in the fishes showing gross lesions. In many of the affected fishes, below the dermal ulcers, the underlying musculature was largely replaced by mycotic granulomas. Sarcolysis and myophagia were also observed at sites far away from the ulcers. In some of the severely affected fishes, the hyphae had penetrated deep into the kidney across peritoneum from the body cavity. However, in fishes without noticeable hemorrhages and ulcerations, the hyphae-associated inflammatory changes were observed only in the epidermis, and the dermis and skeletal musculature were free from pathological lesions. The response was observed in large-sized IMCs.

7.2 DETAILS OF HEMATOLOGICAL, HISTOCHEMICAL, AND ENZYMOLOGICAL STUDIES

The hematological parameters had been estimated using standard procedures and had included determination of total red blood cell (RBC) count, total white blood cell (WBC) count, packed cell volume, and hemoglobin (Hb) content for experimentally infected (EI) and apparently healthy (AH) *Catla*. HP examination, done by them, had

revealed inflammatory responses against the invading organisms, often the formation of granulomas (Chinabut et al., 1995; Wada et al., 1996; Thompson et al., 1999; Saylor et al., 2010). This inflammatory response is said to have been governed by the innate immune defenses of the fish where granulocytes had been thought to play an important role. However, the mechanism by which the granulocytes are known to participate in the inflammatory process against invading oomycetes had not possibly been well understood. Immune functions of the granulocytes such as the reactive oxygen and nitrogen species and lysozyme production had been known to play an important role in eliminating pathogens, especially bacteria (Secombes, 1990). The increase in these parameters in the EUS-infected *Catla* indicated the possible involvement of these mechanisms in inhibiting the replication of *A. invadans*. However, the increased responses that generally occurred at an early stage of any infection might not be effective enough to prevent the spread of infection in later stages. The apparent ability of *A. invadans* to withstand such responses might explain the susceptibility of *Catla* to EUS infection. The EI *Catla* had portrayed increased WBC proliferation in response to the mitogen, concanavalin A; thereby, indicating that leucocytes are polyclonally activated in EUS-infected fish. An enhanced proliferative response was also observed in Atlantic menhaden with ulcer disease syndrome. The possible impact of WBC activation and proliferation on the development of granulomatous reactions associated with the disease had remained to be elucidated in a detailed way.

In addition to the above, the EUS-infected *Catla* were anemic, as evidenced by a decreased total RBC count, packed cell volume, and Hb content. Further, anemia might have been due to hemodilution caused by a loss of body fluid from hemorrhagic/necrotic lesions in the infected *Catla*, as had also been observed in EUS-affected Asian cichlid fish, *Etroplus suratensis* (Pathiratne and Rajapakshe, 1998). Decreased Hb and hematocrit levels in EUS-affected *Channa striatus* had also been attributed to hemodilution (Cruz-Lacierda and Shariff, 1995). The increased total leukocyte counts in EUS-infected *Catla* might be responsible for the enhanced immunological responses. In fact, the WBC count was higher and the Hb content was lower in blood from EUS-affected Indian fish than in healthy fish (Kar et al., 2000).

Concomitant to the above, levels of total plasma protein, serum glutamate oxaloacetate transaminase (SGOT), serum glutamate pyruvate transaminase (SGPT), and serum alkaline phosphatase (ALP) were higher in the EI groups than in the AH fish. One of the reasons for increased total protein concentration in serum might be the formation of anti-*A. invadans* antibodies in the serum. In fact, elevated levels of antibodies to *A. invadans* had been reported from striped snakehead (*Channa striata*;

184 Epizootic Ulcerative Fish Disease Syndrome

Thompson et al., 1997; Miles et al., 2001) and rainbow trout (*Oncorhynchus mykiss*; Thompson et al., 1999) and high values of SGOT, SGPT, and ALP were reported from EUS-affected fish (Kar et al., 2000). Generally, the enzymes SGOT and SGPT were found in different tissues, with a higher concentration in the liver (Nyblom et al., 2006). Increased SGOT, SGPT, and ALP might be due to hepatocyte damage by the *A. invadans* hyphae. The changes in these enzymes were less in fish infected by cohabitation than in intramuscularly injected fish; perhaps, because motile zoospores being directly injected into the muscles where zoospore germination and proliferation were rapid, might be leading to heightened tissue destruction and spread of the organism.

REFERENCES

Anon, 1992. Notes on mycological procedure. In: Workshop on Mycological Aspects of Fish and Shellfish Disease, Bangkok, Thailand, January 1992. Aquatic Animal Health Research Institute.

Anon, 1997. Epizootic Ulcerative Syndrome : OIE Diagnostic Manual for Aquatic Animal Diseases. Source: Chapter 3.1.5, second ed.

Anon, April 1998. Epizooic Ulcerative Syndrome (EUS) in Fishes of Pakistan: A Report Regarding Visit of Mission to Punjab in Connection with Epizootic Ulcerative Syndrome (EUS), pp. 22—25.

Ahmed, G.U., Hoque, A., 1998. Mycotic involvement in epizootic ulcerative syndrome of freshwater fishes of Bangladesh: a histopathological study. In: Proceedings of the Fifth Asian Fisheries Forum, Chiang Mai, Thailand. Asian Fisheries Society, Manila, pp. 11—14.

Anon, 2011. World Animal Health Information Database Interface: Exceptional Epidemiological Events. World Organisation for Animal Health, Paris viewed in 2011.

Bielenin, A., Jeffers, S.N., Wilcox, W.F., Jones, A.L., 1988. Separation by protein electrophoresis of six species of *Phytophthora* associated with deciduous fruit crops. Phytopathology 78, 1402—1408.

Bondad-Reantaso, M.G., Hatai, K., Kurata, O., 1999. Aphanomyces from EUS-infected fish in the Philippines and Bangladesh and MG-infected fish from Japan: II. Pathogenicity studies. In: Book of Abstracts, Fourth Symposium on Diseases in Asian Aquaculture, November 22—26, 1999, Cebu City, Philippines. Fish Health Section, Asian Fisheries Society, p. 36.

Baruah, A., Saha, R.K., Kamilya, D., 2012. Inter-species transmission of the epizootic ulcerative syndrome (EUS) pathogen, *Aphanomyces invadans*, and associated physiological responses. Israeli J. Aquac. — Bamidgeh 696, 9.

Baidya, S., Prasad, A., 2013. Prevalence of epizootic ulcerative syndrome (EUS) in carps in Nepal. Nepal. J. Zool. 1 (1), 33—41.

Callinan, R.B., Fraser, G.C., Virgona, J.L., 1989. Pathology of red spot disease in sea mullet, *Mugil cephalus* L. from eastern Australia. J. Fish Dis. 12, 467—479.

Chinabut, S., 1990. The histopathology of EUS in Asia. In: Phillips, M.J., Keddie, H.G. (Eds.), Proc. Regional Research Programme on Relationship between Epizootic Ulcerative Syndrome in Fish and the Environment: A Report on the 2nd Technical Workshop, 13—26 August, 1990. Network of Aquaculture Centres in Asia-Pacific, Bangkok, p. 75.

Chen, W., Hoy, J.W., Schneider, R.W., 1991. Comparisons of soluble proteins and isozymes for seven *Pythium* species and applications of the biochemical data to *Pythium* systematics. Mycol. Res. 95548—95555.

Cruz-Lacierda, E.R., Shariff, M., 1995. Experimental transmission of epizootic ulcerative syndrome (EUS) in snakehead, *Ophicephalus striatus*. In: Shariff, M., Arthur, J.R., Subasinghe, R.P. (Eds.), Diseases in Asian Aquaculture II. Fish Health Section, Asian Fisheries Society, Manila, Philippines, pp. 327—336.

Chinabut, S., Roberts, R.J., 1999. Pathology and Histopathology of Epizootic Ulcerative Syndrome (EUS). Aquatic Animal Health Research Institute, Department of Fisheries. Royal Thai Government, Bangkok, Thailand. ISBN 974-7604-55-8, 33pp.

Chinabut, S., Roberts, R.J., Willoughby, G.R., Pearson, M.D., 1995. Histopathology of snakehead, *Channa striatus* (Bloch), experimentally infected with the specific *Aphanomyces* fungus associated with epizootic ulcerative syndrome (EUS) at different temperatures. J. Fish Dis. 18, 41—47.

Dykstra, M.J., Noga, E.J., Levine, J.F., Moye, D.W., Hawkins, J.H., 1986. Characterization of the *Aphanomyces* species involved in ulcerative mycosis (UM) in menhaden. Mycologia 78, 664—672.

Egusa, S., Masuda, N., 1971. A new fungal disease of *Plecoglossus altivelis*. Fish Pathol. 6, 41—46.

Fraser, G.C., Callinan, R.B., Calder, L.M., 1992. *Aphanomyces* species associated with red spot disease:an ulcerative disease of estuarine fish of eastern Australia. J. Fish Dis. 15, 173—181.

FAO, 2009. What You Need to Know about Epizootic Ulcerative Syndrome (EUS) — An Extension Brochure. FAO, Rome, pp. 33.

Gayathri, D., Shankar, K.M., Mohan, C.V., Devaraja, T.N., 2008. Determination of specific sampling site on EUS affected fish for diagnosis of *Aphanomyces invadans* by immunodot test using monoclonal antibodies. Res. J. Agric. Biol. Sci. 4 (6), 757—760.

Hatai, K., Egusa, S., Takahashi, S., Ooe, K., 1977. Study on the pathogenic fungus of mycotic granulomatosis — I. Isolation and pathogenicity of the fungus from cultured-ayu infected with the disease. Fish Pathol. 12, 129—133.

Hatai, K., 1980. Studies on the pathogenic agents of saprolegniasis in freshwater fishes. In: Special Report of Nagasaki Prefectural Inst. of Fisheries, No. 8, Matsugae-Cho, Nagasaki, Japan.

Hatai, K., Nakamura, K., Rha, S.A., Yusa, K., Wada, S., 1994. *Aphanmyces* infection in dwarf gourami (*Colisa lalia*). Fish. Pathol. 29 (2), 95—99.

Hanjavanit, C., Hiroki, S., Hatai, K., 1997. Mycotic granulomatosis found in two species of ornamental fishes imported from Singapore. Mycoscience 38, 433—436.

Kar, D., Dey, S.C., 1988. A critical account of the recent fish disease in the Barak valley of Assam. Proc. Reg. Symp. Recent Outbreak of Fish Dis. North East India 1, 8.

Kar, D., Dey, S.C., 1990a. Fish disease syndrome: a preliminary study from Assam. Bangladesh J. Zool. 18, 115—118.

Kar, D., Dey, S.C., 1990b. A preliminary study of diseased fishes from Cachar district of Assam. Matsya 15—16, 155—161.

Kar, D., Dey, S.C., 1990c. Epizootic ulcerative syndrome in fishes of Assam. J. Assam. Sci. Soc. 32 (2), 29—31.

Kar, D., Dey, S.C., Roy, A., Mandal, M., 2000. Epizootic ulcerative syndrome in fish at Barak valley, Assam, India. In: Ramachandra, T.V., Murthy, C.R., Ahalya, N. (Eds.), Lake 2000: Int. Symp. Restoration of Lakes and Wetlands, Indian Inst. Sci., Bangalore, India, pp. 317—319.

Lilley, J.H., Callinan, R.B., Chinabut, S., Kanchanakhan, S., MacRae, I.H., Phillips, M.J., 1998. Epizootic Ulcerative Syndrome (EUS) Technical Handbook. The Aquatic Animal Health Research Institute, Bangkok, pp. 88.

Miyazaki, T., Egusa, S., 1972. Studies on mycotic granulomatosis in freshwater fish I. Mycotic granulomatosis in goldfish. Fish. Pathol. 7, 15–25 (In Japanese).

Miyazaki, T., Egusa, S., 1973a. Studies on mycotic granulomatosis in freshwater fish II. Mycotic granulomatosis prevailed in goldfish. Fish. Pathol. 7, 125–133 (In Japanese).

Miyazaki, T., Egusa, S., 1973b. Studies on mycotic granulomatosis in freshwater fish III. Bluegill. Mycotic granulomatosis in bluegill. Fish. Pathol. 8, 41–43 (In Japanese).

Miyazaki, T., Egusa, S., 1973c. Studies on mycotic granulomatosis in freshwater fish IV. Mycotic granulomatosis in some wild fishes. Fish. Pathol. 8, 44–47 (In Japanese).

Miles, D.J.V., Polchana, J., Lilley, J.H., Kanchanakhan, S., Thompson, K.D., Adams, A., 2001. Immunostimulation of striped snakehead *Channa striata* against epizootic ulcerative syndrome. Aquaculture 195, 1–15.

Nyblom, H., Bjornsson, E., Simren, M., Aldenborg, F., Almer, S., Olsson, R., 2006. The AST/ALT ratio as an indicator of cirrhosis in patients with PBC. Liver Int. 26, 840–845.

Palisoc, F.P., 1990. Histopathology of the epizootic ulcerative syndrome (EUS) positive snakehead, *Ophicephalus striatus*, from Laguna Lake, Philippines. In: Abstracts from the Symposium on Diseases in Asian Aquaculture. 26–29 November 1990. Bali, Indonesia. Fish Health Section, Asian Fisheries Society, Manila, p. 22.

Pathiratne, A., Rajapakshe, W., 1998. Hematological changes associated with epizootic ulcerative syndrome in the Asian cichlid Fish, *Etroplus suratensis*. Asian Fish. Sci. 11, 203–211.

Pradhan, P.K., Rathore, G., Sood, N., Swaminathan, T.R., Yadav, M.K., Verma, D.K., Chaudhary, D.K., Abidi, R., Punia, P., Jena, J.K., 2014. Emergence of epizootic ulcerative syndrome: large-scale mortalities of cultured and wild fish species in Uttar Pradesh, India. Curr. Sci. 106 (12), 1710–1717.

Roberts, R.J. (Ed.), 1978. Fish Pathology. Bailliere Tindall, London, pp. x + 318.

Roberts, R.J., Macintosh, D.J., Tonguthai, K., Boonyratpalin, S., Tayaputch, N., Phillips, M.J., Millar, S.D., 1986. Field and Laboratory Investigations into Ulcerative Fish Diseases in Asia-Pacific Region. Technical Report, FAO Project TCP/RAS/4508, pp. 214.

Roberts, R.J., Wootten, R., MacRae, I., Millar, S., Struthers, W., 1989. Ulcerative Disease Survey: Bangladesh Final Report to the Govt. of Bangladesh and the Overseas Development Administration. Institute of Aquaculture, Stirling University, UK.

Roberts, R.J., Willoughby, L.G., Chinabut, S., 1993. Mycotic aspects of epizootic ulcerative syndrome (EUS) of Asian fishes. J. Fish. Dis. 16, 169–183.

Rha, S.A., Sinmuk, S., Wada, S., Yuasa, K., Nakamura, K., Hatai, K., Ishii, H., 1996. Pathogenicity to ayu (*Plecoglossus altivelis*) of *Aphanomyces* sp. isolated from dwarf gourami (*Colisa lalia*). Bull. Nippon Vet. Anim. Sci. Univ. 45, 9–15.

Scott, W.W., 1961. A monograph of the genus Aphanomyces. Virginia Agric. Exp. Stat. Tech. Bull. 151.

Secombes, C.J., 1990. Isolation of salmonid macrophages and analysis of their killing activity. In: Stolen, J.S., Fletcher, T.C., Anderson, D.P., Robertson, B.S., van Muiswinkel, W.B. (Eds.), Techniques in Fish Immunology. SOS Publ, Fair Haven, pp. 137–154.

Saylor, R.K., Miller, D.L., Vandersea, M.W., Bevelhimer, M.S., Schofield, P.J., Bennett, W.A., 2010. Epizootic ulcerative syndrome caused by *Aphanomyces invadans* in captive bullseye snakehead (*Channa marulius*) collected from south Florida, USA. Dis. Aquat. Org. 88, 169–175.

Tangtrongpiros, M., Luanprida, S., Nukwan, S., Srichumpoung, V., 1990. Breeding and fry nursing of the Thai silver carp (*Puntius gonionotus*) in earthen ponds. In: Proc. Second Asian Fisheries Forum, Tokyo, Japan, pp. 133–135.

Thompson, K.D., Lilley, J.H., Chinabut, S., Adams, A., 1997. The antibody response of snakehead, *Channa striata* Bloch, to *Aphanomyces invaderis*. Fish Shellfish Immunol. 7 (5), 349–353.

Thompson, K.D., Lilley, J.H., Chen, S.-C., Adams, A., Richards, R.H., 1999. The immune response of rainbow trout (*Oncorhynchus mykiss*) against *Aphanomyces invadans*. Fish Shellfish Immunol. 9, 196–201.

Virchow, R., 1858. Rudolf Virchow Endorses Cell Division and its Role in Pathology.

Vishwanath, T.S., Mohan, C.V., Shankar, K.M., 1997. Mycotic granulomatosis and seasonality are the consistent features of epizootic ulcerative syndrome of fresh and brackishwater fishes of Karnataka, India. Asian Fish. Sci. 10, 155–160.

Vishwanath, T.S., Mohan, C.V., Shankar, K.M., 1998. Epizootic ulcerative syndrome associated with a fungal pathogen in Indian fishes-histopathology: a cause for invasiveness. Aquaculture 165, 1–9.

Vogelbein, W.K., Shields, J.D., Haas, L.W., Reece, K.S., Zwerner, D.E., 2001. Skin ulcers in estuarine fishes: a comparative pathological evaluation of wild and laboratory-exposed fish. Environ. Health Perspect. 109, 687–693.

Wada, S., Yuasa, K., Rha, S., Nakamura, K., Hatai, K., 1994. Histopathology of *Aphanomyces* infection in dwarf gourami (*Colisa lalia*). Fish. Pathol. 29, 2299–3237.

Wada, S., Rha, S.A., Kondoh, T., Suda, H., Hatai, K., Ishii, H., 1996. Histopathological comparison between ayu and carp artificially infected with *Aphanomyces piscicida*. Fish. Pathol. 31, 71–80.

Chapter 8

Methodologies of Different Types of Studies

8.1 PHYSICOCHEMICAL CHARACTERISTICS OF WATER

Water is a basic substance on which the dynamics of a water body, and thereby the life of the biota in it, depend. To a limnologist, fishery scientist, and fish farmer, the study of water is a prerequisite for the welfare of the water body and the fish and to know their environment. Further, agencies like public health engineering departments (waterworks), pollution control boards, and various industrial organizations require the analysis of water/wastewater samples. The methodology recommended by APHA (1995) is largely followed and is given below.

The main purposes of physical and chemical examination of water are as follows:

1. Know the exact composition of the sample at the particular point of time of sample collection
2. Classify water with regard to the general level of gaseous and mineral constituents
3. Ascertain the presence or absence of constituents affecting various beneficial uses of water
4. Determine the level of organic impurities
5. Determine the degree of clarity and to ascertain the nature of matter in suspension

Analytical results may be interpreted to suit different purposes for different organizations (according to the aim of the work)—for example, surveillance of water quality, effluent quality, and farming.

Among physicochemical characteristics, the following are generally studied in a water body by a limnologist:

8.1.1 Physical Characteristics

Temperature, turbidity, and transparency

8.1.1.1 Temperature

In limnological studies, water temperature as a function of depth is often required. Water temperature readings are related to the study of pH, DO, salinity, conductivity, etc. However, heated water discharges into the water body generally have detrimental effects.

Water Temperature Measurement

Principle Water temperature could be measured with the help of a good-quality mercury-in-glass precision thermometer (-5 to $50 \times 1/10\,°C$) having minimal thermal capacity, to facilitate rapid equilibration. It is checked and calibrated (if required) against a precision thermometer certified by the National Bureau of Standards (ordinary thermometers may have errors of up to $3\,°C$).

Temperature of both air and surface water could be measured directly. However, bottom water temperature could be measured by collecting the water samples from various depths with the help of a 5-L capacity Kemmerer sampler. Alternatively, water temperature at depths may be most conveniently and accurately measured with the help of a thermistor or a less expensive reversing thermometer or thermophone.

8.1.1.2 Turbidity

Turbidity is a significant parameter in determining the opaqueness of water, which affects photosynthetic processes and fish life. It mainly depends on particulate matter present in the water.

Suspension of particles in water interfering with passage of light is called "turbidity." The suspended particles may be clay, silt, finely divided organic and inorganic matter, plankton, and other microscopic organisms. Turbid water should be unfit for drinking and is undesirable for food products, beverages, etc. Turbidity is a measure of performance of water treatment plants.

The following methods are generally used for determining the turbidity of a water sample:

1. By Jackson's Candle turbidimeter (JCT)
2. By electrical/electronic turbidimeters, commonly called nephelometers

The JCT is generally used for highly turbid waters.

Jackson's Candle Turbidimeter

Principle It is based on the transmittance of light from a flame of a "standard candle" (made of beeswax and spermaceti designed to burn at 114–126 grains per hour) through sample column of certain path length, such that the

Epizootic Ulcerative Fish Disease Syndrome. http://dx.doi.org/10.1016/B978-0-12-802504-8.00008-0

flame becomes indistinguishable against background illumination. Turbidity is inversely proportional to the path length. In brief, it is based on a comparison of the intensity of light scattered by the sample and a standard reference under comparable conditions. The higher the intensity of scattered light, the higher the turbidity.

The lowest turbidity that could be measured with a JCT is 25 units (JTU). As such, indirect secondary methods are required for measuring turbidities in the range of 0—5 units. However, the results obtained with different types of secondary instruments do not match with one another because of fundamental differences in the optical systems, even though the instruments are all precalibrated against JCT.

This method is not in much use today mainly because of difficulties in getting the "standard candle."

Nephelometric Method

Principle Formazin polymer is the turbidity standard suspension preference for water. It is easy to prepare and more reproducible in its light scattering properties. A given concentration of formazin suspension having 40 NTU has approximately a turbidity of 40 JTU. Therefore, turbidity based on formazin will approximate units derived from JCT, but it will not be identical.

Materials **Turbidimeter**: It consists of a light source for illumination of the sample and one or more photoelectric detectors with readout device to indicate the intensity of light scattered at 90° to incident light. The sensitivity of the instrument permits detection of turbidity difference of 0.02 NTU or less in water having turbidity <1.0 NTU. The instrument measures generally from 0 to 40 NTU. Various ranges are necessary to obtain both adequate coverage and sensitivity. However, meters having wider ranges are also available today.

Sample tubes: These should be of clear colorless glass, scrupulously cleaned both inside and outside without scratches. The tubes should be sufficiently long so that they do not need to be touched where light strikes.

Reagents:

(a) Turbidity-free water: It is obtained by passing DW through membrane filter having pore size <100 μm. If filtration does not reduce turbidity, DW itself may be used.

(b) Stock turbidity suspension:

1. Solution I: It is prepared by dissolving 1.0 g of hydrazine sulfate in DW and diluted to 100 mL in a volumetric flask.
2. Solution II: It is prepared by dissolving 10.0 g of hexamethylene tetramine in DW and diluted to 100 mL in volumetric flask.
3. In a 100-mL volumetric flask, mix 5.0 mL of Solution I with 5.0 mL of Solution II. Allow to stand for 24 h at 25 ± 3 °C. This is diluted to the 100 mL mark of the flask and shaken properly. Turbidity of this suspension is 400 NTU.

NB: Stock solutions are prepared monthly.

(c) Standard turbidity suspension:
10 mL of stock turbidity suspension (as prepared above) is diluted to 100 mL with turbidity-free water. The turbidity of this suspension is defined as 40 NTU. Such suspensions are prepared weekly.

Procedure **(a) Turbidimeter calibration**: Generally, these meters are kept calibrated by the manufacturer before being sold out. Further, the manufacturer's operation manual could be useful in calibration, if required.

(b) Measurement of turbidities <40 NTU: The sample is thoroughly shaken. The air bubbles are allowed to escape. The sample is poured into the turbidimeter tube and the turbidity is read directly from the scale or from the calibration curve.

(c) Measurement of turbidities >40 NTU: The sample is diluted with turbidity-free water until its turbidity falls between 30.0 and 40.0 NTU. Now the turbidity of the original sample is computed from the turbidity of the diluted sample and dilution factor. The stock turbidity suspension of 400 NTU is used for continuous monitoring. High turbidities determined by direct measurement are likely to differ appreciably from those determined by dilution technique.

Calculation

$$NTU = \frac{A \times (B + C)}{C},$$

where,

A = NTU found in the diluted sample
B = Volume of the dilution water
C = Sample volume taken for dilution

Interpretation of Results. The turbidity readings are reported as follows:

Turbidity range (NTU)	Record to the nearest NTU
0–1	0.05
1–10	0.10
10–40	1.0
40–100	5.0
100–400	10.0
400–1000	50.0
>1000.0	100.0

8.1.1.3 Transparency

It is a measure of the depth to which one may see into the water. Obviously, this is variable with the day's condition and the eyesight of the observer.

The Secchi disk is a simple device used to estimate this depth, and this parameter could also be called "Secchi disk visibility." The Secchi disk is a simple device used to estimate this depth. It consists of a weighted circular plate, 20 cm in diameter, with the surface painted with opposing black and white quarters. It is attached to a calibrated line by a ring at the center, so that when held by the line, it hangs horizontally. The procedure for determining the Secchi disk visibility follows.

Procedure

1. The Secchi disk is slowly lowered into the water body until it disappears. This depth is noted.
2. The disk is lowered a few more cm and then is slowly raised again until it reappears. This second reading of depth is also noted.
3. The average of these two readings is taken as the final Secchi disk visibility depth or, in short, transparency.

8.1.2 Chemical Characteristics (Water Chemistry)

These are pH, dissolved oxygen (DO), free carbon dioxide (FCO_2), total alkalinity (TA), specific conductivity, carbon, productivity, and nutrients.

8.1.2.1 Hydrogen-Ion Concentration (pH)

The term pH is the negative logarithm of hydrogen-ion activity. Most natural waters have a pH that falls in the range of 4.0–9.0. However, the majority of water is slightly basic because of the presence of carbonates and bicarbonates. A departure from the normal pH could be due to influx of acidic or alkaline wastes, mainly from the industries. Adjustment of pH in effluents as well as in water treatment plants is a common practice. Whereas acidity and alkalinity are measures of total resistance to pH change or buffering capacity of a sample, pH represents the free hydrogen-ion activity not bound by carbonate or other bases.

Hydrogen-ion concentration, expressed as pH, may be determined by the following methods:

1. Colorimetric method
2. Electronic method

Colorimetric Method

Colorimetrically, pH is determined on the spot with the help of a Hellige comparator using BDH (Qualigens) indicator solution and color disks of the ranges from 2.0 to 10.5.

However, the colorimetric method is said to suffer from interferences due to color, turbidity, salinity, colloidal matter, oxidants, and reductants. Indicators themselves may alter the pH of poorly buffered liquids unless preadjusted to nearly the same pH as the sample. Hence, the colorimetric method is suitable only for rough estimation.

Electronic Method

This method involves operation of an electronic pH meter using a standard "glass electrode" along with a "reference calomel electrode" that produces a change of 59.1 mV (milli volt) per pH unit at 25 °C. Glass is an ion-exchange material with preference for H^+ ions that are adsorbed on one surface and their positive charge is transmitted through the glass by Na^+ ion displacement. The H^+ ions are desorbed on the other side. Therefore, glass is a membrane specific for H^+ ions. The glass electrode consists of a glass bulb with dilute HCl in it that is connected to the internal circuit through AgCl.

Materials Required

1. Electronic pH meter with temperature compensation
2. Glass electrode
3. Reference electrode
4. Magnetic stirrer

Standard Solutions (Reagents)

1. Standard buffer prepared by dissolving required amount of substances in DW having conductivity <2.0 microsiemens at 25 °C

NB: The buffer solutions are preserved in polythene or pyrex bottles and the solutions are replaced every month.

2. Saturated potassium—hydrogen—tartrate solution is prepared by dissolving finely crystalline 5—10 g of potassium—hydrogen—tartrate in 100—300 mL of DW at 25 °C in a glass-stoppered bottle. The clear solution is decanted and preserved by adding a thymol crystal
3. Saturated KCl solution to is prepared by dissolving AR grade KCl in DW

Procedure

1. The pH meter is connected to 230 V AC having good earthing and switched on. It is kept as such for a few minutes to warm up.
2. Before use, both the glass and reference electrodes are kept dipped in DW for at least 24 h.

NB: The tips of the electrodes are kept immersed in water when not in use.

At the time of using the electrodes, they are rinsed with DW and wiped with tissue paper and never with hand.

pH measurement

1. The electrodes are connected to the pH meter.
2. Buffer solutions of pH 7.0 and 4.0 or 9.2 are prepared by grinding and dissolving standard buffer tablets (of respective pH) in DW.

3. The electrodes are washed with DW and dried with tissue paper.
4. The temperature of the buffer solution of pH 7.0 is noted.
5. The temperature control switch of the pH meter is adjusted according to the temperature of the buffer.
6. The electrodes are dipped into buffer solution of pH 7.0.
7. The meter will display some reading that may be nearly 7.0.
8. The "standardize control" of the meter is operated to make the meter reading exactly 7.0.
9. The buffer solution of pH 7.0 is removed from the table and the buffer solution of pH 4.0 is brought in.
10. The electrodes are washed and rinsed as before.
11. Temperature of the buffer solution of pH 4.0 is noted.
12. The temperature control switch of the pH meter is adjusted according to the temperature of the buffer.
13. The electrodes are dipped into buffer solution of pH 4.0.
14. The pH meter could display exactly 4.0 or could display some reading that is nearly 4.0.

Titrimetric Method: Alsterburg Azide Modification of Winkler Method

Principle In this method, the amount of DO is computed in mg/L by the quantity of liberated iodine from a mixture of manganous sulfate and alkaline iodide sodium azide neutralized by a known volume of 0.025 N sodium thiosulfate solution. In other words, there is formation of brown hydrated oxide of manganese (IV) by oxidation of manganese (II) hydroxide by the DO when a solution of potassium iodide—sodium azide is added to the sample.

$$MnSO_4 + 2\,KOH \rightarrow Mn(OH)_2\downarrow + K_2SO_4$$
$$2Mn(OH)_2 + O_2 \rightarrow 2\,MnO(OH)_2\downarrow$$
(Brown ppt)

The floc of manganese (II) hydroxide is said to act as a gathering agent for the DO and settles down. On subsequent acidification by concentrated sulfuric acid, the hydrated oxide of manganese (IV) reverts to the original state of manganese (II) while oxidizing the iodide quantitatively into iodine.

$$2\,MnO(OH)_2 + 2\,KI + H_2O \rightarrow Mn(OH)_2 + 2\,I_2 + 2\,KOH$$

Manganese Oxyhydroxide

15. If it is nearly 4.0, the standardize control of the meter is operated to make the meter reading exactly 4.0.
16. The buffer solution of pH 4.0 is also removed from the table. The pH meter is now said to be standardized.
17. The sample water of unknown pH is brought in.
18. The electrodes are washed and rinsed as before.
19. The electrodes are dipped into the water sample.
20. The pH of the water sample is recorded directly from the pH meter.
21. The procedure is repeated for other samples.

8.1.2.2 Dissolved Oxygen

It refers to the oxygen that is dissolved in water. It plays a key role in the dynamics of the water body, particularly in the respiration of the aquatic biota. DO content in water results from:

1. Photosynthetic activity of the green plants
2. Diffusion gradient at the air—water interface
3. Wind-driven mixing

Following are broadly two methods for the estimation of DO in water:

1. Titrimetric (iodometric)
2. Electrometric (electronic)

The liberated iodine is equivalent to the original DO content of the sample. The amount of liberated iodine is estimated by titration with standard solution of sodium thiosulfate using starch as an indicator:

$$I_2 + 2S_2O_3 \rightarrow S_4O_6 + 2I$$

Interference The Alsterburg azide modification of the Winkler method, in which sodium azide is used, is applicable to most natural waters. Interference by higher concentrations of Fe (III) is reduced by this Alsterburg modification, with azide playing the pivotal role.

Sampling Procedure Collection of water samples is done very carefully so that the sample does not mix with air or get agitated. Samples from any depth in streams, lakes, wetlands, or reservoirs need special precautions to eliminate changes in pressure and temperature.

Surface water samples are collected in narrow mouth glass stoppered BOD bottles of 300-mL capacity with tapered ground glass, pointed stoppers, and flared mouths. The bottle is overflown by two or three times its volume and the stopper replaced without air bubbles being allowed to enter. Samples from depths of >2.0 m are collected by Kemmerer type of sampler. The bottle is filled to overflowing for nearly 10 s but without turbulence or formation of bubbles.

Sample Preservation DO in water is determined immediately after collection of the sample. However, samples that are supposed to have low DO may be stored by adding manganous sulfate, alkaline iodide sodium azide, and sulfuric acid, shaken properly and kept away from strong sunlight.

Procedure for DO Determination

Reagents Required. **(a) Manganous sulfate**: It is prepared by dissolving 364 g of $MnSO_4 \cdot H_2O$ in DW. Then it is filtered and diluted to 1 L. This solution is not to produce color with starch when added to acidified solution of KI. Alternatively, 18.2 g of $MnSO_4 . H_2O$ is dissolved in DW, filtered, and made up to 50 mL.

(b) Alkaline iodide sodium azide solution: It is prepared by dissolving 500 g of NaOH or 700 g of KOH and 150 g of KI or 135 g of NaI in DW. This is diluted to 1 L with DW. Next, 10 g of NaN_3 (sodium azide) is dissolved in 40 mL of DW, and this is added to 950 mL of the solution prepared above. Alternatively, if less quantity of reagent is required, it is prepared by dissolving 25 g of NaOH and 7.5 g of KI in DW. This is diluted to 50 mL with DW. Next, 1 g of NaN_3 is dissolved in 4 mL of DW, and this is added to 45 mL of the solution prepared above.

(c) Starch indicator: To 0.25 g of starch powder, a few milliliters of cold water is added and the mixture is rubbed to a paste. This is added, with constant stirring, to 1 L of boiling water, so slowly that boiling never ceases. Boiling is continued for an additional minute or two and then it is allowed to cool. It is preserved in a glass bottle. A fresh solution is prepared as soon as mold appears or when a strong color is not given by very dilute iodine solution.

(d) Concentrated sulfuric acid: Must be AR grade.

(e) Stock sodium thiosulfate (hypo) solution (0.1 N): It is prepared by dissolving 24.82 g of $Na_2S_2O_3 \cdot 5H_2O$ in boiled and cooled DW. It is then diluted to 1 L.

(f) Standard sodium thiosulfate titrant (0.025 N): It is prepared by diluting 250 mL of stock solution to 1 L boiled and cooled DW. Alternatively, 1 L of 0.025 N hypo solution is prepared by dissolving 6.2047 g (approximately 6.3 g) of sodium thiosulfate (AR grade) in 1 L of DW. This standard sodium thiosulfate solution (which is a secondary standard) is to be standardized with a primary standard (say, potassium dichromate solution) before every set of titration for DO estimation, for that any standard book of chemistry dealing with volumetric analysis could be consulted.

Procedure for Estimation of DO

1. To the 300 mL/250 mL of sample collected in a 300 mL/250 mL BOD bottle is added 2 mL of $MnSO_4$ solution followed by 2 mL of alkaline iodide−sodium azide reagent.
2. The bottles are to be stoppered carefully and immediately to exclude air bubbles.
3. The contents are mixed by inverting the bottle at least 15 times. When the brown ppt. settles, leaving a clear supernatant above the manganese hydroxide floc, it is shaken again and the floc is allowed to settle for the second time.
4. After at least 2 min, settling should have produced minimum 100 mL of clear supernatant. Now, the stopper is removed carefully, and immediately 2 mL of concentrated sulfuric acid is added in such a way that it flows down the neck of the bottle. The bottle is restoppered and inverted gently until dissolution of the above floc is complete. A brown-colored solution is formed in the bottle.
5. This brown-colored solution is titrated against standard 0.025 N sodium thiosulfate solution (popularly called hypo) taken in a 50-mL burette. The brown color goes on fading until a straw-colored solution is obtained near the end point.
6. Now, two to three drops of starch indicator solution is added. An intense blue colored solution is obtained.
7. The titration is continued and hypo solution is added drop by drop from the burette. The blue color continues to fade until the solution becomes completely colorless at the end point.
8. The burette reading is noted.
9. The titer value is the amount of DO present in mg/L in the sample water.

The data may be tabulated in the format given below:

No. of obs	IR of burette	FR of burette	Difference	DO in mg/L
1				
2				
3				

Average value of DO (mg/L):
IR = Initial Reading.
FR = Final Reading.

Electrometric (Electronic) Method of DO Estimation by Dissolved Oxygen Meter

The DO meter is generally a battery-operated portable instrument for estimation of DO and determination of temperature of the sample water at the same time. DO is measured in ppm (mg/L) over a range of 0−20 ppm with 0.1-ppm resolution, and temperature is indicated in degrees Celsius from −5 to +45 °C. The DO measurement is automatically temperature-compensated for solubility of oxygen in water and permeability of the probe membrane. Salinity compensation is manual. Measurements are displayed by a three-digit LCD.

Principle The probe supplied with the unit (instrument) is a special Clark-type membrane-covered polarographic sensor. A thermistor probe is also supplied for automatic temperature compensation and measurement. A thin permeable membrane stretched over the sensor isolates the sensor elements from the environment but allows oxygen to enter. When a polarizing voltage is supplied across the sensor, oxygen, which has passed through the membrane, reacts at the cathode and causes a current to flow.

Panel features in brief

1. The following are the principal switches:

Power, salinity, zero, set zero, mode, DO meas/cal, and temperature

2. LCD display

Power Supply This unit is battery-operated. Hence, it does not require any mains outlet. It uses one 9-V small battery—Eveready 216, Hi-Watt 6F22, etc. Correct polarities are ensured while inserting fresh batteries.

Operating Steps

Probe Preparation. The DO probe consists of five main parts: main stem, electrode cover, sleeve, membrane, and Bakelite nut.

The main stem is made of Perspex with a gold electrode at the lower end and a silver ferrule fully surrounding the gold. The upper end carries a connecting cable and a connector. The main stem has threads so that the electrode cover along with the membrane assembly can be fitted to it by means of the Bakelite nut.

The electrode cover is a thin cylinder that fits into the lower half of the main stem. A Teflon oxygen-permeable membrane is held to the electrode cover by a rubber sleeve.

Electrolytic Solution. A 7.5% solution of KCl is prepared by dissolving 7.5 g of AR grade KCl in 100 mL of DW. AgCl crystals are slowly added until AgCl stops being dissolved.

Assembling of the Probe. First, the Bakelite nut is slid over the probe cover. Next, a piece of the membrane (1″ × 1″) is put over the tip of the probe. Then, the sleeve is push-fitted over the electrode cover in such a way that the membrane is flat and tight. There should not be any wrinkles on the membrane. The membrane is removed and refitted if any wrinkles are detected. A few drops of electrolyte solution is poured and checked for any leakage through the membrane or near the sleeve. Now, the electrode cover (probe cover) is filled completely with the solution. It is kept vertical and the main stem of the probe is

slowly inserted into the probe cover that is filled with the solution. The excess solution is let out. Next, the Bakelite nut over the main stem is tightened. It should be ensured that there is no air bubble. In the event of any bubbles being detected, the probe is opened and refilled until no bubbles are visible.

The gold cathode should always be bright and untarnished by wiping with a clean lint-free cloth or hard paper. No form of abrasive or chemical should ever be used.

Preparing the Instrument.

1. The "Mode" switch is set to "Set Zero."
2. The meter is switched to "On."
3. The "Set 20" control is adjusted until the meter displays 20.0.
4. The mode switch is switched to "Zero" and the "Zero" control is adjusted to read 0.0 (this is the zero electrical setting of the meter).
5. The prepared DO probe is attached to the probe connector.
6. A 15-minute period is allowed for the probe to polarize.

Air Calibration. Experience has shown that air calibration method is quite reliable, yet far simpler than other methods. The air calibration method is given below:

1. The probe is placed in moist air. For this, the probe could be wrapped loosely in a damp cloth, taking care that the cloth does not touch the membrane. About 10 min is allowed for temperature stabilization.
2. The mode switch is changed to "Temp" and the temperature is read and recorded. The table showing solubility of oxygen in freshwater against temperature is consulted from the manual supplied with the instrument in order to determine the calibration value. Also, the altitude and the atmospheric pressure correction factor are determined from a different table in the same manual. The calibration value from the first table is multiplied by the correction factor from the second table. An example is given below:

Temperature assumed to be 25 °C

Altitude assumed to be 1000 ft

From Table 1, the calibration value for 25 °C is, say, 8.4 ppm.

From Table 2, the correction factor for 1000 ft is about 0.96.

Therefore, the correct calibration value is 8.4 ppm × 0.96 = 8.1 ppm.

3. The "Salinity" knob is set to "Zero"
4. The mode switch is brought to "Cal."
5. The "CAL Control" is adjusted until the meter reads the correct calibration value as indicated in Step 2 above. Two minutes are allowed to elapse in order to verify calibration stability. Readjustment can be done if necessary.

The DO meter is now calibrated. Calibration should be checked periodically.

DO Measurement.

1. With the instrument calibrated, the probe is dipped in the water sample (whose DO is to be measured) and the sample water is constantly stirred. Stirring is continued with a stirrer during the process of measurement of DO in order to prevent depletion of DO near the membrane of the probe.
2. The "Salinity" control switch is set to the salinity value of the sample under test. Sufficient time is allowed for the probe to stabilize.
3. The DO value is read and recorded.

8.1.2.3 Free Carbon Dioxide (FCO$_2$)

Less than 10 mg/L of FCO$_2$ is generally present in surface waters. However, groundwater may contain a greater amount of FCO$_2$.

FCO$_2$ may be estimated by:

1. Nomographic method
2. Titrimetric method

The nomographic method gives a closer estimation of FCO$_2$ when the pH and alkalinity are determined immediately after sampling. The pH is determined with an electrometric pH meter calibrated with standard buffer solution in the pH range of 7.0—8.0. The TA is determined titrimetrically.

The titrimetric method of estimation of FCO$_2$ is done using phenolphthalein as an indicator.

The titrimetric method is discussed below in detail:

Titrimetric Method of Determination of FCO$_2$

Principle FCO$_2$ reacts with Na$_2$CO$_3$ or NaOH to form NaHCO$_3$. Completion of the reaction is indicated by development of pink color at the end point (pH 8.3). In short, FCO$_2$ is estimated by neutralizing the dissolved CO$_2$ with 0.022 N aqueous sodium hydroxide solution using phenolphthalein as an indicator.

Sampling and Storage Surface water sample may be collected by dipping a collecting bottle with double lid just below the water surface, and after the bottle is filled-in, its inner lid is closed while the mouth of the bottle is still underwater. A deep-water sample may be collected with the help of a water sampler bottle. More precisely, sample water is collected by rubber tubing discharging at the bottom of a 100-mL graduated cylinder or Nelson's tube. The sample water is allowed to overflow several times. The tube is withdrawn while the sample water is flowing. The measuring cylinder is flicked (jerked) to throw off the excess sample water above the 100-mL mark. Some loss of FCO$_2$ may be expected in storage and on transit of the sample, even with a careful collection technique. Hence, field determination immediately after sampling is desirable.

Alternatively, the sampling bottle is completely filled with the sample water and is preserved at a temperature lower than room temperature and FCO$_2$ is estimated as soon as possible.

Reagents

1. A 0.1 N stock NaOH solution is prepared by dissolving about 4 g of AR-grade NaOH pellets in 1 L of DW. It is standardized against 0.1 N H$_2$SO$_4$ using phenolphthalein as an indicator. A 100-mL amount of this 0.1-N solution is diluted to 440 mL with DW to give N/44(0.022N)NaOH.

Alternatively, 0.4735 g (accurately) of NaOH (96% pure) dissolved in 500 mL of DW gives 0.022 N NaOH.

Procedure

1. Sample water is collected following the above procedure and 100 mL is taken in a conical flask.
2. 10 drops of phenolphthalein indicator solution is added.
3. If the sample turns pink, it indicates that FCO$_2$ is absent.
4. If the sample remains colorless, it is titrated rapidly with 0.022 N NaOH solution (taken in a 50-mL burette), stirring gently with a glass rod until a definite pink color is produced that persists for 30 s when viewed through the depth of the sample.

NB: For checking loss of FCO$_2$ during titration, a second sample is taken. After adding 10 drops of phenolphthalein indicator, the full amount of the titrant used in the first titration is rapidly poured-in. If the sample turns pink, no FCO$_2$ should be lost. If it remains colorless, an additional amount of the titrant is added until pink color appears. The result of the second titration is considered.

5. The burette reading is recorded.

The data are recorded in the following table:

Obs. no	Initial burette reading (mL)	Final burette reading (mL)	Difference (mL)
1			
2			
3			

Average value of FCO$_2$ (mg/L):

Calculation :   mg/lit FCO$_2$

$$= (A \times N \times 44{,}000)/\text{mL of sample,}$$

where,

A = mL of NaOH titrant

N = Normality of the NaOH used above

8.1.2.4 Alkalinity

Alkalinity of water is its capacity to neutralize a strong acid to a designated pH, or stated in another way, it is the quantity and kind of compounds present, which collectively shift the pH to the alkaline side of neutrality. Although the alkalinity of natural waters is generally due to bicarbonates, it is usually expressed in terms of $CaCO_3$. The three kinds of alkalinity are indicated as OH^- (hydroxide), normal CO_3^- (carbonate), and HCO_3^- (bicarbonate). The three are summed as TA. Carbonate and bicarbonate are common to most waters because carbonate minerals are abundant in nature. The presence of OH^- can usually be attributed to water treatment or contamination. The expected TA in nature usually ranges from 45 to 200 mg/L of $CaCO_3$.

Alkalinity is significant in interpretation and control of water and wastewater treatment processes. For industrial wastes, measurement of alkalinity could indicate a change in quality if the source of the sample is known to have generally stable levels of alkalinity.

Measurement of Alkalinity

Principle Hydroxyl ions present in a sample as a result of dissociation or hydrolysis of solutes are neutralized by titration with standard acids. Titration to pH 8.3; that is, decolorization of phenolphthalein indicator, show complete neutralization of OH^- and half of CO_3^-, while titration to pH 4.4—that is, a sharp change from yellow to pink of the methyl orange indicator—indicates alkalinity due to OH^-, CO_3^-, and HCO_3^-.

$$OH^- + H^+ \rightarrow H_2O$$

Titration to pH8.3

$$CO_3^- + H^+ \rightarrow HCO_3^-$$
$$HCO_3^- + H^+ \rightarrow H_2O + CO_2$$ Titration to pH 4.4
(from CO_3 or
original HCO_3)

The end point may be determined empirically by titration and is that pH where the derivative of $\Delta pH/\Delta mL$ titrant is the greatest.

Materials Required The required materials are a 50-mL burette, 250-mL Erlenmeyer flask, ordinary filter paper (for watching and comparing the color during titration), 250-mL reagent bottle (for storing 0.02 N H_2SO_4), 60-mL dropping reagent bottles (two—for storing phenolphthalein and methyl orange indicator solutions), 0.02 N H_2SO_4 solution, phenolphthalein indicator solution, and methyl orange indicator solution.

A preparation of 0.02 N H_2SO_4: 0.28 mL of AR-grade concentrated H_2SO_4 added to 500 mL of DW gives 500 mL of 0.02 N H_2SO_4. This being a secondary standard,

it is to be standardized with 0.02 N Na_2CO_3 using phenolphthalein as an indicator.

Sampling and Storage Sample water is collected in polythene or borosilicate glass bottles and is stored at a low temperature. The bottles are filled completely with sample water. Samples are analyzed without delay, preferably within a few hours. Samples are kept sealed until ready for analysis, because loss of CO_2 results in conversion of HCO_3^- to CO_3^-.

Procedure

1. A 100-mL sample of water (collected following the standard procedure) is collected in a 250-mL Erlenmeyer flask.
2. Four drops of phenolphthalein indicator solution are added to it.
3. If the solution remains clear, "0" ppm of TA is recorded.
4. If the solution becomes pink, it be titrated with 0.02 N H_2SO_4 (taken in a 50-mL burette) until a clear solution is obtained.
5. The volume (mL) of 0.02 N H_2SO_4 required in phenolphthalein titration is recorded.
6. Next, two to four drops of methyl orange indicator solution are added to the same solution in the Erlenmeyer flask.
7. The solution is titrated again with the same 0.02 N H_2SO_4 until the solution turns pink-orange.
8. The volume (mL) of 0.02 N H_2SO_4 required in the methyl orange titration is also recorded.

The data are recorded in the following table:
Phenolphthalein titration.

Observation number	Initial burette reading (mL)	Final burette reading (mL)	Difference (mL)
1			
2			
3			

Methyl orange titration.

Observation number	Initial burette reading (mL)	Final burette reading (mL)	Difference (mL)
1			
2			
3			

Result

1. 10 times the volume (mL) of 0.02 N H_2SO_4 required in phenolphthalein titration = phenolphthalein alkalinity (PA) in ppm (mg/L)
2. 10 times the volume (mL) of 0.02 N H_2SO_4 required in methyl orange titration = methyl orange alkalinity (MOA) in ppm
3. Total alkalinity is the sum of PA and MOA when both have values to contribute to the same

8.1.2.5 Specific Conductivity

Conductivity of a water sample is its ability of carrying electric current that depends on total concentration of ionized substances, mobility of each of such ions, their valences, etc. On the other hand, specific conductivity means conductivity of a segment of a particular solution having a length of 1 cm and a cross-sectional area of 1 cm^2. Organic compounds contribute very little to conductivity because they are mostly nonionic. Freshly prepared DW has conductivity of 0.5–2.0 micromhos/cm, which may increase to about 4.0 micromhos/cm on storage due to absorption of CO_2 from the air. Conductivity reflects the characteristics of the water supplied.

Conductivity of a flowing stream could be recorded continuously in situ along with other parameters, for example, DO, pH, and temperature. However, the monitoring equipment is required to be frequently checked to avoid electrode misleading.

There are different kinds of conductivity bridges manufactured by various companies. The digital direct reading conductivity meter is designed for measuring the specific conductivity of a solution by using a conductivity cell. It enables to measure the conductivity, without manual balancing and the specific conductivity is read directly on a digital panel.

Measurement of Specific Conductivity

Principle A conductance cell and a Wheatstone bridge are used for measuring the conductivity. The conductance of a solution increases with temperature at a rate of approximately 2% per degree Celsius. KCl has a lower temperature coefficient of conductivity and hence is used in the cell. For precise works, conductivity is determined at 25 °C.

The main feature of a standard digital conductivity meter is that it avoids the conventional method of manual balancing by employing a self-balancing ratio-transformer bridge technique. A highly stable oscillator powers the bridge. The current through the conductivity cell passes through one winding of the radio transformer. The other winding output is fed to a selective high gain amplifier, the output of which drives a current in the third winding in such a way that the flux in this winding opposes the flux in the first winding. This goes on until both the fluxes cancel each other and consequently the detector output will be almost zero. The output of the amplifier is now a measure of the conductivity of the solution being measured.

Materials Required

1. Conductivity meter with conductivity cell
2. Beaker
3. Stirrer
4. KCl

Operating Procedure

1. Calibrate the instrument.
2. All controls are marked on the instrument.
3. The instrument is switched on. The indicator lamp glows.
4. The CAL/READ switch is kept in CAL position and the CAL control is adjusted to full-scale reading (100) on the meter.
5. After thoroughly cleaning, the "cell" (conductivity cell) is immersed in a standard KCl solution, say 0.1 N KCl, with a specific conductivity of the order of 0.01412 millimhos (14.12 micromhos)/cm at 30 °C.
6. The Cal/Read control is put in the Read position. The meter needle should deflect.
7. The range control is turned in a counterclockwise direction starting from 300 micromhos to get a range where the needle will show maximum deflection. The maximum deflection of the needle can be obtained in the range of 0–30 micromhos.

The standard table supplied with the instrument is referred to in order to determine the conductivity of the solution at 30 °C.

8. Then, the cell constant control is adjusted so that the meter reading corresponds with the table value. In other words, the meter reading is adjusted to read 14.12 micromhos.

Now, the instrument is calibrated for measuring the conductivity of the unknown sample.

The standard 0.1 N KCl solution is prepared by dissolving 7.459 g of dried AR-grade KCl in 1 L of DW.

Determination of Conductivity

1. The instrument is switched on.
2. The "Scale Selector" knob is brought to the 200-millimhos position (maximum).
3. The "Cal/Read" knob is set to "Cal."
4. The "Cal" is operated to display 100 in the display panel.
5. The cell is dipped in N/10 KCl (the specific conductivity of N/10 KCl at 30 °C is 14.12 micromhos/cm).
6. The "Scale Selector" knob is brought to the 20-millimhos position.
7. The "Cell Constant" knob is brought to the 0.95 position.

8. The "Cal/Read" knob is set to the "Read" position. The meter should display a value that is a little less than 14.12.
9. The "Cell Constant" knob is adjusted so that the meter reads 14.12.
10. The "Cal/Read" knob is set to the "Cal" position.
11. The cell is rinsed with DW.
12. Next, the cell is dipped in sample water.
13. The "Cal/Read" knob is brought to the "Read" position.
14. The "Scale Selector" knob is rotated counterclockwise from the 200-millimhos position (highest value) gradually stage by stage to the 20-micromhos mark (lowest value), until the meter displays 1.0.
15. Now, the "Scale Selector" knob is rotated clockwise stage by stage and is stopped at that point where we get the highest value.
16. This value is recorded as the conductivity of the sample water.

8.1.2.6 Carbon

Carbon atoms can be linked to other molecules or atoms on four sides in long chains to form complex organic molecules. Carbon is the structural basis of all living materials on the earth. The amount of carbon fixed in a given time is an important measure of productivity of a lake.

8.1.2.7 Productivity

Lentic bodies can be productive or unproductive in almost the same way as agricultural land. When the input of nutrients in a lentic body increases (either through natural processes or by human activities), changes that occur in the lentic body are known as eutrophication. Conversely, lakes that have a naturally low supply of nutrients are called oligotrophic. Productivity refers to amount of new living matter formed in a certain period of time.

8.1.2.8 Nutrients

Nutrients act as fertilizers for plants, and their supply determines the quantitative and qualitative growth of plants. The most common nutrients are nitrates and phosphates.

Summary

1. Physicochemical characteristics of water refer to the abiotic status of the water body.
2. It is essential to have a quantitative estimation of abiotic environment in order to assess their relationship with the biota.
3. Various tools are available today for quantitative estimation of various physicochemical parameters of water. However, the standard methodology recommended by APHA is generally largely followed. The physical parameters include temperature, transparency, and

turbidity. These are determined by thermometer, Secchi disk, and nephelo turbidimeter, respectively. The chemical parameters include pH, DO, FCO_2, TA, and conductivity in addition to nutrients. They play significant roles in the productivity of the water body.

8.2 PHYSICOCHEMICAL CHARACTERISTICS OF SOIL

8.2.1 Physical Characteristics

These include temperature, texture, color, moisture content, bulk density (BD), and infiltration rate.

8.2.1.1 Soil Temperature

The main materials required are a soil thermometer and a notebook.

Procedure

1. The soil thermometer is inserted into the soil to approximately 3″ at the study point.
2. Wait approximately 1 min for the alcohol/mercury within the thermometer to become steady.
3. The soil temperature is recorded from the dial of the thermometer.
4. The temperature is recorded at different study points.

NB: The thermometer should not be pressed too hard into the soil.

5. In the case of bottom soil in a water body, the thermometer is inserted into the collected soil immediately after collection and the temperature is recorded.

The data are recorded in the following table:

Locality:	Date:	Time:
Site no./Sample no.	Temperature (°C)	
1		
2		
3		

8.2.1.2 Soil Texture

The main materials required are sieves of various mesh size (mm)—0.02, 0.2, 2.0, 5.0, and 7.0; soil sample(s); and graph paper.

Procedure

1. A soil sample is sieved through a series of different meshes as indicated above, beginning with 7.0 mm (stone), and proceeding gradually through 5.0 mm (gravel), 2.0 mm (coarse sand), 0.2 mm (fine sand), and 0.02 mm.

2. This exercise will separate particles of different sizes.
3. With the help of a balance, the quantity of soil of each particle size is weighed.
4. The data are put in a textural triangle.

The data are recorded in the following table:

Locality:	Date:			Time:	
Site no.	Weight of the different-sized particles in the soil sample (g)				
	7.0 mm	5.0 mm	2.0 mm	0.2 mm	0.02 mm
1					
2					
3					

8.2.1.3 Soil Color

Soil exhibits a variety of colors. Soil color may be inherited from the parental material (i.e., lithocentric) or sometimes may be due to soil-forming processes (acquired or genetic color). The color of soil renders a valuable key for the identification of soil type.

Main Materials Required

The main materials required are: (1) a soil auger (for collection of soil sample); (2) a Munsell's soil color chart; (3) a bucket for keeping the collected soil—1 × 10-L capacity; (4) one sheet of white foolscap paper; and (5) one 50 × 50 cm cardboard sheet .

Munsell's Soil Color Chart

It is a booklet formed from the complete edition of the *Munsell Book of Color* (vide *US Department of Agriculture Handbook No. 18: Soil Survey Manual*) and consists of seven soil color charts popularly called "Soil Collection

"hue, value, and chroma." The "hue" notation of a color indicates its relation to red, yellow, green, blue, and purple. The "value" notation indicates its lightness. The "chroma" notation indicates its strength (intensity). The colors displayed on the individual soil color chart are of constant hue designated by a symbol in the upper right hand corner of

the card. Vertically, the colors become successively lighter from the bottom of the card to the top by visually equal steps—their value increases from bottom upward. Horizontally, they increase in chroma to the right and become grayer to the left. The value notation of each chip is indicated by the horizontal scale across the bottom of the chart.

Procedure

1. The soil sample is collected following the standard procedure.
2. The dried soil sample is spread uniformly over a cardboard sheet and is held directly behind the apertures separating the closest matching color chips.
3. The particles of the soil are matched with the color chips of Munsell's Soil Color Chart.
4. After matching the sample soil color with the color chips of Munsell's chart, the color name and the Munsell's notation are ascertained from the chart.
5. Field observations are recorded in the following table.

Locality:	Date:			Time:	
Sample no.	Hue	Value	Chroma	Color	Expression
Example:					
1.	5 YR	5	6	Yellowish red	5 YR 5/6

Display 199." Each soil color chart (in the form of a loose leaflet) consists of different standard color chips systematically arranged according to Munsell's notations. The arrangement is by three simple variables that combine to describe all colors and are known in Munsell's system as

8.2.1.4 Soil Moisture Content

Soil moisture is the water present in the space between the soil particles. Soil moisture influences the physical, chemical, and biological characteristics of the soil. Soil moisture is the percentage of moisture (on a weight basis) in a soil sample at any given time. It shows general moisture content of the soil. Soil moisture helps to understand the water movement and aeration of the soil.

The main materials required are soil digging tools, polythene bags, beakers, a hot plate, and a weighing balance.

Procedure

1. Soil samples are collected in polythene bags following the standard procedure.
2. A 100-mL beaker is weighed empty (W_1).
3. From the stock soil sample, c 100 g of fresh soil is poured out into the preweighed beaker.
4. The beaker with the soil is now weighed (W_2).
5. The weight of the soil is determined by subtracting $W_2 - W_1$.
6. The beaker with the soil sample is first kept on a sand bath and then placed on a hot plate maintained at 105 °C (approximately).
7. The soil sample will be seen to become dry gradually.
8. After the soil sample is dried, the beaker with the soil is removed from the hot plate and cooled.
9. The beaker with the soil sample is now weighed again.
10. After this, the soil sample in the beaker is heated, cooled, and weighed repeatedly until constant weight is obtained (W_3).

Calculation

Percentage moisture content is calculated as follows:

Weight of the empty beaker = say Y g.

Weight of the empty beaker + fresh soil sample = say X g.

Fresh weight of the soil sample = $X - Y$ = say A g.

Final weight of the beaker + dry soil (after getting constant weight) = say Z g.

Dry weight of soil sample = $Z - Y$ = say B g.

Therefore, weight of the moisture in the soil = $A - B$ = say M.

Moisture % = {[Weight of moisture (M)]/[Dry weight of soil sample (B)]} × 100.

8.2.1.5 Soil Bulk Density

The BD or volumetric mass of the soil is the weight of the oven dry soil per unit volume. It includes the weight of the solid soil particles and the pore space in the soil. The dry mass, conversely, is the mass of the dry soil that has no pore space. BD is expressed as grams per cubic centimeter. It is said to increase with the depth of the soil due to the compaction occurring from the pressure on the top layer and also due to the low organic matter content. BD of soil is influenced by density of the mineral particles, quantity of organic matter, soil compaction, and burrowing activity of the soil animals. In other words, BD is the specific gravity of the soil with its pores. BD helps in determining the degree of compaction of the given soil. BD also indicates the soil aeration status.

The main materials required are a 25-mL volumetric flask, an oven, a weighing balance, and a soil sample.

Procedure

The procedure varies according to the nature of the site—disturbed or undisturbed. The procedure for both kinds of sites is described below.

For a Disturbed Site The portion of the earth that is regularly plowed is called a "disturbed site." The procedure for this site type is:

1. The soil sample is oven-dried.
2. A 100-mL volumetric flask is weighed (W_1).
3. This preweighed volumetric flask is filled with a soil sample collected by standard procedure.
4. The flask is now tapped, and tapping is continued until the level of soil poured in reaches the 100-mL mark of the volumetric flask.

Note: It is assumed that by packing the flask with soil followed by tapping the flask, the compactness of soil in nature is represented.

5. The weight of the volumetric flask with the soil is noted (W_2).
6. The weight of the soil is determined by subtracting $W_2 - W_1$.
7. The flask is now emptied and water is poured in to the 100-mL mark of the volumetric flask.
8. The volume of the water (say X mL) is noted by pouring it out into a measuring cylinder.

Calculation (with Model Example)

Weight of the volumetric flask (W_1) = say 71.0 g

Weight of the volumetric flask + soil (W_2) = say 203.5 g

Volume of water in the volumetric flask (X) = 100 mL

Weight of the soil ($W_2 - W_1 = Y$) = say 132.5 g

Now, we know that 1 mL = 1 cc

Result:

BD in g/cubic centimeter = (Weight of the soil in the volumetric flask)/(Volume of the water in the volumetric flask) = Y/X = 132.5/100 = 1.325 g/cc

For an Undisturbed Site The portion of the earth that has not been plowed for some years is referred to as an undisturbed site. The procedure for this site type is:

1. A soil sample from an undisturbed plot of land is collected by following the standard procedure.
2. The external surface of the soil clod is coated with wax; i.e., the soil clod is immersed in liquefied wax.

3. The volume of the clod is determined by finding the volume of water displaced by it. The method is given below:
   a. A known volume of water is taken in a beaker (V_1).
   b. The wax-coated clod is put in the water in the beaker.
   c. The increase in water level is marked and noted.
   d. The clod is removed and water is added to the marked level.
   e. The volume of water is increased, and the marked level is noted (V_2).
   f. The difference between V_2 and V_1 ($V_2 - V_1$) gives the volume of the clod.
4. The external wax coating from the clod is carefully removed, the clod is broken, and the soil is dried in a preweighed empty beaker (W_1).
5. The beaker with the soil is weighed (W_2).

Calculation and Result (with Model Example)
Volume of water taken (V_1) = 100 mL
Volume of water after inserting the clod into it (V_2) = 127 mL
Weight of the empty beaker (W_1) = 46.7 g
Weight of the beaker + soil (W_2) = 102.0 g
Volume of the soil clod ($V_2 - V_1$ = say X mL) = 27 mL
Weight of the soil ($W_2 - W_1$ = say Y g) = 55.3 g
Bulk Density = Weight of the soil/Volume of the soil = Y/X = 55.3/27 = 2.05 g/cc

8.2.1.6 Water Infiltration Rate of Soil

Main materials required

1. Two iron rings of height 25 cm, thickness 2 mm, and diameter of first ring 30 cm and second ring 60 cm
2. Scale
3. Soil sample

Procedure

1. The two rings are inserted (hammered in) about 10 cm inside the soil.
2. Both rings are filled with water up to the brim.
3. The system is left as such for c 10 min.
4. The height of water within the inner ring is measured after c 10 min.
5. The system is left as such again for c 10 min.
6. The height of water within the inner ring is measured again after c 10 min
7. In this way, the height of the water column within the inner ring is measured repeatedly at an interval of c 10 min until constant height (depth) of the water column inside the inner ring is obtained.
8. A graph is plotted with depth (height) of the water column in the y-axis and time interval in the x-axis.

The data are recorded in the following table:

Number of observations	Height (cm) of the water column inside the inner ring after each 10 min
At 10-min intervals	Interval 1, Interval 2, Interval 3, Interval 4, etc.

8.2.2 Chemical Characteristics

Organic matter may fail to oxidize and accumulate at the soil surface while the deeper mineral soil is wholly or partly anoxic. Thus, microbial oxidative metabolism has to depend on electron acceptors other than molecular oxygen. The most frequent of these are NO_3^{-1}, Mn^{3+}, Mn^{4+}, Fe^{3+}, SO_4^{-2}, and CO_3^{-2}. The characteristic gray-green color of the gley horizon is due to the conversion of Fe^{3+} to Fe^{2+} by this process, with loss of the normal red or brown oxide coloration. Fe^{2+} compounds are more water soluble than Fe^{3+} and may be washed out by percolating water, thus bleaching the mineral soil. The concentrations of iron and manganese increase in soil solution are due to electron acceptance by these high-valency ions. Divalent compounds are more soluble in water than are those of oxidized forms.

In soils rich in organic compounds, and especially at high temperatures, sulfate may be used as an electron acceptor by sulfate-reducing bacteria, producing free H_2S. In Fe^{2+} rich soils, the sulfide is immediately removed from solution by the precipitation of the insoluble FeS, which gives the characteristic black color to the *sapropel* sediments of eutrophic freshwaters and coastal mudflats.

In many flooded soils, dissolved iron and manganese reach phytotoxic concentrations. Sulfide is less commonly toxic, primarily because it is scavenged as FeS. However, in soils that have been leached of Fe^{2+} by laterally percolating water, plant growth may be affected. The hot organic-rich circumstances of rice culture are particularly conducive to sulfide formation. Hence, a range of diseases in rice are attributable to sulfide toxicity, although *Oryza sativa* is an extremely waterlogging-tolerant species.

Anaerobic microorganisms in waterlogged soil also liberate organic products that influence plant growth. When the soil first becomes anoxic, some fungi produce ethylene (C_2H_4), a potent plant growth regulator. Organic acids such as acetic and butyric may also act as phytotoxins in anaerobic soils. Conversely, oxidizable organic compounds may compete with plant respiratory oxygen need. Extremely reduced soils and sediments generate copious amount of inflammable methane microbially and a lesser amount of hydrogen, so that the soil atmosphere above the water table of some paddy soils may contain equal volumes of methane and nitrogen.

8.2.2.1 pH

The main materials required are a pH meter, a soil sample, buffer tablets, a beaker, a stirrer, and DW.

Procedure

1. A 20-g sample of powdered soil is taken in a 100-mL beaker.
2. DW (40 mL) is added to it.
3. The suspension is stirred at regular intervals for 30 min.
4. The electrodes of the electronic digital pH meter are dipped into the supernatant (suspension) just above the soil in the beaker, and care must be taken so that the electrode does not touch the soil below the supernatant.
5. The pH value is recorded.
6. The suspension is stirred well every time just before the electrodes are immersed.

The data are recorded in the following table:

Locality:	Date:	Time:
Sample no.	Soil pH	
Average soil pH		

8.2.2.2 Conductivity

The main materials required are a conductivity meter, a beaker, KCl, DW, and a soil sample.

Procedure

1. A 20-g sample of powdered soil is taken in a 100-mL beaker.
2. DW (40 mL) is added to it.
3. The suspension is stirred at regular intervals of 1 h. Alternatively, the soil suspension in the beaker is kept on an electrical shaker for 1 h.
4. The conductivity cell of the electronic digital conductivity meter is dipped into the supernatant (suspension) just above the soil in the beaker, and care be taken, so that the electrode does not touch the soil below the supernatant.
5. The conductivity value is recorded.
6. The suspension is stirred well every time just before the electrodes are immersed.

The data are recorded in the following table:

Locality:	Date:	Time:
Sample no.	Conductivity (micromhos/cm)	
Average value of conductivity (micromhos/cm):		

8.2.2.3 Organic Carbon

Method of Walkley and Black

Walkley and Black (1934) devised a method for the estimation of organic carbon (OC) in soils.

Principle

OC is estimated by titration in which the organic carbonaceous matters are oxidized by a known volume of a standard dichromate solution and then titrating back the volume of dichromate left unused with a standard solution of ferrous (ammonium) sulfate using diphenylamine as an indicator.

Main Materials Required

1(N) potassium dichromate (AR) solution: 49.4 g of AR grade $K_2Cr_2O_7$ is dissolved in 1 L of DW

0.5 (N) (approximately) Ferrous Ammonium Sulfate: 196 g of the hydrated salt is dissolved in 1 L of DW containing 20 mL of H_2SO_4

Diphenylamine indicator: 0.5 g of diphenylamine indicator is dissolved in 20 mL of DW and 100 mL of concentrated sulfuric acid

Concentrated sulfuric acid (sp.gr. 1.84) containing 1.25% silver sulfate

Orthophosphoric acid (85%)

Required glassware such as beakers, conical flasks, volumetric flasks, measuring cylinders, burettes, pipettes, and test tubes

Procedure

1. Soil weighing 1.0 g (ground and sieved through 0.2 mm sieve) is taken using a dry 500-mL conical flask.
2. Next, 10 mL of 1 (N) $K_2Cr_2O_7$ is pipetted into the conical flask and is swirled a little.
3. The conical flask is kept on an asbestos sheet.
4. Then, 20 mL of H_2SO_4 (containing 1.25% of silver sulfate) is added into the conical flask and the flask is swirled again two or three times.
5. The conical flask is allowed to stand for 30 min.
6. Then, 200 mL of DW is added into it.
7. After adding 10 mL of phosphoric acid and 1 mL of diphenylamine indicator, the contents in the conical flask are titrated with ferrous ammonium sulfate until the color changes from blue-violet to green.
8. Similarly, a blank is run without soil.

NB: If more than 7 mL of ferrous ammonium sulfate solution is consumed, the whole procedure of determination is repeated with a smaller quantity of soil.

Calculation and Result

$$\text{Organic Carbon}(\%) = [10(B - T)] \times 0.003$$
$$\times\ 100/(B - \text{weight of soil})$$

Where

B = Volume of ferrous ammonium sulfate solution required for blank (B)

T = Volume of ferrous ammonium sulfate solution required for test soil sample (T)

8.2.2.4 Available Phosphorus

It indicates the amount of phosphorus available in the soil. Available phosphorus (AP) could be estimated by two methods: (1) Olsen's method (Olsen et al., 1954) and (2) Dickman and Bray's method (Dickman and Bray, 1940).

Dickman and Bray's method is briefly described below:

Principle

The AP content in the soil is evaluated by rapidly extracting it with suitable reagents. In this method, the orthophosphate and molybdate ions are said to condense in acidic solution to give molybdophosphoric acid (phosphomolybdic acid), which upon selective reduction (say, with stannous chloride) produces a blue color due to molybdenum blue of uncertain composition.

Main Materials Required

1. Klett-Summerson (USA) photoelectric colorimeter (say model No.800-3)
2. Dickman and Bray's reagent: 15 g of ammonium molybdate is dissolved in 300 mL of DW and warmed to about 60 °C, and cooled and filtered if necessary. To this, 350 mL of 10 (N) HCl is added followed by an excess 50 mL of 10 (N) HCl and made to 1 L. The normality of the HCl used is adjusted correctly by titration.
3. Stannous chloride solution: 10 g of stannous chloride (crystalline) is dissolved in 26 mL of concentrated HCl by warming and then stored in an amber-colored bottle. This is the 40% $SnCl_2$ stock solution. Just before use, 0.5 mL is diluted to 66 mL with DW.
4. Glassware like beakers, conical flasks, volumetric flasks, measuring cylinders, burettes, pipettes, and test tubes

Procedure

1. First, 1.5 mL of soil extract is pipetted into a 25-mL volumetric flask
2. Next, 5 mL of Dickman and Bray's reagent is added.
3. The neck of the flask is washed down and the contents in the flask are diluted to c 22 mL.
4. To this, 1 mL of diluted stannous chloride solution is added, shaken, and filled to the 25-mL mark.
5. The intensity of the blue color is measured just after 10 min. This is important, as the color starts to fade after the stipulated time.

6. Preparation of standard curve for phosphorus: 0.439 g of AR-grade potassium dihydrogen orthophosphate (KH_2PO_4) is weighed accurately (after keeping it in a hot air oven at 40 °C for 1 h) and is dissolved in about 500 mL of DW. To this, 25 mL of 7 (N) H_2SO_4 is added and made up to 1 L with DW. This gives a 100-ppm stock solution of phosphorus (100 μg of P per mL). From this, a 2-ppm solution is made. For preparation of the standard curve, different concentrations of P (1, 2, 3, 4, 5, and 10 mL of 2-ppm solution) are taken using 25-mL volumetric flasks. The color is developed by following Dickman and Bray's method and the reading taken with the help of a Klett-Summerson photoelectric colorimeter against 660 nm. The standard curve is plotted by considering the colorimeter reading on the y-axis and the amount of P (in ppm) on the x-axis.
7. The concentration of phosphorus in the soil water is extrapolated from the standard curve.

8.2.2.5 Available Potassium

The readily exchangeable and water-soluble potassium (K) determined in neutral ammonium acetate extract represents the available potassium in the soil. The estimation is done with the help of a flame photometer (Jackson, 1973).

Principle

The amount of potash is estimated with the help of a flame photometer by comparison with a standard curve of known potassium oxide concentration determined by the same photometer.

Main materials required:

1. Flame photometer
2. Digital pH meter
3. Neutral normal ammonium acetate: Equal volumes of 2(N) acetic acid and 2 (N) ammonium hydroxide are mixed, and the pH adjusted to 7.0 with acid or ammonia
4. KCl (AR grade)

Procedure

1. First, 5.0 g of soil is shaken with 25 mL of neutral normal ammonium acetate (pH 7.0) for 5 min, and the contents are filtered immediately through a dry filter paper.
2. The first few mL of filtrate are rejected.
3. K is estimated in the extract with the help of the flame photometer after necessary calibration of the instrument.

Preparation of Standard Potassium Curve

1. A stock solution of 1000-ppm K is prepared by dissolving 1.9084 g of AR-grade KCl (dried at 60 °C for 1 h) in DW and the volume is made up to 1 L.
2. Required portions are diluted with tiemonium acetate solution to give 10–40 ppm of K.

3. After attaching the appropriate filter and adjusting the gas and air pressure, the flame photometer reading is set at zero for the blank (ammonium acetate) and 100 for 40-ppm K.
4. The curve is obtained by plotting the readings of the flame photometer against the different concentrations of K (10, 15, 20, 25, 30, and 35 ppm).
5. Potassium content in the sample is estimated through extrapolation.

8.2.2.6 Consumption, Decomposition, and Elemental Cycling

All living beings are composed of similar types of chemical elements. If populations of organisms are to grow, multiply (reproduce), and survive, and also in order to maintain a constant biomass, there should preferably be a continuing supply of all the chemical elements that are essential constituents of protoplasm. It is known that if the supply of any one essential element is depleted, this may limit the growth of a population. Further, the supply of elements may limit growth as well as shape the structure of the entire community. It should be noted that the supply of essential elements to populations and communities may come from outside the system, such as terrestrial runoff into streams, stream input into lakes (Godwin, 1978) and estuaries, river input into oceans, or vertical advection of deep water to the surface of the ocean. Nevertheless, much of the flow of energy and materials is said to occur through the detritus food web. Also, the supply of elements through recycling is intimately involved with the fate of detritus. On the other hand, populations of denitrifying bacteria in the anaerobic sediments of estuaries and other coastal waters are perhaps major sinks for available nitrogen (Vedpraskas, 1995).

8.2.2.7 Accumulation of Organic Matter and Their Export

Notwithstanding the above, the elemental cycles of wetland environments differ to some extent from those of terrestrial habitats in the ability of flowing water to import or export particulate and dissolved materials and also in the modification of decomposition and recycling processes. It is said to occur in the presence of low redox soil or sediment that causes loss of nitrogen but assists in the trapping of carbon, sulfur, and some heavy metals. Further, import and export by moving water has a substantial effect on the nitrogen cycle.

Along with the above, the annual nitrogen uptake of wetland vegetation is slightly similar to that of other productive habitats, as the elemental content of herbaceous tissues differs slightly between wet and dry ecosystems. Further, the denitrification rate of eutrophic wetlands is potentially enormous, particularly when they are enriched with sewage effluent or fertilizer.

Summary

1. Soil is the natural medium for growth of biota in both land and water and is a heterogeneous mixture.
2. The physical and chemical characteristics of soil in any area determine the different types of biota it can sustain
3. The methodology recommended by Jackson (1973) and ICAR (1960) is largely followed.
4. Lentic soils are important media in which many chemical transformations take place.
5. Soil may be of different kinds depending on the composition.
6. The remains of plants in various stages of decomposition primarily compose the organic wetland soil.
7. Soil that is generally dominated by minerals (when flooded for extended periods), constitutes the mineral wetland soil. The characteristics of its identification are collectively called redoximorphic features. Mineral wetland soils sometimes contain oxidized matter. Further, soils are categorized according to zonal types that broadly correspond with geographical and climatic conditions.
8. Soil parameters like temperature, pH, conductivity, OC, and NPK are generally estimated. In addition to these, soil moisture, BD, and infiltration rate are sometimes estimated mainly in land soils.

8.3 MORPHOLOGICAL AND ANATOMICAL STUDIES

Fish: Although the word "fish" is used here, the terms "cases" and 'noncases" should be used to embrace the wider definitions where "cases" might be ponds (Moore and Bellamy, 1973), farms, etc. For simplicity, the discussion could be restricted to individual fish only.

Fish are cold-blooded or poikilothermic animals, with their body temperature varying passively in accordance with the temperature of the surrounding water. Although fish as a group are tolerant of a wide range of temperatures, individual species have a preferred range or optimum temperature, which significantly affects the biology of fish, with the rates of all chemical reactions and processes within their bodies showing 50% increases for each 5 °C rise in temperature. The body shape of fish is usually streamlined, an important prerequisite for successful aquatic life, due to the 800-fold higher density that water has than air, and most predatory fish (such as bass and salmonids) are ovoid in cross section and torpedo-like or fusiform in shape. Species, age, sex, and geographical origin are often associated with varying risks of disease. However, it should be kept in mind that fish-level patterns could be closely linked to temporal and spatial patterns of the disease.

The steps in morphological and anatomical works are briefly listed below:

1. Lay the fish on side and remove the abdominal body wall, operculum, and integumental wall over the cardiac cavity.
2. Expose viscera and to identify esophagus, stomach, pyloric caeca (if present), intestine, liver, gall bladder, swim bladder, spleen, kidney, etc.
3. Remove heart and identify two main chambers, viz., the triangular ventricle and the soft atrium, and additional structures like the elastic bulbus arteriosus.
4. Cut a cross section through the fish anterior to the caudal peduncle and examine for red and white muscle, vertebral column, and spinal cord.
5. Remove the top of the cranium carefully and expose the brain. Section through the anterior olfactory nerves and the spinal cord so that the entire brain can be removed.

Concomitantly, the morphology and anatomy of a fish may be studied in a more detailed manner:

1. Examine the inside of the **mouth** and feel for teeth and gill rakers.
2. Examine **scales** under the microscope and look for chromatophores.
3. The **skin** of fish is very important for various reasons. Fish skin could be viewed as having two main layers, viz., the outer epidermis and the underlying dermis. Outside the epidermis is the cuticle or mucous layer that, in addition to providing lubrication, makes the skin less permeable and prevents the entry of pollutants and microorganisms. The mucous is secreted from mucus cells that reside in the fragile epidermis. The epidermis is composed of living cells to the outermost layers. The scales, which are calcified flexible plates, grow out from the dermis. In higher teleosts, the scales have spicular processes from their external posterior edge. Growth rings or annuli are visible on scales of wild fish, similar to the rings seen in the main trunks of trees. Pigment cells or chromatophores are highly developed in fish and the melanophores are the dark brown or black pigmented cells; iridophores are silver; and there is a range of lipophores that contains the organic solvent-soluble pigments (reds, yellows, etc.). The decision on the color the fish is said to be depends to some extent on what it sees outside and further on its health status. If it is sick or starving, the fish releases the melanin concentrated in the melanophores and this results in a dark fish.
4. Study the **paired and unpaired fins** of the fish. It is known that the dorsal and the caudal fins are vital for locomotion, but may be subject to erosion or damage in a crowded farm situation. Dorsal fin rot or erosion is often seen where stocking levels are too high,

nutrition is marginal, or water temperatures have been at the low end of normal ranges, for that fish species, for a prolonged period. Aggression, especially at feeding time, could result in fin and tail nipping, which will result in erosion or tail rot. Too high a stocking level may also damage pectoral fins or when the tank or pond sides are constructed from an abrasive material. The adipose fin acts like the spoiler of a car and is only present in some fish species. Most fish swim by passing a wave of increasing amplitude along the body and this is generated by sequential contraction from head to tail of the muscle blocks or myomeres.

5. The **digestive systems** of fish may vary in a number of aspects in accordance with species. Herbivorous fish usually have long intestines and little or no stomachs (e.g., grass carp and silver carp), whereas carnivorous fish generally have larger stomachs and short intestines (e.g., salmonids and striped bass). Other variations include dentition and presence and numbers of diverticula. The stomach is usually sigmoid and highly distensible. Pyloric caeca, blind ending diverticula from the distal pyloric valve region of the stomach and from the anterior intestine, are found in many species, most notably in salmonids, where they may number 70 or more. Histologically these are said to resemble intestine.
6. The teleost **liver** is relatively large and the color depends on diet. In wild fish, it is usually reddish brown in carnivores and lighter brown in herbivores but seasonal variations may occur. In farmed fish, the color is usually a reflection of the dietary lipid levels, and is normally lighter than in the wild fish. In some species, the liver could be a compound organ, called the hepatopancreas, where exocrine pancreatic tissue is located around hepatic portal veins. The fish liver is not lobulated like those of mammals. The biliary system also differs, in that the intracellular bile canaliculi occur that eventually anastomose to form typical bile ducts. The bile ducts fuse and ultimately are said to form the gallbladder. The pancreatic tissue in fish may generally show more variations in their location in fish, than the other abdominal viscera. However, the most common location is in the mesenteric fat interspersed between the pyloric caeca.
7. The **pancreas** is located as a subcapsular layer in the spleen in some species, but in most species, the main elements of the spleen are the ellipsoids, the pulp, and the melanomacrophage centers.
8. Locate the nostrils, operculum, lateral line sense organ, eyes, anus, and urinogenital opening. Locate the nostrils of the fish. The nostrils have epidermal flaps in some species and are blind ending pits, which house nerve endings and mucus cells. The nerves run directly to the forebrain. Pollutants can damage these sensitive

surfaces, which are relied on greatly by migratory species such as salmon and eels. The lateral line is the main vibration sense organ in fish and may be damaged by pollutants, chemicals, or parasites. It runs as a paired canal along the flanks, has an integumental cover, which is punctuated by sequential pores along its length. The mechanoreceptors or neuromasts are located basally in the canal and are stimulated by changes in the external milieu, in terms of displacement or vibration.

9. Locate the **operculum** of the fish. The operculum or gill cover provides physical protection for the gills but is also an actual component of the respiratory mechanism. Foreshortened opercula are a problem in many species and can be either genetic or environmental in origin.

10. Study the **gills** of the fish. The gills are located beneath the opercula and consist of four white bony or cartilaginous arches and the red or pink gill lamellae. The fifth gill or pseudobranch is an embryonic red gill-like structure located on the underside of the operculum but is not present in all species (e.g., eels). Its function remains to be defined in full but it has an endocrine and regulatory function as well as a hyperoxygenation function for the retinal blood supply. The gills undertake the tasks of the uptake of oxygen and associated loss of carbon dioxide. Secondary lamellae branch off the primary lamellae and the numbers present reflect the fish's lifestyle; i.e., slow moving bottom dwellers may have only 10 lamellae per mm of filament, whereas fast swimming predators will have 30 to 40. A complex of capillary channels is present in the secondary lamellae and the thin lamellar walls (usually only one cell layer thick) readily allows for respiratory exchange between the blood and the surrounding water. Blood flow is arranged, so that the direction of flow is opposite that of the water crossing the gills; thereby, increasing the efficiency of respiratory exchange. The spikes on the gill arches are the gill rakers and these prevent food materials entering the gill chambers. They are particularly well developed in plankton or filter feeding fish. As well as a respiratory function, the gills are also responsible for regulating the exchange of salt and water and play a major role in the excretion of nitrogenous waste products (ammonia). Even slight structural damage could render fish very susceptible to osmoregulatory as well as respiratory difficulties.

With low oxygen levels in the water column or gill damage that reduces the respiration efficiency, a direct consequence will be an increased ventilation rate, which could be observed clinically as an increased rate of opercular movement. This could be further observed as the water temperature rises, because less oxygen can dissolve in the water at higher temperatures. Gill epithelium is very prone to damage from parasites, water borne irritants or toxins, and high levels of suspended solids.

11. Notwithstanding the above, the gas-filled **swim bladder** is a characteristic feature of many teleost fish, although absent in bottom dwelling fish and some fast-swimming pelagics. Its primary function is to offer buoyancy, but it is also believed to be used for sound and pressure reception. Further, in some species, it is equipped with drumming muscles for sound production. The embryonic connection between the gut and the swim bladder is often retained as a pneumatic duct in the more primitive fish species (physostomes), but has been lost in most of the spiny-rayed fish (physoclists). Moreover, in many of the physostomes, the swim bladder is two-chambered, separated by a diaphragm. In such cases, the anterior chamber is associated with gas reception and retention, while the posterior chamber is involved with gas reabsorption.

12. The **eyes** have fixed spherical lenses that are virtually free floating and are vulnerable to parasites, environmental damage, and nutritional deficiencies. The lens protrudes partially through the iris to provide a very wide angle of view and the iris is limited in reaction to light intensity, having a poorly developed sphincter and dilator muscle. The cornea may be tinted in some species.

13. The **spleen** is located in the peritoneal fat, near the greater curvature of the stomach or the first flexure of the intestine. It is usually single, although in some species, it may be paired.
In circulation, the venous deoxygenated blood enters the thin-walled cardiac atrium, is then pumped into the muscular ventricle, and from there into the fibroelastic bulbus arteriosus. Coronary vessels run over the outside of the ventricle, supplying the compact muscle. Heart rates vary considerably according to temperature, from 15/min in trout at 5 °C to 100/min at 15 °C. The ventral aorta runs from the heart and distributes blood to the gills via the afferent branchial arteries.

14. Concomitant to the above, there are two types of bones in teleosts, viz., (1) cellular as in other vertebrates and (2) acellular, which may usually be found in the advanced teleosts such as the perches and the sunfishes. However, the majority of the fishes are believed to have no hemopoietic tissue in their bone spaces and vascular canals.

15. Concomitantly, the piscine brain is similar to the brain of higher animals in its basic components. However, there are significant differences in form and

complexity. The fish brain could be divided into five main chambers:

a. the telencephalon or forebrain (olfaction, color vision, and memory);

b. the diencephalon (thalamus, epithalamus, and hypothalamus);

c. the mesencephalon (optic lobes);

d. the metencephalon or cerebellum; and

e. the medulla oblongata, which joins with the spinal cord. The pituitary gland, which incorporates the neurohypophysis, is located at the ventral base of the brain in a bony cupula. Its function is similar to that in other vertebrates—that is, it conducts the body orchestra.

16. The **thymus** is a paired organ, an ovoid pad of primary lymphoid tissue, located subcutaneously in the dorsal commissure of the operculum.

17. The **excretory** section of the fish kidney varies dramatically depending on whether the fish is marine, euryhaline, or freshwater, reflecting the significant differences in their respective functions. The major work in this field is that of Hickman and Trump (1969). In freshwater, fish drink very little but produce copious amounts of dilute urine; few salts appear in the urine because the kidneys reabsorb them. Salts are also gained from the surrounding water by the active uptake through the gills by special chloride cells found at the base of the secondary lamellae. In seawater, fish drink a lot (up to 15% of body weight per day); selectively excrete monovalent ions (Na^+, Cl^-) through the gills, and produce small amounts of concentrated urine. The chloride cells in the gills are responsible for removing the excess salt from the blood and passing it out to the water. Salt and water regulation and excretion require that gills and kidneys are in healthy condition, as damage to either or both of these organs could result in a fish's inability to respond to osmotic changes. Hence, kidney damage, from diseases such as, bacterial kidney disease (BKD) or nephrocalcinosis, may not be apparent in salmonids until such fish are moved to seawater where they will suffer high mortalities.

18. Concomitant to the above, teleosts generally depict more diversity in their **reproductive** patterns than any other group in the Animal Kingdom. Sexes are separate in most of the teleosts. However, hermaphroditism and bisexuality may also occur. Also, both parthenogenesis (development from an unfertilized ovum) and gynogenesis (development from an ovum stimulated to divide by penetration from a sperm that does not contribute genes) are also recorded. In fact, the gonads develop as paired organs and generally lie just below the kidneys. In immature fish, these are rudimentary thread-like structures. But in matured fish, the ovaries could constitute up to *c* 70% of the body weight. The testes have a vas deferens or collecting duct, which transports mature spermatozoa to the excretory meatus at the urinary papilla. The ovaries pass the ova to the outside through the oviduct or into the abdominal cavity in the more primitive species, for evacuation via the genital opening. Live-bearers store eggs in a pouch referred to as a uterus, but this is in essence a simple storage space.

8.3.1 Details of Investigation of EUS Outbreaks

8.3.1.1 Procedure of Investigation

The procedure for an outbreak investigation follows nine basic steps. Not all the steps are necessarily included in every investigation, nor do they always follow the same sequence. In practice, several steps will be undertaken simultaneously.

The nine basic steps are:

8.3.1.2 To Establish a Diagnosis

The initial provisional diagnosis in an EUS outbreak is usually based on species of fish affected, clinical signs, gross pathology, and perhaps seasonality (Lilley et al., 1998). Whenever possible, laboratory tests should be undertaken to verify the provisional diagnosis. Since some laboratory procedures (e.g., HP, fungal isolation) may take weeks, the implementation of control measures is often based on the provisional diagnosis.

8.3.1.3 Diagnostic Methods (OIE, 2009)

Diagnostic methods comprise (1) field diagnostic, (2) clinical, and (3) agent detection and identification methods.

Field Diagnostic Methods

There is mass mortality of various species of freshwater fish in the wild and in farms during periods of low temperatures and after periods of heavy rainfall. Loss of appetite and fish had become darker. Infected fish may float below the water surface, and become hyperactive with a very jerky pattern of movement.

Clinical Methods

Development of red spots followed by small-to-large ulcerative lesions on the body surface, head, operculum or, caudal peduncle. In later stages, large red or gray shallow ulcers, often with a brown necrosis, are usually observed.

8.3.1.4 To Define a "Case"

Depending on the type of investigation, an EUS case might be an individual affected fish or an aggregation of individuals such as the population in an affected pond (Roberts et al., 1994). A useful case definition at the individual animal level might be "a fish with necrotizing granulomatous dermatitis and myositis associated with highly invasive nonseptate fungal hyphae."

8.3.1.5 Confirm That an Outbreak Is Actually Occurring

This step may seem unnecessary but in many instances it is required, particularly in areas where EUS is already endemic. The disease may be expected to occur at low prevalence at certain times, but even a moderate prevalence increase, especially if ulceration is severe and/or toxigenic *Aeromonas hydrophila* is present, will lead to substantial production losses if not recognized early. Moreover, dermal ulcerations caused by other agents may be common in fish populations and are often macroscopically very difficult to distinguish from EUS. Laboratory confirmation of a diagnosis of EUS will usually be necessary.

8.3.1.6 Analysis of Data

Factor-specific attack rates for such factors, such as species, age, sex, pond, and management system are calculated and arranged in an "attack rate table." A theoretical example is shown below for EUS where size indicating nutritional stress is suspected.

In the table, attack rates are usually expressed as percentages. The second-to-last column is the difference in attack rates (sometimes called the "attributable risk") and the last column is the "relative risk," which is the ratio of attack rates.

The higher the attack rate difference and the relative risk, the more important the specific factor is in increasing the risk of the disease. The analysis becomes more complicated when trying to sort out interactions and confounding among factors. Stratified and multivariate analyses are used to investigate these phenomena.

8.3.1.7 Go for Reporting

This may take the form of a brief discussion with the farm manager for small outbreaks, outlining the important features and actions required to prevent further occurrences. However, it could be a wise effort to always produce some form of written report so that a permanent record of events exists for future use. For large outbreaks, findings should be published in the scientific literature.

For substantial investigations, the report may contain the following sections: background, methods, results, hypotheses, financial impact (where appropriate), recommendations, and appendices containing laboratory reports, etc.

8.3.1.8 EUS Field Sampling Data Sheets

To make information available for the region-wide collaborative programs on EUS:

8.3.1.9 EUS Sampling Data Sheet

Name of site	Date of record:	Collector:
Current EUS outbreak, if any:		
If yes, is the present outbreak:		1. restricted to sampling site? 2. occurring throughout the local area? 3. a national problem?

Tentative date on which the present outbreak started:

Estimated number/weight of fish lost from present outbreak:

Value of losses from present outbreak:

What were the conditions 3–12 days before the outbreak (temperature, rainfall, other important physico-chemical features, etc.)?

Fish market price:

Species	Price/kg of unaffected fish	Price/kg of affected fish
Site description		
Country:		
Province:		
District:		
Town:		
Village:		

Type of water body at site:

Farm	Lake	Reservoir
Canal	River	Rice field
Swamp	Others	

Short description of the site:
Previous history of EUS at the site:

8.3.1.10 Sampling Point

How and where the fishes were sampled on the site:
Was selection of sampling point random?
If not, give reason:

Size of pond: Depth of pond: Fish species:

Stocking rate:

Are there wild fish in the pond?

Were fish releases made just before the outbreak?

What is the source of the water to this pond?

Describe attempted treatments/control strategies:

Perceived importance of the problem to local farmers/fishers:

Water quality data:

Fish population data:

8.3.1.11 Sampling for Disease Diagnosis

The ideal specimens for disease investigation or health monitoring are live fish. Samples could be taken from these either on site or transported live to the appropriate laboratory.

Infected fishes could preferably be collected endemic and intensely affected regions. The author of this treatise has been collecting EUS-infected fishes from various locations in Assam and other Northeastern regions of India. The tissue samples would be aseptically removed. Standard methods would be followed for the isolation of viruses using cell lines. The methodology for isolation of viruses is given below:

8.3.1.12 Transport of Live Fish

Place the affected fish in a plastic bag filled approximately one-third with water and two-thirds with oxygen. Seal the bag and then place in another bag and seal again. Then, place the bag on ice or cold packs in an insulated box—for example, polystyrene—and place more ice on top and seal the container.

The maximum transport time depends on water temperature and the ratio between biomass, water volume, and oxygen. As a rough guide, the transport time may not exceed 12 h, and also the biomass may not exceed one-third the water volume. Transport time is to be significantly reduced if oxygen is not used.

Live fish weighing >300 g (approximately) may preferably be sent not by normal goods transport (e.g., by air, rail, or road), but rather may be either sampled on site or sent via specialized forms of transport.

8.3.1.13 Transport of Fresh Material

Unopened fish, reproductive materials, viral samples, and fish heads (for *Myxobolus*) may be dispatched for laboratory investigation in fresh state without freezing. Pack samples in ice and in an insulated containers and then dispatch. The maximum transport time may not be >24 h.

8.3.1.14 Microscopic Pathology

Early EUS lesions are said to be caused by erythematous dermatitis with no obvious oomycete involvement. *Apha-nomyces invadans* hyphae are usually observed growing in skeletal muscle as the lesion progresses from a mild chronic active dermatitis to a severe locally extensive necrotizing granulomatous dermatitis with severe floccular degeneration of the muscle. The oomycete elicits a strong inflammatory response, and granulomas are formed around the penetrating hyphae. Lesion scrapes from fish body or ulcers generally show secondary fungal, bacterial, and/or parasitic infections.

8.3.1.15 Microscopic Methods

The squash preparation may be carried out as follows:

1. Remove the ulcer surface using a sharp scalpel blade.
2. Cut the muscular tissue at the edge of the ulcer.
3. Place the pieces of the tissue on a cutting board; then, make thin slices using a sharp scalpel blade.
4. Place the thinly sliced tissue between two glass slides and squeeze gently with fingers.
5. Remove one of the glass slides and cover the tissue with a cover slip. View under a light microscope in order to find the nonseptate hyphae structure of *A. invadans* (12–25 μm in diameter).

The chemicals used should be reagent (analytical) grade. To avoid contaminating the stock pimaricin suspension, dispense a 1-mL aliquot for day-to-day use aseptically. All antibiotics should be kept at 4 °C.

8.3.2 Materials for Laboratory Work

8.3.2.1 Glassware

All glassware used throughout this procedure must be scrupulously clean. Minimum requirements are as follows:

1. Soak in laboratory detergent.
2. Rinse three times in tap water.
3. Rinse five times in DW.
4. Mark 1 L glass bottles at 200 mL capacity.
5. Sample collection bottle: The sample collection bottle used is a modified 100-mL glass Schott bottle. The inlet (a small glass filter funnel) is covered with nylon mesh of 200-mm aperture size. The collection bottle is secured to a 2-m long stick, so that when the stick touches the bottom sediment, the inlet will be about 5 cm above the sediment. The stick is used to push the bottle to the bottom as rapidly as possible. Plastic bottles and tubing are not suitable.

Pipettes

By preference, all pipetting is to be done with air displacement pipettes with disposable tips. Two micropipettes are used, one capable of 40–200 μL and the other 200–1000 μL. Where this may not be possible, glass capillary plunger-type pipettes (Socorex) could suffice, if capillaries are discarded after use. If glass pipettes have to be used, they must be carefully washed. Where small volumes are required—for example, pimaricin—it may be necessary to prepare intermediate dilutions.

Filtered Pond Water

Collect 3 L or more of pond water from one of the untreated ponds to be tested. For each pond, a pooled water sample of *c* 250 mL of filtered pond water (FPW) will be required. Filter through Whatman No. 1 filter paper and autoclave at 121 °C for 15 min. Store at room temperature. Before use, this water sample is to be tested using the FPW positive control system described below in order to ensure that 100% of *A. invadans* colonies sporulate satisfactorily.

For stock metal ion solutions, use sterile glassware to make up solutions. Keep the solutions in separate sterile disposable plastic containers at 4 °C.

8.3.3 General Guidelines for Sampling

8.3.3.1 Materials and Methods

General Field Surveys

On arrival at the sampling site of the water body, information is usually gathered from the farmers. This is generally followed by the estimation of water quality parameters and examination of some susceptible fish.

Selection of Individual Specimens

Scoop net, cast net, and seine net represent the best choices for catching EUS-infected fish in natural waters or fishponds. For outbreak investigations, diseased fish with ulcerative lesions or red spots on the body should be sampled.

After collection, they were immediately brought to the laboratory and acclimatized in 1000-L indoor circular tanks supplied with dechlorinated and aerated tap water at ambient temperature (23–26 °C). The fish were fed twice daily (rice bran and mustard oil cake in a ratio of 1:1). Water was maintained at optimum levels for the fish species throughout the experiment: DO 6.88 ± 0.56 mg/L, pH 7.14 ± 0.77, ammonia 0.029 ± 0.011 mg/L, and nitrite 0.016 ± 0.01 mg/L. Water was periodically exchanged to remove waste feed and fecal material.

Preservation of Samples

Fish specimens should be transported to the laboratory live or in ice-cooled boxes for further diagnosis. Fish collected from remote areas should be anesthetized and can be fixed in normal 10% formalin or 10% phosphate-buffered formalin (PBF) for at least one to two days. The fixed specimens are then transferred to double-layer plastic bags with formalin-moistened tissue paper. The bags are sealed and sent to the laboratory in semidry conditions.

Pooling of Samples

Ten diseased fish specimens are sampled from the EUS-infected site. Diagnosis is achieved using the histological technique and oomycete isolation on individual fish or a group of a few fish.

Best Organs or Tissues

Fish with minor clinical signs are recommended, and the muscle tissue next to or underneath the ulcer is best for oomycete isolation. The best tissue for HP examination is muscle tissue at the edge of the ulcers.

Sample/Tissues That Are Not Suitable

Severely diseased or dead fish are not suitable for oomycete isolation.

Diagnostic Methods

Diagnosis of EUS is based on clinical signs and confirmed by HP. Diagnosis of EUS in clinically affected fish may be achieved by HP or by oomycete isolation. Positive diagnosis of EUS is made by demonstrating the presence of mycotic granulomas in histological sections or isolation of *A. invadans* from internal tissues.

Field Diagnostic Methods

EUS outbreaks have been associated with mass mortality of various species of freshwater fish in the wild (including rice—fields, estuaries, lakes, and rivers) and in farms during periods of low temperatures and after periods of heavy rainfall.

Clinical Signs

Fish usually develop red spots or small to large ulcerative lesions on the body.

Behavioral Changes

The early signs of the disease include loss of appetite and fish become darker. Infected fish may float near the surface of the water, and become hyperactive with a very jerky pattern of movement.

8.4 HISTOPATHOLOGICAL AND HEMATOLOGICAL STUDIES

8.4.1 Method of Sampling Fish for Histological Examination

Histology deals with the study of tissues and also their pathology called histopathology. HP can therefore provide information on the processes and changes occurring in tissues. In many cases, it forms the basis of disease diagnosis and prognosis. Accurate sampling of tissues for histology is thus a vital part in the diagnostic procedure (Chinabut and Roberts, 1999). However, it is necessary to note any behavioral abnormalities or visible external lesions before sampling any fish.

Notwithstanding the above, histologically and biochemically, the muscle could be classified into two types: (1) the red, aerobic, slow contracting muscle fibers and (2) the white, anaerobic, fast contracting muscle fibers.

In some species of fish, there are also pink fibers, which are sandwiched between the two types and these appear to be intermediate in function as well as location. The well-vascularized red muscle is best observed as the triangles of darker muscle located over the lateral line and midline dorsally. The majority of the body muscle is the white muscle and is usually used under escaping or chasing situations.

Notwithstanding the above, it is always preferred to follow the steps given below:

1. Complete the EUS sampling data sheet for each site, recording full details of each fish sampled. As with other diagnostic procedures, a mixture of sick (moribund) and healthy (control) fishes are to be sampled. Dead fish (recent mortalities) could provide little accurate information and are said to be useful for histology.
2. Sample only live specimens of diseased fish. If fish with clinical signs of EUS are readily available, several samples of each species are to be collected, preferably at different stages of infection. To sacrifice the chosen fish by stunning the fish by a blow on the head, or by using anesthesia.
3. Dissect large fish and to take samples of skin, muscle (<1 cm^3), spleen, kidney, and liver. The muscle sample may include the lesion and the surrounding tissues. In brief, remove (excise) a small piece of tissue (<1 cm^3) from the organs, and include any areas showing gross abnormalities.
4. Fix the tissues immediately in cold 10% formalin. The amount of formalin in the jar should be 15–20 times the volume of the tissue to be fixed. Often, PBF is an effective tissue fixative for the samples. The usual concentration is 10% formalin (formalin being 40% formaldehyde in water). Care should be taken, as formalin is an irritant, especially to the eyes.
5. Gently agitate the fixative two to three times over the first hour after adding the tissue.
6. The selected site could be sampled repeatedly during the period of outbreak, and the specimens are to be sent to a centralized diagnostic facility. If an appropriate facility is not available in the country, it could be better to make the information available to the Regional Collaborative Program on EUS for further necessary works.
7. Live samples, if available, are considered best for laboratory examinations. The fish be packed in double

plastic bags; filled with water to one-third of their capacity; with the remaining two-third volume inflated with air/oxygen. Bags are tightly sealed with rubber band or tape.
8. If live fish, which could be transported to the laboratory, is not available, recently dead or moribund fish with clinical lesions could be used.
9. Take samples of skin/muscle sections using a scalpel or a blade.
10. The samples are to be immersed in formalin as quickly as possible after the fish is sacrificed (killed). In fact, all tissues should preferably be sampled within 5 min of killing, as postmortem changes occur rapidly in these cold-blooded animals.
11. Sample gill, heart, spleen, liver, pyloric caeca, pancreas, kidney, skin, muscle (preferably at lateral line), and brain. Small fish or fry may be sampled whole with the abdomen opened to allow proper fixation of the internal organs.
12. Ensure that the amount of formalin in the container is about 10 times the amount of tissue.
13. Record any details about the fish in a piece of paper; e.g., body condition, length, weight, feeding, internal appearance, etc.
14. Label the sampling container with a waterproof marker and to ensure that each container will not leak.
15. Dispatch/deliver samples with relevant details and clinical history and also to contact the diagnostic service over telephone or fax:

8.4.2 Brief Examination of EUS-Affected Samples

1. The fixed tissues could be wrapped into formalin-moistened tissue paper and placed into small plastic bags to prevent leakage or smell during transport.
2. Make sure that the samples are properly labeled with the following information:
   a. Date of samples; type of tissue samples (e.g., skin, gills, muscle, kidney, other internal organs), collected; locality (place of collection); species of fish; weight and length of the affected fish, if possible; name of the collector; type of fixative used (10% formalin); etc.
   b. Samples could be packed into a padded envelope or container and sent by post, if no courier service is available.
   c. The laboratory to be informed about the kind of samples collected and when they are expected to arrive or to be delivered.
   d. Processing the fixed tissue involves dehydration through ascending alcoholic grades; clearing in a wax-miscible agent, and impregnation with wax.

The blocks of fish tissue are cut at about 5 μm and mounted on a glass slide. Before staining, the section is completely dewaxed and stained in haematoxylin and eosin (H&E). H&E and general fungal stains (e.g., Grocott's stain) could demonstrate typical granulomas and invasive hyphae.

8.4.3 Emergency Disease Hotline

There could a national disease hotline number. This number is supposed to put us in contact with the appropriate state or territory agency.

Hematology and blood biochemistry of fish are areas, which have been utilized only to a limited extent for clinical investigations. Normal values for many species remain to be established and these will vary according to season, age, temperature, genetic strain, physiological status, nutrition, and sampling methodology.

8.4.4 Sampling of Blood

Fish blood may be sampled for disease monitoring or health status analyses in the following areas:

1. Hematology (examination of blood cells and blood cell indices); for example, red blood cell count, hematocrit
2. Blood biochemistry; for example, hormones, enzymes, etc.
3. Plasma parameters for example, plasma chlorides (for salt water challenge tests in salmon)
4. Serology for pathogen antibodies; that is, salmon pancreas disease virus antibody screening
5. Virology and Bacteriology: Some microbes could be screened for using frank blood. For biochemistry and hematology, an anticoagulant be used in the blood collection tube and this is normally heparin (heparinized tubes may be purchased). However, the laboratory where samples would be submitted, should be consulted for their normal requirements. For serology, bacteriology, and virology, blood should be taken without anticoagulants. Where fish are large enough, blood samples could be taken under anesthesia; that is, nonlethal sampling. The normal sampling site is from the caudal vein and the easiest approach is ventrally at midline toward the vertebral column.

8.4.5 Hematological Parameters

For blood sampling, the preferred sites are the caudal vein (from either a ventral or a lateral aspect), cardiac puncture (ventrally), or the brachial plexus (caudal to the gills). Notably, the blood volumes in fish are small compared with those of mammals, being c 5% of the body weight.

Collection of blood and separation of serum and plasma: Blood from experimentally infected and apparently healthy catla was drawn from the caudal vein using a sterile 1-mL hypodermal syringe and a 24-gauge needle. The blood was allowed to clot at room temperature and the serum was separated and stored in sterilized vials at −20 °C, until further use. Prepare plasma using blood that was collected from the fish, using EDTA as an anticoagulant, and centrifuge at 1000 rpm for 10 min. The supernatant (plasma) is collected and stored in sterile tubes at −20 °C, until further use.

Notwithstanding the above, hemopoietic tissue in the fish is predominantly located in the stroma of the spleen and the interstitium of the kidney. It is also found in the periportal areas of the liver, the intestinal submucosa, and the thymus to a lesser extent. The kidney of fish is usually located in a retroperitoneal position, up against the ventral aspect of the vertebral column. It is usually divided by function and histology into the anterior or head kidney, which is predominantly hemopoietic, and posterior or excretory kidney. In the salmonid kidneys, the corpuscles of Stannius could be seen as paired white nodules in the mid-kidney. These are endocrine glands and appear to be involved with calcium metabolism, although their exact role is unclear.

Hematological parameters be measured using standard procedures and may include total red blood cell count, total white blood cell count, packed cell volume, and Hb content for experimentally infected and apparently healthy fish.

8.4.6 Biochemical Parameters

Biochemical parameters (Blazevic and Ederer, 1975) could be measured by using a biochemical test kit (Crest Biosystems, India) per the manufacturer's instructions, and could include serum glutamate pyruvate transaminase (SGPT) activity, serum glutamate oxaloacetate transaminase (SGOT) activity, serum alkaline phosphatase (ALP) activity, and total plasma protein for all experimentally infected and apparently healthy fish.

8.5 BACTERIOLOGICAL STUDY

8.5.1 Methodology

1. Take a clean slide, and using a sterile loop, emulsify a small amount of material in a drop of sterile saline or DW.
2. Allow the slide to air-dry, then heat fix it by passing it through a flame three times, then allow the slide to cool.
3. Flood the slide with crystal violet for 1 min.
4. Hold the slide at a steep angle, wash off the stain with Gram's iodine, and allow it to sit for 1 min.
5. Tip off the iodine and pour the alcohol/acetone mixture over the slide from the upper end so as to cover its whole surface. Repeat until no more color runs off. Then, wash gently in water.

8.5.1.1 Examination and Direct Inoculation of Solid Media

1. Before examining a fish internally, the external body surface, including gills, tail, and fins, is to be examined for the presence of any lesions. Observations are always recorded on papers.
2. Samples from these sites are taken by searing the surface with a hot scalpel blade followed by insertion of a sterile bacteriological loop or swab. Material from the loop/swab is then plated out onto suitable agar medium by the spread plate technique.
3. Once external examination or sampling has been carried out, the body surface is opened to expose the internal organs. Care is taken not to puncture the gastrointestinal tract. In the absence of any visible internal lesions, a sample of kidney is taken and inoculated onto suitable agar medium. The surface of the kidney (or other organ) is seared with a hot scalpel blade before insertion of the sterile loop.
4. Agar plates containing the streaked out samples be incubated and examined daily for any evidence of growth. The majority of bacterial fish pathogens should grow on Tryptone Soya Agar (TSA) within seven days. Media such as marine agar or TSA along with 1.5% NaCl could be used for marine pathogens.

Kidney Smears

1. Direct examination of Gram's solution—stained kidney smears may give an indication of bacterial septicemia and is especially useful in the examination of fish for BKD.
2. A small portion of kidney is emulsified in a loopful of sterile physiological saline on a microscopic slide, allowed to air-dry, and then fixed by passing the slide through a Bunsen flame several times.
3. Direct observation of the slide is carried out using the X40 and X100 (oil immersion) objectives after being stained using Gram's method.
4. Bacteriological plates or swabs may be despatched to the lab for investigation, but preferably should be clearly identified and labeled. A written note regarding the history and clinical appearance of the disease, the fish specimen, and the details of the samples taken are important.

8.5.1.2 Agar Plate Inoculation

1. Petri dishes containing solid medium (agar) are used to provide a large surface of media for the cultivation of microbes. Inoculation of an agar plate is often carried out using the streak plate technique as indicated before.
2. This involves diluting the culture or other sample—for example, the kidney material—by smearing it across the surface of the agar. Organisms present in the sample would be separated, and after suitable incubation, each organism present will give rise to a colony. Although this colony contains many millions of organisms, they will have all originated from one. Therefore, all organisms in one colony are said to be identical.
3. By using this method, organisms could be cultured in the laboratory, and if a mixed culture is present, it will become apparent on plating out. This is essential before starting identification procedures, because methods are only valid when carried out on pure cultures; that is, cultures containing one type of organism.
4. Care should be taken that the plates used for this purpose are dry. Also, unnecessary exposure of the agar surface to potential contamination from the environment throughout the procedure should be avoided.

8.5.1.3 Streak Plate Technique

1. Sterilize a bacteriological loop and allow it to cool.
2. Pick up a small amount of sample using the sterile loop.
3. Inoculate the sample on a segment of the surface of the culture medium.
4. Sterilize the loop to allow it to cool and, touch the edge of the uninoculated area of the medium to ensure coolness.
5. Spread part of the sample over about a quarter of the plate by making three to four parallel streaks with the loop.
6. Repeat streaking procedure, sterilizing and cooling the loop between each sequence.
7. Label the underside (not the lid) of the plate, writing only at the edge, and then, to incubate or despatch with related clinical notes and history. Ensure that the dispatched samples are adequately padded to protect against physical damage.

8.5.1.4 Gram Staining

1. This is a differential staining method and is the first test for the characterization of all bacterial isolates.
2. It tries to distinguish bacteria of different types:
   a. Gram-positive: those that resist decolorization and stain blue/purple
   b. Gram-negative: those that are decolorized and stain red/pink

It may be noted here that the difference in color reaction is due to the different chemical composition of the cell wall and the membrane. This stain is also said to reveal the general shape of the bacteria.

The method could be useful in the diagnosis of BKD and rainbow trout fry syndrome, by using kidney material taken directly from the fish.

8.6 MYCOLOGICAL STUDIES

The genus *Aphanomyces* is a member of a group of organisms commonly known as the water molds (Lilley et al., 1997). Although long regarded as a fungus because of its characteristic filamentous growth, this group, the Oomycetida, is not a member of the Eumycota, but is classified with diatoms and brown algae in a group called the Stramenopiles or Chromista.

8.6.1 Survival Outside the Host

How *A. invadans* survives outside the host is still unclear. If the motile zoospore cannot find suitable substrates, it will encyst (Willoughby et al., 1995). There is no suitable method to recover or isolate the encysted zoospore in EUS-infected fishponds. How long the encysted spore can survive in water or on a nonfish substrate is still unclear (Roberts et al., 1993). In an in vitro experiment, the encysted zoospore had survived for at least 19 days.

8.6.2 Stability of the Agent (Effective Inactivation Methods)

Aphanomyces invadans is said to grow best at 20–30 °C. It does not grow in vitro at 37 °C. Water salinity of >2 parts per 1000 (ppt) could stop spread of the agent. Preparing fishponds by sun drying and liming are effective disinfection methods for EUS. Similar to other oomycetes or water molds, general disinfection chemicals are said to effectively destroy any *A. invadans* that might contaminate farms, fishponds or fishing gears.

8.6.2.1 Life cycle

Aphanomyces invadans (Saprolegniales, Oomycetes) has an aseptate fungal-like mycelia structure. This oomycete has two typical zoospore forms. The primary zoospore consists of round cells that develop inside the sporangium. The primary zoospore is released to the tip of the sporangium where it forms a spore cluster. It quickly transforms into the secondary zoospore, which is reniform with laterally biflagellate cells and can swim freely in the water. The secondary zoospore remains motile for a period that depends on the environmental conditions and presence of the fish host or substratum. Typically, the zoospore encysts and germinates to produce new hyphae, although further tertiary generations of zoospores may be released from cysts (polyplanetism).

8.6.3 Target Organs and Infected Tissue

The motile zoospore plays an important role in the spread of the disease. Once the motile spore attaches to the skin of the fish, the spore will germinate under suitable conditions and its hyphae will invade the fish skin, muscular tissue, and reach the internal organs. Fish skeletal muscle is the target organ and exhibits major EUS clinical signs with mycotic granulomas.

8.6.4 Persistent Infection with Lifelong Carriers

There is no information to indicate that fish could be lifelong carriers of *A. invadans*. Generally, most infected fish die during an outbreak. Some mild or moderately EUS-infected fish could recover, but they are unlikely to be lifelong carriers.

Vectors: Not much information is perhaps available about the possible vectors of EUS.

Known or suspected wild aquatic animal carriers: Not much information is perhaps available about the Known or possible wild aquatic animal carriers of EUS.

8.6.4.1 Isolation of A. invadans from EUS-Affected Fish

Pale, raised lesions, which have not yet completely ulcerated, are said to be very suitable for fungal isolation attempts. Conversely, yellow to red focal skin lesions or healing ulcers are unsuitable. The fish specimen may be killed by decapitation and be pinned, with the lesion uppermost, to a dissecting board. The scales around the periphery of the lesion are removed and underlying skin seared with a red-hot spatula, so as to sterilize the surface. If possible, the fish and the dissecting board are then removed to a laminar flow cabinet containing filtered air, free of fungal elements. Then, one could cut through the stratum compactum underlying the seared area by using a sterile scalpel blade and sterile, fine pointed, rat tooth forceps. Then, the underlying muscle is exposed by cutting horizontally and reflecting the superficial tissues. It is to be ensured that the instruments do not contact the contaminated external surface and otherwise contaminate the underlying muscle. Four pieces of affected muscle, approximately 2 mm^3 in size, are carefully excised up using aseptic techniques, and to place them on a petri dish containing the isolation medium.

Where a suitable lesion is found on the tail of a small fish, the fish could be cut into two pieces using a sterile scalpel. Then, a cross section is sliced through the fish at the edge of the lesion. The scalpel is flamed, until red-hot and this could be used to sterilize the exposed surface of the muscle. A small-bladed sterile scalpel could be used to cut out a circular block of muscle (2–4 mm^3) from beneath the lesion. Used sterile, fine pointed pair of forceps to remove the block and placed it on the isolation medium. In this way, it could be easy to prevent the instruments contacting the contaminated external surface of the fish.

Two different isolation media are generally in use to obtain cultures of *A. invadans*. The use of Czapek Dox agar with penicillin G (100 units/mL) and oxolinic acid (100 μg/mL) was reported by Fraser et al. (1992), and an adapted version of Willoughby and Roberts (1994). GP-PenOx broth could be another medium for mycotic isolation. Inoculated media are incubated at approximately 25 °C and examined under a microscope (preferably an inverted microscope) within 12 h. Emerging hyphal tips may be repeatedly transferred to fresh plates of GP-PenStrep agar, until cultures are free of bacterial contamination. Then, they may be subcultured on GP agar at intervals of not more than five days.

The fungus is subcultured by aseptically cutting a block of agar, 3–4 mm in diameter, from the periphery of a colony and placing this upside down onto a Petri dish of fresh agar. Agar dishes may be inoculated within 24 h of preparation and the surface may not be dried before use.

8.6.4.2 Count Method for A. invadans Propagules in Pond Water

The method involves collection of a 1-L pooled sample (10 × 100 mL aliquots) from the pond and performance of a plate count on a 200-mL sample. Gelatin and selected metal ions are added to the subsample as nutrient supplements. Antibiotics are added to reduce the growth of contaminating bacteria and zygomycete fungi.

The method is a modification of aquatic fungal propagule count methods described by Anon (1992), Willoughby et al. (1984), and Celio and Padgett (1989). These previously described methods was said to have proved unsatisfactory for *A. invadans* as they failed to support growth of isolates while allowing abundant growth of other fungi and bacteria. The modifications are based on the observations that *A. invadans* is one of a limited number of aquatic fungi that could utilize complex protein sources and that a large proportion of colonies attach to plastic surfaces. During the counting procedure, most non-*A. invadans* colonies fail to attach and are removed by washing. Bacterial contamination is usually minimized by the low concentrations of complex protein, absence of glucose and addition of antibiotics.

After incubation of plate count cultures, colonies consistent with *A. invadans* are identified based on their attachment to the plastic surface, and hyphal and sporangial morphology. Representative *A. invadans*-like colonies are subcultured for further characterization.

The technique had perhaps been tried successfully in several artificially infected pond waters, but to date, it had perhaps been possible to try the method only once during a natural EUS outbreak. Counts of *A. invadans* propagules in pond water during that outbreak had a range of 10–30/L.

8.6.5 Counting and Identification

Examine the positive plates, and then (1) gently discard the fluid and any floating fungal colonies in the plates and fill the dishes with sterile DW, and (2) allow the dishes to stand for 5 min before discarding all but about 5 mL of wash fluid.

8.6.5.1 Count Method for A. invadans

1. Mark colonies resembling *A. invadans* on the plastic dish, with a felt tip pen, close to, but not obscuring the colony. The colonies may loosely adhere to the bottom and measure 3–5 mm in diameter.
2. Examine the marked colonies with the 10x and 20x objectives of an inverted microscope.
3. Mark colonies that have the characteristic right angle branching, 10 μm hyphal diameter and the rounded tip, with a circle. Hyphal diameter could be estimated with a calibrated eyepiece graticule.
4. At this stage, the hyphal tips from the representative colonies in the test samples could be removed with a sharp scalpel and inoculated deep into agar plates (Czapek Dox with 100 units/mL penicillin and 100 mg/mL oxolinic acid, or GP-PenStrep).
5. Examine the plates daily with a stereo microscope for at least five days, and subculture fungal tips as soon as possible. The recovered fungi can be identified by sporulation features, hyphal diameter, growth rate at 22 °C and failure to grow at 37 °C. Count only those colonies that are typical of *A. invadans*.
6. Spores per liter = *A. invadans* colony count (total for all seven plates) × 5. If <100% of colonies sporulate satisfactorily in the FPW positive control, then it may be necessary to repeat the counts after adjusting ionic strength of FPW by diluting up to one part FPW with two parts DW. Determine the optimum dilution by titration in the FPW positive control system.

8.6.5.2 Recording

1. Record the colony count for each sample and for the FPW and pond water positive control.
2. For the latter, subtract any *A. invadans* that were detected in the test sample.
3. Record the number of colonies subcultured, and for each subculture, to record the results of sporulation, hyphal diameter and 22 °C and 37 °C growth tests.
4. Maintain axenic cultures of representative probable *A. invadans* colonies for possible pathogenicity studies.

8.6.6 Maintenance of Cultures of A. invadans

Aphanomyces invadans cultures could be maintained in flasks of 200 mL GP broth at 10 °C for only six weeks before subculturing is required. This is due to the fast staling of the growth medium (Willoughby et al., 1995). The advantage of this technique, however, is that any bacterial contamination could be easily recognized as clouding of the medium.

Cultures could be maintained for longer periods on agar slopes in universal tubes, with sterile light paraffin oil covering the entire slope, as described by Smith and Onions (1994). Particular care may be taken to avoid contamination, as bacterial growth is not readily apparent in these cultures (Dobos et al., 1995 a,b). GPY agar could be used for this procedure, but *A. invadans* could be sustained for longer periods (may be for >6 months at 20 °C) using a buffered medium developed for *Aphanomyces astaci*.

8.6.7 Bioassay

Fish can be experimentally infected by intramuscularly injecting a 0.1 mL suspension of 100+ motile zoospores in EUS-susceptible fish (preferably *Channa striata* or other susceptible species) at 20 °C, and demonstrating histological growth of aseptate hyphae, 12–25 μm in diameter, in the muscle of fish sampled after seven days, and of typical mycotic granulomas in the muscle of fish sampled after 10–14 days.

8.6.8 Inducing Sporulation in *A. invadans* Cultures

The induction of asexual reproductive structures is necessary in order to identify fungal cultures as members of the genus *Aphanomyces*. Induce sporulation, it is required to place an agar plug (3–4 mm in diameter) of actively growing mycelium in a Petri dish containing GPY broth and incubate for four days at about 20 °C. Wash the nutrient agar out of the resulting mat by sequential transfer through five Petri dishes containing autoclaved pond water (APW), and leave overnight at 20 °C in APW. After about 12 h, the formation of achlyoid clusters of primary cysts and the release of motile secondary zoospores could be apparent under the microscope. There are features that distinguish sporulating cultures of *Aphanomyces* from *Saprolegnia* and *Achlya*

8.6.9 Identification of Saprolegniacean Fungal Cultures

Concomitant to the above, the Saprolegniaceae are aseptate, eucarpic fungi that typically demonstrate two zoospore forms. The secondary zoospores are characteristically reniform and laterally biflagellated. The two flagellae differ in type (heterokont), with one anteriorly directed tinsel-type flagellum and one posteriorly directed whiplash-type flagellum.

Saprolegniacean genera are distinguished primarily by asexual characters, particularly zoosporangial shape, method of zoospore release, and method of zoosporangial renewal. The production of asexual characters could also be induced.

There are variations in the characters between the three main saprolegniacean fungi associated with fish disease (*Aphanomyces*, *Achlya*, and *Saprolegnia*). Identification of these fungi to the species level usually depend on the production of sexual structures, but these are said to be commonly absent from fish-parasitic species, and unknown from *A. invadans*.

Concomitantly, the zoosporangia of *Aphanomyces* spp. may typically be no wider than the hyphae. A single row of primary zoospores could be formed within a zoosporangium and released from an apical tip, or from lateral evacuation tubes, at that time, they immediately encyst and form achlyoid clusters. The primary zoospore is therefore not fully released from the sporangium. The main free-swimming stage of *Aphanomyces* spp. is perhaps the secondary zoospore that is discharged from the encysted primary zoospores. The secondary zoospore possibly remains motile for a period depending on environmental conditions and location of a host or substratum. Typically, the zoospore undergoes encystment and later germinates to produce new hyphae, although further tertiary generations of zoospores may be released from cysts (polyplanetism).

Concomitant to the above, the *Achlya* spp. zoosporangia are usually formed from terminal hyphal swellings that differentiate into the primary zoospores. These encyst, as with *Aphanomyces,* in an achlyoid manner, but only at the apical tip of the zoosporangium. Zoosporangial renewal is typically sympodial, branching from the hypha below the basal septum delimiting the spent zoosporangium.

In addition to the above, the zoosporangia of *Saprolegnia* spp. are, as with *Achlya* spp., said to possess short terminal hyphal swellings. However, in saprolegnians, the primary zoospore is fully released from the zoosporangium and remains motile for a short period before encysting and releasing secondary zoospores. However, polyplanetism could be particularly pronounced among fish-parasitic *Saprolegnia* spp. Zoosporangial renewal is typically by internal proliferation, that is, the secondary zoosporangium develops within the previously emptied primary zoosporangium.

It is opined by some workers that that some epidermal insult (such as rhabdovirus infection) precedes infection with *A. invadans*. Examination of samples of fish at different times during the outbreak may attempt to characterize the different types of lesions present and the rhabdovirus-infection status of individual fish. Thus, the "spectrum" of disease is important here.

Describing the outbreak in terms of place may lead to identification of the cause. For farmed fish, this may involve looking at the pattern in different ponds. It is often useful to consider time and place together. This could be done by drawing a plan of the ponds and recording the dates when cases may occur. Such a diagram may also give a lead to whether the outbreak is a common source or propagating. For larger scale epidemics, spot maps are useful.

8.7 VIROLOGICAL STUDY

8.7.1 Virus Isolation and Characterization

8.7.1.1 Isolation of Viruses

The isolation of viruses (Chen et al., 1993) from EUS-diseased specimens could be very difficult, even with access to specialized virological facilities (Mahy and Kangro, 1996). Two factors are said to be critical for successful isolation:

1. first, the diseased fish need to be collected during the early period of an outbreak (i.e., within one to two weeks of the disease being noticed), and
2. second, the specimens need to be alive just before collection of the tissue samples.

8.7.1.2 Virological Sampling

For virological sampling (Burleson et al., 1992), the fish organs and transport medium/conditions will be determined by the virus/es suspected. The laboratory where samples will be submitted is to be contacted before sampling.

8.7.1.3 Tissue Sampling and Handling

Diseased fish with early skin lesions are to be collected. Fish are usually sacrificed and wiped clean with tissue paper. c 1 g of each tissue sampled is taken. For muscle samples, tissue debris and surface fungus on the ulcerated lesions are to be removed using a clean razor blade. Pieces of muscle tissue are taken from beneath the lesions. For internal organ samples, the abdomen is carefully opened using a clean pair of scissors. Small pieces of tissue from kidney, spleen, intestine, and pancreas are to be taken and pooled. If the fish are very small, the entire viscera could be taken. Tissue samples from up to five fish could be pooled and processed as one tissue extract. Tissue samples could be stored up to 48 h in HBSS supplemented with 2% FCS, 500 IU/mL penicillin, 500 mg/mL streptomycin, and 10 mg/mL amphotericin B (fungizone) at 4 °C.

Samples are then homogenized using a sterile, precooled pestle and mortar until a smooth paste is obtained. Sterile fine sand is added to facilitate homogenization. Samples are diluted 1:10 by the addition of 9 mL HBSS containing 2% FCS. After mixing well, the samples are transferred to sterile centrifuge tubes and spun at 1000 g at 4 °C for 15 min in order to separate the cell debris, sand, and possibly contaminating microbes from the fluidextract. A further 1:5 dilution is carried out by filling 5-mL sterile disposable syringes with 4 mL HBSS (with 2% serum) and then drawing up 1 mL supernatant. These 1:50 final dilutions are mixed well and then filter-sterilized through 0.45 mm disposable filter units. The filtrates or tissue extracts are kept in 5 mL sterile bottles at 4 °C which are ready to be inoculated directly onto fish cell lines or, if necessary, transported to the fish virology laboratory.

8.7.1.4 Virus Isolation

Simultaneous cell culture (Wolf and Quimby, 1969; Wolf et al., 1966) and sample inoculation can be carried out using susceptible fish lines such as BF-2 and/or SSN-1 cell lines. Tests are sometimes conducted in 24-well plates. Each plate is first seeded with a single cell suspension of the indicator cell line in L-15 medium containing 2% FCS and 1x antibiotics (100 IU/mL penicillin and 100 mg/mL streptomycin). Each well receives 1.3—1.4 mL of cell suspension. Cell density should be sufficient to produce an 80%—90% confluent monolayer one day after seeding. Tissue extracts (1:50 dilution) are immediately inoculated into two replicate wells. The inoculum volume is 200 µL/well. An equal number of inoculated wells as negative control wells is to be allocated for each plate. Cells are incubated at 23—25 °C and observed daily for CPE for at least 14 days. The first blind passage of culture fluids is performed between days 7 and 10 by transferring 200 µL of supernatant from each well to fresh culture wells and observing the plates for a further 14 days. A second and third blind passage should also be carried out.

Samples showing CPE (Amend et al., 1984) in which the cell monolayer changes (e.g., disintegrates, sloughs off the surface of the tissue culture wells, or results in cell lysis) are to be passaged to provide larger quantities of the suspected virus. Supernatant measuring 200 µL from a single well-exhibiting CPE is inoculated into 25 cm^2 flasks containing an 80%—90% confluent cell monolayer. The suspected virus is allowed to be adsorbed for 1 h. The cells are washed once with 5 mL phosphate buffered saline (PBS), and then 7 mL of maintenance medium (L-15 with 2% FCS) is added. Flasks are incubated at 23—25 °C together with uninoculated control flasks for comparison. When the cells show complete CPE, they are spun at 1000 g at 4 °C for 15 min. The supernatant is collected and divided into 1 mL aliquots. Some tubes are kept at 4 °C for further characterization within six months and others stored at −20 °C or −70 °C for long-term storage.

For virus isolation and characterization, pooled tissues are placed in small volumes of antibiotic-supplemented viral transport medium or Hank's Balanced Salt Solution (HBSS) (without Ca^{++} Mg^{++}) plus 10% fetal calf serum (FCS) and kept on ice without freezing until being further processed. Pooled tissue samples are weighed, homogenized, and diluted to 1:10 (w/v) with HBSS. Homogenates are then clarified by centrifugation at 1500 g for 10—15 min. The supernatant is further diluted to 1:5 with HBSS to give a final dilution of 1:50 before filtration through a 450-nm membrane filter or through a 220-nm syringe filter. Freshly grown 60%—70% confluent 25 cm^2 monolayer cultures of different warm water fish cell lines (e.g., SSN1, SNKD2A, SNKD3, EPC, BB, etc.) are prepared in Leibovitz L-15 medium supplemented with 10% FCS, 100 U penicillin, and 100 μg streptomycin/mL (Sigma), and inoculated with 1 mL tissue extract. Tissue extracts would be allowed to adsorb for 30—60 min before overlaying the monolayers with 5 mL of fresh medium with 2% FCS and incubated at 28 °C. CPE at varying degrees of severity would develop between 1 and 14 days postinoculation. It may not always be possible to distinguish between toxic and virus-specific effects. Aliquots of medium from all cultures, which developed a CPE, also a number of which appeared normal, would be passed into fresh susceptible fish cell line. A repeatedly transmissible CPE, if obtained with the tissue extracts of the diseased fishes, would reveal isolation of the virus (Frerichs et al., 1991). The CPE would be processed with a blind passage in cases of negative samples or second passage for positive samples for confirmation of viral etiology. The viral titer would be determined in susceptible fish cell line by end-point infectivity titration or end-point dilution assay (TCID 50) (Rovozzo and Burke, 1973). The cell would be incubated for seven days at 27 °C. An end point of 50% (TCID 50) would be calculated using Spearman Karber method (Karber, 1931) or by the method of Reed and Muench (1938). Stock virus suspension would be prepared by propagation of the viral strain in susceptible fish cell line using Leibovitz L15 medium supplemented with 10% FCS for cell growth and maintenance. Following complete development of CPE, cell culture fluids would be harvested and cell debris would be removed by centrifugation at 1000 g for 10 min.

For virus replication, fish cell line would be seeded in 96 well plates. Ten-fold serial dilution of the virus would be inoculated in six wells of each line. The plates would be incubated for seven days at 27 °C, and the viral titer on each cell line would be computed (Frerichs et al., 1989; Lilley and Frerichs, 1994).

8.7.1.5 Virus Purification for the Analysis of Their Structural Polypeptide

For virus purification for the analysis of their structural polypeptide, 100 mL of the virus culture lysate would be clarified by the centrifugation of the cell debris for 20 min at 20,000 g, and the supernatant centrifuged in aliquots for 20 min at 21,000 g on to 0.3 mL 50% sucrose TNE pads (TNE: 20 mM Tris, 1 mM EDTA, 100 mM NaCl, pH 7.5). Virus bands would be collected, pooled, and centrifuged again on to a 50% sucrose TNE pad. The resulting virus bands would be collected, diluted to 6 mL with TNE, and pipetted on to a discontinuous sucrose gradient of 5 mL 40% sucrose TNE and 5 mL 20% sucrose TNE. After centrifugation at 90,000 g for 90 min, 1 mL fraction would be collected, samples of which would be titrated on to susceptible fish cell lines to confirm the presence of virus.

The refractive index of each fraction would be checked and the interface between 20% and 40% sucrose layers would be retained. This would be concentrated by centrifugation in TNE at 1,00,000 g for 90 min, and the resulting pellet would be resuspended in 20 μL TNE and stored at −20 °C (Lilley and Frerichs, 1994).

Virus proteins (Laemmli, 1970) would be denatured by dilution to approximately 1:4 (v/v) in reducing buffer (62.5 mM, Tris pH 6.8, 10% glycerol, two SDS, 2-B-mercaptoethanol, and 0.00125% Bromophenol blue) and heating to 95 °C for 5 min. Polypeptides would be separated by discontinuous polyacrylamide gel electrophoresis using a 4% stacking gel and 10% resolving gel on a BRL-mini-V 8-10 system at a constant 200 V for 40 min. The bands would be visualized using Coomassie brilliant blue (Lilley and Frerichs, 1994).

For characterization of virus-specific proteins, standard immune-blotting procedure could be followed (Lilley and Frerichs, 1994).

8.7.1.6 Extraction of Viral Nucleic Acids

Viral nucleic acids would be extracted from purified viruses and the nucleic acid analysis would be carried out using standard protocols depending on the nature of the virus genome (Feldman and Wang, 1961). Depending on the presumptive identification the viruses, selected and reported primers pertaining to specific viruses or random primers would be used for amplification of portions of viral genome and would be sequenced and sequence analysis would be carried out (Sambrook et al., 1989).

8.7.1.7 Antibody-Based Antigen-Detection Methods

Polyclonal antibodies against *A. invadans* or *Aphanomyces* saprophyte had been reported to show cross-reactivity using protein gel electrophoresis and Western blot analysis and immunohistochemistry. However, specific monoclonal antibody (MAb) against *A. invadans*, developed later, was found to have high specificity and high sensitivity to the *Aphanomyces* pathogens of the EUS using immunofluorescence. This MAb could detect the *Aphanomyces* hyphae at the early stage of the EUS infection.

8.7.1.8 Test for the Cross-Neutralizing Activity of the Antisera

The alpha procedure (Rovozzo and Burke, 1973) would be used to test the cross-neutralizing activity of the antisera against the virus(es). Antiserum would be prepared appropriately in HBSS after preliminary assay against homologous virus would be added to four replicated rows of serial virus dilution in 96-well plate and the reaction mixture held at room temperature for 1 h. The mixture would be assayed for infectivity by adding susceptible cell suspension to each well and incubating at 25 °C until a full CPE had developed. Log 10 neutralization indices would be determined according to Rovozzo and Burke (1973).

8.7.1.9 Serological Methods

Serological methods for detection and identification of *A. invadans* in EUS specimens are said to be not practical. If necessary, the MAb is said to offer a better specificity and sensitivity than polyclonal antibody for serological detection or identification of *A. invadans* in diseased specimens or in pathogen isolates.

8.7.1.10 Tests Recommended for Targeted Surveillance to Declare Freedom from Epizootic Ulcerative Syndrome

The test for targeted surveillance to declare freedom from EUS is examination of gross signs. Targeted surveillance is conducted twice a year to cover the range of seasonal variation, preferably at least once during the season that favors EUS occurrence or when water temperatures are about 18–22 °C or below 25 °C. Biosecurity measures are to be implemented in order to maintain disease-free status in controlled aquaculture facilities or compartments.

Using the gross sign for targeted surveillance, a large number of the fish are to be examined without killing them. Fish on farms or in natural water bodies are to be sampled carefully using suitable gears or nets. The suitable number of fish specimens to be examined, could be based on details described in the *OIE Guide for Aquatic Animal Health Surveillance* (2009).

Once fish show similar gross signs of EUS, they should be categorized as suspect EUS fish, and the location/farm/zone could be considered suspect. Suspect specimens are to be further tested through presumptive diagnosis followed by confirmative diagnosis.

8.7.2 Molecular Techniques

8.7.2.1 Polymerase Chain Reaction Amplification of the DNA of A. invadans

DNA Preparation from A. invadans Isolate

DNA is extracted from an actively growing colony of *A. invadans* culture in GY broth at about four days or when young mycelia reach 0.5–1.0 cm in diameter. The mycelia are transferred to sterile 100-mm Petri dishes, washed twice with PBS, and then placed on tissue paper for liquid removal. Hyphal tips (*c* 50–250 mg) are excised with a sterile scalpel blade and transferred to a 1.5-mL microcentrifuge tube for DNA extraction. Commercial DNA extraction kits could be used successfully.

DNA Preparation from EUS-Infected Tissue

Small pieces of EUS-infected tissue, 25–50 mg, are suitable for DNA extractions.

Diagnostic PCR Technique

Two techniques are said to be more specific to *A. invadans* than the previous one.

Method 1: The species-specific forward primer site is located near the 3′ end of the SSU (small subunit) gene and a species-specific reverse primer site is located in the ITS1 region for Ainvad-2F (5′-TCA-TTG-TGA-GTG-AAA-CGG-TG-3′) and Ainvad-ITSR1 (5′-GGC-TAA-GGT-TTC-AGT-ATG-TAG-3′). The polymerase chain reaction (PCR) mixture contained 25 pM of each primer, 2.5 mM each deoxynucleoside triphosphate, 0.5 U of Platinum Taq DNA polymerase and 20 ng of genomic DNA (either from an *Aphanomyces* isolate or from infected tissue) for a total volume of 50 μL. DNA is amplified in a thermocycler machine under the following cycle conditions: 2 min at 95 °C; 35 cycles, each consisting of 30 s at 95 °C, 45 s at 56 °C, 2.5 min at 72 °C, and a final extension of 5 min at 72 °C. The PCR product is analyzed by agarose gel electrophoresis and the target product may be 234 bp.

Method 2: The species-specific primer sites are located in the ITS1 and ITS2 regions. The forward primer is ITS11 (5′-GCC-GAA-GTT-TCG-CAA-GAA-AC-3′) and the reverse primer is ITS23 (5′-CGT-ATA-GAC-ACA-AGC-ACA-CCA-3′). The PCR mixture contains 0.5 μM of each primer, 0.2 mM each deoxynucleoside triphosphate, 1.5 mM $MgCl_2$, 0.6 U of Taq DNA polymerase and 20 ng of genomic DNA (from an *Aphanomyces* isolate) for a total volume of 25 μL. The DNA is amplified under the following cycle conditions: 5 min at 94 °C; 25 cycles, each consisting of 30 s at 94 °C, 30 s at 65 °C, 1 min at 72 °C, and a final extension of 5 min at 72 °C. The PCR product is analyzed by agarose gel electrophoresis and the target product could be 550 bp. PCR amplification using the DNA template from the infected tissue is often similar to the above protocol except that 5 ng of the DNA template is used for 35 cycles.

Method 3: The species-specific primer sites are located in the ITS1 and ITS2 regions. The forward primer is BO73 (5′-CTT-GTG-CTG-AGC-TCA-CAC-TC-3′) and the reverse primer is BO639 (5′-ACA-CCA-GAT-TAC-ACT-ATC-TC-3′). The PCR mixture contains 0.6 μM of each primer, 0.2 mM of each deoxynucleoside triphosphate, 1.5 mM $MgCl_2$, 0.625 units of Taq DNA polymerase, and *c* 5 ng of genomic DNA (or 2.5 μl of DNA template extracted from 25 mg of infected tissue and suspended in 100 μL buffer) in a 50-μL reaction volume. The DNA is amplified under the following cycle conditions: 96 °C for 5 min; 35 cycles of 1 min at 96 °C, 1 min at 58 °C and 1 min at 72 °C; followed by a final extension at

72 °C for 5 min. The PCR product is analyzed by agarose gel electrophoresis and the target product could be 564 bp.

All *A. invadans* isolates found so far belong to a single genotype. This is said to facilitate the identification process. Alternatively, sequencing of the PCR products could be performed and the results could be compared with the sequence deposited in the public gene data banks. *A. invadans* is said to have similar characteristics as *A. astaci,* which is the etiological agent of crayfish plague. Both pathogenic oomycetes could be differentiated using molecular tools (8, 17, 28, and 32).

Maintaining *A. invadans* in the axenic culture is necessary. As it is characteristically slow growing, it easily becomes contaminated with other microorganisms, such as bacteria and other fast-growing oomycetes and fungi. Attempts to purify or isolate *A. invadans* from contaminated cultures usually fail.

Isolation of Head Kidney Leukocytes Head kidney leukocytes could be isolated following the method described by Kamilya et al. (2008) with some modification. Following blood collection and euthanasia, the head kidney may be removed aseptically from experimentally infected and apparently healthy catla. A cell suspension could be obtained by teasing the head kidney tissues with forceps in complete RPMI-1640 medium containing penicillin (100 IU/mL), streptomycin (100 μg/mL), and 10% FCS (Hi−media, India). The cell suspension be washed by centrifugation for 10 min at 1000 rpm. The pellet is resuspended in RPMI-1640 medium, carefully layered on top of the leukocyte isolation medium, HiSep (Hi−media, India), and centrifuged at 1500 rpm for 30 min. Cells at the medium-HiSep interface are transferred to clean tubes and washed twice by centrifugation at 1000 rpm for 10 min. Purified leukocytes are counted using a hemocytometer and their number adjusted to 1×10^6 cells/mL in RPMI-1640 medium. Cell viability is determined by the trypan blue exclusion test.

8.7.3 Nitric Oxide Assay

Production of nitric oxide could be assessed by measuring absorbance of nitrite from the supernatant using the Griess reaction (Green et al., 1982). Head kidney leukocytes (1×10^6/mL) be cultured in complete RPMI-1640 medium in 96-well microtiter plates for 24−72 h at 25 °C in a humidified 5% CO_2 incubator. After incubation, the plate is centrifuged and 100-μL supernatant is removed from each well. Next, 100 μL of 1% sulfanilamide in 2.5% phosphoric acid is added to each sample followed by 100 μL 0.1% N-naphthyl-ethylenediamine in 2.5% phosphoric acid, and absorbance was measured at 540 nm.

8.7.4 Superoxide Anion Production

Production of superoxide anion (O^{2-}) by head kidney leukocytes could be evaluated following the method of Kamilya et al. (2008). Leukocytes (1×10^6 cells/mL) are seeded into the wells of a 96-well microtiter plate. Nitro-blue tetrazolium (NBT) is dissolved in RPMI-1640 medium to a final concentration of 2 mg/mL. Medium containing NBT be filter-sterilized and 50 μL NBT be added to the leukocyte culture. After incubating the plate for 25 min at room temperature, the supernatant is removed from each well and the cells are fixed by adding 200 μL 70% methanol for 1 min. Unreduced NBT is removed by washing cells several times with 70% methanol. Reduced NBT is dissolved by adding 120 μL 2 M KOH to each well, followed by 140 μL dimethyl sulfoxide (DMSO). Optical density is measured at 595 nm.

8.7.5 Leukocyte Proliferation Assay

The proliferative response of the head kidney leukocytes be determined by the MTT [3-(4,5-dimethyl thiazol-2-yl)-2,5-diphenyl tetrazolium bromide] colorimetric assay method (Mosmann, 1983) with modifications. Head kidney leukocytes (1×10^6 cells/mL) are distributed into the wells of a 96-well microtiter plate. The mitogen concanavalin A, at concentrations of 50 μg/mL, is added to each well. The plate is incubated for 72 h at 25 °C. After incubation, 20 μL MTT solution (dissolved in PBS at a concentration of 5 mg MTT/mL and sterilized by filtration) is added to the wells and the plate is incubated at 25 °C for 4 h. After centrifugation of the plate, the media is removed and 200 μl DMSO is added to each well and mixed for 2 min. The microplate is read on a microreader at 595 nm.

8.7.6 Lysozyme Activity

The serum lysozyme activity is determined following a turbidimetric method (Ellis, 1990), modified to a microtiter plate assay. Next, 25 μL of each serum sample is added to a 175 μL suspension of the lysozyme sensitive gram-positive bacterium *Micrococcus lysodeikticus* (0.2 mg/mL in 0.05 M sodium phosphate buffer, pH 6.2). The reaction is carried out with constant shaking, and absorbance was recorded at 450 nm after 1 and 5 min. One unit of lysozyme activity is defined as the amount of sample causing a reduction in absorbance of 0.001/min.

8.8 ELECTRON MICROSCOPIC STUDY

For study using electron microscopy (EM), the virus would be propagated on cell monolayers at 25 °C for 24 h. The culture medium would be decanted and replaced with fixative containing 1.3% paraformaldehyde and 1.6% glutaraldehyde (Karmovsky, 1965). The cells would be centrifuged at 2000 g for 10 min and the pellets stored in a new fixative at 4 °C for 1 h. The fixative would then be replaced by a 0.1 M cacodylate buffer and rinsed for a further period of 4 h before secondary fixation with 0.5% osmium tetroxide in cacodylate buffer. The pellet would then be dehydrated through a graded acetone series, impregnated with epoxy resin and polymerized in an oven at 60 °C for 48 h before sectioning at 70 nm and examination using a Philips EM 30 or any other standard EM (Lilley and Frerichs, 1994).

8.9 FISH SAMPLING FOR PARASITOLOGICAL WORKS

8.9.1 Examination of the Skin

1. First, stun the fish with a sharp blow on the head.
2. Then, take scrapings for microscopic examination using a scalpel and scrape from front to back of the fish or around the fins.
3. Place the scrapings on a clean glass slide with a drop of water and cover the slide with a cover slip.
4. Examination of the gills: After the gross examination of the gills, clip a small portion of gill lamellae with a pair of sharp scissors and to place on a glass slide. Add a drop of water and cover with a cover slip. Examine under the low power of a microscope with high contrast.
5. Examination of other organs: Any other organs suspected of having parasitic infection may go for squash preparations made from small subsamples of tissue and be examined similarly using light microscopy.
6. Record and draw findings.

Biochemical parameters. Biochemical parameters were measured using a biochemical test kit (Crest Biosystems, India) as per the manufacturer's instructions and included SGPT activity, SGOT activity, serum ALP activity, and total plasma protein for all experimentally infected and apparently healthy catla.

8.9.1.1 Statistical Analyses

Data are statistically analyzed using SPSS-15.0 or any other latest version for Windows software (SPSS Inc., Chicago, IL, USA). Results are presented as mean ± standard deviation. Mean values be compared by one-way analysis of variance.

Fish and husbandry. Apparently healthy catla (mean weight 70 g) were collected from a local fish farm. After collection, they were immediately brought to the laboratory and acclimatized in 1000-L indoor circular tanks supplied with dechlorinated and aerated tap water at ambient temperature (23−26 °C). The fish were fed twice daily (rice bran and mustard oil cake in a ratio of 1:1). Water was maintained at optimum levels for the fish species throughout the experiment: DO 6.88 ± 0.56 mg/L,

pH 7.14 ± 0.77, ammonia 0.029 ± 0.011 mg/L, nitrite 0.016 ± 0.01 mg/L. Water was periodically exchanged to remove waste feed and fecal material.

Sampling techniques: The infected fishes were identified by the symptoms of fishes such as loss of normal glaze; spot having hemorrhagic lesions on the body, excess mucus secretions, discoloration of gill filament and damage of gill. For the isolation of fungus fishes were carried to the laboratory immediately after collection.

Serological comparison of BSNV and IPNV was performed by reciprocal β cross-neutralization of BSNV and four classical strains of IPNV (WB, Sp, Ab, and TV-1) by the method of Okamoto et al. (1983) with minor modifications. Antisera against two of the IPNV reference strains WB and TV-1 were provided by P. F. Dixon (CEFAS, Weymouth, UK). Antisera were serially five-fold diluted in microtiter plates (40 µl per well) and mixed with C 100 TCID&! (40 µL) virus. The virus−antiserum suspension was incubated at room temperature (24 °C) for 60 min with frequent mixing on a plate shaker (Titertek, Flow Labs). BF-2 single-cell suspension (100 µL) was added to each well and the plates were incubated at the temperatures given above and observed for 10−14 days for the development of CPE. Neutralizing antibody titer of antiserum was expressed as the highest antiserum dilution protecting 50% of the inoculated cultures, as calculated by the Spearman−Karber method (Karber, 1931). Serological relationships (r) between the IPNV strains and BSNV, calculated from the formula $r = \sqrt{(r1 \times r2)}$, where $r1$ and $r2$ are the titer ratios (Archetti and Horsfall, 1950).

The biophysical and biochemical characteristics, including virus morphology, buoyant density, presence of a bisegmented double-stranded RNA genome, heat and pH stability, and nature of capsid structural polypeptides indicate that the virus belongs to the genus *Aquabirnavirus* in the family Birnaviridae (Dobos et al., 1995). Two of the proteins, VP2 (45−60 kDa) and VP3 (29−35 kDa), together make up 80%−90% of the total protein content of the birnaviruses (Dobos et al., 1979; Dobos, 1996). Corresponding polypeptides (44 and 37 kDa) also formed the most abundant viral proteins of BSNV. The high molecular mass protein (112 kDa) found in BSNV is similar to the 101 kDa polyprotein of IPNV encoded by genome segment A (Duncan and Dobos, 1986). The polyprotein, not previously reported to occur in purified virus preparations, has since been found in purfied preparations of IPNV (Magyar and Dobos, 1994). Although the BSNV structural proteins correspond favorably with those of IPNV, the individual molecular masses of the proteins and migration pattern are considerably deterrent from the structural polypeptides of IPNV. The difference in the protein pattern was unambiguously evident, as the low molecular mass protein of BSNV was 37 kDa compared with IPNV, which has two structural proteins of molecular mass falling below 32 kDa. The molecular masses of BSNV genomic RNA segments (2 ± 56¬10′ and 2 ± 0¬10′ Da) were also found to be widely different from those of the IPNV RNA segments, which have molecular masses of 2 ± 5¬10′ and 2 ± 3¬10′ Da (Dobos et al., 1977). When compared with the reported molecular masses of the genomic segments of birnaviruses (Dobos et al., 1991), BSNV has the largest size difference's yet reported between the two genomic segments.

The type species of the genus *Aquabirnavirus*, IPNV, currently has 10 serotypes identified in two serogroups, A and B, which show some kind of cross-reactivity within each serogroup but not between the two serogroups (Hill and Way, 1996). The present study, involving three classical serotypes of serogroup A and the sole serotype of serogroup B, revealed that BSNV did not have any cross-reaction with strains of either serogroup. SGV, isolated from an ulcerated sand goby in Thailand, is another tropical birnavirus reported to be significantly different from the classical IPNV strains.

This isolate, however, which showed some degree of cross-reaction with IPNV Ab and Sp, has later been recognized as a mixture of IPNV Ab and Sp (Hill and Way, 1996). Apart from serological distinctness, the differences in the molecular masses and migration patterns of capsid proteins and viral RNAs also distinguish BSNV from other reported *Aquabirnavirus* strains. Phenotypically, the virus was different from all the IPNV strains, as it was unable to multiply in CHSE-214, RTG-2, EPC, or FHM cell lines (Macdonald and Gower, 1981). Although BSNV was identified as belonging to the genus *Aquabirnavirus*, biochemical, biological, and serological characteristics demarcated it from reported strains of the type species of the genus, IPNV. Since the virus cannot be compared with any of the existing IPNV strains and also since the virus was serologically distinct from the two reported serogroups of IPNV, BSNV could be classified currently as third serogroup of IPNV (serogroup C) in the genus *Aquabirnavirus*. Further studies would be required, however, to clearly identify whether the virus really forms a new serogroup or a new species altogether under the genus *Aquabirnavirus*. The name blotched snakehead virus is therefore proposed for this agent.

8.9.1.2 Viruses Observed in Electron Microscopy

The presence of virus particles in the tissues of EUS-infected fishes was said to have been first demonstrated by EM. The initial observations of viral association with EUS were noted by transmission electron microscopic studies of the tissues of diseased *Channa striatus* and *Clarias batrachus* specimens obtained from the epizootic in Thailand during 1983 and 84. The presence of viruslike particles had been reported in hepatocytes, spleen, kidney and blood cells, capillary endothelium, and muscle

myofibrils taken from the site of ulceration in diseased specimens. These particles observed in the cytoplasm were of *c* 70 nm diameter. However, they could not be precisely identified, as they were pleomorphic with shapes varying from round, oval, elongate, and kidney-shaped. Further, particles of rhabdovirus-like morphology were noticed in skin lesions of infected barramundi (*Lates calcarifer*). The presence of rhabdovirus and infectious pancreatic necrosis viruslike particles in cell cultures inoculated with tissue extracts from EUS-infected snakeheads was noted by Wattanavijarn et al. Rhabdoviruses were also observed in cell cultures inoculated with tissue extracts from a variety of infected fishes including those of a barramundi that had depicted extensive CPE.

REFERENCES

Anon, 1992. Norwegian Aquaculture: Controlling the Antibiotic Explosion. Animal Pharmacy, P.J.B. Publication Ltd, Surrey.

APHA, 1995. Standard Methods for the Examination of Water and Wastewater. American Public Health Association, USA pp. xxxix + 1193.

Amend, D.F., McDowell, T., Hedrick, R.P., 1984. Characteristics of a previously unidentified virus from channel (*Ictalurus punctatus*). Can. J. Fish. Aquat. Sci. 41, 807–811.

Archetti, I., Horsfall Jr., F.L., 1950. Persistent antigenic variation of influenza A viruses after incomplete neutralization *in ovo* with heterologous immune serum. J. Exp. Med. 92, 441–462.

Blazevic, D.J., Ederer, G.M., 1975. Principles of Biochemical Tests in Diagnostic Microbiology. John Wiley & Sons, Inc, New York, N.Y.

Burleson, F.G., Chambers, T.M., Wiedbrauk, D.L., 1992. Virology: A Laboratory Manual. Academic Press, London, pp. 186–246.

Celio, D.A., Padgett, D.E., 1989. An improved method of quantifying water mould spores in natural water columns. Mycologia 81 (3), 459–460.

Chen, M.M., Kou, G.H., Chen, S.N., 1993. Establishment and characterization of a cell line persistently infected with infectious pancreatic necrosis virus (IPNV). Bull. Inst. Zool. Acad. Sin. 32, 265–272.

Chinabut, S., Roberts, R.J., 1999. Pathology and Histopathology of Epizootic Ulcerative Syndrome. Aquatic Animal Health Research Institute Newsletter, Bangkok, Thailand, 33 p.

Dickman, S.R., Bray, R.H., 1940. Colorimetric determination of phosphate. Ind. Eng. Chem. Anal. Ed. 12, 665–668.

Dobos, K.M., Quinn, F.D., Ashford, D.A., Horsburgh, C.R., King, C.H., 1995a. Emergence of a unique group of necrotizing mycobacterial diseases. Emerg. Infect. Dis. 5 (3), 367–378.

Dobos, P., 1996. The molecular biology of infectious pancreatic necrosis virus (IPNV). Annu. Rev. Fish Dis. 5, 25–54.

Dobos, P., Berthiaume, L., Leong, J.A., Kibenge, F.S.B., Muller, H., Nicholson, B.L., 1995b. Family *birnaviridae*. In: Murphy, F.A., Fauquet, C.M., Bishop, D.H.L., Ghabrial, S.A., Jarvis, A.W., Martelli, G.P., Mayo, M.A., Summers, M.D. (Eds.), Virus Taxonomy. Sixth Report of the International Committee on Taxonomy of Viruses. Springer-Verlag, Vienna & New York, pp. 240–244.

Dobos, P., Hallett, R., Kells, D.T.C., Sorensen, O., Rowe, D., 1977. Biophysical studies of infectious pancreatic necrosis virus. J. Virol. 22, 150–159.

Dobos, P., Hill, B.J., Hallett, R., Kells, D.T.C., Becht, H., Teninges, D., 1979. Biophysical and biochemical characterization of animal viruses with bisegmented double-stranded RNA genomes. J. Virol. 32, 593–605.

Dobos, P., Nagy, E., Duncan, R., 1991. Birnaviridae. In: Kurstak, E. (Ed.), Viruses of Invertebrates. Marcel Dekker, New York, pp. 301–314.

Dobos, P., Berthiaume, L., Leong, J.A., Kibenge, F.S.B., Muller, H., Nicholson, B.L., 1995. Family Birnaviridae. In: Murphy, F.A., Fauquet, C.M., Bishop, D.H.L., Ghabrial, S.A., Jarvis, A.W., Martelli, G.P., Mayo, M.A., Summers, M.D. (Eds.), Virus Taxonomy. Sixth Report of the International Committee on Taxonomy of Viruses. Springer-Verlag, Vienna and New York, pp. 240–244.

Duncan, R., Dobos, P., 1986. The nucleotide sequence of infectious pancreatic necrosis virus (IPNV) dsRNA segment A reveals one large ORF encoding a precursor polyprotein. Nucleic Acids Res. 14, 5934.

Ellis, A.E., 1990. Lysozyme assays. In: Stolen, J.S., Fletcher, T.C., Anderson, D.P., Roberson, B.S., van Muiswinkel, W.B. (Eds.), Techniques in Fish Immunology. SOS Publ, USA, pp. 101–103.

Feldman, H., Wang, S., 1961. Sensitivity of various viruses to chloroform. Proc. Soc. Exp. Biol. Med. 106, 736–738.

Fraser, G.C., Callinan, R.B., Calder, L.M., 1992. *Aphanomyces* species associated with red spot disease; an ulcerative disease of estuarine fish from eastern Australia. J. Fish Dis. 15, 173–181.

Frerichs, G.N., Millar, S.D., Alexander, M., 1991. Rhabdovirus infection of ulcerated fish in South-East Asia. In: Ahne, W., Kurstak, E. (Eds.), Viruses of Lower Vertebrates. Springer-Verlag, Berlin, pp. 396–410.

Frerichs, G.N., Millar, S.D., Alexander, M., 1989. Rhabdovirus infection of ulcerated fish in South-East Asia. In: Ahne, W., Kurstak, E. (Eds.), Viruses of Lower Vertebrates. Springer-Verlag, Berlin, Germany, pp. 396–410.

Godwin, H., 1978. Fenland: Its Ancient Past and Uncertain Future. Cambridge University Press, Cambridge.

Green, I.C., Wagner, D.A., Glogowski, J., Skipper, P.L., Wishnok, J.S., Tannenbaum, S., 1982. Analysis of nitrate, nitrite and (15N) nitrate in biological fluids. Anal. Biochem. 126, 131–138.

Hill, B.J., Way, K., 1996. Serological classification of infectious pancreatic necrosis (IPN) virus and other aquatic birnaviruses. Annu. Rev. Fish Dis. 5, 55–77.

Hickman, C.P., Trump, B.F., 1969. The kidney. In: Hoar, W.S., Randall, D.J. (Eds.), Fish Physiology, vol. 1. Academic Press, New York and London, pp. 91–239.

ICAR, 1960. Hand Book of Fisheries and Aquaculture. ICAR, New Delhi.

Jackson, M.L., 1973. Soil Chemical Analysis. Prentice Hall of India Pvt.Ltd, New Delhi pp. xiv + 498.

Karber, G., 1931. Beitrag zur kollektiven behandlung pharmakologischer Reihenversuche. Arch. Exp. Pathol. Pharmacol. 162, 480–483.

Kamilya, D., Joardar, S.N., Mal, B.C., Maiti, T.K., 2008. Effects of a glucan from the edible mushroom (*Pleurotus florida*) as an immunostimulant in farmed Indian major carp (*Catla catla*). Isr. J. Aquacult. Bamidgeh 60 (1), 37–45.

Karnovsky, 1965. A formaldehyde-gluteraldehyde fixative of high osmolarity for use in E.M. J. Cell. Biol. 27, 137–142.

Lilley, J.,H., Frerichs, G.N., 1994. Comparison of rhabdoviruses associated with epizootic ulcerative syndrome (EUS) with respect to their structural proteins, cytopathology and serology. J. Fish Dis. 17, 513–522.

Lilley, J.H., Hart, D., Richards, R.H., Roberts, R.J., Cerenius, L., Soderhall, K., 1997. Pan-Asian spread of a single fungal clone results in large scale fish-kills. Vet. Rec. 140, 11–12.

Lilley, J.H., Callinan, R.B., Chinabut, S., Kanchanakhan, S., Mac Rae, I.H., Phillips, M.J., 1998. Epizootic Ulcerative Syndrome (EUS) Technical Handbook. The Aquatic Animal Health Research Institute, Bangkok, ISBN 974-7604-58-2, 88 pp.

Laemmli, U.K., 1970. Cleavage of structural proteins during assembly of the head of bacteriophage T4. Nature 227, 680–685.

Macdonald, R.D., Gower, D.A., 1981. Genomic and phenotypic divergence among three serotypes of aquatic birnaviruses (infectious pancreatic necrosis virus). Virology 114, 187–195.

Magyar, G., Dobos, P., 1994. Expression of infectious pancreatic necrosis virus polyprotein and VP1 in insect cells and the detection of the polyprotein in purified virus. Virology 198, 437–445.

Mahy, B.W.J., Kangro, H.O., 1996. Virology Methods Manual. Academic Press, London.

Mossman, T., 1983. Rapid calorimetric assay for cellular growth and survival: application to proliferation and cytotoxic assays. J. Immunol. Methods 65, 55–63.

Moore, P.D., Bellamy, D.J., 1973. Peatlands. Elek Science, London, UK.

OIE, 2009. Manual of Diagnostic Tests for Aquatic Animals, sixth ed. O.I.E, Paris, France.

Okamoto, N., Sano, T., Hedrick, R.P., Fryer, J.L., 1983. Antigenic relationships of selected strains of infectious pancreatic necrosis virus and European eel virus. J. Fish Dis. 6, 19–25.

Olsen, S.R., et al., 1954. Estimation of Available Phosphorus in Soils by Extraction with Sodium Bicarbonate. Circ. U.S. Dep. Agric., p. 939.

Reed, L.J., Muench, H., 1938. A simple method of estimating fifty percent endpoints. Am. J. Hyg. 27, 493–497.

Roberts, R.J., Frerichs, G.N., Tonguthai, K., Chinabut, S., 1994. Epizootic ulcerative syndrome of farmed and wild fishes. In: Muir, J.F., Roberts, R.J. (Eds.), Recent Advances in Aquaculture, vol. 5. Blackwell Science Ltd, Oxford, England, pp. 207–239.

Roberts, R.J., Willoughby, L.G., Chinabut, S., 1993. Mycotic aspects of epizootic ulcerative syndrome (EUS) of Asian fishes. J. Fish Dis. 16, 169–183.

Rovozzo, G.C., Burke, C.N., 1973. A Manual of Basic Virological Techniques. Prentice Hall, Englewood Cliffs, NJ, pp. 126 ± 163.

Sambrook, J., Fritsch, E.F., Maniatis, T., 1989. Molecular Cloning: A Laboratory Manual, second ed. Cold Spring Harbor Laboratory Press, Cold Spring Harbor, New York.

Smith, D., Onions, A.H.S., 1994. The Preservation and Maintenance of Living Fungi. International Mycological Institute. Technical Handbooks No. 2. 122 pp.

Vedpraskas, M.J., 1995. Redoximorphic Features for Identifying Aquic Conditions. Technical Bulletin 301. North Carolina Agricultural Research Service, North Carolina State University, Raleigh, 33 pp.

Walkley, A.J., Black, I.A., 1934. Estimation of soil organic carbon by chromic acid titration method. Soil Sci. 37, 29–38.

Willoughby, L.G., Roberts, R.J., Chinabut, S., 1995. *Aphanomysis invaderis* spp. nov.: the fungal pathogen of freshwater tropical fish affected by epizootic ulcerative syndrome. J. Fish Dis. 18, 273–275.

Willoughby, L.G., Roberts, R.J., 1994. Improved methodology for isolation of the *Aphanomyces* fungal pathogen of epizootic ulcerative syndrome (EUS) in Asian fish. J. Fish Dis. 17, 541–543.

Willoughby, L.G., Pickering, A.D., Johnson, H.G., 1984. Polycell-gel assay of water for spores of Saprolegniaceae (fungi), especially those of the *Saprolegnia* pathogen of fish. Hydrobiologia 114, 237–248.

Wolf, K., Gravell, M., Malsberger, R.G., 1966. Lymphocystis virus: isolation and propagationin centrarchid fish cell lines. Science 151, 1004–1005.

Wolf, K., Quimby, M.C., 1969. Fish cell line and tissue culture. In: Hoar, W.S., Randall, D.J. (Eds.), Fish Physiology, vol. 3. Academic Press, New York, pp. 253–305.

Chapter 9

Epizootic Ulcerative Syndrome Works in Relation to Different Aspects

9.1 HUMAN-INDUCED FACTORS

Humans have been living in and with nature since time immemorial. It has been said that humankind was with nature at the dawn of civilization without much engineering being done. At the noon of civilization, humans started to engineer with nature. This is said to have resulted in different kinds of biotic and abiotic alterations, some friendly and some inimical. Ultimately, human-made alterations are believed to have culminated in often-perilous consequences, possibly at the zenith of civilization. Today, many unfriendly, unwanted, and often harmful results of human-induced factors and direct and indirect human interventions seem to be evident before us. One such outcome could be the outbreak, in epidemic dimension, of a hitherto unknown virulent and enigmatic fish disease variously called epizootic ulcerative syndrome or EUS.

Possibly due to impact on the environment caused by various human activities, for about the last three decades EUS has been seriously damaging a large number of commercially important culturable fish species during the cold season in a large number of water bodies covering a big part of the world in general, and Asia in particular. EUS seems to appear from the end of autumn to early spring and affects large number of fishes in ponds and in other lentic water bodies. EUS is considered, by some quarters, to be an infection due to fungal genera *Aphanomyces* and *Alternaria* (said to be in Pakistan), which are characterized histologically by penetrating hyphae surrounded by granulomatous inflammation. EUS is considered as the same disease as MG and RSD (Blazer et al., 1999). It is believed that <10 zoospores of this fungus, on average, are enough to kill a fish. Further, even infection by just a single zoospore may result in ulcerative lesions that could lead to mortality (Mohan and Shankar, 1995; Lilley et al., 1998; Thompson et al., 1999; Johnson et al., 2004). Fishes infected with *Aphanomyces* are believed to develop ulcerative lesions, which ultimately lead to extensive tissue necrosis (Blazer et al., 1999).

The syndrome had been reported to be broad-based and expanding, occurring in a number of culturable species in South east (SE) and South Asia (Roberts et al., 1993),

Japan (Egusa and Masuda, 1971; Hatai et al., 1977; Wada et al., 1996), and Australia (Fraser et al., 1992; Lilley and Roberts, 1997). The large-scale outbreak of EUS could be due to continued human impact on the environment through various kinds of activities. This most destructive disease, occurring in both fresh and brackish-water farmed and wild fishes, had been causing major fish losses in many countries since >3 decades (Baldock et al., 2005). Similar EUS was reported on the bodies of carps (*Catla catla, Cirrhinus mrigala, Hypophthalmichthys molitrix,* and possibly *Ctenopharyngodon idella*) and snakeheads *(Channa marulius)* in public waters of Sindh, Pakistan, during cold season under low salinity ranges. EUS disease in fishes was reported from waters having low temperature and salinity, and also probably had low light intensity. The spores of *Aphanomyces* and *Alternaria* were said to have developed abundantly in waters due to use of unhygienic organic dung and rotten feed and; perhaps, due to this, the fishes had been affected severely. The *Alternaria* species are commonly found in decayed and decomposed materials. The spores are airborne, but may also be found in the soil and water. Possibly, under the influence of human activities, there had reportedly been human health disorders, caused by *Alternaria*, which may grow on skin and mucous membranes, including on the eyeballs and within the respiratory tract. Almost similar situations had been reported in the carps, in which there was damage to the skin, gills, and eyeballs. Infections caused by motile spores of fungi in the closed water bodies enter the skin of fish and then germinate, forming fungal filaments or hyphae. A variety of saprophytic *Aphanomyces, Achlya, and Saprolegnia* species had been isolated from the surface of EUS-affected fish or from infected waters, and possibly also other oomycete fungi had been isolated from other diseases of aquatic animals (Lilley, 1997). Disease was reported in active form during cold seasons under low temperature during winter to early spring and affecting large numbers of fishes in ponds and stagnant public waters. Disease had also been reported in estuarine fishes like mullets (Mckenzie and Hall, 1976). In fact, epizootic condition in FW fishes is a complex infectious etiology characterized by the presence of invasive *Aphanomyces* infection and necrotizing ulcerative lesions

Epizootic Ulcerative Fish Disease Syndrome. http://dx.doi.org/10.1016/B978-0-12-802504-8.00009-2

typically leading to a granulomatous response (Roberts et al., 1993).

9.2 SOME PERTINENT INFORMATION ABOUT EPIZOOTIC ULCERATIVE SYNDROME

EUS, being an outcome of complex interactions among different parameters, a host of information have been generated through long-standing works in the field.

9.2.1 Similar Reported Diseases

Pathological and epizootiological evidence had indicated that the condition, called red spot disease (RSD) in Australia, is indistinguishable from EUS. Similarly, available evidence further suggests that the condition, known as mycotic granulomatosis in Japan, is also indistinguishable from EUS. RSD and MG might have originated due to human-induced factors as well, and need further study to trace the root cause of their origin.

9.2.2 Expansion of Range

EUS is endemic in many countries, and could be due to human impacts. It had been extending its geographic range even into subtropical, subtemperate, and temperate climates. Experimental evidence had indicated that the *Aphanomyces* involved is capable of causing disease in temperate species, as well.

9.2.3 Relation to Consumption by Humans

All available evidence seemed to suggest that consumption of EUS-infected fish poses no proven specific health problems to man provided that they are properly prepared in sanitary conditions.

9.2.4 Virus-Related Information

Earlier studies had shown the presence of a wide range of viral agents in fishes affected by EUS. It had thus been recommended that further extended work be undertaken to determine more accurately the incidence and distribution of tropical fish viruses throughout the region.

9.2.5 Epidemiological Information

Further studies are likely to produce large volumes of complex data related to EUS outbreaks in fish populations, and relationship and impacts of human-induced influences, if any. It had been suggested that epidemiological expertise be developed within the region to enable these data to be effectively utilized.

9.2.6 Possible Relation with Environment

It is believed that there is evidence to portray relation between the initiation of EUS and environmental factors. Human-induced impacts may have a role in the root cause of its origin, which, of course needs further confirmation. It is usually suggested that further studies on environmental conditions, including physical, chemical, and biological factors in addition to human-induced impacts, are essential for a clear understanding of the role of environment in the outbreak of EUS. There is often concern over the limited understanding of the relationship between fish health and environmental conditions, in general.

9.2.7 Specific Identification of the Fungus

The exact identification of the fungus is essential, may be also by molecular process. The isolates be fully characterized and their relationship made evident. Although much works have been done in this regard, further works may bring to light newer information.

9.2.8 Aspects of Diagnosis

Diagnosis is usually based on a number of clinical and pathological features of the disease. However, it is important that a rapid, specific, accurate, and low-cost diagnostic test be developed.

9.2.9 Development of Resistance

Evidence from some countries seemed to suggest that there is an element of resistance that might develop in a fish, after an initial outbreak of EUS. This resistance may be ecological, genetic, or associated with some kind of acquired immunity. It is important that the mechanism for this be investigated.

9.2.10 Mode of Transmission

Evidence seems to suggest that the spread of EUS disease in the Asian countries could be largely via rivers and other natural waterways. There is further to be said about evidence available of distinct transmission of EUS over marine barriers. It is important that an understanding be gained regarding the mechanism of transmission between the water bodies. It is further essential that attention be given to the development of quarantine measures to prevent the transfer of these via live fish transportation or via transportation of infected materials.

9.2.11 Socioeconomic Impact of EUS

The outbreak of EUS in India, as in other countries, had also left people (fish consumers) panic-stricken. Concomitantly, the fishers had been incurring colossal economic losses. This led to socioeconomic difficulties. In this regard also, humankind could ascertain the role of its own impacts, if any, in the outbreak of this virulent disease, in order to find corrective measures.

9.2.12 Social Effect

Investigations carried out in different places had revealed that c 73% aquaculture operation units were adversely affected by EUS. The outbreak of the disease resulted in a dwindling of the fish consumption rate by c 28.7%, 23.3%, and 20.5% in urban, suburban, and rural sectors, respectively, on an average estimation. Consequently, the fish trade was also affected seriously. Owing to consumer resistance, the traders did not accept such fish for selling. Moreover, in rural markets, diseased fishes were sold at a very low price. Role of human impacts could also be thought of, in this regard.

9.2.13 Organization of the Small-Scale Livestock and Fisheries Sector

Small-scale farmers are significant components and players in the agricultural and aquacultural sectors. The target of a service provider, say an organization, is the "individual farmer," particularly a small-scale farmer in a farming system who may not always be competent enough to handle and tackle all kinds of situations such as floods, droughts, etc. On the other hand, apex bodies such as the government operate on a collective basis like a group of "farmers" rather than "individuals." In fact, the organization of the small-scale livestock sector usually operates in a hierarchical manner beginning with small-scale farmers followed by villages, provinces, and countries. Their issues are also to be dealt with significantly.

Concomitant to the above, c 42.19% of the aquaculturists are said to have suffered c 31%—40% loss of fish in their culture ponds. Further, a section of farmers had to look for alternative jobs. In addition, c 88.9% of fish traders had also suffered losses to some extent during the affected period.

Another study, undertaken in five districts of Kerala specifically, revealed that the spread of EUS had completely paralyzed the inland fish market and threw the fishers out of their occupation. The women fish vendors in particular were subjected to tremendous hardship. They had to seek alternative employment as agricultural laborers, headload and quarry workers, etc., but without much success.

9.3 INFORMATION PERTAINING TO PHYSICAL AND CHEMICAL CHARACTERISTICS

The physicochemical characteristics of the aquatic regime are significant parameters that exert a tremendous amount of influence on the dynamics of interaction among the various factors.

9.3.1 Minor Elements

Various kinds of studies had revealed that the concentration of metals and other potentially toxic minor elements in water samples had usually been low and also had been within permissible guidelines, although there had been occasional peaks above toxic thresholds. But such peaks could not be correlated with specific disease occurrences. Copper, zinc, and mercury had reportedly been found at high residual levels in comparison with "permissible" values in some areas, specifically India (Phillips and Keddie, 1990; Lilley et al., 1992).

Concomitant to the above, Macinthosh and Phillips (1986) had recorded consistently high concentrations of Al and Fe in water samples from affected areas. The former (Al) might have been released from the soils after heavy rains (specifically if such rains had followed a dry spell), resulting in damage to the epidermis and gills of the fish. It may be mentioned here that Al is generally considered to be toxic to fish at a pH range of 5.0—6.0. It is also believed to be a significant toxin in even mildly acidic freshwater (Exley and Phillips, 1988). However, its activity in the outbreak of EUS, if any, is perhaps yet to be fully explored.

9.3.2 Soil Status

It had been believed that no specific characteristics of soil fertility or organic content appear to be related to the outbreak of EUS (Macintosh and Phillips, 1986; Kar, 2007). Nevertheless, the soil condition of a number of affected sites had been reported to be slightly acidic in nature with low calcium content. Such soils could account for the poorly buffered acidic water and high levels of Al and Fe in the water samples from the affected sites. It may be mentioned here that the outbreak of RSD in Australia was believed to have been associated with acid sulfate soils. It may be opined here that high levels of Al and Fe are usually flushed out of acidic soils when rainfall follows a period of drying. Such conditions are sometimes reported to precede EUS outbreaks.

9.3.3 Basic Physicochemical Characteristics of Water

Rodgers and Burke (1981) had reported the rapid seasonal decrease of salinity and temperature as significant factors, which are supposed to play some role in predisposing fish to attack by RSD in Australia. Conversely, Callinan had opined that these factors could also prevail following periods of heavy rainfall, when acidic compounds are washed out of the surrounding soil. The resulting low pH was believed to trigger the outbreak of RSD.

Concomitant to the above, seasonal variable factors had also been implicated in the outbreaks of EUS. Various reports had portrayed that EUS usually occurs during periods of decreasing or fluctuating temperatures. The EUS-affected sites in Bangladesh, China, India, Lao PDR, etc., had been monitored during 1988 and 1989 (Lilley et al., 1992). It had been found that outbreaks had occurred during such months when the average daily temperature was less than the average annual daily temperature (Phillips and Keddie, 1990). Nevertheless, outbreaks of EUS had also been recorded in different places during the warmer months (Phillips and Keddie, 1990; Kar, 2007, 2013), thereby suggesting no clear relationship between temperature and outbreak of EUS. Nevertheless, fluctuations in diurnal temperature to the tune of 10 °C had been recorded during the outbreaks in both Bangladesh and the Philippines (Phillips and Keddie, 1990). Such fluctuations had been believed to be of greater magnitude in shallow paddy field environments. As such, they are perhaps enough to stress the resident fish population. The dynamics of the relationship between temperature and initiation of EUS, if any, is to be studied further.

Notwithstanding the above, low alkalinity, hardness, chloride concentration, and fluctuating pH had also been opined to be related to the outbreaks of EUS. This is believed to be important, because waters with low alkalinity (TA) are often considered vulnerable to fluctuations in water quality. Such waters may also respond poorly to fertilization and may result in increased toxicity of metals and certain pollutants to fish. Further, low concentration of chloride could also make fish less tolerant to environmental toxins. In addition, the epidermis of the skin and the gills are believed to be damaged due to acidic water. This could lead to decreased resistance in fish to infection. Moreover, the significance of TA and chloride had been further emphasized by reports from various countries. There had been reports of treatments of EUS infection through the application of lime and NaCl.

Notwithstanding the above, some of the EUS-affected sites had revealed values of physicochemical characteristics of water, particularly DO and FCO_2, much different from the optimum range. Such values had been considered to be potentially harmful to nonair-breathing fishes, and this could have certainly resulted in "stress" to the fishes (Phillips and Keddie, 1990). However, no definite correlation could be ascertained between the suboptimal levels of these parameters and the outbreak of EUS, because such irregular values tended to persist for a big part of the year.

Concomitant to the above, there had possibly been no record of very high values of NH_3 and NO_2^- (nitrite) in the water before the outbreak of EUS. This might have indicated that toxicity of these substances was unlikely to be a significant factor in the initiation of EUS (FAO, 1986). Further, suboptimal levels of nitrogen and phosphorus compounds seemed to have indicated that the outbreak of EUS might not have been associated with heavily fertilized or organically loaded waters.

9.3.4 The Water Supply

Water is the habitat for the fishes. Fish farms that receive water supply from the irrigation canals or paddy fields generally have a potential risk of getting infection, which may lead to outbreak of EUS. As such, it could be better, if the ponds in the fish farms could be filled with tube well water. If this is not possible, then, the pond water may not be changed during an outbreak of EUS. Also, the embankments of the ponds may preferably be raised in order to prevent the entry of floodwater. On the other hand, if it is at all essential to use surface water, this water may be filtered to remove the wild fishes, if any. Further, the organic contaminants and the pathogens in the water may also be removed through filtration (e.g., sand filter, charcoal bed, etc.). Also, the general water quality may be improved through stocking of herbivorous fishes and shellfishes, and the planting of grasses in the source water body before it enters the culture facility.

9.3.5 Season and Climate

It had been opined by various quarters at different times that limnological parameters such as low temperature, alkalinity, and hardness appear to correlate best with the initiation and outbreak of EUS. However, such situations generally follow the period of intense rainfall at the beginning of the dry season. As reported (Lilley et al., 1992), c 72% of the outbreaks occurred during the dry season and normally stopped with the onset of the next rainy season (Phillips and Keddie, 1990). However, there seems to be tremendous amount of variations in the pattern of rainfall in SE Asia. It may be mentioned here that a big portion of Thailand generally experiences a dry spell from September to March, whereas the equatorial regions of SW Malaysia and Indonesia may get rains during most part of the year.

In contrast to the above, it had been reported by FAO (1986) that some incidences of outbreaks of EUS had occurred in the estuarine fishes of Singapore and Thailand after phytoplankton blooms. Such blooms are much related to climate. Blooms also occur in still bright weather conditions following a period of rainfall. However, there is possibly a need for more detailed analysis evaluating the potential significance of rainfall patterns in the outbreaks and epidemiology of EUS.

9.4 ASPECTS OF HYGIENE

Proper hygiene prevents contaminations. Maintenance of hygiene is essential for a healthy life. Proper hygiene could also prevent diseases among the fishes.

9.4.1 Aspects of General Husbandry

Good general husbandry practices are very much essential for healthy, disease-free management of the water body. Recurrence of EUS could perhaps be largely prevented through good general husbandry practices. Menasveta (1985) had reported that c 50% of the EUS-affected snakeheads had recovered when they were shifted to ponds with improved water quality. Further, Das et al. (1992) observed that poorly fed overstocked fish are more susceptible to infection by EUS, particularly under poor sanitary conditions. Almost all the reports had emphasized the significance of good fish husbandry practices in managing the problems of EUS. However, the fish farmers are to prevent the entry of pathogens into their water bodies from the affected areas. There is perhaps no conclusive information about the exact route or path of EUS transmission. As such, specific care must be taken to keep healthy ponds away from the contaminated water bodies, equipment, and affected fishes.

9.4.2 Disinfection

Disinfection is an important aspect of preventing the recurrence of EUS. A stringent means of eliminating EUS is to destroy the stock, to sterilize the rearing facilities, and then, to restart the work with the disease-free fish. This could be a significant means of eliminating EUS, particularly in small, well-controlled, and manageable environments like hatcheries. Concomitantly, the EUS-affected fishes are to be disposed of through incineration or burial with disinfectants such as slaked lime. Further, the nets, fishing gears, and other equipment are to be dipped in or sprayed with disinfectants, such as slaked lime (before they are moved to other water bodies) to prevent reintroduction of the disease after sterilization. Hygienic practices involve the requirement of workers and visitors to wash their hands and feet with disinfectants such as Densol, Lysol, or Dettol before entering the fish farm. This is in addition to the maintenance of general cleanliness while in the farm. Moreover, the wheels of the vehicles may also be disinfected, if there is a potential risk of infection from the adjacent water bodies.

9.4.3 Public Health

EUS has been a hitherto unknown enigmatic virulent disease among the fresh and brackish-water fishes in the world and occurring on a semiglobal dimension today. As such, this epidemic had possibly not been known to the living memory of humans. Hence, being undefined, its modus operandi for management, tackling, and dealing had been shrouded with mysteries, misbeliefs, fears, and confusion. In view of this, people, not only were averse to eating fishes during the EUS outbreak period, but also were reluctant to use the waters obtained from the EUS-affected water bodies. In fact, there had been reports of moralities among the cattle, birds, and even human beings, attributed to the so-called consumption of EUS-affected fishes. This had been found to be totally baseless and unfounded (Kar and Dey, 1988). Also, Rahman et al. (1988) could not induce any disease symptoms in ducks fed with EUS-affected fishes. Moreover, injection of *Aeromonas hydrophila* cultures could not induce any symptoms of EUS.

Concomitant to the above, patients reportedly suffering from diarrhea in Thailand were believed to have contracted their condition from *A. hydrophila* appearing through EUS outbreaks during 1982–83 (Chumkasien, 1983). However, Reinpraryoon et al. (1983) had reported a fairly high carrier rate of the said bacterium in some rural areas. As a stringent public health measure, the EUS-affected dead fishes are not being disposed of near human habitations. Otherwise, the bacteria or the toxins present in any dead fish may cause illness in human beings.

In addition to the above, the use of chemotherapeutants, notably antibiotics, should be discouraged in view of public health hazards. As an example, it may be cited here that with Chloramphenicol, which is used for treating typhoid in man, there is apprehension that the built-up bacterial resistance in treated fish (Poonsuk et al., 1983) could be transferred to humans. Further, there may be possibility of severe allergic reactions being generated among the farm workers who could come in contact with the drug. Moreover, there could also be the concern that consumers may be exposed to drug residues in marketed fish that had been hastily harvested before the recommended period of withdrawal had been completed.

In view of the foregoing account, there is need for legislations to protect the fishes, the fish consumers, and also the water bodies from indiscriminate use of chemotherapeutants, notably the antibiotics.

9.5 ASPECTS OF FOOD AND MANAGEMENT

Fish happened to be one of the potential entities of nutrition and avocation. It becomes a matter of concern when the fishes in a water body are affected by a disease. The ailments are to be managed judiciously so that the fishes as well as the people are not much affected in the long run.

9.5.1 The Fish Stock

The fishes to be stocked are to be disease-free and healthy. It is to be ensured that wild fishes and diseased fishes do not get an entry into the fish farm on any account. Further, the hatcheries are to maintain the movement records of the fishes, including the fry, fingerlings, and adults covering the species, number of individuals, size range, and the destination of the consignments, so that the sites at risk could be traced in the event of outbreak of EUS. Seed suppliers are encouraged to provide verification certificates of the disease-free status of the juveniles. Concomitantly, the fish farmers could insist on the practice of giving fish juvenile health certificates with each consignment. Further, a prophylactic treatment of salt or malachite green could be administered to new stock, and they could be isolated from the rest of the farm stock for two to three weeks.

Concomitant to the above, high stocking density are often the major causes of stress to cultured fishes. However, optimum density also varies with the species of fish. The water quality could deteriorate, if the stocking density is very high, and this may stress the fish further. Importantly, the EUS-affected dead fishes are to be removed immediately to prevent further contaminations.

Concomitant to the above, the reduction in the severity of outbreaks during later periods seemed to suggest that resistance to the disease gradually builds up over successive generations. Nevertheless, stocking of resistant fish stock could offer some amount of protection. However, such a step could be adopted only in such areas, where the disease might have occurred earlier, and the pathogens could remain in a latent state. In addition, stocking of resistant fishes; e.g., Tilapia, during the EUS-outbreak season, had been encouraged by some quarters.

9.5.2 Nutrition and Feeding

Nutrition is an important parameter of metabolism of the body. Optimum nutrition maintains a healthy mind in a healthy body, as the proverb goes. Nutrition is also equally important in the physiology of a fish life. On the other hand, mal nutrition could lead to bad consequences. Malnutrition is said to result in profound changes in the immune response of a fish (Landolt, 1989). It may be noted here that suboptimal deficiencies in balanced food could reduce the ability of the fish to resist occurrence of disease. In this regard, Menasveta (1985) had opined with anecdotal evidence that cultured snakeheads fortified with vitamin C had depicted less chance of being infected by EUS. He also recommended this procedure to be followed prior to the onset of cold season.

Concomitant to the above, good quality pelleted feed, stored in a cool dry area, and fed at an appropriate dose, could provide with all the basic nutritional requirements of an intensively farmed fish. This could serve as an alternative to feed given as trash fish. In this regard, Hardy (1990) had argued that the improved nutritional profile and water quality conditions presented to fish on a fixed pelleted diet could result in a higher yield for the farmer under more intensive farm conditions.

Notwithstanding the above, a drastic decrease in feed levels could be recommended at the initial sign of infection. There may not be any other input of organic matter, such as the application of fertilizers, during an EUS outbreak. It is because these may lead to additional stress to the farm stock. In this regard, Goswami et al. (1988) recorded that grass fed to cultured *C. idella* had usually been collected from submerged adjacent areas, and thus these might have served as potential sources of infection.

9.5.3 Rice Field and Pond Management Practices

Rice is one of the staple food items of the people of South and SE Asia. Fish is said to add a delicious taste to the dish. As such, people cultivate paddy on a large scale. Concomitantly, they go for fish culture as well, along with rice paddy through the technology of "paddy-cum-fish culture," in order to harvest two kinds of crops at the same time. The fishes commonly grown in the paddy fields include mainly murrels, anabantids, clariids, and some of the minor carps. These fishes generally fall victim of EUS attack and cause colossal loss to the farmers. It had been a general observation that the occurrence of EUS in the rice field ecosystem had been usually coinciding with the later part of paddy cultivation period. During this time, the water level is usually low, there could be accumulation of certain types of fertilizers, and there is usually decomposition of organic matter, notably aquatic macrophytes, in the water body. Such a situation could render the aquatic biotope little uncolonizable by the aquatic biota including the fishes, which could then become susceptible and vulnerable to various kinds of pathogens, especially those causing EUS fish disease. Such a situation is *in vogue* in Asia in general, notably in India, Bangladesh, Myanmar, Thailand, China, Malaysia, Indonesia, the Philippines, Lao PDR, Kampuchea, Vietnam, and so on (Phillips and Keddie, 1990; Kar, 2007, 2013).

Notwithstanding the above, surveys (Phillips and Keddie, 1990) had revealed that the common types of fertilizers used in the region were urea, TSP, NPK, Ammophos, etc. In addition, various types of unspecified and undefined inorganic fertilizers are also used by the farmers with the expectation of a bumper harvest. In addition, the application of certain types of pesticides is also quite common, notably dieldrin, DDT, Malathion, endrin, Dipterex, paramine, 2-4-D, Gramoxone, Ker Tex 200, Azodrin, diazinon, Machete, etc. However, no precise correlation could be revealed between application of agrochemicals and outbreak of EUS. This had been known after examination of the cropping patterns and timing of application of the agrochemicals. Further, the herbicides had probably accounted for *c* 75% of the chemicals used, and moreover, they had been applied at the time of preparation of the agricultural plot, and had only been used occasionally for clearing of dykes before the period of cultivation. No doubt, certain subacute effects might result from the residual accumulation of these chemicals, but no direct effects seemed to relate to the outbreak of EUS.

Concomitant to the above, it had also been a common experience that the major outbreaks of EUS in the pond ecosystem had been occurring toward the end of the fish culture period. Such situations had been reported from Myanmar, Lao PDR, Vietnam, Indonesia, etc. (Phillips and Keddie, 1990). Excess of organic loads may have some role in the initiation of EUS outbreak (Kar et al., 2007). Organic matter decomposition, to some extent, may result from fertilizers and manures, notably the poultry manure, cow dung, pig dung, night soil, etc. Moreover, supplementary feeding given to the cultured fishes could add further organic load, if remained unutilized. Such organic loads are said only to add to the existing "stress" of the fishes. As such, incidence of EUS outbreaks could be reduced through adoption of proper hygienic measures, adequate fish husbandry measures; maintenance of good water quality through proper application of lime, fertilizers, etc., and above all, healthy pond management.

9.6 CONCLUSIONS

EUS has been a challenge to ichthyologists. Extensive research and intensive studies could not as yet possibly find an answer to its effective management and control. As such, further continuous and comprehensive studies are to be conducted, if necessary, collaborating with different expertise, in order to find a definite clue and cure of the ailment.

Notwithstanding the above, Atlantic menhaden are the species of estuarine fish most severely affected by ulcerative lesions in the United States, which are characterized by

solitary, typically peri-anal, focal, deep, granulomatous lesions containing oomycete hyphae, primarily those of *Aphanomyces* (Noga et al., 1996).

Human-induced factors may have some role in the initiation and pathogenesis of a fish disease. But concomitantly, human beings may in turn be at risk of being infected and affected by the disease. However, evidence for human health risks from exposure to *Pfiesteria* toxins was largely based on limited cognitive testing. Further, there had been no reports of health effects on mammals exposed to estuarine waters or on avian populations that fed on menhaden from active fish kills containing diseased fish in *Pfiesteria*-infested waters. The etiology of ulcerative mycotic lesions in menhaden and other fish was said to be unclear, though it is most probably multifactorial in nature (Lilley et al., 1998). The association of ulcerative fish lesions with viral, bacterial, fungal, and other parasitic agents (Noga and Dykstra, 1986; Noga, 1988; Sindermann, 1988; Roberts et al., 1989; Lilley et al., 1998; Marty et al., 1998) certainly does not preclude the involvement of biotic or abiotic toxins, including those from *Pfiesteria*-like dinoflagellates. These toxins could predispose fish to epithelial lesion-initiating events. In addition, other predisposing factors (stressors) may include suboptimal or rapidly changing chemical and physical water quality (Wedemeyer, 1974; Noga et al., 1993), sublethally depressed DO (Scott and Rogers, 1980), trauma (Kane et al., 1998), and the presence of irritants (Murty, 1986; Heath, 1987; Anderson and Zeeman, 1995). Further, human-induced factors might be having some role in such alterations in the physicochemical characteristics of the environment.

Further needed research includes discerning the factors that cause fish ulcer initiation and progression, as well as species-specific sensitivities, and perhaps above all, possible role of human-induced factors, if any, as the possible root cause in the initiation of the condition. Further, we need to characterize the toxins from *Pfiesteria* and other closely related dinoflagellate species, determine the mechanisms of biological activity of these toxins, and develop a widely utilitarian, toxin-specific probe to identify and quantitate the presence of toxic *Pfiesteria* activity.

REFERENCES

Anderson, D.P., Zeeman, M.G., 1995. Immunotoxicology in fish. In: Rand, G.M. (Ed.), Fundamentals of Aquatic Toxicology, Effects, Environmental Fate and Risk Assessment, second ed. Taylor and Francis, Washington, D.C., pp. 371–404.

Blazer, V.S., Vogelbein, W.K., Densmore, C.L., May, E.B., Lilley, J.H., Zwerner, D.E., 1999. *Aphanomyces* as a cause of ulcerative skin lesions of menhaden from Chesapeake Bay tributaries. J. Aqua. An. Heal. 11, 340–349.

Baldock, F.C., Blazer, V., Callinan, R., Hatai, K., Karunasagar, I., Mohan, C.V., Bondad-Reantaso, M.G., 2005. Outcomes of a short expert consultation on epizootic ulcerative syndrome (EUS): re-examination of casual factors, case definition and nomenclature. In: Walker, P., Lester, R., Bondad-Reantaso, M.G. (Eds.), Diseases in Asian Aquaculture V. Fish Health Section, Asian Fisheries Society, Manila, pp. 555–585.

Chumkasien, P., 1983. *Aeromonas Hydrophila* in Man, Special Reference to Diarrheal Cases during the Fresh Water Fish Epidemic: 1982–1983, pp. 547–549.

Das, P., Mishra, A., Kapoor, D., 1992. Some observations on epizootic ulcerative syndrome (EUS) around Allahabad. Adv. Biosci. 12, 1–8.

Egusa, S., Masuda, N., 1971. A new fungal disease of *Plecoglossus altivelis*. Fish Path. 6, 41–46.

Exley, C., Phillips, M.J., 1988. Acid rain: implications for the farming of salmonids. In: Muir, J.F., Roberts, R.J. (Eds.), Recent Advances in Aquaculture, vol. 3. Croom Helm, London.

FAO, 1986. Report of the Expert Consultation on Ulcerative Fish Diseases in the Asia-Pacific Region.

Fraser, G.C., Sallinan, R.B., Calder, L.M., 1992. *Aphanomyces* species associated with red spot disease: an ulcerative disease of estuarine fish from eastern Australia. J. Fish Dis. 15, 173–181.

Goswami, U.C., Dutta, A., Sarma, D.K., Sarma, G., 1988. Studies on certain aspects of prevention and treatment of fish suffering from epizootic diseases syndrome. In: Proceedings of the Symposium on Recent Outbreak of Fish Diseases in North Eastern India Guwahati, Assam, pp. 16–17.

Hatai, K., Egusa, S., Takahashi, S., Ooe, K., 1977. Study on the pathogenic fungus of mycotic granulomatosis – I. Isolation and pathogenicity of the fungus from cultured ayu infected with the disease. Fish Pathol. 12, 129–133.

Heath, A.G., 1987. Pollution and Fish Physiology. CRC Press, Boca Raton, Florida.

Hardy, R.W., 1990. Fish health managemant and nutrition in the Asia-Pacific region. In: Fish Health Management in Asia-Pacific. Report on a Regional Study and Workshop on Fish Disease and Fish Health Management.ADB Agriculture Department Report Series No.1. Network of Aquaculture Centres in Asia-Pacific, Bangkok, Thailand, pp. 425–434.

Johnson, R.A., Zabrecky, J., Kiryu, Y., Shields, J.D., 2004. Infection experiments with *Aphanomyces invadans* in four species of estuarine fishes. J. Fish Dis. 27, 287–295.

Kar, D., Dey, S., 1988. Preliminary electron microscopic studies on diseased fish tissues from Barak valley of Assam. Proc. Annu. Conf. Electron Microsc. Soc. India 18, 88.

Kane, A.S., Oldach, D., Reimschuessel, R., 1998. Fish lesions in the Chesapeake Bay: *Pfiesteria*-like dinoflagellates and other etiologies. Md. Med. J. 47, 106–112.

Kar, D., 2007. Fundamentals of Limnology and Aquaculture Biotechnology. Daya Publishing House, New Delhi pp. vi + 609.

Kar, D., Mazumdar, J., Halder, I., Dey, M., January 25, 2007. Dynamics of initiation of disease in fishes through interaction of microbes and the environment. Curr. Sci. Bangalore 92 (2), 177–179.

Kar, D., 2013. Wetlands and Lakes of the World. pp. xxx + 687. Springer, London. Print ISBN: 978-81-322-1022-1; e-Book ISBN: 978-81-322-1923-8.

Landolt, M.L., 1989. The relationship between diet and the immune response of fish. Aquaculture 79, 193–206.

Lilley, J.H., Callinan, R.B., Chinabut, S., Kanchanakhan, S., MacRae, I.H., Phillips, M.J., 1998. Epizootic Ulcerative Syndrome (EUS) Technical Handbook. The Aquatic Animal Health Research Institute, Bangkok, Thailand.

Lilley, J.H., Phillips, M.J., Tonguthai, K., 1992. A review of epizootic ulcerative syndrome (EUS) in Asia, Aquatic Animal Health Research Institute and Network of Aquaculture Centres in Asia-Pacific. Bangkok, Thailand, pp. Ix +73.

Lilley, J.H., Roberts, R.J., 1997. Pathogenicity and culture studies comparing the *Aphanomyces* involved in epizootic ulcerative syndrome (EUS) with other similar fungi. J. Fish Dis. 20, 135–144.

Lilley, J.H., 1997. Studies on the comparative biology of *Aphanomyces invadans*. (Ph.D. thesis). University of Stirling, Scotland. 228 pp.

Mckenzie, R.A., Hall, W.T.K., 1976. Dermal ulceration of mullet (*Mugil cephalus*). Aust. Vet. J. 52, 230–231.

Menasveta, P., 1985. Current fish disease epidemic in Thailand. J. Sci. Soc. Thailand 11, 147–160.

Marty, G.D., Freiberg, E.F., Meyers, T.R., Wilcock, J., Farverd, T.B., Hinton, D.E., 1998. Viral hemorrhagic septicemia virus, *Ichthyophonus hoferi*, and other causes of morbidity in Pacific herring *Clupea pallasi* spawning in Prince William Sound, Alaska, USA. Dis. Aquat. Org. 32, 15–40.

Macintosh, D.J., Philips, M.J., 1986. The Contribution of Environmental Factors to the Ulcerative Fish Diseases Condition in Southeast Asia, pp. 175–208.

Murty, A.S., 1986. Toxicity of Pesticides to Fish, vol. 2. CRC Press, Boca Raton, Florida.

Mohan, C.V., Shankar, K.M., 1995. Role of fungus in epizootic ulcerative syndrome of fresh- and brackishwater fishes of India: a histopathological assessment. In: Shariff, M., Arthur, J.R., Subasinghe, R.P. (Eds.), Diseases in Asian Aquaculture II. Fish Health Section, Asian Fisheries Society, Manila, pp. 299–305.

Noga, E.J., Dykstra, M.J., 1986. Oomycete fungi associated with ulcerative mycosis in menhaden, *Brevoortia tyrannus* (Latrobe). J. Fish Dis. 9, 47–53.

Noga, E.J., 1988. Determining the relationship between water quality and infectious disease in fishery populations. Water Resour. Bull. 24, 967–973.

Noga, E.J., Johnson, S.E., Dickey, D.W., Daniels, D., Burkholder, J.M., Stanley, D.W., 1993. Determining the Relationship between Water Quality and Ulcerative Mycosis in Atlantic Menhaden. Albemarle–Pamilico Estuarine Study, Report 92–15. North Carolina Department of Environmental, Health, and Natural Resources, Raleigh.

Noga, E.J., Khoo, L., Stevens, J.B., Fan, Z., Burkholder, J.M., 1996. Novel toxic dinoflagellate causes epidemic disease in estuarine fish. Mar. Pollut. Bull. 32, 219–224.

Poonsuk, K., Saitanu, K., Navephap, O., Wongsawang, S., 1983. Antimicrobial Susceptibility Testing of *Aeromonas Hydrophila*: Strain Isolated from Fresh Water Fish Infection. Chulalongkorn University, Bangkok, pp. 436–438.

Phillips, M.J., Keddie, H.G., August 13–26, 1990. Regional research programme on relationships between epizootic ulcerative syndrome in fish and the environment. In: Report on the Second Technical Workshop. NACA, Bangkok, p. 133.

Rodgers, L.J., Burke, J.B., 1981. Seasonal variation in the prevalence of "red spot" disease in estuarine fish with particular reference to the sea mullet, *M. cephalus* L. J. Fish Dis. 4, 297–307.

Reinprayoon, S., Lertpokasombat, K., Chentaratchada, S., 1983. The incidence of *Aeromonas hydrophila* in asymptomatic people at Bang-Toi. In: The Symposium on Fresh Water Fishes Epidemic. Chulalongkorn University, Bangkok, pp. 535–547.

Rahman, T., Choudhury, B., Barman, N., 1988. Role of UDS affected fish in the health of ducks in Assam: an experimental study. In: Proceedings of the Symposium in Recent Outbreak of Fish Diseases in North Eastern India Guwahati, Assam, p. 10.

Roberts, R.J., Wootten, R., MacRae, I., Millar, S., Struthers, W., 1989. Ulcerative Disease Survey, Bangladesh. Final Report to the Government of Bangladesh and the Overseas Development Administration. Institute of Aquaculture, Stirling University, Scotland, 85 pp.

Roberts, R.J., Willoughby, G., Chinabut, S., 1993. Mycotic aspects of epizootic ulcerative syndrome (EUS) of Asian fishes. J. Fish Dis. 16, 169–183.

Scott, A.L., Rogers, W.A., 1980. Histological effects of prolonged sublethal hypoxia on channel catfish *Ictalurus punctatus* (Rafinesque). J. Fish Dis. 3, 305–316.

Sindermann, C.J., 1988. Epizootic ulcerative syndromes in coastal/estuarine fish. NOAA Technical Memorandum NMFS-F/NEC-54. Smayda, T.J., 1992. A phantom of the ocean. Nature 358, 374–375.

Thompson, K.D., Lilley, J.H., Chen, S.C., Adams, A., Richards, R.H., 1999. The immune response of rainbow trout (*Oncorhynchus mykiss*) against *Aphanomyces invadans*. Fish Shellfish 9, 195–210.

Wedemeyer, G.A., 1974. Stress as a Predisposing Factor in Fish Diseases. U.S. Fish and Wildlife Service Fish Disease Leaflet, 38.

Wada, S., Rha, S., Kondoh, T., Suda, H., Hatai, K., Ishi, H., 1996. Histopathological comparison between ayu and carp artificially infected with *Apahnomyces piscicida*. Fish. Path. 31, 71–80.

Chapter 10

Control (Treatment) of Epizootic Ulcerative Syndrome

10.1 INTRODUCTION

In the last three decades (approximately), a serious and severely damaging fish disease has been sweeping through countries of the Asia–Pacific region, and then those of America and Africa, with dangerous consequences to fish resources and serious threats to the nutrition of the people and the avocations of inland fisherfolk. For a disease that is about 30 years old, it is a worry that the scientific community has not yet been able to pinpoint its causative agents.

In this regard, there have been a number of regional, national, and international seminars, symposia, and workshops to discuss various aspects of epizootic ulcerative syndrome (EUS) in different parts of the world. The NACA-ADB Regional Study and Workshop on Fish Disease and Fish Health Management has been one such significant initiative. Subsequently, a Regional Technical Cooperation Program (TCP/RAF/311[E]) of the Food and Agriculture Organization (FAO) entitled "Emergency Assistance to Combat EUS in the Chobe-Zambezi River" was approved in October 2007 for implementation in seven participating Southern African countries, viz., Angola, Botswana, Malawi, Mozambique, Namibia, Zambia, and Zimbabwe (Anon, 1992). There had also been programs for the control of EUS in other countries, wherever there had been EUS outbreaks. The programs had stressed the importance of enhancing surveillance and diagnostic capacities as well as the formulation of regional emergency response strategies to enhance educational status and awareness among the people, to promote responsible trade in aquatic animals in both affected and unaffected areas, and so on. Further, the Office d'Epizootie (OIE) had since organized a number of Regional training seminars to heighten awareness of aquatic animal diseases among the OIE focal points and people, notably in Africa.

EUS has been reported as one of the most destructive diseases for both farmed and wild fishes of fresh and brackish-water origin. After its report of possible origin in cultured ayu (*Plecoglossus altivelis*) from Japan in 1971, EUS had been known to occur in >94 species of fishes spread across 26 countries. EUS was considered a reportable disease by the Network of Aquaculture Centers in Asia–Pacific (NACA) and World Organisation for Animal Health (OIE) after recognizing its potential impact on cultured and wild fisheries. Therefore, OIE member countries had been obliged to report any new incursions of EUS, whether it is in a new species or in a new area. Once an outbreak of EUS occurs in a region, it generally recurs with less severity over the next 2–3 years, and generally with a reduced frequency thereafter. In India, EUS was first reported in 1988 from Assam and Tripura, and by 1999, it had spread to several other regions, resulting in large-scale fish mortalities. After the initial severe outbreaks during the late 1980s and the 1990s, mortalities due to EUS, in equivalent epidemic proportions, had not been widely reported in spite of the endemic status of the disease. Incidentally, a thorough survey of fish farms was carried out in different provinces, notably Uttar Pradesh (UP), with the objectives to confirm the cause of large-scale mortalities and to document the severity and prevalence of the disease. Incidentally, there had been reports of heavy mortalities of cultured and wild fish species in wetland districts of Uttar Pradesh (UP), India during 2011.

Investigations into the potential causative factors have focused on viral, fungal, and bacterial agents. Environmental parameters have also been studied. These abiotic factors had been believed to cause sublethal stress to the fish, initiating disease outbreaks. Potential causes of stressful environmental conditions include temperature, eutrophication, sewage, metabolic products of fishes, industrial pollutants, and pesticides.

The quality of water also appears significant from an etiological point of view. Parameters like salinity, alkalinity, temperature, hardness, and chloride concentration (many of which are seasonally variable) are known to predispose fishes to attacks of EUS. Infected fishes showed signs of improvement when transferred to clean freshwater ponds.

EUS is believed by a section of workers to be a **fungal disease of freshwater and brackish fish** affecting >100 species of fishes, and is said to be caused by the fungal species *Aphanomyces invadans*. The organism is said to require a specific combination of factors in order to germinate within the dermis of the fish. The disease causes lesions in both the skin and the visceral organs.

Epizootic Ulcerative Fish Disease Syndrome. http://dx.doi.org/10.1016/B978-0-12-802504-8.00010-9

EUS is present in parts of the **Asia—Pacific region as well as Australia**. A similar disease has also been reported in the eastern United States. EUS occurs commonly during periods of **low temperature and heavy rainfall in tropical and subtropical waters**. These conditions favor sporulation, and cold temperatures **delay the inflammatory response** of the fish to infection. Some species have been shown to be resistant to EUS, notably Chinese carps, milkfishes, and tilapia.

Further, little is known of the means by which EUS spreads within and between regions. However, movements of subclinically affected fish are probably important in transmitting *A. invadans* infection. Until effective control and prevention measures are implemented, it is likely that the disease will continue to spread, and that outbreaks will continue to recur, particularly in endemic areas. Comprehensive investigations of initial outbreaks in previously EUS-free areas, as well as recurrent outbreaks in previously endemic areas, are urgently needed. Findings from such studies could contribute important information on causal factors for EUS (Lilley et al., 1992).

In the initial stages of EUS, **red spots or small hemorrhagic lesions** are generally found on the surface of the fish. These lesions progress to form **ulcers and eventually large necrotic tissue erosions**. Fungal **mycelium** is often visible on the surface of ulcers. **Death** then follows rapidly due to **visceral granulomata**, septicemia, and failure of osmoregulatory balance. In this regard, positive diagnosis is made by **analysis of histological sections demonstrating mycotic granulomas and isolation of the causal fungus**. Concomitantly, keeping diseased fish in **good-quality clean water** may lead to recovery, but only if the lesions are not too extensive. Moreover, **dark scars** are often left behind upon healing. There is **no effective treatment for advanced stages of the disease**. Quarantine and health certification practices for movement of live fish between and within countries is the method of preventing spread of EUS to the EUS-free areas. In the endemic areas, **eradication, exclusion, management, surveillance, and treatment** are all required in order to gain control.

10.2 THERAPEUTIC APPLICATION IN DISEASED FISH

It seems quite well known that the therapy of fish diseases is still relatively unsophisticated (Singh et al., 1993). Even knowledge of the etiology of diseases is still at a rudimentary stage. The literature seems quite unlinked, varied, relatively rudimentary, and fragmentary. On the other hand, aspects of treatment are generally reported incidentally to the main theme of papers, with relatively less detail attached to them. Moreover, there have been few concerted and cohesive research efforts devoted exclusively to establishing definitive criteria (i.e., toxicity, pollution, etc.) for the etiology of an ailment, and not much effort has been directed toward the synthesis of treatment compounds. Further, there have not been many attempts at establishing exact and friendly *modus operandi* for the correct application of chemicals and drugs. Nevertheless, Roberts and Shepherd (1974) and Wood (1974) seemed to have brought out useful practical guides to be used at the grassroots level.

Similarly, the present trend in intensive fish farming seems to lay emphasis on prevention of the ailment rather than cure and treatment. This also seems to be the case with a majority of the intensive animal husbandry practices. However, most outbreaks of disease among cultured fishes are due to inadequate husbandry or management (Shepherd and Poupard, 1975). But in contrast to terrestrial environments, the aquatic domain seems to present unique difficulties with regard to husbandry and disease outbreaks.

With the foregoing in mind, therapeutic application in fish could be applied in three ways:

1. External chemical treatments
2. Systemic treatments via the diet
3. Parenteral treatment

While discussing all of the above briefly, attempts will be made in the following discussion specifically to mention various drugs and chemicals for the treatment of diseases among fishes.

10.2.1 External Chemical Treatments

This type of treatment could be applied in a variety of ways. All methods are said to involve immersion in a chemical solution. However, swabbing is an exception. Swabbing is a useful method for dealing with individual fish with localized external infections. However, immobilization is usually required before removal of the fish from the water. However, maintenance of the environment for fish seems to be a prerequisite for treatments involving immersion. In this regard, the dissolved oxygen (DO) levels of the water body are to be maintained above the minimum acceptable level. This is mainly because DO is considered the prime factor that is said to limit survival in the short term. On the other hand, NH_3 concentrations may not affect the survival of fishes directly during the treatment process, but chemicals may pose additional burdens on diseased fish. As such, all-out efforts should be made to ensure that the poor quality of water does not endanger survival. In addition, the toxicity of the chemicals or drugs and their efficacy against the disease-causing organisms are to be balanced, possibly in order to ensure the success of the treatment. In this connection, the gills are the most susceptible organs in terms of chemical toxicity. As such, any disease affecting the gills, could itself reduce the ability of the fish to survive the treatment process. Further, the concentration of the active ingredient of a chemical or a drug vis-à-vis its efficacy may be reduced due to the input of suspended materials (mainly solids), especially organic materials, from the floods or arising due to poor tank hygiene. In this regard, the pH and the hardness of the water are the most influenced parameters. As a rule of thumb, it is believed that as pH becomes lower, the water becomes softer and more acidic.

In addition to the above, immersion treatments can be further classified as (1) baths, (2) prolonged immersion, (3) dips, (4) flushes, and (5) flowing treatments.

Protozoan parasites are often quite abundant and sometimes pose a potential threat to the health of fishes. *Ichthyophthirius multifiliis*, an external protozoan parasite (white spot), is quite difficult to treat in its encysted form. Hence, mixed treatment with malachite green and formalin is often recommended. Conversely, the monogenetic trematodes, notably the *Gyrodactylus* and the crustacean parasites; for example, the *Lernaea* and the *Argulus*, can be treated with organophosphates, mainly because they do not seem to be susceptible to formalin. In this regard, Hoffman and Meyer (1974) comprehensively reviewed the treatment of external parasites. On the other hand, Kabata (1970) reviewed the treatment process of crustacean parasites. Concomitantly, Saring (1971) described the requisite treatment techniques in warm-water aquaculture.

10.2.2 Mycotic Infection

In addition to parasites, bacteria, and viruses, fungi are often considered potential threats to the health of fishes and other aquatic biota. It should be noted that malachite green, the fish farmers' choice against fish parasites, is also considered the chemical of choice against most external fungal infections, especially *Saprolegnia* sp. A concentration of 2 mg/L for 1 h is recommended as a prophylactic treatment during egg incubation.

Malachite green is available in a number of different forms. Of these, the oxalate, chloride, and sulfate salts are generally acceptable. However, only the double salt of $ZnCl_2$ (which is said to contain toxic levels of zinc) may be avoided. Incidentally, Steffens et al. (1961) had reviewed the use of malachite green in the treatment of fish disease.

10.2.3 Bacterial Infection

Bacterial infection could be classified into three kinds for treatment purposes and where external treatments may be used. These are (1) myxobacterial etiology, (2) endotoxic gram-negative bacteria, and (3) mixed infection.

Among the above, myxobacterial gill disease has traditionally been treated with quaternary ammonium compounds (QACs). Recommended concentrations range from 1 to 4 mg/L for 1 h applied as a bath. Further,

nifurpirinol had been reported as showing great promise for the treatment of all myxobacterial diseases at an active concentration of 1 mg/L for 1 h applied as a bath.

In addition to the above, $CuSO_4$ is considered a useful alternative to QACs, although its use is often restricted (for reasons of toxicity) to dip method at a concentration of 500 mg/L for 1 min. Further, proflavine hemisulfate is a good all-round treatment for external lesions due to gram-negative bacteria. Bath at 20 mg/L for 30 min is generally recommended (Roberts, 1978).

Incidentally, Bullock et al. (1971) produced an extensive review of bacterial diseases of fishes and their treatment.

10.3 SYSTEMIC TREATMENTS VIA THE DIET

Contrasting problems to those outlined for external treatments are presented through incorporation of a drug into the food to treat systemic bacterial diseases or gut parasites. In fact, the fish may eat the medicated food, and a lowering of the appetite is one of the first signs of disease in fish. This portrays the need for rapid diagnosis. Diagnostic techniques such as immunodiffusion and immunofluorescence, etc., had been introduced. However, the fish farmer may not usually expect a specific identification of the responsible pathogen in fewer than three to four days, even if the required diagnostic facility is available. Under said situation, a simple answer is to use drugs with a wide spectrum of Gram-negative antibacterial activity. Conversely, there are fewer populations of gram-positive bacteria of fish. They are associated, however, with chronic diseases in which treatment is infeasible. This is in contrast with the situation in higher animals, where many gram-positive microbes such as the streptococci are significant causes of often-acute infections.

Concomitant to the above, it is often considered important to achieve a correct rate of dosage in the quantity of medicated feed offered to the fish. In fact, a balance has preferably to be reached between toxic and therapeutic levels in the bloodstream. Nevertheless, the feeding rate is said to depend on a number of factors, including the activity of the drug, water temperature, and the time of year. Usually, healthy fish are reluctant to accept medicated diets. There could also be variations in the acceptance of medicated diets by the diseased fish. It could sometimes be difficult to correctly ascertain, how much diet the fish could actually consume. Moreover, the availability of drugs could prove difficult in certain countries, and drugs might be prescribed by registered veterinary practitioners only. In fact, such regulations and restrictions are considered significant because they help in reducing the risk of development of resistant strains of pathogenic bacteria. Likewise, physical incorporation of drugs into the diet could also present practical difficulties. Concomitantly, there could be strict limitations of marketing of fishes following a course of medicated feeding. There are of course no legislative requirements, but still the producers are expected to be under a moral obligation not to harvest such fish for at least four weeks after treatment, in order to allow the antibiotic to be cleared from the tissues of the fish. Otherwise, such fish could be responsible for developing antibiotic resistance in the bacterial flora of the person consuming them.

10.4 PARENTERAL TREATMENT

Injection, like swabbing, is of limited use in intensive culture. However, the fish has to be immobilized before removal from the water body for injection. This is a labor-intensive process. Moreover, parenteral antibiotics, such as oxytetracycline, are usually expensive. Hence, this technique is generally reserved for small number of valuable fish; that is, brood stock. However, an improved system is said to have been developed for the mass injection of carps using automatic multidose syringes. This technique is in wide use in Eastern Europe in order to premedicate populations of cyprinids susceptible to superior vena cava (SVC). This technique is generally carried out just before rise in Spring temperature, which usually initiates the epizootics of SVC, complicated by invasion of secondary gram-negative bacteria. Chloramphenicol is now generally used as a drug for this. Nevertheless, it may be stated here that antibiotic injection techniques in treatment of fish disease may be regarded as experimental in nature.

10.5 ADDITIONAL INFORMATION

Historically, chemotherapy for fish diseases was developed in the United States during the mid-1940s. Conventional sulfonamides were used to combat outbreaks of furunculosis in FW fishes (Snieszko and Friddle, 1952). In addition, a range of nitrofurans, including furazolidone, is available for treatment of fishes. Further, potentiated sulfonamides have become available for the treatment of fishes. Furthermore, oxytetracycline is the drug of choice for vibriosis and ulcer disease, and nifurpirinol can be used for the treatment of systemic vibriosis. However, therapy of corynebacterial kidney disease (BKD) is said to pose a special problem due to its chronic nature, and erythromycin could be a successful medication (Snieszko and Griffin, 1955; Roberts, 1978). Concomitantly, the secondary aeromonad infection of the organs of cyprinids associated with SVC and cauda equina syndrome had been treated by parenteral injection of antibiotic. In addition, systemic diseases are often associated with clinically obvious external lesions, and in such cases,

recovery is often facilitated by combining medicated treatment with a suitable external treatment. Incidentally, Bullock et al. reviewed alternative treatments for bacterial fish diseases (Roberts, 1978). Further, the role of phytoflagellates such as the Chrysophyta in fish kills, especially *Prymnesium parvum*, is quite well known. The organisms seemed to be of continuing significance in Israel, and methods of control were developed utilizing the lytic action of NH_3 on the *Prymnesium* cell >15 °C and pH 7.0. It had sometimes been opined that disease prevention had not been given much importance, nor had it been treated as a priority area of research. Rather, research in this field had been much oriented toward the sport, restocking, and ecological aspects of fisheries (Roberts, 1978). Further, in the United States, legislation, often, said to have imposed stringent limits on chemical treatment. Concomitantly, the flourishing of fish farming industry in Japan had been accompanied by the production of new therapeutic compounds by the pharmaceutical industries. Further, a thorough analysis of the traditional techniques for the management of bacterial gill diseases, may provide valuable new inputs. On the other hand, it may be noted that formalin and malachite green, which are considered the fish farmers' two traditional drugs, have been in use for more than 50 years and will probably remain preeminent and in vogue for the foreseeable future. In addition, application of external chemical treatments had been improved by the concept of flowing treatment. On the other hand, systemic treatments are said to have benefited from the progress achieved in veterinary medicine; for example, the recent introduction of potentiated sulfonamides. It had been believed that their use might offer an alternative to the appetite problems faced with other types of medications fed via the diet, although such things are still at the experimental stages. Moreover, the development of fast handling and injection techniques might also prove useful for correct treatment of systemic fish diseases. Nevertheless, development of methods of vaccination (using oral route, wherever possible) is considered one of the most promising and hopeful areas of advance. However, in spite of some amount of progress being made in this field, there seemed to be not much strong indication of any immunological protection being given to the vaccinated fish. Thus, there is a need for further works in this regard (Summerfelt and Warner, 1970).

10.6 PROPHYLAXIS (PREVENTION)

Prophylaxis revolves around good general husbandry practices including disinfection, opting for water from tube wells rather than irrigation canals or paddy fields, and ensuring disease-free stock and healthy fry. Apart from not overstocking ponds, other preventive measures include the use of antibiotics and chemicals. However, the use of antibiotics to control and treat EUS has been strongly opposed by FAO.

Further, successful prophylactic and therapeutic treatments have generally involved the addition of quicklime, which is considered a relatively simple and inexpensive way of enhancing water quality. This fact only reinforces the need to overcome the environmentally degrading conditions that may predispose fish to disease.

In addition to the above, the following preventive (prophylactic) and control measures were suggested to farmers at the "Consultation on Epizootic Ulcerative Syndrome vis-à-vis the Environment and the People" under the aegis of the International Collective in Support of Fishworkers.

1. Better water-quality management, as in aquaculture practices
2. Application of unslaked lime in unaffected ponds at 200 kg/ha once a month for three consecutive months, depending on the pH of the water and soil
3. Restrictions on transferring the use of nets and other potentially transmitting agents from affected to unaffected water bodies
4. Fish from infected waterways, especially those with lesions of EUS, are not to be relocated to other water bodies

However, as a prophylactic measure, salt, potassium permanganate, **bleaching powder, and malachite green** may be recommended as alternative or additional prophylactic measures. Others measures include formalin, iodine, and peroxide disinfectant. There have been claims of prophylactic success against EUS coming from "traditional" homespun remedies such as the application of crushed tamarind or banana leaves or turmeric powder to the affected ponds (Rungratchatchawal, 1999). These methods, however, had probably not been scientifically tested. Antibiotics have been found to be useful in controlling secondary bacterial/fungal infections.

10.7 CONTROL

Control of EUS in wild populations is probably impossible. However, in small closed water bodies, mortality could be reduced by liming the water and by improving the water quality, together with removal of infected fish. Incidentally, EUS has been listed by OIE under the category "Other Significant Diseases" (OSD); that is, diseases that are of current or potential international significance in aquaculture, but have not been included in the list of diseases notifiable to the OIE. This is because of factors related to their importance or geographical distribution, current inconclusive knowledge about them, or lack of knowledge of approved diagnostic methods about them.

However, the following points may be considered for an attempt at effective control of EUS:

1. Control of EUS in natural waters (e.g., rivers and wetlands) is very difficult, as indicated.
2. Fish farmers whose farmed fish have been affected by EUS may be encouraged to culture non-EUS susceptible fishes, or to avoid farming susceptible species during the EUS outbreak period.
3. A strict ban on the movement of fish from the affected water bodies to other unaffected water bodies is recommended.
4. Diseased fishes are not to move from one fish farm to another.
5. Properly dried, salted, and iced fish may not be potential carriers of EUS. As such, trade with such fishes and their products could be allowed to continue.
6. Further, a number of simple biosecurity measures could minimize or prevent the spread of EUS. These may include the following:
   a. All possible carriers or vectors such as recently dead fish, birds, or terrestrial animals, as well as contaminated fish-catching devices, fish-transport containers, etc., are to be prevented from getting into unaffected water bodies.
   b. In the case of outbreaks occurring in small closed water bodies, liming of water, improvement of water quality, and removal of infected fish are often effective in reducing mortality.
   c. Increasing salinity in lentic water bodies may also prevent outbreaks of EUS in water bodies such as aquaculture ponds.
   d. During the dry and cold seasons, wild fishes are to be closely monitored to detect the presence of EUS-affected fishes in neighboring tanks or canals, in which case the exchange of water may be avoided.
   e. The EUS-infected fish are not be thrown back to the open waters, but are to be disposed of properly by burying them into the ground or incineration.
   f. In addition to the above, additional practical aquaculture biosecurity measures could include:
      - Good farm hygiene
      - Following procedures such as separation of nets, tanks, and stocks, and regular and correct disinfection methods
      - Good husbandry practices
      - Good water-quality management
      - Proper handling of fish
      - Regular monitoring of fish health
      - Good record keeping (gross and environmental observations and stocking records including movement records of fish in and out of the aquaculture facility, etc.)
      - Early reporting or notification to concerned authorities

Moreover, What could one do in the event of a disease outbreak?

1. Report immediately to concerned authorities (nearest fisheries or veterinary authority) about a suspected EUS outbreak, and ask for guidance.
2. Take note of simple observations such as:
   a. Abnormal fish behavior (e.g., fish swimming near the surface, sinking to the bottom, loss of balance, flashing, corkscrewing or air gulping (for non-air-breathers), or any signs that deviate from normal behavior.
   b. Date and time of observed outbreaks
   c. Estimate of mortalities
   d. Species of fish affected and estimation of mortalities; pattern of mortality; a small number of fish dying every day or a large number of fish dying at one time; etc.)
   e. Any unusual events

Further, what could be done in the event of a disease outbreak?

The following measures were recommended to reduce the incidence of EUS and the mortality of fish in affected ponds:

• Application of 25 kg turmeric powder and 100 kg unslaked lime, thoroughly mixed and sprinkled over the pond surface, in 1-m depth of water for every hectare. This should be done once a week for three weeks.
• The use of 200 kg unslaked lime per hectare three times in 14 days (i.e., on Day 1, Day 7, and Day 14), provided the water pH does not exceed 8.0.

10.7.1 Control Activities of EUS in Different Provinces: Regional Reports

There had been efforts to control and contain the unimpeded and unabated spread of the virulent, enigmatic, and hitherto unknown fish disease EUS by different provinces and nations. The following is a brief discussion about this:

Ulcerative disease rhabdovirus (UDRV) was found in the diseased fish.

Aphanomyces, *Achlya*, and *Saprolegnia* had been isolated and identified from the surface of ulcerated fish. However, the exact etiology of EUS is yet to be identified. While the causes of EUS are not clearly identified, Incidentally, Nontawith Areechon of the Faculty of Fisheries, Kasetsart University, Bangkok, as did many others, pointed out that the outbreak of EUS might be related to stressful conditions in the water. Stress could be caused by, among other things, environmental changes and toxic substances. These could weaken the fish and make them more vulnerable to pathogens.

Treatment for therapeutic purposes includes antibiotics and chemicals. Though these possibly brought some satisfactory results, they have many undesirable side effects, too. These may include residues, cost enhancements, development of bacterial resistance, negative impacts on the environment, and so on. Based on a number of experiences, the keys to a successful crop in fish culture are healthy fry, proper pond preparation, and good management. However, these measures may not be applicable to wild fish populations. However, prevention of any condition, which may be stressful to aquatic animals, could be considered very important. This could hinge around good water management, optimum stocking rates and proper proportion of fish species. Later it was found that EUS-affected fish had improved in health when removed and put into a normal freshwater pond.

10.7.1.1 India (in General)

In India, the first outbreak of EUS had occurred in the northeastern provinces of Assam and Tripura during the monsoon month of July 1988. After many deliberations, the Indian scientific investigations on remedial measures recommended quicklime at 200–600 kg/ha as showing encouraging results in controlling and containing the EUS disease. In fact, limited success had come from antibiotic therapy. CIFAX—a drug formulated by the Central Institute of Freshwater Aquaculture (CIFA)—at a dose of 1 kg/ha m water area had been helpful in containing EUS. Similarly, bleaching powder at 1 ppm was said to be effective.

10.7.1.2 West Bengal

In West Bengal (WB), the socioeconomic impact of EUS has been staggering. A sample survey of 500 affected fish farmers in an EUS-afflicted area revealed that a large number of the respondents (c 30–40%) had suffered a loss of fish. Moreover, c 44.4% had consumed fish before the outbreak of the disease.

Later, the amount of consumption had been reduced to c 15%. In WB, the medium of radio was the main instrument for disseminating information about EUS for public awareness.

10.7.1.3 Manipur

Manipur has the biggest freshwater lake in NE India, viz., Loktak Lake with a water spread area of about 28,000 ha. In fact, fish is an item of staple food of the inhabitants of Manipur. Approximately 90% of them eat fish in either dried or fresh form. Incidentally, Manipur had received the first information on the spread of EUS in the NE India, during June 1988. The provincial government had geared up its machinery in the form of opening separate fish disease cells, organizing workshops, creating public awareness campaigns, etc. Among the measures recommended were:

(1) disinfection of nets by sun-drying, boiling in water, etc.; (2) treatment of ulcerated fish in 3% solution of common salt for 5—10 min or in 500—1000 ppm of potassium permanganate for 1 min; and (3) disinfection of the affected ponds with unslaked lime (quicklime) at 150—200 kg/ha, depending upon soil pH.

10.7.1.4 Odisha

In this province, EUS had first been reported during February 1989 from ponds in Balasore and Mayurbhanj, and then from Cuttack. The disease had affected around 80 development blocks, but the lotic systems and reservoirs had not been affected. This belies the assertion and conventional wisdom that EUS usually spreads from wild waters to human-made ponds, or at least that it occurs simultaneously in both areas. As such, it was quite unclear how the EUS in Odisha had been confined to ponds only.

The province had lost an estimated 186 tonnes of fish worth a little more than Rs 3 million. In view of this, the provincial government had targeted *c* 5500 farmers in four districts for free supply of lime, and 11 of the 13 districts were covered between 1989 and 1991. The government had also tried to disseminate information on EUS at the panchayat level.

Among the measures recommended were:

1. Disinfection of nets by sun drying, boiling in water, etc.
2. Treatment of ulcerated fish in 3% solution of common salt for 5—10 min or in 500—1000 ppm of potassium permanganate for 1 min
3. Disinfection of the affected ponds with unslaked lime (quicklime) at 150—200 kg/ha, depending upon soil pH

It should be noted that the control measures adopted by the department of fisheries were found to be effective, and there were no further reports of the spread of EUS in the province. This seemed to be exceptional, since almost all other provinces of India reported a second or even multiple outbreaks of EUS.

10.7.1.5 Tamil Nadu

In Tamil Nadu (TN), only four of the province's 22 districts were affected by EUS, according to the government's fisheries department. However, according to Tamil Nadu Agricultural University, seven districts of the province had been hit by the EUS disease. The disease was first reported from the Chengai MGR, Thanjavur, and South Arcot Districts during January, February, and March 1991 respectively. The major species affected belonged to the indigenous varieties of channids, mastacembelids,

cyprinids, and bagrids. The intensity of infection was mild among major carps, exotic carps, and the Tilapia species.

A collaborative work with the School of Tropical Medicine, Calcutta, had identified the causative agents to be *Citrobacter intermedius* and *Klebsiella aerogenes* from samples collected during February and April 1991. Further, the affected fishes were treated with the poultry drug Bifuran (nitrofurazone 100 mg; furazolidone 14.5 mg/100 mg) at a concentration of 25 ppm in water, based on the drug-sensitivity pattern of the isolate during investigation. Ulcerative wounds had healed in five to seven days of therapy. However, cost considerations prevented this from being tried out in larger areas.

Notwithstanding the above, the following preventive and control measures had been suggested to the farmers, with encouraging results:

1. Better water quality management, as in aquaculture practices.
2. Application of unslaked lime in unaffected ponds at 200 kg/ha once a month for three consecutive months, depending on the pH of the water and soil.
3. Restriction on transferring the use of nets and other potentially transmitting agents from infected to unaffected water bodies.

The following measures were recommended to reduce the incidence of EUS and the mortality of fish in the affected ponds:

1. Application of 25 kg turmeric powder and 100 kg unslaked lime, thoroughly mixed and sprinkled over the pond surface, in 1 m depth of water for every hectare. This should be done once a week for three weeks.
2. The use of 200 kg unslaked lime per hectare, three times in 14 days (i.e., on Day 1, Day 7, and Day 14), provided the water pH does not exceed 8.5.

10.7.1.6 Andhra Pradesh

EUS was reported first during October 1990 from the Kolleru Lake of West Godavari and Krishna Districts. Outbreaks had occurred in all water areas like irrigation canals, drains, swamps, ponds, and lakes. It had affected the following genera: (1) *Channa*, (2) *Clarias*, (3) *Anabas*, (4) *Heteropneustes*, (5) certain other types of catfishes, (6) *Puntius*, (7) and IMC such as Catla, Rohu, and Mrigal. In wild waters, the first to succumb to the disease were the IMCs. The disease then spread to fish in cultured ponds.

The following preventive measures had been adopted against EUS in Andhra Pradesh:

1. Application of lime at 50 kg/acre
2. Application of salt at 10 kg/acre through gunny bags hanging from feed poles in the fish tank, and 2 kg of salt/100 kg feed

3. Dip treatment with 0.5%−2% potassium permanganate
4. Application of antibiotics like oxytetracycline, doxycycline, and terramycin at 5 g/100 kg of fish for 10 days. As already indicated, however, FAO strongly opposed the application of antibiotics in controlling and treating EUS infections.
5. Preventing the entry of diseased fish into the tanks by the use of a mesh
6. Addition of mineral and vitamin mixtures to the feed
7. Avoiding the exchange of water when neighboring tanks and canals have been affected
8. Periodic monitoring of the health of the fish

The following treatment was advised:

1. Application of malachite green
2. Use of antibiotics like oxytetracycline, doxycycline, and Terramycin at 10−20 g/100 kg of fish for 10 days or erythromycin at 60−100 mg/1 kg of fish feed
3. Addition of mineral mixture at 2% and 100 g of vitamin mixture per tonne of feed may be applied to the feed
4. Stopping the use of manure in the tank during the disease period

The Commissioner of fisheries of the province directed the field staff to publish and distribute pamphlets on EUS among the fish farmers in the areas affected by the disease.

10.7.1.7 Kerala

Kerala has rich fishery resources, and fisheries are an important sector of the economy of the province. The sector is said to employ c 3% of the province's population and contributes to c 2.5% of its net domestic product.

The province of Kerala had also been affected by EUS. EUS was first reported from Pookote Lake in the Banasurasagar Reservoir area in the northern district of Wynad in June 1991. The disease took the department of fisheries by surprise because it had affected even fingerlings that the department itself had cultured in the lake. In three weeks, it had spread to wells and ponds in Wynad. EUS had spread to the fresh and the brackish waters of Kuttanad and Vembanad Lakes, and the rivulets in Kottayam, Alappuzha, and Pathanamthitta in the south, by the end of August 1991.

Subsequently, the provincial government of its own volition, during May 1992, proposed to sanction funds for liming. However, the implementation of relief measures was not without shortcoming. In fact, the government was slow to react to the reassurance by the expert opinion that EUS could subside naturally with changed weather conditions. Meanwhile, EUS had recurred during October 1992. But despite government admittance that EUS had recurred in contrast to its expectation, fisherfolk had not begun to receive cash relief, and there were not enough government outlets to purchase the diseased fish.

Interestingly enough, none of the unions demanded prophylactic measures to combat EUS, despite its evident recurrence. Once the relief measures had been accepted, the demands for control of the epidemic subsided.

Other studies had also brought another dimension to the hypotheses on the origin of EUS. The intricately interlinked riverine network of Kerala had been strongly highlighted. Starting from Wynad in the north, a continuity of water bodies could be established between the Kaveri and Kaverretty Rivers, which flowed through Tamil Nadu and Karnataka, and the three Kerala rivers, viz., Pamba, Achenkoil, and Meenachil, are said to drain the province. Given that EUS had been reported in Tamil Nadu during early 1991 and in Karnataka during 1990, it is likely that the EUS disease spread from these neighboring provinces through the riverine network, possibly aided by floods.

10.7.2 Prevention and Control of EUS Abroad

EUS had been sweeping unabated and unhindered across seas and oceans to different parts of the world. It had spread to both the United States and Africa. In Africa, it had affected a large diversity of fishes; thereby, threatening the nutrition and avocation of the people. As such, effective measures are required for the prevention and control of EUS in different overseas countries.

10.7.2.1 Prevention and Control of EUS in Africa

Control of EUS in natural waters is probably impossible. Perhaps the most effective way of controlling the spread of the disease is to restrict the movement of fish (and/or contaminated material) from infected areas to uninfected areas. In outbreaks occurring in small, closed water-bodies, liming the water and improving water quality, together with the removal of infected fish, was often found to be effective in reducing mortalities and controlling the disease.

10.7.2.2 Measures for Preventing EUS Outbreak in Pakistan

The following measures had been suggested in Pakistan:

1. Ichthyospecies resistant to EUS are the only fish to be stocked in regions already affected by EUS.
2. The stocking ponds are to be drained, dried, and limed before stocking with fish.
3. Disease-free hatchery-reared fry (3″−4″ in size and pretreated with 1% salt bath for 1 h) are preferred for stocking.

4. Tube well water or treated incoming water is to be used.
5. Wild fishes are to be excluded from culture ponds.
6. Farm equipment and shoes of visitors are to be disinfected with hypochlorite or iodophor disinfectants.
7. Good water quality with high DO is to be maintained.
8. Stress to the fish is to be minimized. This could include low stocking density, healthy food (nutrition), and no excess handling or netting of stocked fishes.
9. Special care should be taken during the periods of low water temperature (winter) with regard to monitoring of potential "risk factors" such as water quality, parasitic infestations, etc.
10. The farmers are to report immediately to the fishery experts in the event of a suspected EUS outbreak, in order to take steps for reducing losses.

The following measures had been suggested:

1. Better water quality management as in aquaculture practices
2. Application of unslaked lime in unaffected ponds at 200 kg/ha once a month for three consecutive months, depending on the pH of the water and soil
3. Restriction on transferring the use of nets and other potentially transmitting agents from infected to unaffected water columns

10.8 TREATMENT

With regard to **treatment and control**, Humphrey and Pearce (2004) opined that there had been no specific control measures in fish for EUS in natural environments. In captive fish, early "red spot" lesions may respond to topical treatment with an antiseptic iodophor solution. Increasing the salinity of water may prevent outbreaks of EUS in aquaculture ponds.

Keeping diseased fish in **good-quality clean water** may allow recovery, but only if lesions are not too extensive, and **dark scars** are often left behind on healing. However, there is possibly **no effective treatment for advanced stages of the disease.**

Attempts at effective treatment and control of EUS are impeded by a number of significant difficulties, notably widespread prevalence of the disease among the wild fish populations and its still unknown etiology. In fact, it is usually appreciated that there is much difficulty in controlling pathogens in large bodies of water containing wild freshwater fish. Fish farms are usually designed for the easy application of chemical treatments. These treatments appear to have met with limited success, particularly in the major carp farms of India and the snakehead farms of Thailand. Both prophylactic and therapeutic treatments, generally involving the addition of quicklime (CaO), had been reported as achieving satisfactory results. But no large-scale comparative assessment has been done involving diverse types of affected species. Even if the primary disease agent is a virus, most of the compounds used may be just as effective in limiting the severity of the ulcerative condition by treating secondary fungal or bacterial infections. The success of lime applications illustrates the need to improve environmental conditions so that fishes may not be predisposed to the disease. Further, Haniffa et al. (2013) worked on the in vitro and in vivo antimicrobial effects of *Wrightia tinctoria* (Roxb.) R. Br. against EUS in *Channa striatus*. The herbal paste applied topically on the lesions showed a positive effect by controlling and curing the lesions. Thus, the active extract of *W. tinctoria* could be used to treat fish diseases.

10.9 REMEDIAL MEASURES

Humans are believed to have some role in the initiation of EUS. Such roles are said to have brought about changes in the form of climate and other environmental factors. The prophylactic and therapeutic remedial measures so far tried in India for controlling or containing EUS, however, are applicable only to manageable water bodies. The remedial measures developed so far may not be applicable to large open waters such as rivers, reservoirs, lakes, and big beels (perennial wetlands in India), which are usually >30 ha in area and may not be applicable to the backwaters where EUS outbreaks have occurred (Boyd, 1990). The difficulty encountered in countering the disease outbreak at present is primarily the lack of knowledge on the primary causative agent, and occurrences of the disease in large water bodies affecting wild populations. However, the chemicals used for therapeutic and prophylactic treatments in manageable water areas are lime, $KMnO_4$, NaCl, bleaching powder, and antibiotics. The chemical treatment is primarily aimed at controlling the external pathogens observed, such as bacteria and the fungi. Perhaps above all, a clean environment must be maintained so that we may have disease-free aquatic biota.

10.9.1 Application of Lime

Lime is generally available from suppliers of construction materials and in hardware shops. It is a relatively simple and inexpensive means of enhancing water quality. This is largely accomplished by increasing the alkalinity and appears to be effective in helping to control losses in fishponds. Agricultural lime ($CaCO_3$ or $CaMg(CO_3)_2$) stabilizes the pH of the water but has no sterilizing effect. Burnt lime (quicklime—CaO; or slaked lime—$Ca(OH)_2$) could have a therapeutic effect against disease, but has a very high pH. Hence, if applied in large quantity, it inhibits bacterial decomposition of organic matter, but an overdose may kill the fish. Overdosing could easily cause problems in ponds that are naturally alkaline.

Along with the above remedies, ponds should ideally be allowed to completely dry out as a general husbandry procedure. Then, lime (preferably slaked lime) is evenly spread over the bottom. Water is then supplied to a depth of 10 cm for 7–10 days before filling to the normal depth and stocking with fish. Emergency prophylactic and therapeutic applications may also be made directly to stocked ponds but doses of treatment could vary considerably. Amounts of 150–600 kg/ha slaked lime in 1-m deep ponds at two- to four-week intervals had usually been recommended for treatments, depending mainly on the pH of the pond. If the alkalinity of the water is <50 mg/L CaCO$_3$, a high dose is required, but if the alkalinity is between 150 and 200 mg/L CaCO$_3$, the pH would be considered fairly high and less lime would be required. Low levels can be applied for prophylaxis but Purkait (1990) had noted that liming <100 kg/ha fails to control EUS disease.

Depending upon the pH, quick lime at 100–600 kg/ha is believed to be effective in manageable water areas. In areas having alkalinity <40 mg/L, a higher dose of lime is applied, and in areas with higher alkalinity, a lower dose of lime is applied at one-month intervals during the outbreak period. It is observed that CaO applied at 50 kg/ha in the disease-prone water areas during the postmonsoon period and just prior to the outbreak of EUS disease might have either arrested the occurrence of the disease, or if outbreak had occurred, lowered its intensity. The information collected from the different provinces of India through a questionnaire developed by CIFRI of Barrackpore and distributed to all provinces had gathered that lime treatment had given encouraging results in checking the intensity and spread of EUS disease. Lime treatment at 20 kg/acre/ft of water is said to control EUS at least to some extent.

10.9.2 Salt (NaCl)

Common salt (NaCl) is a less expensive nonpolluting and effective agent, particularly when used in association with lime. Salt usually controls some ectoparasites, and is known to temporarily inhibit the activities of some bacterial and fungal flora. It also helps to lower the toxicity of nitrite and ammonia to fish. Nevertheless, the tolerance of fish to salinity appears to be species-specific. In fact, certain species of the IMC had been reported to be successfully treated with a short dip in 3–4% NaCl solution. Further, 1% NaCl solution had been reported to be used in unparticularized fishponds in Thailand. Dip treatment of other affected species of fishes, at a concentration of c 3–4% NaCl, had given fairly effective result in healing up of ulcers at the initial stage of the disease. Some other important measures are as follows:

1. Successful prophylactic and therapeutic treatments have generally involved the addition of quick lime, a

relatively simple and inexpensive way of enhancing water quality. This fact only reinforces the need to overcome the environmentally degrading conditions that may predispose fish to disease.
2. Salt, potassium permanganate, bleaching powder, and malachite green can also be recommended as alternative or additional prophylactic measures.
3. Other substances include formalin, iodine, and peroxide disinfectant.

The application of salt at a concentration of 3–4% for dip treatment of affected fishes has given fairly effective results in healing up ulcers at the initial stage of the disease.

10.9.3 Potassium Permanganate

Application of this chemical as a deterrent for EUS seemed to be quite widespread in India. An application rate ranging from 1 to 10 ppm had given fairly encouraging results in different provinces. While the application rate for bath treatment of fish is 1–6 ppm, the pond treatment rate is 5–10 ppm. This rate had been found to be effective in containing EUS and healing up of initial stage of ulceration of fish.

In some cases, concentrations of KMnO$_4$ at 5 mg/L, had been recommended for bath treatments. It could also be applied at 3–5 kg/ha in ponds, in order to control the secondary infections in the EUS fish. As reported by Tachushong and Saitanu (1983), KMnO$_4$ was said to have taken about 120 min to inactivate *Aeromonas hydrophila* in purified water. This could possibly be due to the permanganate ion (MnO$_4$) reacting with organic matter and other reduced substances before it could possibly react with the bacteria. Further, organic load is usually high in most of the fishponds. The DO levels in the water could be enhanced due to a decrease in the BOD level. This could be a positive consequence of the breakdown of the organic matter and deactivation of zooplankton. However, it is believed that KMnO$_4$ might result in some amount of decrease in DO levels because DO is removed during oxidation of organic matter. Nevertheless, the chemical is often expensive for many poor fish farmers. As such, its use is believed to be restricted mainly to dipping valuable fishes, such as the brood stock in the hatchery.

10.9.4 Bleaching Powder

This chemical could serve as an alternative or additional prophylactic measure to liming. By its application, the ponds could be disinfected with 50 mg/L of bleaching powder (25–30% chlorine). The system may be left for 7–10 days before stocking. It should be noted that chlorine may act as a virucidal agent at a sufficient level of contamination (Frerichs, 1990), but such levels may not be used in fishponds. Bleaching powder at 1 mg/L or 5–10 kg/ha is usually used in therapeutic treatments in

India, and had been reported to be effective in healing early lesions of EUS. Further, this chemical at 3—5 ppm is usually in use for disinfecting all fishery equipment meant for fishery activities in EUS-affected areas. Investigations at CIFRI showed that EUS could be contained in manageable water areas, by applying a prophylactic dose of 50 kg/ha of CaO, and after one week, bleaching powder at 0.5 ppm in disease prone water areas. Further, according to others, therapeutic dose of 100 kg/ha CaO, and after one week, bleaching powder at 1 ppm could be effective, when initial symptoms of EUS are seen.

In brief, bleaching powder at 5—10 kg/ha was reported to be useful in healing up of initial lesion of EUS-affected fishes and at 3—5 ppm for disinfecting all fishery equipment used for fishery activities in EUS-affected areas. Investigations at CIFRI showed that EUS could be contained in manageable water areas, by applying a prophylactic dose of 50 kg/ha of CaO, and after one week, bleaching powder at 0.5 ppm in disease prone water areas.

Therapeutic dose of 100 kg/ha CaO, and after one week, bleaching powder at 1 ppm could be effective, when initial symptoms of EUS are seen.

10.9.5 Malachite Green

It is a very efficient fungicide and ectoparasiticide. It may be toxic to some Ichthyospecies at high temperature. Its use is restricted in the United States because of its suspected carcinogenic properties. Hence, it may be used in the fish farms, but with due care. Dip treatment of affected fish at a dose of 1 mg/L had been reported to eliminate mycotic infections in carps of Bangladesh, which allowed lesions to heal spontaneously (Roberts et al., 1989). Further, there had been reductions in the rate of mortality among EUS-affected *Trichogaster pectoralis* through daily spray application of 0.1 mg/L of malachite green (Tangtrongpiros et al., 1983). It should be emphasized that the use of malachite green may be restricted to hatcheries or smaller ponds for commercial purposes.

10.9.6 Antibiotics

Antibiotics are a significant but expensive means of controlling secondary bacterial infections, particularly of *A. hydrophila*. Importantly, tetracycline had been found to be efficacious in oral treatment of *Puntius gonionotus* at 500 mg/500 g of feed. It is important to note that EUS-affected fishes are afflicted by a wide variety of bacteria and often by fungi. A microencapsulated feed containing *c* 30% protein, nalidixic acid, and erythromycin along with vitamin A and C, formulated by CIFRI trial to cure diseased fishes by the use of pelleted feed, had shown the fishes to be recovering. In general, it was found that

antibiotics—erythromycin, oxytetracycline, or terramycin at 60—100 mg/kg of feed for 7 days—have cured the ulcers of EUS-affected fishes. However, FAO has strongly opposed using antibiotics in controlling and treating EUS-affected fishes. Attempts to cure EUS have also yielded good results by the injection of major carp brood stock (Roberts et al., 1989). In addition, an expensive fish could further benefit from an additional antibiotic bath. Nevertheless, oral treatments are the easiest to apply. In view of this, Jhingran (1990) had recommended the use of erythromycin, nalidixic acid, oxytetracycline, or Terramycin at doses of 60—100 mg/kg feed for a week. Antibiotic cocktails with a combination of three or more drugs are to be discouraged, even though a cocktail could be more efficient and effective than a single antibiotic (Saitanu and Poonsuk, 1983)—this could be mainly because of the increased risk of bacteria developing resistance to all of these drugs. In fact, Poonsuk et al. (1983) had already recorded an increase in resistance to chloramphenicol, sulfadiazine, sulfathiazole, and tetracycline in strains of *A. hydrophila* isolated from farmed fish in Thailand. Further and more importantly, a sufficient withdrawal period must be allowed before harvesting treated fish in order to prevent risks to public health due to the presence of drug residues in the marketed fish. Further, FAO (1986) has cautioned the use of antibiotics in the treatment of EUS-affected fish in view of antibiotic resistance being developed.

Pending the knowledge of the definitive primary causative agent of the disease, what is apparent is that the EUS-affected fishes are afflicted by a wide variety of bacteria, and in acute cases, by fungi. A microencapsulated feed containing 30% protein, nalidixic acid, and erythromycin along with vitamin A and C had been formulated by CIFRI trial, with pelleted feed to diseased fishes, and showed the fishes recovering. In general, it was found that antibiotics—erythromycin, oxytetracycline, or Terramycin at 60—100 mg/kg of feed for 7 days—cured the ulcers of EUS-affected fishes. However, FAO has strongly opposed the use of antibiotics in controlling and treating EUS-affected fishes.

10.9.7 Others

In addition to others, formalin may also lead to substantive toxic conditions to fish in the tropical region. Frerichs (1990) had found that formalin could inactivate the snakehead rhabdovirus. However, this could be effective only at concentrations that could be unsuitable for fish stocks. Reportedly, it had been used at a concentration of 25—50 mg/L. However, it had sometimes been used in conjunction with Dipterex (0.01 mg/L). However, the principal effect could be reflected mainly in the control of unrelated parasites.

In addition to the above, iodine has been used, sometimes under experimental setup. Although it may be toxic to fish under certain circumstances, its lethal effects may be decreased by the presence of proteinaceous matters, which are abundant in most fishponds. Notably, iodine killed *A. hydrophila*, but catfish could survive in 12.5 mg/L of iodine, as was opined by Temeyavanich (1983). Further, the peroxide disinfectant "Virkon S" exhibited effectiveness in inactivating suspension of rhabdovirus in murrels, but subject to in vivo tests. Moreover, the use of tamarind or banana leaf based on indigenous traditional knowledge could be useful in managing EUS in fishes.

10.9.7.1 CIFAX

In order to regulate EUS, a drug formulated by CIFA for application in EUS-affected captive waters, was reported to show encouraging results in controlling EUS. The drug was applied at 1 L/ha meter of water area to fishes with symptoms of EUS, which were said to have been cured within 7 days.

10.10 VACCINATION

There is possibly no protective vaccine available against EUS. However, snakehead fishes that had been immunized with a crude extract of *A. invadans* elicited humoral immune response as detected by SDS-PAGE and Western blot analysis (Thompson et al., 1997).

10.11 CHEMOTHERAPY

There is no effective treatment for *A. invadans*-infected fish in the wild and in aquaculture ponds (Cooper et al., 1997). To minimize fish losses in infected fishponds, water exchange should be stopped, and lime or hydrated lime and/or salt should be applied (Lilley et al., 1998). Attempts at using ash, lime, and neem seeds or branches (*Azadirachta indica*) for prophylactic treatments of *A. invadans*-infected fish in fishponds gave variable results.

10.12 RESTOCKING WITH RESISTANT SPECIES

Some important culturable species including Nile tilapia, milkfish, and Chinese carp have been shown to be resistant to EUS attack and infection with *A. invadans*, and thus they could be cultured in endemic areas. Introducing resistant indigenous fish species is recommended.

10.13 DISINFECTION OF EGGS AND LARVAE

There is routine disinfection of fish eggs and larvae against water molds to make them resistant against *A. invadans*. It should be noted that there has been no report of the presence of *A. invadans* in fish eggs or larvae.

10.14 CONCLUSION

Investigations done on EUS are expected to contribute to the termination of outbreaks. Information gained could ensure that the risk of similar occurrences is reduced. Strategies to stop the epidemic are to be put in place as soon as possible and may soon be undertaken in the absence of conclusive findings. Also, further works are required before effective methods for treating EUS outbreaks in ponds can be developed. Detailed investigations of a number of outbreaks could provide valuable information about possibly important "component" causes.

Correct diagnosis of EUS focusing on symptoms of behavior, external signs, and HP is the prelude to treatment. Both prophylactic and therapeutic treatments, usually involving the addition of quicklime, have reported satisfactory results. Yet without large-scale comparative assessments across a variety of affected species, liming cannot be unequivocally advocated. In fact, lakes in Kerala with high levels of natural deposits of lime have also been sites of EUS outbreaks.

REFERENCES

Anon, 1992. Consultation on Epizootic Ulcerative Syndrome vis-à-vis the Environment and the People, International Collective in Support of Fish Workers, May 25–26, 1992. Institute of Management in Government, Vikas Bhavan, Trivandrum 695 033, Kerala (India).

Boyd, C.E., 1990. Water Quality in Ponds for Aquaculture. Alabama Agricultural Experiment Station. Auburn University, Alabama.

Bullock, G.L., Conroy, D.A., Snieszko, S.F., 1971. Bacterial diseases of fishes. In: Snieszko, S.F., Axelrod, H.R. (Eds.), Diseases of Fishes. T.F.H. Publishers, Neptune, NJ.

Cooper, J.A., Pillinger, J.M., Ridge, I., 1997. Barley straw inhibits growth of some aquatic saprolegniaceous fungi. Aquaculture 156, 157–163.

FAO, August 5–9, 1986. Report of the Expert Consultation on Ulcerative Fish Disease in the Asia Pacific Region. (TCP/RAS/4508). FAO, Regional Office for Asia and the Pacific, Bangkok, Bangkok, Thailand.

Frerichs, G.N., 1990. Efficacy of chemical disinfections against snakehead Rhabdovirus. J. Appl. Icthyol. 6, 117–123.

Haniffa, M.A., Sheela, P.J., Milton, M.J., deBrito, J., 2013. *In vitro* and *in vivo* anti-microbial effects of *Wrightia tinctoria* (Roxb.) R.Br. against Epizootic Ulcerative Syndrome in *Channa striatus*. Int. J. Pharmacy Pharmaceutical Sci 5 (3), 219–222.

Hoffman, G.L., Meyer, F.P., 1974. Parasites of Freshwater Fishes: A Review of Their Control and Treatment. T. F. H. Publishers, Neptune. NJ.

Humphrey, J.D., Pearce, M., 2004. Epizootic Ulcerative Syndrome (Red-Spot Disease). Fishnote, No. 1: pp. 50–58. www.fisheries.nt.gov.au.

Jhingran, A.G., 1990. Status of research on epizootic ulcerative syndrome: strategy for containing the disease. In: The National Workshop on Ulcerative Disease Syndrome in Fish, March 6–7, 1990, Calcutta, India.

Kabata, Z., 1970. Crustacea as enemies of fishes. In: Snieszko, S.F., Axelrod, H.R. (Eds.), Diseases of Fishes. T.F.H. Publishers, Neptune, NJ. Book 1.

Lilley, J.H., Phillips, M.J., Tonguthai, K., 1992. A Review of Epizootic Ulcerative Syndrome (EUS) in Asia. Aquatic Animal Health Research Institute and Network of Aquaculture Centres in Asia-Pacific, Bangkok, Thailand.

Lilley, J.H., Callinan, R.B., Chinabut, S., Kanchanakham, S., MacRae, I.H., Phillips, M.J., 1998. Epizootic Ulcerative Syndrome (EUS) Technical Handbook. The Aquatic Animal Health Research Institute, Bangkok, 88 p.

Poonsuk, K., Saitanu, K., Navephap, O., Wongsawang, S., 1983. Antimicrobial susceptibility testing of Aeromonas hydrophila: strain isolated from fresh water fish infection. In: The Symposium on Freshwater Fishes Epidemic: 1982−1983, June 23−24, 1983. Chulalongkorn University, Bangkok, pp. 436−438.

Purkait, M.C., 1990. Case studies on the epizootic ulcerative syndrome of fishes in Chantitala-1 of Hooghly district. In: The National Workshop on Ulcerative Disease Syndrome in Fish, March 6−7, 1990, Calcutta, India.

Roberts, R.J., Shepherd, C.J., 1974. Handbook of Trout and Salmon Diseases. Fishing News, London.

Roberts, R.J., 1978. Experimental Pathogenesis of Lymphocystis in the Plaice (Pleuronectes platessa). Wildlife Diseases. Plenum, London.

Roberts, R.J., Frerichs, G.N., Tonguthai, K., Chinabut, S., 1989. Epizootic ulcerative syndrome of farmed and wild fishes. In: Muir, J.F., Roberts, R.J. (Eds.), Recent Advances in Aquaculture. Blackwell Science, UK.

Rungratchatchawal, N, 1999. Acute Toxicity of Neem Extract on Some Freshwater Fauna (MSc thesis). Kasetsart University, Bangkok.

Saitanu, K., Poonsuk, K., 1983. The prophylactic and therapeutic studies of antimicrobial agents to the Aeromonas hydrophila infection of snakehead fish (Ophicephalus striatus). In: The Symposium of Freshwater Fish Epidemic: 1982−1983, June 23−24, 1983. Chulalongkorn University, Bangkok, pp. 461−514.

Saring, S., 1971. Diseases of Warm Water Fishes. T.F.H. Publishers, Neptune, N.J. Book 3.

Shepherd, C.J., Poupard, C.W., 1975. Veterinary aspects of salmonid fish farming: husbandry diseases. Vet. Rec. 97, 45−70.

Snieszko, S.F., Friddle, S.B., 1952. Further studies on factors determining tissue levels of sulfamerazine in trout. Trans. Am. Fish. Soc. 81, 101−110.

Snieszko, S.F., Griffin, P., 1955. Kidney disease in brook trout and its treatment. Prog. Fish-Cult. 17, 3−13.

Singh, G., Upadhyay, R.K., Narayanan, C.S., Padmkumari, K.P., Rao, G.P., 1993. Chemical and fungitoxic investigations on the essential oil of Citrus sinensis (L.). Z. Pflanzenk. Pflanzen. 100, 69−74.

Steffens, W., Leider, U., Mehring, D., Hattop, H.W., 1961. Possibilities and dangers of the use of malachite green in the fishing industry. J. Fisch. 10, 745−771.

Summerfelt, R.C., Warner, M.C., 1970. Incidence and intensity of infection of Plistophora ovariae, a microsporidian parasite of the golden shiner, Notoemigonas crysoleucas. In: Snieszko, S.F. (Ed.), A Symposium on Diseases on Fishes and Shellfishes. American Fisheries Society, Washington, DC, pp. 142−160. Special publication no 5.

Temeyavanich, S., 1983. Acute toxicity of iodine, chlorine and malachite green on three fresh water fishes and its treatment on snake head fish pond. In: The Symposium of Fresh Water Fishes Epidemic: 1982−1983, June 23−24, 1983. Chulalongkorn University, Bangkok, pp. 415−427.

Tachushong, A., Saitanu, K., 1983. The inactivation of Aeromonas hydrophilla by potassium permanganate. In: The Symposium of Fresh Water Fishes Epidemic: 1982−1983, June 23−24, 1983. Chulalongkorn University, Bangkok, pp. 394−402.

Tangtrongpiros, J., Thirapatsakun, T., Wongsattayanont, B., 1983. The examination and treatment of ectoparasite in Trichogaster pectoralis affected from an epidemic during December 1982 to February 1983. In: The Symposium of Fresh Water Fishes Epidemic: 1982−1983, June 23−24, 1983. Chulalongkorn University, Bangkok, pp. 428−435.

Thompson, K.D., Lilley, J.H., Chinabut, S., 1997. The antibody response of snakehead fish, Channa striata (Bloch), to Aphanomyces invaderis. Fish Shellfish Immunol. 7, 349−353.

Wood, J.W., 1974. Diseases of Pacific Salmon, Their Prevention and Treatment, second ed. Department of Fisheries, Hatchery Division, State of Washington, Olympia, Washington.

Chapter 11

Monitoring of Epizootic Ulcerative Syndrome

11.1 INTRODUCTION

There has been an explosion of anecdotal information on outbreaks of EUS, as well as extrapolated data regarding fish losses due to EUS-related causes. But actual field survey data that could reveal the actual randomized counts of dead fish due to EUS are largely unavailable. Also, only some studies have defined the lesions caused by EUS and could confirm the diagnosis in each case. OIE defined an EUS-affected fish as a fish with necrotizing granulomatous dermatitis and myositis associated with hyphae of *Aphanomyces invadans*. This definition differs slightly from the one given by Roberts et al. (1994). The OIE definition requires HP processing of fish tissues with H&E and Grocott's stains in order to make positive diagnosis of EUS.

In addition to EUS, there have been reports of occurrences of lesions on the fish body that may be due to various causes and do not fit with the conventional views of EUS, and are also not associated with large-scale fish mortalities. Nonetheless, surveys had been conducted in various places to ascertain the correct etiology of EUS. For example, a cross-sectional survey conducted in Bangladesh during the winter of 1998–99 revealed that c 80% of the 471 fish with clinical lesions sampled from 84 sites had been diagnosed as EUS-positive (Khan and Lilley, 2001). Similar results might also be obtained from other places. It thus portrayed that EUS is probably still the principal cause of lesions produced on FW fish, and further that these surveys do not seem to overestimate the prevalence of EUS. In view of the alarming nature of EUS, it is highly essential to monitor this hitherto unknown fish disease.

11.2 DISEASES AMONG FW FISHES

It is essential to have a glimpse about the diseases among the fishes in order to monitor them effectively. Fish diseases among the carp-rearing farms in Asia, notably in Bangladesh, had been reported in c 31% of the extensive farms and in c 24% of the semi-intensive farms (Chowdhury, 2002). One of the most common and virulent disease problems among the FW fishes is EUS, which has been causing large-scale mortalities among FW fish since the early 1970s, rendering many Ichthyospecies endangered. Nevertheless, some other diseases among FW fish are briefly given below:

1. There are ailments caused by Protozoan parasites, notably *Trichodina, Chilodonella, Ichthyobodo, Ichthyophthirius multifiliis, Myxobolus*, and certain other myxosporideans (Banu et al., 1993).
2. There are diseases caused by Metazoan parasites, notably *Dactylogyrus, Gyrodactylus, Argulus, Piscicola*, and *Lernaea*.
3. There are diseases caused by bacteria, notably infectious dropsy of carps (Hossain et al., 1994); edwardsiellosis in *Pangasius hypophthalmus, Anabas testudineus,* and *Channa punctata*; bacterial gill disease caused by *Flavobacterium branchiophilum*; Columnaris disease; and so on.
4. There is RSD in *Barbodes gonionotus* and *Clarias batrachus*, caused by *Pseudomonas* sp. (Banu et al., 1999).
5. There are fungal diseases caused mainly by *Saprolegnia* and *Branchiomyces*.
6. There are viral diseases caused by the viruses.
7. Nutritional deficiency disease is a main hindrance in FW fish farming.

Notwithstanding the above, the occurrence of diseases in farmers' ponds and other associated difficulties are recorded and taken care of regularly by the government fisheries institutes.

11.3 DISEASE INFORMATION AND ITS SOURCES

Genuine information from authentic sources is a prerequisite for tackling a disease. In the event of a disease outbreak, reported surveys revealed that c 57.7% (75 out of 130) of affected farmers had consulted a government fishery extension officer for guidance and help. Respondents indicated that they had also contacted other farmers (c 44.6%, $n = 58$), other hatchery owners (c 52.3%, $n = 68$), drug and chemicals salespersons (c 11.5%, $n = 15$), and so on.

Epizootic Ulcerative Fish Disease Syndrome. http://dx.doi.org/10.1016/B978-0-12-802504-8.00011-0

Ranking of information support revealed that government extension officers had provided affected farmers with the most useful information (c 42.3%, $n = 55$), followed by other hatchery owners (c 27.6%, $n = 36$), other farmers (c 16.9%, $n = 22$), and drug and chemicals salespersons (c 6.2%, $n = 8$).

11.4 CARP HATCHERIES AND NURSERIES: THEIR STATUS

Hatcheries and nurseries are significant components of aquaculture. These are the places where fish begin their lives. Before 1990, carp seed production in the Indian subcontinent, notably in India and in Bangladesh, was based on spawn collected from natural sources; i.e., the rivers. The situation changed in 1990, after a large number of hatcheries had been established in various parts of the country. As reported, >500 hatcheries had been established and contributed c 97.6% of the total spawn production in the country (Chowdhury, 1993). Further, out of the total hatchery-produced spawn, c 94.6% had been contributed by private hatcheries and c 3% by public-sector hatcheries (Banik, 1999). Thus, the private hatcheries play an important role in the fish production of Bangladesh.

In the Indian subcontinent, notably in India and in Bangladesh, AQC is mostly carp-based, with IMCs (*Labeo rohita*, *Cirrhinus mrigala*, and *Catla catla*) and exotic carps (*Cyprinus carpio*, *Hypophthalmichthys molitrix*, and *Ctenopharyngodon idellus*) being the main species cultured in different types of AQC systems, including culture-based fisheries in the oxbow wetlands (Hasan et al., 1999), as well as in enhanced fisheries in floodplains through open-water stocking (Ali and Islam, 1998).

11.5 DISEASE OCCURRENCES IN HATCHERIES AND NURSERIES

"Disease" is one of the major constraints to intensification of AQC, and eventually could become a limiting factor to the economic success of the industry. Hatcheries and nurseries are more vulnerable to fish diseases because they deal with more juveniles, and hence they assume some importance in this matter. Nursery operators sometimes have a tendency to overstock ponds with fish in the improved farming systems, which ultimately may lead to diseases. It should be noted further that high stocking densities of fry and fingerlings in nurseries usually lead to increased stress, and subsequently the fish become more prone to infectious diseases (Snieszko, 1974). In fact, juveniles are usually more susceptible to pathogenic attacks mainly because of their less developed immune systems.

Occurrence of diseases and incidences of droughts and floods happen to be the major management problems faced by the hatcheries and nurseries. However, reports had revealed that diseases are less prevalent among the former than the latter. However, among the major ailments reported from nurseries are the following: white spot, tail and fin rot, gill rot, sudden spawn mortality, dropsy, malnutrition, and EUS. Conversely, the major diseases reported from hatcheries were fish lice and sudden spawn mortalities. As reported, the economic loss due to disease had been about 7.6% of the profit in Bangladesh. Nevertheless, gill rot probably caused the highest financial losses to affected farms, followed by sudden spawn mortality, fish lice, EUS, and malnutrition. The results of this case study in Bangladesh indicated that disease is a significant issue in hatcheries and nurseries. The issue of disease and health management must be cared for with due emphasis. But in spite of adversities, hatchery and nursery operations somehow remain as profitable enterprises in Asia in general, and the Indian subcontinent in particular, including India and Bangladesh.

Notwithstanding the above, parasitic diseases in nurseries are considered significant factors that could limit the growth and survival of fry and fingerlings. Some episodes/instances/examples of parasitic attacks in fishes (as exemplified from Bangladesh) are briefly given below:

Gill myxoboliasis is caused by *Myxobolus* sp. and *Henneguya* sp. These parasites cause severe losses to IMC fisheries, notably to *C. catla* in Bangladesh. Hossain et al. (1994) had reported that c 61% of the carp fry in the nurseries of the greater Mymensingh district had been infected with ectoparasites. But the highest mortalities of the carp fingerlings were due to infestation by *Trichodina*, *Myxobolus*, and *Dactylogyrus*. Chandra et al. (1996) further reported a high prevalence of myxosporeans in the juveniles of IMCs, notably *L. rohita* and *C. mrigala*, in the nursery ponds of Mymensingh. They had detailed severe gill infections caused by five species of the genus *Myxobolus*. In fact, large-scale mortalities of carps associated with gill myxoboliasis had raised concerns among Bangladeshi fish farmers (Chandra et al., 1996). In addition to the above, as reported in Bangladesh, fry had also been affected by gas bubble disease (Ahmed, 2004). This could be caused by water that is supersaturated with dissolved gases (notably N_2 and O_2). Further, one of the most serious diseases of eggs in hatcheries is caused by the fungi *Saprolegnia* and *Achlya* (Ahmed, 2004). The fry are also usually highly susceptible to microbial diseases.

In view of the significant role played by hatcheries and nurseries in providing a sustainable supply of quality fish seed for increased AQC and enhanced fisheries, surveys are being initiated from time to time to study the different aspects of management issues in small-scale carp hatcheries and nurseries, with special reference to health management.

11.6 DEVELOPMENT IN AQUACULTURE AND PROBLEMS OF DISEASES

Aquaculture (AQC) plays a significant role in the nutrition and avocations of the people. Hence, efforts are being made today toward development in the AQC sector different countries in the world. Concomitantly, disease and health problems in AQC are also in the rise mainly due to intensification in AQC. There is enhancement in AQC in the countries of the region, mainly through increased stocking densities. This may, rather, invite disease and health problems among the intensely stocked fishes (mainly due to overcrowding) and may lead to large-scale mortalities and consequently poor production, rather than so-called enhanced production.

11.7 SIMILARITIES AND DIFFERENCES: LIVESTOCK PRODUCTION AND AQUACULTURE

Livestock production and AQC are both practiced by human beings, and both are related to the nutrition and avocations of the people. Hence, there are obviously similarities and differences between the two.

11.7.1 Similarities

1. They are organized similarly.
2. The same farmers are often involved.
3. Both serve as nutrition and avocation.
4. Both may be highly impacted by diseases.

11.7.2 Differences

1. Living habitats are significantly different.
2. Developing countries are said to export more aquatic animals than livestock.
3. A wider range of diagnostic tests is available for livestock than for aquatic animals.
4. Management systems may not be very well developed in AQC in many countries.

11.8 AQUATIC ANIMAL HEALTH: POSSIBLE LESSONS FROM LIVESTOCK

There could be pitfalls in dealing with both livestock production and AQC. Lessons learned from one could serve as a check for the other. Notwithstanding the above, the care of livestock may impart the following lessons that could be emulated for AQC:

1. Systematic research to identify factors other than pathogens that could contribute to the occurrence and severity of diseases
2. Research and extension to formulate, develop, and implement on-farm quality assurance measures

3. Systematic approach for disease surveillance, with the focus on early detection followed by rapid response
4. Improved understanding of the theory and practice of risk analyses, with the focus on early detection followed by quick response
5. Strengthening of quarantine arrangements

11.9 LIVESTOCK AND AQUACULTURE HEALTH SYSTEMS: HOW ARE DISEASES AND OTHER ISSUES CHANGING?

Health and disease are two dynamic and closely related entities and are subject to changes based on various factors and interactions. It may sometimes be difficult to ascertain what cause leads directly to what result. Alterations may also take place due to rapid, often unpredictable changes in the environment, notably the climate. However, the scenario could have a significant impact if the following steps are adopted:

1. Increasing emphasis on good disease surveillance systems
2. Wider application of inexpensive screening tests

11.10 INTEREST AMONG CONSUMERS

People will generally associate with others if they have an interest in a subject, or if something is related to their health, disease, nutrition, and avocation. Awareness among consumers is on the rise with regard to major calamities among livestock due to disease outbreaks, notably "mad cow" disease in the UK, the Nipah virus in Malaysia, the EUS epidemic among fishes, and so on. These have strengthened the move to on-farm management of livestock.

11.11 DISEASE MONITORING AND THE GOVERNMENT

Disease monitoring needs a big work force to cover every nook and corner. As such, government machineries are to be involved in the process. However, experience has revealed that differences may exist in the capabilities of governments at different levels for improving the health of animals in the small-scale sector. The responsibilities of livestock and aquatic animal health are entrusted with the same governmental department in countries like Lao PDR and the Philippines. Thus, marvelous opportunities exist for sharing experience, expertise, and resources. Further, livestock health surveillance and control systems for smallholders are quite well developed in some Asian nations. This could be expanded to other nations through network collaboration.

Concomitant to the above, the governments in different nations now seem to be under increasing pressure to supply better information about the health status of livestock and aquatic animals. This could become increasingly more important over the coming decades for those countries that export aquatic animal products, particularly live animals and their products. The *OIE Aquatic Animal Health Code* of the Office International des Epizooties has guidelines about the modus operandi of surveillance to be conducted and concomitant reporting of diseases, if any. Improvements in the reporting system could be an ongoing process, with inputs in the form of tables, figures, maps, plates, etc., so that targeted goals could be achieved. Nevertheless, it may be appreciated that the modus operandi of disease surveillance for livestock and fisheries and AQC may not be the same, and thus they may call for different strategies.

11.12 DISEASE PROBLEMS AND THEIR IMPACTS

A disease occurring in a living being hampers with the physiology of the host, and this is said to exert impacts on the life of the organism. The impact could be serious dimension to the juveniles in the hatcheries and nurseries. The impact of disease in hatcheries and nurseries could be measured by:

(a) fish mortality; (b) economic loss; and (c) extent of change of attitude, if any, of farmers due to disease problems and so on.

Surveys in general had reported that diseases may cause at least partial loss of fry and fingerlings in most facilities (*c* 66%), and only *c* 7% of the hatcheries and nurseries reported total loss of their stock. Interestingly, *c* 27% of farmers had reported no loss of their stock due to disease. Nevertheless, farmers sometimes reported varying degrees of losses for individual types of disease outbreaks. The approximate average rates (percentage) of mortalities along with the causes, as reported in the case of hatchery and nursery operations, are as follows:

Malnutrition (55%), air gulping (43%), dropsy (42%), EUS (39%), sudden spawn mortality (38%), gill rot (34%), tail and fin rot (34%), argulosis (23%), and white spot (21%).

11.13 IMPACTS OF EPIZOOTIC ULCERATIVE SYNDROME ON SOCIETY

Any disease is expected to exert some impact on society, and the impact may become profound if the disease is a hitherto unknown one and spreads unabated in an epidemic dimension. So, EUS fish disease, being a hitherto unknown virulent, enigmatic fish disease, had been exerting significant impact on society since its inception, particularly with regard to its effect on fish consumers, specifically humans. Notwithstanding the above, there had been other types of impacts of EUS on society. For instance, due to EUS in fishes, as reported, *c* 91% of the farmers (*n* = 149) had faced reduced price of their fish in the markets, and consequently, fall in their family income.

In addition, EUS had paved the way for an increased debt of a substantive number of fish farmers, as reported by *c* 42.95% (64 out of 149 respondents) of the affected farmers. Nevertheless, the majority of the fish farmers had wanted to retain their faith in the traditional trade with AQC even though EUS had been a severe blow to their economy. As such, there had appeared to have not much supporting information about the change in attitude of the fish farmers (with regard to practice and species selected) toward AQC, in spite of their hardships in the event of outbreaks of EUS.

11.14 CATEGORIZATION OF IMPACTS

The impacts of a disease on society could be of various kinds, and hence the impacts may often need categorization. Categorization of the impacts of ailments in fish revealed that gill rot had caused the highest losses, followed by sudden spawn mortality, fish lice, EUS, and malnutrition. In addition, diseases like tail/fin rot, dropsy, red spot, air gulping, and white spot had also been reported to be causing considerable losses in terms of fish. Numerically, as an example, in Bangladesh, taka (Tk) 8062.0/ha/year was reported to have been lost on average due to the outbreak of diseases (including EUS). Consequently, the total loss, due to diseases, in 174 fish farms, in Bangladesh, could be extrapolated as Tk 6275.0/ha/year. Concomitantly, the reported cost of treatment of the sick fish in hatcheries and nurseries could be Tk1669.0/ha/year. However, in spite of adversities, the average profit of the fish farm was Tk 104, 575/ha. Also, the incurred losses had amounted to *c* 7.6% of the profit.

11.15 FISH FARMERS

A fish farmer is the significant instrument in dealing with the fish for livelihood. As such, a fish farmer may adopt various steps to monitor fish stock along with the related environmental conditions in order to get a complete idea of the state of health of the fish. Thus, monitoring of the fish farm could enable the farmer to take urgent steps to prevent any outbreak of disease in its early stages itself, particularly if any deviation is noticed in the aquatic domain.

Concomitant to the above, attention should be given to any physical signs of injury or illness such as discoloration of the body or gills based on regular observations. Moreover, individual and group behavior are to be monitored, specifically in terms of swimming and feeding. In fact, lethargic swimming close to the water surface, away from the main body of the water, particularly in the corners of the pond, indicates a warning of illness. A suitable opportunity to observe the fish is during the feeding time, if the water is murky. Further, loss of appetite could be due to the initiation of the disease.

Up-to-date records of feed and stock are to be monitored regularly in the intensive fish farms. In this way, a progressive loss of appetite could be determined by comparing the records of the daily ration. Dead fish are to be quickly disposed off, as they could serve as potential sources of diseases. Further, some kind ailment could be guessed in the event of large-scale mortality. A tentative postmortem diagnosis could be obtained through examination of the dead fish. In addition, fish farmers and fishers are familiar with the external symptoms of the disease, notably the EUS. In the case of culture of valuable fish with a long growth cycle, weighing of the sample periodically may be considered. This could serve as a general indicator of health and condition of the population.

Notwithstanding the above, fundamental monitoring of the environment is specifically relevant in the prediction of EUS outbreaks. This could range from simple observations of weather and rainfall patterns to recording of potentially stressful water quality parameters, such as temperature and pH. Evidence had revealed a higher risk of EUS, particularly when rain follows a dry spell. As such, specific attention might be given to the different aspects of rainfall. Level of DO in a water body is significant. Hence, DO could be estimated before sunrise, when readings are at the lowest. However, the depth of water in the pond is to be monitored during the dry season. It is mainly because there could be rapid diurnal fluctuations of different parameters at low water level. It should be noted that Secchi disks are simple devices to quantify turbidity. and turbidity may be related to other parameters.

11.15.1 Disease Problems and the Farmers' Response

A farmer is worried and often hard hit by disease problems with crops, and is often upset by crop losses. This situation may occur anywhere in the world. As an example, in Bangladesh, out of c 174 fish farmers surveyed, c 166 had responded that they had faced disease outbreaks in their farms. With regard to treatment of the sick fishes in the event of a disease outbreak, nearly all the farmers (c 161 responses out 166 interviewed: 97%) had responded that they had used some kind of treatment. In fact, majority of the farmers had been found to attempt preventive and curative treatment measures, and only a negligible percentage of farmers (c 1.2%) had harvested and marketed their products. A few farmers (c 1.8%) took no measures during the disease outbreaks in hatcheries and nurseries, probably because of incapability or other unavoidable reasons.

Notwithstanding the above, certain other farmers in the study areas are reported to have used different treatments, such as chemicals and antibiotics, water exchange, and manipulation of feeding and fertilization. Nevertheless, use of chemicals had ranked highest ($n = 158$, 98.1%); followed by stopping of fertilization ($n = 51$, 31.7%) and feeding ($n = 46$, 28.6%), water exchange ($n = 44$, 27.3%)

and use of antibiotics ($n = 25$; 15.5%). Further, insecticides and pesticides had often being used as part of chemical treatment, for example, Sumithion and Malathion had commonly being used to treat against fish lice on brood stock. As reported, c 15% of the brood fishes had completely recovered as a result of the treatment, while c 31% of the fishes had partial recovery, and c 4% had no recovery. In fact, farmers engaged in hatchery and nursery trades appeared to be well informed about their problems. Further, as reported, many of the farmers had good understanding of the disease problems, as well as apparently, a number of them ($n = 142$; 87.7%) had the ability to identify some diseases, although 16 (9.9%) farmers had the expertise and experience of identifying most of the diseases in their fishes.

11.15.2 Farmers and the Problems Faced by Them

It is an accepted fact that sound health management practice is a key to success in any hatchery or nursery operation. However, both the hatchery and nursery operators have been faced with various kinds of problems in their operational exercises. With regard to ranking of the problems, reported surveys indicated that c 51 (29.3%) had mentioned the occurrence of diseases as their major problem. Conversely, c 48 (27.6%) had pointed out shortage of water as their major problem. Some other problems faced by the farmers included flooding, theft, and extremes of temperature. In addition, some farmers had expressed difficulties with regard to the lack of financial support, frequent "bandhs," and bad roads for commuting. Nevertheless, as a group, c 118 (67.8%) and c 133 (76.4%) of the farmers opined that disease and a shortage of water, respectively, were their main problems.

11.16 FISH HEALTH WORKERS

A medicine doctor treats a human patient along using a team of human health workers. Likewise, a sick fish also needs treatment by a knowledgeable fishery scientist. In this exercise, the scientist could also involve a team of fish health workers. However, it had been emphasized that a standard and uniform approach is necessary for monitoring EUS fish disease throughout the region. With this aim in view, a standard monitoring pro forma had been drawn at NACA's second technical workshop on EUS and the environment. It had called for a general basic site description and details of management practices of the paddy fields and the pond ecosystems. It suggested that proper monitoring exercises could be conducted with regard to certain aspects such as the application of fertilizers, pesticides, etc.; occurrences of the disease itself and the fish species affected; data on physicochemical characteristics of the water; and economic impact assessments. In addition, HP examination of the ulcerative lesions could be a specifically useful diagnostic tool for investigating EUS.

Nevertheless, it could require a great deal of experience and knowledge, on the part of the fish farmers and fish health workers, to be able to recognize the characteristic features.

11.17 MONITORING EXERCISE OF EPIZOOTIC ULCERATIVE SYNDROME

Monitoring is an ongoing exercise. It is performed to track something. Monitoring is an essential exercise in tracking the spread of a disease. Tracking the spread of a disease is highly essential to know the exact status of a disease. This could help in predicting the outbreak of a disease in a locality, as well as help to take effective measures for its control. Monitoring becomes a little difficult task, when a disease is hitherto unknown and virulent, and sweeps the countries unabated, thereby giving little scope and time to combat it effectively. EUS among the fish is one such disease. Little is known of the means whereby, EUS spreads within and between the regions, although movements of subclinically affected fish are probably important in transmitting *A. invadans* infection. Until effective control and prevention measures are implemented, it is likely that the disease will continue to spread and that outbreaks will continue to recur. Comprehensive investigations of initial outbreaks in previously EUS-free areas, as well as of recurrent outbreaks in previously endemic areas are urgently needed. This could contribute important information to the existing knowledge on causal factors for EUS.

From an epidemiological point of view, it is important to characterize the outbreak in terms of certain variables. This characterization must be done in such a way that hypotheses could be developed regarding the source, mode of transmission and duration of the outbreak. The information is organized in an attempt to find answers to the following kinds of questions:

1. What is the probable date of initiation of the EUS outbreak?
2. What is the exact period of the outbreak?
3. Given the diagnosis, what is the probable period of exposure?
4. Is the outbreak most likely from a common source, propagated source, or both?
5. Are there any characteristics about fish for that specific attack rates vary?
6. Which groups have the highest and lowest attack rates?
7. What are the significant features of the geographical distribution of cases?
8. What are the relevant attack rates?
9. In addition to the above, there are three basic time spans used to describe disease temporal patterns, viz., (a) the epidemic period, which is of variable length depending on the particular epidemic, (b) a 12-month period to describe seasonal patterns, and (c) an indefinitely long period of years to identify long-term trends.

11.17.1 Regional Monitoring Program

NACA implemented a regional research program on the relationship between EUS and the environment beginning in 1987. Such a research program established a region-wide monitoring system for monitoring of the status of EUS in the region. Such a network provides a basis for an early warning system to make people aware of EUS.

Case studies related to monitoring of fish diseases are discussed below.

11.18 CASE STUDY

Notwithstanding the detailed account of monitoring of the EUS fish disease as described above, there have been a number of case studies reflecting different aspects of the monitoring of fish and livestock diseases in Asia. Some significant case studies related to EUS, notably in Bangladesh, are summarized below.

A case study was conducted by Khan and Lilley (2002) pertaining to risk factors and socioeconomic impacts associated with EUS in Bangladesh.

11.18.1 Materials and Methods

The cross-sectional interview-based survey was conducted in a thana (subdistrict) selected randomly from each of the 64 districts of Bangladesh from December 1998 to April 1999, because this period is generally said to be the "EUS occurrence season" for fishes. Three MSc students from the Bangladesh Agricultural University interviewed a fish farmer and a fisher at random in each thana, and examined 100 fishes for EUS-type ulcers (lesions).

11.18.2 Areas of Survey

One thana known to have adequate fishery resources had been randomly selected from each of the 64 administrative districts in Bangladesh. In fact, a letter had been sent to thana fisheries officers (TFOs) during August 1998, requesting a list of different categories of fish farms (both registered and unregistered) and wild fisheries areas in their respective thanas. One fish farm and one wild fishery had been randomly selected from these lists.

11.18.3 Development of Questionnaire

The procedure for questionnaire development had been developed after the methods described by Thrusfield (1995). Both the fish farmers' and fishers' questionnaires had been designed to record information in a standard format with inbuilt error checks. Closed questions had been used wherever possible, to give data in a "Yes/No/Do not know" format, or in a categorical format, to facilitate ease of coding and analyses. Further, attempts had been made to keep the language brief, nontechnical, polite, and doubtless. Both types of questionnaires were bilingual (prepared in both English and Bengali), and the Bengali version was used for the interviews. It should be noted that before starting the final survey and the actual final interviews, the questionnaires had been rehearsed twice by interviewing target interviewees in order to identify and remove ambiguous and irrelevant questions.

11.18.4 Conducting the Interviews and Concomitant Sampling of EUS Fish

The three interviewers had been trained together in order to minimize differences in their technique of implementing the works. The training had also included examination of fish for EUS-type ulcers, as well as sampling of EUS-affected fish tissues for HP. Each interviewer covered one-third of the total number of districts in Bangladesh. TFOs had been requested to help and cooperate with interviewers. Each interviewer carried the following things:

(a) the required number of questionnaires, (b) fish sampling sheets, (c) photographs of EUS-affected fish, (d) 10% buffered formalin, and (e) stationery items such as vials, scalpels, marker pens, cast nets, hapa, etc.

After completion of the interview, each interviewer examined at least 100 susceptible fish from each fish farm and at least 100 susceptible fish from each wild fishing site in order to record any EUS-type lesions among the examined fishes. Further, one fish of each species recorded with lesions had been sampled for HP. The tissue samples so collected had been fixed in 10% buffered formalin. In this case, a sampling net was not available, so the interviewer's net was provided for catching fishes. In order to avoid recounting of the same individuals during farm visits and on-farm activities, fish that had been examined for lesions were isolated (separated) into the hapa until 100 individuals had been examined. The nets were disinfected after having been used once. A fish farm or wild fishery could be categorized as "EUS-affected" if one or more fish of any species presented characteristic mycotic granulomas that could be confirmed through HP.

11.18.5 Preparation of the Database and Analyses of the Data

Two MS Access™ databases (for the fish farmer and fisher data) and two MS Excel™ spreadsheets (for the fish species data) were used to key in the information. Univariate analyses was done using the software Epi Info™ to examine the association between the occurrence of EUS and putative risk factors using crude relative risk (RR) as the measure. The data on fish farm and wild fishery were analyzed separately.

11.18.6 Processing for Histopathology (HP)

Formalin-fixed tissues of underlying muscles containing EUS lesions were processed, embedded in paraffin wax, and sectioned at 5 μm. The tissue sections had been stained with hematoxylin and eosin (H&E) in order to see the granulomas. Grocott's silver stain had been used in order to confirm the mycotic involvement.

11.18.7 Results and Discussion

Variables were analyzed for their effects on RR of EUS. Incidentally, RR > 1 indicated that the variable is a putative causal factor of EUS; RR = 1 depicted no association between the variable and EUS; and RR < 1 indicated that the variable is a sparing factor for EUS (i.e., it might reduce the chance occurrence). It should be noted that where the lower confidence limit (CL) is >1, there is 95% confidence that the variable is a risk factor for EUS. Conversely, where the upper CL is <1, there is 95% confidence that the variable is a sparing factor for EUS.

11.18.8 Connections with Pond

As reported, there may be >10 times the chance of EUS occurrence in culture ponds that contain wild fish. This was probably the highest RR measured out of the variables examined. The data also reflected that there could have been a significantly lower RR (0.39) of EUS occurrence among farmed fish, had pond embankments been sufficiently high to prevent entry of incoming waters. Likewise, the ponds that had been inundated that year portrayed a significantly higher RR (2.33). Further, a significantly higher RR (2.63) was also revealed in fish farms directly linked to water bodies and that had received wild fishes in them. Furthermore, each type of connecting water body (e.g., rice field, ditch, and beel) provided a similar level of risk. Ponds containing water sourced only from underground wells or rain were at much lower risk of EUS (RR = 0.91, 0.52) than ponds that sourced their water from rice fields (RR = 2.36). These results corroborate with those of Hossain et al. (1992), which when recalculated for RR portrayed a significantly lower risk (RR = 0.65) of EUS-type lesions occurring in fishes in fish farms containing rain-sourced water rather than flooded or irrigated water.

Concomitant to the above, it may be pointed out here that floodwater and entry of wild fish are considered "risk factors" because they could serve as routes of entry for the pathogens (Kabata, 1985). Incidentally, Roberts et al. (1989) had opined that floodwater is an effective means of spreading EUS to different places. Changes in the water quality of a fish habitat due to mixing water with floodwater and agricultural runoff could cause some amount of stress to farmed fish, and hence could serve as an additional reason for the outbreak of EUS. It is possible that underground water and the rainwater do not contain as many parasites and pathogens, and hence more use of these types of water could possibly reduce the risks of EUS occurrence in AQC (Munro and Roberts, 1989).

11.18.9 Preparation of Pond

Studies revealed that complete draining of pond water, removal of bottom mud, drying, and liming during the period of pond construction resulted in low RRs of 0.55, 0.41, 0.17, and 0.42, respectively. Concomitantly, fertilization during pond construction seemed to result in a low but insignificant RR (0.50).

It should be pointed out that construction and preparation of a fish culture pond, if done in the above manner, could exclude fungal pathogens like *A. invadans* and many other pathogenic microbes from the pond. Removal of bottom mud led to a low RR. Further, the *A. invadans*, unlike other oomycete fungi, does not seem to show strong "geotaxis," and may possibly accumulate on the pond bottom. Though soil assays could not succeed in isolating *A. invadans* (Willoughby, 1999), nonetheless the fungus *A. invadans* could possibly thrive in the thick bottom mud of old derelict ponds during the warmer months of summer when the water temperature could be around 31 °C. However, the fungus might be activated to grow with declining temperatures or with disturbances in the pattern of rainfall. Such a hypothesis could get support from Ahmed and Rab (1995) who had opined that fishes cultured in previously derelict ponds had a higher probability of EUS-occurrence.

11.18.10 Poststocking Management and Hygienic Practice in the Habitats

As reported, poststocking liming gave a significantly low RR of 0.46. Ponds with a high content of phytoplankton and/or zooplankton as revealed by pond water color also usually depicted low values of RR. But ponds with high levels of organic wastes as revealed by black-colored pond water also portrayed high values of RR (2.21).

Liming increases the pH, alkalinity (TA), hardness, and buffering system of pond water. It also reduces stress for fish and thereby reduces the risk of EUS. In fact, exposure of fishes to water having low pH may lead to skin damage, which is necessary for the entry of fungi and cause disease. EUS lesions have been induced in fishes, but they were first exposed to acidified water taken in an aquarium, and then spores of *A. invadans* were released into the same aquarium water. Thus, the aquarium study mentioned above tried to prove tentatively that the two factors (exposure of the fish to acidified water and concomitant release of spores of *A. invadans*) in combination might be sufficient to cause the initiation of EUS (Callinan et al., 1996). Further, it is also possible that increased Ca and Mg content in pond water might have a more direct effect by

benefiting fish skin and inducing encystment of the fungal zoospores and thereby making them ineffective in causing disease.

11.18.11 Other Diseases vis-a-vis Adoption of Hygienic Measures

A high value of RR (2.90) had also been reported from water bodies where cattle are washed and drink pond water after plowing or grazing in other areas, possibly because of transport of pathogens with the cattle. On the other hand, netting the fishes with dried and disinfected nets revealed lower values of RR (0.59 and 0.14 respectively). Further, chances of infection might increase during the transport of implements between farms, and hence drying and disinfection is recommended.

In addition to the above, a high value of RR (2.65) was reported from fish farms in which the fishes had been affected by parasites. In fact, a number of parasites had been isolated from EUS-affected fish (Tonguthai, 1986). The parasites could have served either as vectors for the pathogens or at the very least as stress-inducing factors during EUS outbreaks. Subasinghe et al. (1993) demonstrated such an association between the level of infection by *Trichodina* sp. and the susceptibility of *Channa striata* to EUS infection. The mechanism of attachment of these parasites to the body of the host might in fact cause rupture of the skin of the host's body that could facilitate infection by pathogens, notably fungi.

11.18.12 Climate vis-a-vis Seasonality

Farmers who reported the occurrence of EUS in their ponds during the previous season were said to at higher risk of an EUS occurrence even during the current year (RR = 3.00). Unusually low air temperature and/or incessantly heavy rainfall for approximately 3–15 days (prior to the start of the present survey), as opined by the fish farmers and fishers, might also be correlated with occurrences of EUS.

Furthermore, occurrences of EUS had generally been associated with low temperature and had often occurred after periods of heavy rain. It had been observed by Phillips and Keddie (1990) from 1988–1999 data that EUS outbreaks had occurred during the months when the average daily temperature was less than the annual average temperature in India, Bangladesh, China, and Lao PDR. Nevertheless, EUS outbreaks in the Philippines and Thailand had also been recorded during warmer months. In addition, Chinabut et al. (1995) had challenged stripped snakehead (*C. striata*) by injecting zoospores of *A. invadans,* and had found a weak inflammatory response. There had been a higher rate of mortality and more extensive fungal invasion among fishes maintained at 19 °C, compared with fishes held at 26 °C and 31 °C.

11.18.13 Data Obtained from the Interviews with the Fishers

Among the different types of fish habitats sampled, the "haors" had displayed the highest value of RR (1.33), and the rivers had depicted the lowest RR (0.54). A "haor" is a seasonal floodplain wetland in NE India (Kar, 2007, 2013). But in Bangladesh, a "haor" is the biggest natural depression between two or more rivers (Khan, 1997; Khan et al., 1997). Chemicals, wastes, pathogens, etc., might enter the haor through the riverine networks. With the onset of the dry season, the water level of the haor decreases, and the aquatic biota (the collective plants and animals) becomes crowded. This could often result in stressful conditions for fishes. Moreover, the presence of a wide range of EUS-susceptible fishes under these circumstances makes the haor a susceptible area for EUS outbreaks. On the other hand, the movement of water in lotic water bodies, notably rivers, is said to lessen the chances of fungal pathogens attaching to the body of the fish host, thereby resulting in fewer occurrences of EUS and consequently lower values of RR. Further, as reported, there had been no significant relationship between artificial stocking of natural water bodies and occurrences of EUS. As mentioned earlier, water bodies in which EUS-occurrence had been reported during the previous year are at more risk and prone to be affected by EUS, even during the current season, as also reported by the fishers (RR = 2.19).

11.18.14 Observations in General

The general observations and opinions of the farmers, as recorded by the interviewers during the period of survey, are summarized below. Some of these points had been corroborated by the current study, while others may require further study for confirmation.

1. EUS outbreaks had been occurring almost every year since 1988, affecting a large number of species of FW fishes. The intensity of the disease was quite high during the initial period of its inception, which had gradually subsided with often a decreasing trend with the passage of time.
2. The wild fisheries had sometimes been more affected than the farmed fisheries, although the reverse was also sometimes true. In fact, the farmers and fishers from the wetland areas had reported that an outbreak of EUS occurs almost every year in their wetlands as the temperature begins to fall. It had affected mainly the snakehead murrels, some eels, notably the *Mastacembelus* sp., *Macrognathus* sp., etc. Later, the fish farms, which are situated quite close to the affected sites of wild fisheries, become affected, and then the process of transmission of the disease continues to distant farms.
3. As reported, EUS often occurs in culture ponds that are directly connected to rice fields through drainages. But other adjacent ponds that are not linked to rice fields through drainages generally remain unaffected.
4. Some farmers and fishers had opined that aquatic birds (particularly the piscivorous ones), reptiles, mammals, etc., might transmit the disease from one place to another mainly by preying upon easy-to-catch EUS-affected sick fishes, and in this process, drop the affected fish (either in part or in full) into the water bodies that had remained unaffected by EUS.
5. It is a common belief among many farmers and fishers that floods play a significant role in spreading EUS far and apart.
6. As reported, the affected snakeheads, walking catfishes, and the climbing perches might transmit the EUS disease by entering an unaffected water body.
7. Some farmers had opined that their unaffected farms had become affected after the introduction of duckweed from a wild water body.
8. As reported, a large number of old ponds, situated in shade and having no connection with the wild or affected water bodies, but containing a large amount of bottom deposits, had been repeatedly affected by EUS over several years. This could therefore reveal that the fungus is able to thrive in isolated water bodies under particular environmental conditions.
9. As reported, ponds with high stocking density are usually more prone to be affected by EUS.

Similar to the above, another set of observations was recorded during a separate case—control study of ponds in the Mymensingh district of Bangladesh. This study was undertaken concurrently with the cross-sectional study. The salient features of this latter study are briefly given below:

1. A series of contiguous and adjacent fish farms were affected with EUS during the late winter. Surprisingly, it was difficult to find an unaffected "control" fishpond in that area, as was reported.
2. It was reported that ponds with a common embankment, situated in proximity to a wetland (say, a "beel"), had all been affected. On the other hand, newly constructed ponds with quite high embankments, but situated near the same beel, remained unaffected.
3. In view of the above, on a number of occasions, the affected ponds were separated from the unaffected ponds (with similar cultural characteristics and practices) by a highway or high embankments.
4. It has been reported, perhaps based on summary observations, that the unaffected ponds of fish farms situated adjacent to EUS-affected beels usually become EUS-affected within a week. It was thought that this could be due to washing cattle and implements used in the beel in the unaffected pond. Some farmers further reported that after fishing in EUS-affected beels, rice fields, etc., EUS had

occurred in the ponds of their fish farms, possibly due to the washing of the affected nets and other implements in their unaffected ponds.

REFERENCES

Ahmed, A.T.A., 2004. Development of Environment Friendly Medicant for the Treatment of Argulosis in Carp Brood Stock Pond. SUFER-DFID Final Report. 57 pp.

Ahmed, M., Rab, M.A., 1995. Factors affecting outbreaks of epizootic ulcerative syndrome in farmed and wild fish in Bangladesh. J. Fish Dis. 18, 263–271.

Ali, M.L., Islam, M.Z., 1998. As assessment of the economic benefits from stocking seasonal floodplain in Bangladesh. In: Peter, T. (Ed.), Inland Fishery Enhancements. FAO Fish Tech paper 374, pp. 289–308.

Banik, R.C.K., 1999. Some information about fisheries of Bangladesh. In: Fish Week 99 Compendium. Department of Fisheries, Dhaka, pp. 96–104.

Banu, A.N.H., Hossain, M.A., Khan, M.H., 1993. Investigation into the occurrence of parasites in carps, catfish and tilapia. Progr. Agricult. 4 (1–2), 11–16.

Banu, A.N.H., Islam, M.A., Chowdhury, M.B.R., Chandra, K.J., 1999. Pathogenicity of *Aeromonas hydrophila* artificially infected to Indian major carp (*Cirrhinus mrigala* Ham.). Bangladesh J. Fish. Res. 3 (2), 187–192.

Callinan, R.B., Paclibare, J.O., Reantaso, M.B., Lumanlan-Mayo, S.C., Fraser, G.C., Sammut, J., 1996. EUS outbreaks in estuarine fish in Australia and the Philippines: associations with acid sulphate soils, rainfall and *Aphanomyces*. In: Shariff, M., Arthur, J.R., Subasinghe, R.P. (Eds.), Diseases in Asian Aquaculture II. Fish Health Section. Asian Fisheries Society, Manila, pp. 291–298.

Chandra, K.J., Begum, A.A., Ahmed, G.U., Wooten, R., 1996. Infection of myxosporean ectoparasites of juvenile carps in nurseries of Mymensingh, Bangladesh. Bangladesh J. Aquacult. 18, 39–44.

Chinabut, S., Roberts, R.J., Willoughby, L.G., Pearson, M.D., 1995. Histopathology of snakehead, *Channa striatus* (Bloch), experimentally infected with the specific *Aphanomyces* fungus associated with epizootic ulcerative syndrome (EUS) at different temperatures. J. Fish Dis. 18, 41–47.

Chowdhury, M.B.R., 1993. Research priorities for microbial fish disease and its control in Bangladesh for fish health. In: Tollervey, A. (Ed.), Disease Prevention and Pathology, pp. 8–11.

Chowdhury, M.B., 2002. Cross-sectional survey of epizootic ulcerative syndrome (EUS) in Bangladesh, December 1998–March 1999. In: Lavilla, P.C.R., Cruz-Lacierda, E.R. (Eds.), Diseases in Asian Aquaculture IV. Fish Health Section, Asian Fisheries Society, Manila.

Hasan, M.R., Bala, N., De Silva, S.S., 1999. Stocking strategy for culture based fisheries: a case study from Oxbow lakes fisheries project. In: Middendorp, H.A.J., Thompson, P., Pomeroy, R.S. (Eds.), Sustainable Inland Fisheries Management in Bangladesh, ICLARM Conf Proc. 58, pp. 157–162.

Hossain, M.A., Banu, A.N.H., Khan, M.H., 1994. Prevalence of ectoparasites in carp nursery of greater Mymensingh. Prog. Agric. 5, 39–44.

Hossain, M.S., Alam, M., Majid, M.A., 1992. Survey on fish disease epizootic ulcerative diseases in Chandpur district. Bangladesh. J. Train. Dev. 5, 55–61.

Kabata, Z., 1985. Crustacea as enemies of fishes. In: Snieszko, S.F., Axelrod, H.R. (Eds.), Diseases of Fishes. T. F. H. Publishers, Neptune, NJ. Book 1.

Khan, M.A., 1997. Ecology of floodplains in the northeast region of Bangladesh. In: Tsai, C., Ali, M.Y. (Eds.), Open Water Fisheries of Bangladesh, Chapter 11. The University Press Ltd, Dhaka, Bangladesh, pp. 153–172.

Khan, M.H., Marshall, L., Thompson, K.D., Campbell, R.E., Lilley, J.H., 1997. Susceptibility of five species (Nile tilapia, rosy barb, rainbow trout, stickleback and roach) to intramuscular injection with the Oomycete fish pathogen, *Aphanomyces invadans*. Bull. Euro. Assoc. Fish Pathol. 18 (6), 192–197.

Khan, M.H., Lilley, J.H., 2001. Risk factors and socio-economic impacts associated with epizootic ulcerative syndrome (EUS) in Bangladesh. In: Proceedings of DFID/FAO/NACA Asia Regional Scoping Workshop "Primary Aquatic Animal Health Care in Rural, Small-scale Aquaculture Development in Asia". Dhaka, 26–30 September 1999.

Khan, M.H., Lilley, J.H., 2002. Risk factors and socio-economic impacts associated with EUS in Bangladesh. In: Arthur, J.R., Phillips, M.J., Subasinghe, R.P., Reantaso, M.B., MacRae, I.H. (Eds.), Primary Aquatic Animal Health Care in Rural, Small-scale, AQC Development, pp. 54, FAO Fisheries Tech. Pap No. 406, pp. 27–39.

Kar, D., 2013. Wetlands and Lakes of the World. pp.xxx + 687. Springer, London. Print ISBN 978-81-322-1022-1; e-Book ISBN: 978-81-322-1923-8.

Kar, D., 2007. Fundamentals of Limnology and Aquaculture Biotechnology, vi + 609. Daya Publishing House, New Delhi.

Munro, A.L.S., Roberts, R.J., 1989. The aquatic environment. In: Roberts, R.J. (Ed.), Fish Pathology, second ed. Bailliere Tindall, London, pp. 1–12.

Phillips, M.J., Keddie, H.G., 1990. Regional research programme on relationships between epizootic ulcerative syndrome in fish and the environment. In: A Report on the Second Technical Workshop. 13–26 August 1990. NACA, Bangkok.

Roberts, R.J., Wootten, R., Macrae, I., Miller, S., Struthers, W., 1989. Ulcerative Disease Survey, Bangladesh. Final report to the government of Bangladesh and the overseas development administration. Institutes of Aquaculture, Stirling University, Scotland, 104 pp.

Roberts, R.J., Willoughby, L.G., Chinabut, S., 1994. Mycotic aspects of epizootic ulcerative syndrome (EUS) of Asian fishes. J. Fish Dis. 16, 169–183.

Snieszko, S.F., 1974. The effects of environmental stress on outbreaks of infectious disease of fishes. J. Fish Biol. 6, 197–208.

Subasinghe, R.P., Arthur, J.R., Shariff, M., 1993. In: Proceedings of the Regional Expert Consultation on Aquaculture Health Management in Asia and the Pacific Serdang, Malaysia, 22–24 May 1995. FAO Fish. Tech. Pap. No. 360.

Thrusfield, M., 1995. Veterinary Epidemiology, second ed. Blackwell Science Ltd, London.

Tonguthai, K., 1986. A Preliminary Account of Ulcerative Fish Diseases in the Indo-Pacific Region (A Comprehensive Study Based on Thai Experiences). National Inland Fisheries Institute, Bangkok, Thailand, pp. 39.

Willoughby, L.G., 1999. Mycological aspects of a disease of young perch in Windermere. J. Fish Biol. 2, 113–116.

Chapter 12

Quarantine and Health Certification

12.1 INTRODUCTION

EUS has been a hitherto unknown, virulent, and enigmatic disease among FW fish ever since its inception in SE Asia (Kar, 1990). It has been inflicting colossal losses on fish farmers mainly by sweeping through rice fields and fish farms. As usual, there had been widespread fear of and psychosis about potential EUS disease transmission to human consumers, which has been found to be totally unfounded and baseless (Kar, 2007, 2013). But this consequently led to a drastic decrease in the market demand for fish (Dey and Kar, 1990), including marine species that were not in fact affected by EUS.

Notwithstanding the above, EUS has been considered a virulent disease of epidemic dimension that has been affecting FW fish in the Indian subcontinent (including India and Bangladesh) since 1988 (Kar and Dey, 1990a,b). However, there is evidence to suggest that initiation of EUS outbreaks happen when a number of determinants or causal precipitating and predisposing stress factors amalgamate to cause the effect. The principal effect is the exposure of the dermis. These exposed sites could serve as potential points of entry for both primary and secondary etiological agents, viz., bacteria, fungi, and viruses, not only to primarily initiate the disease (EUS), but secondarily to cause havoc in fish during later periods. There had been several hypotheses about the dynamics of initiation of EUS. According to Lilley et al. (1998), the exposed sites could provide the points of attachment and entry for the spores of the fungus *Aphanomyces invadans*. According to Callinan et al. (1996), outbreaks of EUS in the estuarine fish of Australia were due to acid-sulfate soils, and they were said to have reproduced EUS in estuarine fish by exposing susceptible fish to acid water and spores of *A. invadans*. To the contrary, Kanchanakhan (1996) had further shown that EUS could be reproduced when susceptible snakeheads (*Channa* sp.) had been injected with a particular strain of Rhabdovirus and bathed in spores of *A. invadans*. Thus, it appeared that there could be a number of sufficient causes for the initiation (outbreak) of EUS, although every set of sufficient causes of EUS could be different from one another. However, each combination might have the common result of exposing the dermis and allowing entry of the etiological agents. Nevertheless, the presence of the fungus *A. invadans* is common in almost all cases of EUS

outbreaks, whether as a primary or secondary etiological agent. The highly invasive ability of the *A. invadans* to enter into the tissues of bones, gizzard, and spinal cords has been reported, and it could indicate that under certain circumstances, the fungus is able to invade the healthy skin of the fish (Vishwanath et al., 1998).

12.2 SURVEY OF WORKS

Notwithstanding the above, there had been various kinds of works on different aspects of EUS in different places since its inception. As an example, a review of some earlier surveys on EUS in Bangladesh is briefly given below. It could be evident that these surveys provide information on the impacts of EUS on the social, economic, and biodiversity aspects during that period.

12.2.1 Survey of Farmed EUS Fish in Bangladesh

1. Approximately 68% of the 200 ponds in Chandpur District were affected with EUS as noted during March—April 1998 (Hossain et al., 1992).
2. Approximately 50% of the 234 carp ponds had suffered from EUS-type outbreaks in 1991—92 (Ahmed and Rab, 1995).
3. Approximately 13% of the 96 extensive and *c* 7% of the 522 intensive/semi-intensive carp farmers had reported EUS during 1992—95 (ADB/NACA, 1991).
4. Approximately 16% of 6401 farmed fish had lesions during the period November 1998 to March 1999 (Khan and Lilley, 2001).

12.3 TRANSBOUNDARY ANIMAL DISEASES

It is a common notion that diseases know no boundaries and generally spread in an epidemic form like wild fire. Nevertheless, they have enough potential to affect both aquatic (piscian and nonpiscian) as well as the human population adversely (Baldock et al., 2005). Concomitantly, transboundary animal diseases (TADs) may be defined as "those epidemic diseases that are severely contagious and have the potential to spread very fast irrespective of national boundaries or borders, causing serious

Epizootic Ulcerative Fish Disease Syndrome. http://dx.doi.org/10.1016/B978-0-12-802504-8.00012-2

socioeconomic and public health consequences." Such high risks of morbidity and mortality among the susceptible animals (including fish) pose serious threats to the livelihoods of livestock and fish farmers.

12.3.1 Some Examples of Transboundary Animal Diseases (among Livestock and Fish)

1. **Foot and mouth disease (FMD)**: The disease had seriously affected the commercial pig industry in Taiwan during 1997. It has been absent from SE Europe. Efforts had been made to eradicate it from South America. It is now endemic in some parts of Africa, the Middle East, and Asia.
2. **Rinderpest**: It was introduced into Africa during the late nineteenth century, and within a decade had covered the whole country. It has been possibly eliminated from West Africa.
3. **Rift Valley fever**: It is said to have occurred in Egypt during 1977 with *c* 200,000 affected persons and *c* 600 deaths.
4. **Contagious bovine pleuropneumonia**: It has caused catastrophes in Africa over the last few years, and it is said by now to have covered 27 countries.
5. **Hog cholera (or classical swine fever)**: A recent serious outbreak of this disease had led to the death of *c* 12 million pigs. It had occurred significantly in Europe, notably in The Netherlands, Germany, Spain, Italy, and Latin America.
6. **African swine fever**: It had occurred for the first time in Cote d'Ivoire during 1996, where it had killed *c* 25% of the pig population. It is endemic in much of sub-Saharan Africa.
7. **Highly pathogenic avian influenza (HPAI)**: In Pennsylvania, USA, during 1983–84, the cost of eradication of HPAI was US $64 million, and the indirect loss to the consumers was US $500 million.

Concomitant to the above, TADs have the potential to threaten food security through serious loss of animal protein. TADs cause major production loss of both livestock and fish products; hamper trade of livestock, fish, and their products; cause public health consequences; and so on.

8. **Epizootic ulcerative fish disease syndrome (EUS)**

A significant addition to the above list is the hitherto unknown, virulent, enigmatic, and epidemic fish disease known as EUS, which has been causing large-scale mortality among FW and brackish-water fish on what is now a semiglobal scale. It appears to have crossed all national and international boundaries, rendering a number of species endangered as well as posing serious threats to the nutrition and avocations of the people (Kar, 2007, 2013).

12.3.2 Trends that Could Affect Transboundary Animal Diseases

TADs exhibit a great deal of dynamism. New diseases are said to emerge; old diseases are said to reemerge. They display a great propensity for sudden and unexpected spread to new regions. Significantly, the spread and impact of TADs may be seriously hampered due to global warming and climate change, possibly due to changes in rainfall patterns, etc., thereby possibly affecting the distribution of insect vectors.

12.3.3 Transboundary Animal Diseases to be Combated (Fight Against TADs): Application of Appropriate Technology

An effective national animal and fish quarantine system (FAO, 2009) could be the first line of defense against the entry and establishment of TADs. There could be a second line of defense as well. Some points could be as follows:

1. Disease surveillance and animal health information systems
2. Devise various alternative methods for studying epidemiology of disease outbreaks
3. Disease diagnosis and characterization of the etiological agents
4. Production of vaccines for disease control and eradication

12.4 FISH AND LIVESTOCK HEALTH: MAJOR ISSUES

International trade at the global level has enhanced the risks of disease emergence to almost an exponential scale. The situation has been aggravated with the raising of GMOs, crossbreedings, and other modern applications, sometimes without considering the consequences. As such, there has been an emergence of global bodies such as the WTO, which is responsible for (1) administering a number of multilateral agreements, notably the GATT, Agreement on Agriculture, Agreement on Sanitary and Phytosanitary Measures (SPS Agreement), and so on; and (2) taking care of frameworks that facilitate trade, along with evolving mechanisms to reduce the risks associated with free trade. The SPS Agreement provides core principles, while the *International Animal Health Code of the OIE* (OIE, 1997) provides standards for the management of these risks. Concomitantly, "The Animal Health Service" of the FAO appreciated and recognized the international spread of diseases as one of the potentially greatest downsides in the emerging changes to world trade. The FAO therefore consolidated its activities under the Emergency Prevention System for Animal and Plant Pests and Diseases to control and contain serious and severe epidemics among fish and livestock such as TADs, including fish diseases, notably Rinderpest, FMD, Newcastle disease, and others. Aspects of EUS could also be considered under this scheme.

Concomitant to the above, global production from AQC is increasing at approximately 13% per year, and production is expected to double every six years if this rate is sustained along with uniform rates of production from different countries. Conversely, fast growth may also help potential pathogens to find a greater number of susceptible hosts. Also, rapid unchecked growth might boost unhealthy competition among neighboring farms, thereby leading to the spread of infectious diseases in epidemic dimensions leading to devastation among not only the fish, but also other types of aquafauna in aquafarms of both the large- and the small-scale sectors.

Notwithstanding the above, it is often appreciated that similarities exist between smallholder livestock and AQC systems with regard to application of similar principles of health management and outcomes. Similar problems often occur with increased stocking densities coupled with commercial pressures for making production efficiency gains continually. This often may lead to enhanced chances of infectious diseases. A glaring example in this regard is the poultry industry. During the early 1960s in Australia, there had been small poultry farms of small number of batches, which had been run mainly on familial efforts. Gradually, avaricious commercial natures transformed the small avocations into huge commercial establishments, thereby inviting numerous diseases and associated problems. Consequently, management had to reconvert "multiage" farms into "all-in all-out" farms, with mandates for reduced stocking density.

12.5 ADDRESSING DISEASE PROBLEMS IN SMALL-SCALE AQC

Rural small-scale farmers are generally resource-poor with usually limited knowledge of disease and health management among the fish. Thus, it is essential to understand how the rural small-scale AQC sector is being managed by the farmers and their associates engaged in various sectoral activities. This could, concomitantly, help us in framing appropriate interventions through better health management practices. Some significant health management practices could be as follows:

1. Framing of effective quarantine measures at the national level
2. Access of the farmers to basic aquatic animal health services
3. Basic health management measures incorporated into fishery enhancements and rural livelihood programs (Kar and Kar, 1994)

4. Strengthening extension services (FAO, 2009) and accelerating communication exchange, which could bring about a letup in the situation through faster responses to disease situations

12.6 RULES AND REGULATIONS

Rules and regulations are part of systematic management. Ailments in fish may not occur, if the fish farm could be maintained following stringent rules of hygiene and strictly scientific norms of management.

12.6.1 Quarantine and Health Certification

The EUS fish disease had been sweeping fast and unimpeded across watersheds, oceans, and country borders for a few decades. Such unhindered spread of EUS points to the possibility that human agencies might be responsible for its irresistible spread. Further, the unrestricted transport of fish and piscine products, along with their pathogens, if any, within and across national and international boundaries might have played some role in the initiation and spread of EUS. It is sometimes believed that the trade of some EUS-susceptible species such as carps, catfish, murrels, and gouramies, which are of interest to aquarists, is often responsible for the initiation and outbreak of EUS. A number of recommendations were recorded at the ADB-NACA Regional Workshop on Fish Disease and Fish Health Management (Anon, 1991) emphasizing the establishment of centralized quarantine facilities and a system of fish health certification. The compelling need for such a strategy had arisen in view of the severe economic losses suffered by the fish farmers and fishers due to the devastating outbreaks of EUS. In fact, such protective systems had been operational in some countries of the region for quite some time.

12.6.2 Farm Registration

Concomitant to quarantine measures, there had been a growing need for regulations for controlling movement of live farm consignments in connection with national and international trades. Such steps, if effectively implemented, could reduce the impact of infectious diseases within the regions and countries. Ideally, all farms and hatcheries are expected to possess a valid license for responsible trade. They are also expected to maintain an up-to-date records of the movement of consignments of fish and piscine products, along with the health status of the fish. The arrival of diseased fish to unaffected sites could then be controlled and restricted wherever necessary. In addition, commercial pressure from the farm owners and fishers to the suppliers of fish and fish seeds, could go a long way in supplying the farms with "certified" disease-free fish. Such a practice, if followed honestly and regularly, could eliminate the need for legislation.

12.6.3 Chemotherapeutants

The chemotherapeutants are to be used judiciously. Chemotherapeutants, like antibiotics, are often used indiscriminately, perhaps due to ignorance or bad advice. Such erroneous use may cause damage to the aquaculture industry, and could damage useful flora through the assumption that many microbes could become antibiotic resistant. Many antibiotics may not be directly harmful to humans. However, they may perturb their intestinal flora and could pave the way for the development of resistant bacterial strains. This could make the treatment of fish diseases more difficult. There are a few regulations already in vogue. However, there are no clear guidelines regarding the withdrawal periods after their administration. However, such guidelines could be indispensable to ensure and safeguard both the aquaculture industry and its workers, as well as ensure the safety of consumers of aquaculture products.

12.7 RESEARCH

Continuous and comprehensive research work is a prerequisite to knowing about a disease in detail. Such efforts could give an answer to the etiology of the ailment, particularly if the illness is hitherto unknown, as well as help in targeting the actual measures for its treatment and control. It should be noted that intensive research work has been ongoing, based on extensive field and laboratory work, to determine the correct etiology of the hitherto unknown, enigmatic, and virulent EUS (Kar, 1991a,b).

12.7.1 Etiology

As indicated earlier, there is still a knowledge gap regarding the exact etiology of EUS. As such, more studies are needed to fill this void. Particular attention is required to establish the primary etiology and the associated secondary infections. It is very much essential to confirm the exact roles of bacteria, fungi, and viruses, and solve the riddle for good. HP studies (based on both light microscopy and EM) along with comparisons between HP of EUS fish and other non-EUS fish, could also be attempted.

12.7.2 Control

Aspects of control with particular reference to EUS have already been discussed in detail. At the cost of reiteration, it may be emphasized here that an integrated comprehensive approach to fish health and disease is highly essential in today's scenario, utilizing available techniques of management. The lessons learned from the repeated devastating episodes of EUS have paved the way for this new kind of thinking. The application of lime could be a pragmatic approach in the management and control strategy of EUS, but there is need for more detailed studies in order to quantify the dosages of treatment, and for more effective prophylactic and therapeutic preventive measures. In addition to the application of lime in other kinds of water bodies, such measures could also be applicable in managing EUS in cultured fish populations, particularly in the wetlands (notably the "beels" in eastern India and Bangladesh).

12.7.3 Environment

There had always been a reported (Kar, 2013) close relation between the initiation of EUS and changes in the environment, particularly in the limnological parameters (Kar and Dey, 1991, 1992, 1993, 1996) of the water bodies in which the fish live. A sudden alarming decline in TA and also of temperature of water had often been found to trigger the initiation of EUS under the Indian scenario, and therefore TA and water temperature have often be regarded as "stress factors" in the initiation of EUS (Kar et al., 1990a,b, 1993, 1994, 1995a,b, 2003, 2007a,b).

The Regional Research Program of NACA, along with other studies, emphasized the importance of environmental factors in the outbreak of EUS. Even so, the critical factors remained elusive (FAO, 1986). NACA further opined that (FAO, 1986) declining or unstable temperatures, low alkalinity and hardness, and fluctuating pH appear to be significant in the outbreak of EUS. But rigorous experimental works are required to confirm these roles. In addition, high levels of Al and other elements might also be mobilized under low pH conditions. Significantly, greater understanding of the dynamics of these factors might allow the possibility to manipulate the environment positively. Alternatively, such understanding could also permit avoidance of certain developing conditions, which could assist in controlling EUS in both wild and cultured fish populations. The exercise of region-wide monitoring of selected environmental parameters under the guidelines of the NACA is expected to be ongoing—the dissemination of more information on EUS is expected, particularly on the dynamics of interaction among various parameters. This could help in unraveling the exact etiology of EUS.

Similarly, regional studies have portrayed that pollution of water bodies from inorganic and organic pollutants is a growing concern and a significant constraint to fisheries and aquaculture development in the Asia—Pacific region. It is sometimes opined from certain quarters that our understanding of the complex relationship between EUS in fish, and the environment, had been hindered by an acute lack of data on the related aspects. During the study of EUS, on a number of occasions the practical benefit to be obtained from improving the pond environment was the yield of disease-free fish. Further, a more comprehensive understanding of interactions between pollutants, fish production, and disease could assist in developing strategies to protect fish from ailments, get high fish yield, and help in making up fish losses in both wild fish populations and aquaculture.

12.7.4 Improved Facilities

Facilities are essential for doing research, and such facilities should be improved types when the topic of research is a hitherto unknown one. In this regard, it may be opined that EUS has been a hitherto unknown, enigmatic, and virulent fish disease that has had devastating effects on freshwater fish with a semiglobal reach. As such, research aimed at solving the riddle of this unknown illness in fish has been requisitioning for improved facilities—at both laboratory and field levels—in the form of infrastructure and expertise, including tissue culture, virology, EM, elemental and pesticidal analyses, analyses of different water-quality parameters, and so on. EUS has further highlighted the need for improved fish disease diagnostic and research facilities in developing countries, particularly in the SE Asian region, mainly because EUS is almost endemic in this region today. In addition, the socioeconomic benefits of developing the resources and the infrastructure to investigate such challenging tasks could outweigh the disadvantages and the cost of such facilities. In fact, millions of dollars have probably been spent on and lost in tackling EUS. In view of the pressing need for an advanced and improved facility, the Royal Thai government, with the support of the Overseas Development Organisation (ODA), established the Aquatic Animal Health Research Institute (AAHRI) at the Kasetsart University campus in Bangkok. This institute gives facilities and technical expertise for sophisticated research on EUS in regional, national, and international arenas.

Realization of the objectives indicated above for the prevention, control, cure, and eradication of EUS requires the acceptance, adoption, and implementation of correctly and adequately planned policies by concerned governments. Moreover, there is need for a high level of organization, and national and international cooperation and coordination, when handling and dealing with such a varied and widespread problem like EUS, which evidently takes no heed of national and international boundaries.

12.8 CONCLUSION

Welfare to animals has of late been a major issue in livestock and fisheries management systems. Outbreaks of EUS have been subsiding in many regions. New occurrences are being reported from previously unaffected areas, however,

as well as in newly developed fisheries and farming management systems. It was recently revealed that EUS may not always occur as seasonal outbreaks and may not always cause high mortalities, but instead may prevail at low levels throughout the year. It could thus have some effect on productivity that may not be measurable in terms of mortalities alone. On one hand, communities that are dependent on local fisheries have been affected greatly by outbreaks of EUS everywhere. On the other hand, societal impacts may sometimes extend beyond persons who are directly affected by the loss of fish. There have been reports of direct economic losses due to EUS-caused deaths in various affected nations, and further estimates are being prepared based on recent survey data. But losses due to lower productivity could be of greater significance. Naturally, during the period of severe EUS outbreak, reduced aquaculture and fish production could be demonstrated. Declines in fish production positively due to EUS have not been positively ascertained, as even during periods of severe EUS outbreaks, it may be difficult to locate the susceptible species as revealed from anecdotal evidence. To date, there has not been much information available on the long-term effects of EUS on aquatic ecology.

REFERENCES

ADB/NACA, 1991. Fish Health Management System in Asia-Pacific. Report on a Regional Study and Workshop on Fish Disease and Fish Health Management. ADB Agriculture Department Report Series No.1. Network of Aquaculture Centres in Asia-Pacific, Bangkok.

Ahmed, M., Rab, M.A., 1995. Factors affecting outbreaks of epizootic ulcerative syndrome in farmed and wild fish in Bangladesh. J. Fish Dis. 18, 263–271.

Anon, 1991. Nepal Status Report. ADB/NACA Fish Health Management in Asia-Pacific. Report on a Regional Study on Workshop on Fish Disease and Fish Health Management. ADB Agriculture Department Report Series No.1. Network of Aquaculture Centres in Asia-Pacific, Bangkok, Thailand, pp. 89–92.

Baldock, F.C., Blazer, V., Callinan, R.B., Hatai, K., Karunasagar, I., Mohan, C.V., Bondad-Reantaso, M.G., 2005. Outcomes of a short expert consultation on epizootic ulcerative syndrome (EUS): reexamination of causal factors, case definition and nomenclature. In: Walker, P.J., Lester, R.G., Bondad-Reantaso, M.G. (Eds.), Diseases in Asian Aquaculture V. Fish Health Section, Asian Fisheries Society, Manila, pp. 555–585.

Callinan, R.B., Sammut, J., Fraser, G.C., 1996. Epizootic ulcerative syndrome (red spot disease) in estuarine fish: confirmation that exposure to acid sulphate soil runoff and an invasive fungus, *Aphanomyces* sp., are causative factors. In: Robert, J. (Ed.), Proceedings of the National Conference on Acid Sulphate Soils. Smith and Associates and ASSMAC, Australia.

Dey, S.C., Kar, D., 1990. Fish yield trend in Sone, a tectonic lake of Assam. Matsya 15–16, 39–43.

FAO, 1986. Report of the Expert Consultation on Ulcerative Fish Diseases in the Asia-Pacific Region (TCP/RAS/4508), Bangkok, 5–9 August 1986. FAO Regional office for Asia and the Pacific, Bangkok.

FAO, 2009. What You Need to Know about Epizootic Ulcerative Syndrome (EUS) — an Extension Brochure. Food and Agricultural Organisation of the UN, Rome, pp. 1–36.

Hossain, M.S., Alam, M., Mazid, M.A., 1992. Survey on fish disease, epizootic ulcerative syndrome in Chandpur district. Bangladesh. J. Tran. Develop. 5, 55–61.

Kanchanakhan, S., 1996. Field and Laboratory Studies on Rhabdoviruses Associated with Epizootic Ulcerative Syndrome (EUS) of Fishes (Ph.D. thesis). University of Stirling, Scotland. 278 pp.

Kar, D., 1990. Limnology and Fisheries of Lake Sone in the Cachar District of Assam (India), pp. viii + 201 (Ph.D. thesis) (Published). University of Gauhati, Assam (1984), Matsya. vols. 15-16, pp. 209–213.

Kar, D., 1991a. Fish disease syndrome affects the people of Barak valley. Fish. Chimes 10 (11), 63.

Kar, D., 1991b. EUS still prevails in South Assam. Fish. Chimes 10 (11), 63.

Kar, D., 2007. Fundamentals of Limnology and Aquaculture Biotechnology. Daya Publishing House, New Delhi pp. vi + 609.

Kar, D., 2013. Wetlands and Lakes of the World. pp. xxx + 687. Springer, London. Print ISBN: 978-81-322-1022-1; e-Book ISBN: 978-81-322-1923-8.

Kar, D., Baishya, A.K., Mandal, M., Arjun, J., Laskar, B.A., 2003. A preliminary account of the vegetations in Dargakona locality of Silchar in Assam. Environ. Ecol. 21 (2), 254–258.

Kar, D., Bhattacharjee, S., Kar, S., Dey, S.C., 1990a. An account of 'EUS' in fishes of Cachar district of Assam with special emphasis on microbiological studies. Souv. Annu. Conf. Indian Assoc. Pathol. Microbiol. India NE Chap. 8, 9–11.

Kar, D., Dey, S.C., 1990a. Fish disease syndrome: a preliminary study from Assam. Bangladesh J. Zool. 18, 115–118.

Kar, D., Dey, S.C., 1990b. A preliminary study of diseased fishes from Cachar district of Assam. Matsya 15–16, 155–161.

Kar, D., Dey, S.C., 1991. Gill nets in Lake Sone of Assam with their economics and impact on fishery. J. Appl. Zool. Res. 2 (2), 76–79.

Kar, D., Dey, S.C., 1992. Interrelationship and dynamics of fish population of Lake Sone in Assam. Environ. Ecol. 11 (3), 718–719.

Kar, D., Dey, S.C., 1993. Variegated encircling gears in Lake Sone in Assam. J. Appl. Zool. Res. 4 (2), 171–175.

Kar, D., Dey, S.C., 1996. Scooping gears of Lake Sone in Assam. J. Appl. Zool. Res. 7 (1), 65–68.

Kar, D., Dey, S.C., Michael, R.G., Kar, S., Changkija, S., 1990b. Studies on fish epidemics from Assam, India. J. Indian Fish. Assoc. 20, 73–75.

Kar, D., Dey, S.C., Kar, S., Bhattacharjee, N., Roy, A., 1993. Virus-like particles in Epizootic Ulcerative Syndrome of Fish. Proc. Int. Symp. on Virus-Cell Interaction: Cellular and Molecular Responses 1, 34pp.

Kar, D., Dey, S.C., Kar, S., Roy, A., Michael, R.G., Bhattacharjee, S., Changkija, S., 1994. A candidate virus in epizootic ulcerative syndrome of fish. Proc. National Symp. of the Indian Virological Society 1, 27pp.

Kar, D., Dey, S.C., Kar, S., 1995b. Fish and fisheries of Lake Sone in Assam. J. North Eastern Counc. 15 (2–3), 49–57.

Kar, D., Dey, S.C., Kar, S., Roy, A., 1995a. Epizootic ulcerative disease syndrome among the fishes of North-East India. J. North Eastern Counc. Govt. India 14 (4), 21–24.

Kar, D., Mazumdar, J., Barbhuiya, M.A., 2007a. Isolation of mycotic flora from fishes affected by epizootic ulcerative syndrome in Assam, India. Asian J. Microbiol. Biotechnol. 9 (1), 37–39.

Kar, D., Mazumdar, J., Halder, I., Dey, M., January 25, 2007b. Dynamics of initiation of disease in fishes through interaction of microbes and the environment. Curr. Sci. Bangalore 92 (2), 177–179.

Kar, S., Kar, D., 1994. Health and disease pattern among the Ao Nagas. Soc. Sci. VI (7), 417–422.

Khan, M.H., Lilley, J.H., 2001. Risk factors and socio-economic impacts associated with epizootic ulcerative syndrome (EUS) in Bangladesh. In: Proceedings of DFID/FAO/NACA Asia Regional Scoping Workshop "Primary Aquatic Animal Health Care in Rural, Small-scale Aquaculture Development in Asia". Dhaka, 26–30 September 1999.

Lilley, J.H., Callinan, R.B., Chinabut, S., Khanchanakhan, S., MacRae, I.H., Phillips, M.J., 1998. Epizootic Ulcerative Syndrome (EUS) Technical Handbook. The Aquatic Animal Health Research Institute, Bangkok, p. 88.

OIE, 1997. Epizootic Ulcerative Syndrome: Diagnostic Manual for Aquatic Animal Diseases. Office International Des Epizootics, Paris, pp. 127–129.

Vishwanath, T.S., Mohan, C.V., Shankar, K.M., 1998. Epizootic ulcerative syndrome associated with a fungal pathogen in Indian fishes-histopathology: a cause forinvasiveness. Aquaculture 165, 1–9.

Chapter 13

Epizootic Ulcerative Syndrome and the Economy of the Nation

13.1 INTRODUCTION

There is a common Bengali proverb, *"Machhe Vate Bangalee."* It means that Bengalees may not be able to live without fish in two square meals a day. It is believed that *c* 1.4 million people earn their livelihood from the fisheries sector. Further, *c* 11.0 million people are involved in fishing and its related and other ancillary activities (Mazid, 1995). Furthermore, *c* 80% of the population is rural and generally catches wild fishes from various small and big lentic (lakes, wetlands (beels, haors, baors or anuas, etc.), ponds, marshes, fens, ditches, etc.) and lotic (rivers, streams, brooks, canals, etc.) water bodies. Although fish is the staple food of the people for nutrition, many poor people may not be able to buy fish for every meal in a day.

Even so, the fisheries domain is seen by many as an important sector in the nutrition and avocations of the people. This fact is more pronounced in tropical Asia, notably the Indian subcontinent including India and Bangladesh. In this regard, the fisheries sector contributes *c* 5.3% to the gross domestic product (GDP) of Bangladesh. Also, >7% of the country's population is supported by the fisheries sector and its allied activities. Further, *c* 6% of export-related earnings is contributed by the fisheries sector, thus paving the way to earn the third rank after jute products and leather. Furthermore, fish accounts for *c* 6% of the per capita protein intake and contributes about 60% of the animal protein consumed throughout Bangladesh. Moreover, the fisheries sector provides income to *c* 1.5 million full-time and 11.0 million part-time fishers. The fish sector plays similar roles in India, Thailand, and other countries.

Concomitant to the above, production through aquaculture (AQC) is continuing by leaps and bounds in the developing countries in Asia in order to meet the needs of nutrition and avocation. Burgeoning population pressure vis-a-vis a shortage of alternative employment opportunities in certain countries had increased the attraction of fisheries and AQC as avocations. But these countries, notably Bangladesh and India, had been facing many difficulties in the appropriate management in AQC, mainly due to lack of up-to-date knowledge of management practices and a lack of awareness among growers (producers). Information pertaining to appropriate management of emerging problems should be disseminated among the fisheries and AQC sectors.

Notwithstanding the above, small-scale fish farming has become increasingly popular among unemployed youths as a potential source of income in view of the burgeoning unemployment problem. Some small-scale farmers own ponds but, rather than culturing, lease the ponds to others on an owner–tenant basis. In view of this, in the event of outbreak of a disease such as epizootic ulcerative syndrome (EUS), there could be colossal losses to AQC mainly because the tenant farmer may not always be competent enough to tackle and manage the grievous situations posed by an outbreak of EUS.

13.2 FRESHWATER FISHERIES SECTOR

There is widespread appreciation that the fisheries and AQC sectors play significant roles in meeting national goals in social, economic, and nutritional arenas. In certain Asian countries like Bangladesh, pisciculture is progressing toward semi-intensive culture in a water-spread area of *c* 147,000 ha. As already mentioned, indiscriminate use of fertilizers and feeds has been rendering fishes susceptible to diseases due to foul water, and consequently the effects of the diseases on improved systems of AQC are significant. At the same time, outbreaks of diseases had been insignificantly reported and poorly documented.

13.3 PROMISING AQC AND SOCIOECONOMICS

Notwithstanding the above, statistical figures from the Indian subcontinent, notably Bangladesh, indicated that *c* 1.31 million ton of fish was produced during 1996–1997. Out of this, *c* 1.0 million ton was obtained from inland waters, contributing to *c* 79% of total fish production. Further, open-water capture fisheries

Epizootic Ulcerative Fish Disease Syndrome. http://dx.doi.org/10.1016/B978-0-12-802504-8.00013-4

contributed c 58% of total inland production, and the rest of the c 42% was contributed by closed-water culture fishery. In fact, the quantum of production from culture fishery has been increasing since 1983–1984. Its yearly growth rate over the previous years varied between 4% and 20%. Further statistical figures portrayed that the contribution of culture fisheries to inland fish production in some Asian countries such as Bangladesh had been c 16% during 1983–1984, and increased to c 33% by 1996–1997. The average rate of depletion from capture fisheries, however, was c 1.7% per year during the period 1983–1984 to 1989–1990. Nevertheless, a trend of increasing fish production has been registered since 1991–1992, perhaps due to large-scale open-water stocking programs initiated by some governments neighboring India, notably Bangladesh. The projected target of fish production during 2001–2002 was 2.08 million ton. This goal was probably achieved through intensification of AQC and possibly through enhanced fisheries from floodplains. However, a sustained supply of large-sized quality fish fry and fingerlings could be a prerequisite for both the intensification of AQC and the enhancement of fisheries through open-water stockings.

13.4 PADDY-CUM-FISH CULTURE AND EPIZOOTIC ULCERATIVE SYNDROME

Similarly, and in addition to other species, EUS had been affecting fishes that grow in paddy fields, notably murrels, anabantids, clariids, and some minor carps. As such, the potential socioeconomic impacts of the highly infectious EUS affecting rice-field fishes are immense. As indicated, c 250 million families in the region are dependent upon paddy cultivation as a principal crop, with rice being a staple food item of their diets (Macintosh, 1986). Farmers also go for fish production through the application of the paddy-cum-fish culture technique. They usually produce two crops—rice and fish. They often invest a huge sum of money in such ventures with the expectation of earning a lucrative return. The usual period of harvest of the crops is from September to February, but the outbreak of EUS generally coincides with the harvesting period of crops, as indicated above. As such, fish farmers face difficulties in marketing their fish crops, and in the process they incur colossal losses. As such, estimates of the economic value of fish losses to commercial fish traders are huge. These figures do not, however, account for indirect socioeconomic costs due to market rejection of harvested ulcerated fish or, in some cases, even unaffected fish. In some communities during the 1980s, a widespread but unfounded fear of disease transmission to consumers led to a drastic decrease in market demand for all food fish. Consequently, confidence in freshwater fish farming,

particularly among potential investors and financial agencies, was badly damaged.

13.5 LIVESTOCK VERSUS AQC

Livestock production is generally perceived as possessing high risk due to disease outbreaks, vis-a-vis the high cost of animals and their maintenance. On the other hand, AQC, being comparatively less money-intensive than livestock farming, may not always have severely devastating economic effects in the event of disaster. In this regard, AQC possesses positive features. However, a shortage of juveniles for stocking ponds may invite their import from neighboring areas or countries, and imported juveniles are sometimes of poor quality and often have poor survival rates. Moreover, there is the potential risk of introduction of diseases from the countries of origin. However, the production of juveniles (mainly fingerlings) is usually confined to provincial hatcheries and a few private entrepreneurs. Such a situation is in vogue in Lao PDR, where this activity is on the increase, and the consequences are aggravated by health-related management problems. The latter limit aquaproduction in Lao PDR and constrain development, but does not inflict any direct and severe economic loss. However, this may not be the case for wild fisheries. Ultimately, a lack of baseline information on aquatic animal health issues for Lao PDR may be limiting its ability to assess risks in AQC and fishery operations.

13.6 INPUTS THROUGH AGRICULTURE

Rice bran and cassava are the two agricultural feedstuffs produced by the farming system in Lao PDR, and surplus corn may also be produced in some regions. Rice bran is usually retained by farmers for use as supplementary feed in pisciculture and livestock production, although some is exchanged as the cost of rice milling. On the other hand, cassava is eaten or fed to livestock in upland areas. Chemical fertilizers are not in much use mainly because of financial constraints. Nevertheless, they may be used along with pesticides in irrigated areas, where there is greater promotion of the more intensive methods of rice production.

There is only limited intensive livestock production in the villages mainly due to a lack of animal feedstuffs. Limited production area, coupled with marketing difficulties, could serve as a partial reason for this, and the fragmented nature of subsistence farming could be a further cause. Although livestock production is encouraged, often through bank lending, the agricultural promotion banks in Lao PDR do not seem to be very conversant with lending to the fisheries sector, notably to small-scale fish farmers for pisciculture.

13.7 EPIZOOTIC ULCERATIVE SYNDROME, THE HITHERTO UNKNOWN FISH DISEASE

A disease may usually cause loss of life and property. The loss may be colossal if the outbreak of the disease occurs on an epidemic dimension. Concomitantly, the inhabitants of a nation suffer losses due to outbreaks of disease, and the brunt of a disease is usually faced by the economically poor masses of a country. Poverty-stricken stakeholders at the grassroots level are rendered beggars with enormous losses when the entity that fell ill relates to their nutrition and avocation. Such a situation occurred when several diverse species of fishes were killed by EUS on a semiglobal scale, rendering a number of fish species less abundant and endangered.

EUS has been a hitherto unknown, virulent, and enigmatic disease among FW fishes since its inception in SE Asia, and has inflicted enormous losses on fish farmers mainly by sweeping through rice fields and fish farms. As usual, there was widespread fear of and psychosis about potential EUS disease transmission to human consumers, which has been found to be totally unfounded and baseless (Kar, 2007, 2013; Kar and Dey, 1990a,b,c). But this fear and psychosis consequently led to a drastic decrease in the market demand for fish, including marine species that were not in fact affected by EUS.

13.8 THE RISK FACTORS

A principal strategy in subsistence farming is to minimize "risk factors." This is possibly reflected in the low input vis-a-vis low productivity of rural Lao AQC. The size of Lao ponds ranges from 550 to 1520 m^2 with a water depth of c 50 cm on average. Productivity is low, usually ranging from 417 to 708 kg/ha per year, mainly due to low stocking densities (generally, from 1 to 4 fish/m^2) and limited artificial feeding. It is important to note that low-input AQC systems are generally not very disease-prone, but may become so during the dry season or with increased inputs.

Disease is a less significant risk factor among fishes in Lao PDR, as compared with its significance for livestock. Rather, the significant risk factors are theft, excess use of time and labor, and so on.

13.9 EPIZOOTIC ULCERATIVE SYNDROME AND THE LOSS TO ECONOMY

In addition to the above, there have been reports of economic losses due to fish disease and mortalities caused by EUS in countries such as India, Bangladesh, Nepal, Pakistan, Sri Lanka, Myanmar, Thailand, Lao PDR, Kampuchea, Vietnam (Anon, 1991a), the Philippines, Malaysia, and Indonesia, as well as on the Australasian continent.

Nevertheless, estimates of the huge losses incurred by various nations due to EUS are given below:

Country	Year	Economic loss in US$ due to EUS	References
Thailand	1982 −1983	8.7 million	Tonguthai (1985)
Sri Lanka	1988 −1989	25,000	Anon (1991c)
Eastern Nepal		937.0	Anon (1991d)
Bangladesh		3.4 million	Barua (1990)

13.9.1 Thailand

As reflected above, EUS had devastating effects in most of the countries mentioned. Murrels and catfishes were severely affected. As reported (Tonguthai, 1985; Lilley et al., 1992), Thailand incurred an estimated economic loss of around 200 million Baht (equal to approximately US$ 8.7 million) during the EUS outbreak of 1982−1983 alone. There was a sharp decline in the market demand and sale of food fishes, possibly due to an unfounded rumor of suspected transmission of EUS to human consumers. Consequently, confidence in fish farming badly declined among fish farmers, potential investors, and financial agencies. Tonguthai (1985) estimated the *direct* losses to culture fisheries in Central Thailand, due to EUS, to be c US$ 8.0 million.

13.9.2 Indonesia

There were losses of brood stock of common carp (*Cyprinus carpio*) in Indonesia during 1980. This caused an economic loss of US$ 2.0 million. However, this was probably due to an unrelated bacterial septicemic condition (Priyadi, 1990).

13.9.3 The Philippines

The average daily income of fishers declined from approximately US$ 4.0 to US$ 1.50 in the Philippines during the EUS outbreak period in Laguna de Bay. This was probably due to the rejection of affected fish by human consumers. There was also worry that repeated yearly outbreaks of EUS, specifically in Laguna de Bay, could lead to a significant decline in the natural fish population. This in turn might lead to dwindled stocks of the affected species in the wild (Anon, 1991b).

13.9.4 Australia

Callinan estimated the yearly economic loss of fresh and brackish-water fishes on the east coast of Australia due to ulcerative diseases, notably EUS, as US$ 1,330,000.0. However, this estimate probably did not take into account the potentially irreversible damage caused to fish stocks that in turn affected the potential commercial and recreational fishing industry in the country.

13.9.5 Bangladesh

In addition to the above, Bangladesh had incurred colossal losses of fish and money due to the devastating outbreaks of EUS since 1988−1989. Decreased consumption of fish by humans following the outbreak of EUS resulted in a fall in fish prices of up to c 75%, leading to high losses to fish traders, fish farmers, and fishers.

13.9.6 Nepal

Nepal is a country with hilly terrains. But in spite of this, EUS could prevail with potentially severe impact on the newly established fish farms. This in turn could affect the nutritional welfare of the population. As reported, Nepal had initially incurred a loss of c 15−20% of the total fish production, which equated to US$ 937,000 from a small number of affected water bodies covering an area of only c 328 ha in the Eastern Development Region. Further, the occurrence of EUS in cultured major carp fingerlings gave rise to fears of a potentially crippling effect on the expansion of carp culture in the mountain country (Anon, 1991d).

13.9.7 Pakistan

EUS had affected the Punjab province of Pakistan. The financial losses due to the impact of EUS fish disease in some parts were estimated and reported (Iqbal et al., 1998). It had been reported that fish losses to the tune of Rs 125,000/- (US$ 3125.0) related to the losses incurred in fish farms. In fact, the losses spread over c 10% of fish farms are included in the five affected districts, notably Kausar, Lahore, Gujranwala, Sialkot, and Narowal in the province of Punjab. The total economic loss amounted to approximately US$ 300,000.0 (AAHRI, NACA, IOA and NSW-Fisheries, 1997). Moreover, the wild fish catch had been reported to have declined by c 25% due to EUS. Further, almost complete disappearance of murrels, notably *Channa punctatus,* in portions of the Rivers Chenab, Ravi, and Sutlej, as well as in the associated canals, had also been reported.

Loss of fishes due to EUS also has social and economic impacts on fishing communities. There had been a loss of c 80% in the daily income of fishers in Maliwal village on the bank of River Ravi, which represents a reduction of Rs 1.92

million (US$ 48,000.0) over one fishing season. In addition, there had also been a c 20% reduction in government revenue from auctions of fishing rights during August 1996, perhaps due to EUS. In other words, this represented an economic loss in revenue of c Rs 8.0 million (US$ 200,000.0), which could accumulate to a total of Rs 40.0 million in successive years (AAHRI, NACA, IOA, NSW-Fisheries, 1997).

13.9.8 India

Concomitant to the above, the devastating outbreak of EUS among cultured IMC had a significantly bad impact on and severely bad economic implications for almost the entire Indian subcontinent (Kar and Dey, 1988a,b). This episode was also said to have a potentially crippling effect on the expansion and viability of carp culture in some regions. Bhaumik et al. (1990) reported that greater than 73% of the AQC ponds in **West Bengal** (WB) had been affected by EUS. Further, most of these farms had been reported to incur losses ranging from 31% to 40% of their stock. In addition, Prasad and Sinha (1990) reported that out of the 39 districts in **Bihar**, six recorded a loss of US$ 250,000.0 due to fish kills caused by EUS. Devastating effects of similar magnitude occurred in **Assam** (Kar et al.,1993, 1995a,b) and other provinces in India—this was more so during the initial periods, but the adverse effects on the economy continue even today (Kar, 2007, 2013; Kar and Dey, 1990a,b,c).

Similarly, the province of **Kerala** was affected by EUS during the mid-1990s. Kerala has rich fishery resources, and fisheries are an important sector of the province's economy. According to the fishery department of the province, the sector employs about 3% of the population of the province and contributes to around 2.5% of its net domestic product. Concomitantly, the total annual inland fish production of the province had been estimated to be 36,000 ton from 355,000 ha of inland waters. Around 200,000 people belonging to 33,000 fishers' families depend almost entirely on the inland fishery resources for their livelihood. They are usually the subsistence fisherfolk. Concomitant to the above, EUS was first reported from Pookote Lake in the Banasurasagar Reservoir area in the northern district of Wynad in June 1991. EUS had subsequently spread to the fresh and the brackish waters of Kuttanad and Vembanad Lakes and the rivulets in Kottayam, Alappuzha, and Pathanamthitta in the south by the end of August 1991. It was estimated that EUS had probably affected c 25% of fish in the Vembanad Lake. However, EUS had moved to Thrissur by October and was reported from ponds and canals inside "kole" fields. Later, in November 1991, EUS had inflicted Kuttanad again, appearing in Kumarakom, affecting c 30% of the fish there. Subsequently, EUS spread to southernmost district

Trivandrum in January 1992, where c 15—30% of the catch in Veli Lake was affected. It was then reported from the Achenkoil River, two months later. The various fish types affected were the murrels, *Clarias*, *Etroplus*, *Puntius*, *Wallago*, etc. Fish farmers and fishers were estimated to have incurred a colossal economic loss of Rs 20 million (per government records) to Rs 120 and 200 million. The value of the annual catch from the Vembanad Lake alone had been estimated at Rs 100 million. Further, in their report giving details of losses to inland fishworkers in Kerala, ICSF (1992) quoted the official figure of Rs 20 million, but commented that newspapers reported losses of up to 10 times this figure.

Under the compelling situation in Kerala described above, fishers often had to resort to loans with interest rates as high as 180%, when income from other sources, notably from fishes, had been totally wiped out. The provincial government agreed to buy EUS-affected fish at Rs 2.00 per kg. Three fishworkers' associations took up the matter of relief. They were the Kerala Matsya Thozhilali Aikya Vedi, the Matsya Thozhilali Union, and the Kerala Swatantra Matsya Thozhilali Federation. However, proper developmental works probably began at the panchayat level in order to generate employment. Nevertheless, a useful by-product of the socioeconomic impact of the disease has been the government's fresh attention to conducting research on EUS in Kerala.

Notwithstanding the above, the effects of pesticides and agrochemicals on Kuttanad were debated. According to a very significant Indo-Dutch study on water conditions in Kuttanad, large doses of fertilizers (c 20,000 ton) and toxic pesticides (c 500 ton) had been sprayed over 66,000 ha of paddy fields. Notwithstanding the above, a scientist from the Central Institute of Fisheries Technology (CIFT), ICAR, Cochin, had questioned the conclusion that had been drawn on pesticides. He opined that the residues of pesticides, and also heavy metals like mercury and cadmium, were within acceptable limits of toxicity. He opined further that the first outbreak of EUS could have occurred in Kuttanad, and not in Wynad, if high levels of pesticides and heavy metals were the main causes of EUS.

The scientists from Thailand also had a similar line of thinking. Kamonporn Tonguthai had opined that there should be for more experimentation before a specific pesticide could be pinpointed as a triggering factor. She had further indicated the case of Lao PDR, where EUS had been reported, despite the absence of pesticide pollution. On the other hand, Supranee Chinabut had indicated a study by FAO in which pesticides had been assigned as the possible cause for the outbreak of EUS. Further, studies indicated the occurrence of EUS even in hilly ponds that were not affected by environmental pollution. However, low alkalinity seemed to be the common factor all over SE Asia, although pesticides could also affect water quality.

Similar views had been put forward by the scientists from Kerala, who had opined that the monsoon rains may usually lower the alkalinity of water, leading to outbreak of EUS.

In addition to the above, some of the more farfetched etiological conjectures pointed to infected fishlings brought into Kerala by private agencies, and to the droppings of birds that had eaten diseased fish. Moreover, according to the subjective perceptions of fisherfolk, EUS had been regarded as a curse of nature and a warning against destructive human intervention into the natural aquatic environment.

13.9.9 China

Reports from Eastern China had portrayed significant socioeconomic damage to fishes, fishers, and fish farmers from an ulcerative disease that supposedly showed conditions similar to the symptoms in the Australian condition. However, the dynamics of EUS outbreak, indicating the relationship between Chinese episode, the Australian outbreak, and any other similar outbreak, had not been very clearly ascertained and needs more elucidation.

13.9.10 Riverine Routes and the Spread of Epizootic Ulcerative Syndrome

Notwithstanding the above, works by different workers had brought another dimension to hypotheses on the **origin of EUS**. It has been highlighted that there is a strongly interlinked intricate riverine network of Kerala's lotic water system. A continuity of water bodies could be established between the Kaveri and the Kaverretty rivers, starting from Wynad in the north, which flowed through Tamil Nadu and Karnataka. Some prominent lotic systems in Kerala are the Pamba, Achenkoil, and Meenachil. EUS had been reported in Tamil Nadu in early 1991 and in Karnataka in 1990. It could be probable that EUS might have spread to Kerala from neighboring provinces via the riverine network, possibly aided by floods.

13.9.11 Epizootic Ulcerative Syndrome and Nutritional Deficiency

There had been obvious nutritional deficiency among fish consumers, mainly due to the exclusion of fish, because the Asian nations—including India, Bangladesh, Pakistan, and Sri Lanka—are those where fish is a main source of animal protein available to accompany a rice meal. As indicated earlier, c 250 million families in SE Asia cultivate paddy as a principal crop and eat rice as a staple food item. As such, a big portion of the incidental fish harvest from the paddy fields forms an important part of the family diet (Macintosh, 1986). In fact, it is quite difficult to estimate the economic value of fish losses due to virulent fish diseases, like the EUS. It is mainly because in rural areas, fish is mostly consumed at a local level and only a small portion is marketed. Tonguthai (1985), however, had estimated the direct losses, to the culture fishery in Central Thailand, to be approximately US$ 8.0 million.

13.9.12 Fisherfolk, Epizootic Ulcerative Syndrome, and the Economy of the Nation

Fisherfolk procure fish in various water bodies in the country. Their catch no doubt has a significant impact on the economy of the country. In the event of a disease, notably the hitherto unknown, enigmatic, and virulent EUS, not only are the livelihoods of fisherfolk seriously hampered, but also the economy of the nation is devastated.

13.9.12.1 Fisherfolk in the Region

"Fisherfolk" or "fishers" are the instruments to harvest fish from any water body. They are trained not only in various methods of fishing, but also in the fabrication of different types of fishing implements and devices according to fish, habitat, and seasonal types. Sometimes a fishing implement is fabricated specifically for a particular fishing community and often they feel proud of such identification of a community specialized in operating a particular type of fishing implement or device (Kar, 2000, 2003a,b, 2007, 2013).

In India (particularly its eastern region) and Bangladesh, there are large numbers of fishers belonging to diverse types of fishing communities. They could be broadly classified under different categories based on their communities, intensity of fishing, and so on. Most fishers in this region belong to the Scheduled Caste community (if nontribal) and to the Scheduled Tribe community (if tribal) as notified in the Constitution of India. Fishers, in general, are poor and not very literate. There had not been many ambitious welfare measures for their upliftment, although some programs meant for their benefit are believed to be in the offing now. The cooperative movement has not been well organized among fishers except in the province of Tripura. The Sone Beel Fishermen's Co-operative Society

(SBFCS), one of the biggest of its kind in this region, has not been very active and progressive, although it should have been (Kar, 2003b).

13.9.12.2 In Sone Beel (the Biggest Wetland in Assam)

Sone Beel is the biggest wetland in Assam, often serving as a reservoir of EUS-affected fishes, when there is an outbreak of the disease. Concomitantly, there are about 100,000 fishers in Sone Beel, who derive their nutrition and avocation from the beel. Thus, in the event of an outbreak of a disease among fishes, notably EUS, there is serious economic hardship, mainly because there is no alternative source of earnings. Therefore, the aspects of fish disease, fisherfolk, and the economy of a nation appear to be inseparably associated with one another in the form of a linked chain. Similar situations also occur in other wetlands in the region. Situations in some of these wetlands are briefly discussed below.

Incidentally, villages on the shore of a beel are generally inhabited by fishers belonging to different communities. The activities of fishers, and their socioeconomic condition, have been influencing the development and utilization of fishery resources in a significant way. This aspect has gained importance in view of the so-called inferior social status of fishers vis-à-vis overfishing of the water bodies by these poverty-stricken, so-called less progressive people.

13.9.12.3 Abounding Fishers of the Beel

There are 39 villages around Sone Beel, the biggest wetland (beel) in Assam and one of the biggest in India. Fishers belonging to four principal communities have been recorded in the 39 villages around Sone Beel (Kar, 1990a,b,c, 2003, 2007, 2013; Dey and Kar, 1989). These are the "Kaibarta," "Patni," "Maimal," and "Namasudra." Certain relevant information pertaining to the relative composition of each community in terms of fishers, the total population living in the villages around the beel, and the annual average area exploited by individual fishers of each community, are tabulated below:

Fisher community	Relative abundance (% of total) Total fishers	Total inhabitants	Fishing area (ha/fisher)
Kaibarta	68.14	47.67	0.32
Patni	14.66	10.24	1.51
Maimal	12.29	8.59	1.81
Namasudra	4.91	3.43	4.53
		30.07 (nonfishers)	

Moreover, on the basis of fishing intensity, fisherfolk of Sone Beel have been further classified, after FAO (1962), into three categories, viz., professional, part-time, and occasional, an account of which, with relevant information (Kar, 1990), is tabulated below:

A brief description of the three categories of fishers is given below.

cycle of fish (Welcomme, 1979), and the seasonal needs of agriculture seem to impose cyclicity on such communities. As such, reservation of fishing sites is not practiced by this group. Rather, they generally prefer cultivation to fishing in shallow water. Although their interest lies in hauling the large growing fishes, the yield status is usually poor because of the lack of adequate knowledge of the distri-

| Fisher category | Relative abundance (% of total) | | | | |
	Kaibarta	Patni	Maimal	Namasudra	Total
Professional	55.04	6.91	5.54	2.51	70.00
Part-time	8.00	3.00	2.60	1.40	15.00
Occasional	4.50	4.75	4.75	1.00	15.00
Total	67.54	14.66	12.89	4.91	100.00

13.9.12.4 Lifescape of Fishers and Glimpse of the Economy

Occasional Fishers

They take recourse to fishing only when an occasion arises or when the situation compels. Such a category of fishers is prevalent in Sone Beel generally during the dry days when fishing sites are easily accessible on foot. The relative ease of access coupled with a certain amount of free time between sowing and harvesting of the beel crops (notably paddy) makes this type of fishing attractive and often sportive to the group. This category of fishers constitutes c 15% of the fisher population of the beel. Both affluent and poor sections of fishers belong to this category: those compelled to this profession are the poor, while those who adopt it as a sport are the rich. Certain deeper portions of the beel, locally called "kuri," are found to be traditionally or occasionally reserved for them. These usually fished during certain festivals or feasts commensurate with holidays, where a large section of society generally participates. Most fishes caught in such operations usually go directly into their diet. The individual time spent is short and the gears used are usually simple and less efficacious. As the number of participants is large, it is difficult to assess the contribution of individual efforts to total catch. Migratory habit is not generally seen in this category of fishers.

Part-Time Fishers

This non-nomadic group of fishers constitutes c 15% of the fisherfolk population of Sone Beel. They fish only during part of a year and consider this activity equal or inferior to other activities of the group. The flood cycle, the biological

bution of fish species in the Beel. Nevertheless, part-time fishers try to operate almost all the types of gears operated by professional fishers. In addition, they have a tendency to practice certain aquacultural techniques.

Professional Fishers

This is the largest group among fishing communities in Sone Beel, constituting c 70% of the total fisher population around the beel. Both poor and rich fishers belong to this group. They fish throughout the year with the help of diverse types of fishing gears and devices without any restriction or reservation of fishing sites. However, they have a tendency of operating sophisticated and little-mechanized fishing gears, and they are proud of being identified as operators of certain specific gears; for example, professional fishers belonging only to the Kaibarta community operate the Maha jal, Chat jal, Ghuran jal, and the Haran jal. Similarly, professional fishers belonging only to the Patni community operate the Kuchrung jal and the Dorar jal. Also, professional fishers belonging only to the Maimal community operate the Chhat jal and Rekh jal. Although the bamboo cage traps and the bamboo barricades are used by almost all the categories of fishers, the professional fishers belonging to the Namasudra community operate only the bamboo made cage traps of different kinds; for example, Dori, Paran, Gui, Polo, Runga, etc. However, the Namasudras also sometimes operate the Lar Barshi in Sone Beel. The catches by the professional fishers include a rich diversity and quantity of fishes, notably IMCs, exotic carps, *Puntius* spp., *Salmophasia bacaila*, *Danio* spp., *Botia dario*, *Mystus vittatus*, *Sperata* spp., *Wallago attu*, *Eutropiichthys vacha*, *Channa* spp., *Chanda nama*, *Mastacembelus armatus,* and *Macrognathus* spp.

Their catches also include a large number of juveniles. The professional category of fishers often becomes nomadic. They construct temporary thatched fishing huts and lead camp life on high grounds around the beel. They spend days together on boats and rafts while continuing fishing in the beel as well as in the inlets and outlets of the beel.

Thus, in view of the debilitating economic condition of fisherfolk, it is quite understandable that hitherto unknown, virulent, and enigmatic fish diseases, like the EUS, could cause serious financial constraints among fisherfolk, and this could, consequently, devastate the economy of a nation.

Socioeconomics of the Fisherfolk Population of Sone Beel

In India, fishing is generally considered a low-category profession. However, in Sone Beel it is the principal way of earning a livelihood by poor fishers. The entire water spread area of the beel is used for intensive fishing on both domestic and commercial scales by operating different types of fish-catching devices as mentioned earlier. Of these fishing gears, all but Pelain are commercial types. Jhaki jal and Paran are sometimes found in domestic use as well. CPGH (catch/person/gear/ hour: Dey, 1981) of the principal gears operated by the fishers of Sone Beel, and calculated on the basis of 12 months' moving average method (Coxton and Cowden, 1950), indicated that Dori is the most efficacious and is sequentially followed by Jhaki jal > Haran jal > Ghuran jal > Patan jal > Chat jal > Lar barshi > and Maha jal. Use of fine-meshed nets involving catch of juveniles, which is seen to be, in vogue, in Sone Beel, causes concern. Lack of proper training of fishers regarding professional fishing regulations, a tremendous amount of poverty, and a lack of rehabilitation measures taken for them to date seem to be some causes impelling the use of banned nets and devices by poor fishers. Further, in the absence of any large-scale welfare measures being undertaken for them to date, poor fishers are compelled to remain in the grip of un- scrupulous "intermediaries." Consequently, the per capita income of fishers is low and ranges from Rs 1.30 to Rs 8.25 (mean of Rs 3.83 ± 2.14), according to 1980s statistics (Dey and Kar, 1989; Kar, 2003). Their poverty is further aggravated by their early marriage and big family sizes. A low level of life expectancy (c 55 years) and high rate of child mortality (c 10% per year) also reflect their state of distress. Poor adult literacy (10–20%) may be considered the principal factor behind their financial hardship. Juvenile literacy, though higher (c 60–70%), is soon followed by an increased percent- age of dropouts. Consequently, the number of respon- sible earning members is meager, and the grip of poverty lingers on in the community.

Fishers now realize the gross imbalance that exists between poor income from fish trade and the hard labor put-in. In view of this, there is a burgeoning tendency among quite a large number (c 59%) of fishers to motivate their children against taking up the fishing profession. As a result of this, it is found that only a meager percentage (c 20%) of the youth wish to keep faith with their parental avocation of fishing. Others aspire for varied types of profession like banking, agriculture, defense, medicine, engineering, and teaching. As a remedy against such un- desirable trend, a cross section (c 12%) of their society advocates setting up a local "fish farmers' training institute" to help make the fishing profession more attractive to fishing youth.

During the dry season, when Sone Beel shrinks to 409.09 ha, and fishing becomes less intensive, fishers cultivate "boro" paddy (winter–spring paddy) within the beel itself in order to tide over the adverse situation. The area of their cultivable land ranges from 0.07 to 4.2 ha (Mean 0.02 ha ± 0.105). The Kaibartas possess the biggest quantum of land and the Namasudras the least. Approximately 7% of fishers do not have any cultivable land. Nevertheless, during the off-season some associate themselves with livestock and poultry farming, pottery and carpentry, net and boat making, beverages, etc. Some affluent fishers who possess big areas of cultivable land raise the "sali" paddy (monsoon-autumn paddy) in addition to the "boro." In this way, they involve themselves in paddy trade in a big way. In addition to paddy, they also grow vegeta- bles and sun grass on a commercial scale.

Incidentally, religion forms a part of the daily life of fishers living around Sone Beel. Among Hindus, there are both "Vaishnavites" (c 33%) and "Shaktas" (c 30%). Muslims, however, belong to one sect only. Being largely illiterate, fishers have peculiar beliefs about supernatural things.

It is amazing to note here that there is an ancient hermitage of a Muslim hermit called "Sond Peer" or "Hond Peer" (after whom the beel was named "Sone Beel") at the village of Mokam Tila (Anandapur) along the eastern shoreline of Sone Beel. There is also an equally ancient temple of Goddess Kali situated adjacent to the Muslim hermitage. Both Hindus and Muslims pay equal homage and are equally regardful to both religious institutions.

Inference on the Socioeconomics of Fisherfolk

The fishers of Sone Beel have been struggling against poverty for a couple of decades. In view of the nature of economic constraints faced by fishers, it is strongly felt that elimination of unscrupulous intermediaries by organizing credit facilities, marketing of product and purchase of domestic as well as production requirements through co-operatives would go a long way in the emancipation of fisherfolk. Further, encouragement of IMC culture in the confined "bandhs" at the DSL of the beel could lead to a potential IMC fishery. Concomitantly, fast conveyance of the catch from fishing centers to fish landing stations and the urban markets would accelerate fish trade. Provision of cold storage facilities in the landing stations and the urban markets would be an added advantage to prevent loss of flesh due to decay. Moreover, popularization of the recent fish culturing and fishing techniques, and proper rehabilitation of fishers could enable them to give up nomadism and paddy cultivation in the event of hardship and help gear up the upliftment of the anglers in the long run (Dey and Kar, 1989).

The Sone Beel Fishermen's Co-operative Society

The Sone Beel Fishermen's Co-operative Society (SBFCS) is a registered Society managed by fishers living around the beel. The society has a chairman (elected directly by the fishers), a vice chairman, and a few members. The secretary of the society is deputed by the government. The beels in Assam are under administrative control of the deputy commissioners, as they are considered "Khas" lands. In another context, the beels in Assam are also under administrative control of the Assam Fisheries Development Corporation (AFDC) for their "so-called" management. The AFDC floats tender notices, and in response to the tender notice, the SBFCS as well as certain affluent individuals try to obtain the beel on lease from the AFDC. Except in very rare instances, the Sone Beel Fishermen's Co-operative Society generally gets preference over the individuals in obtaining the beel on lease from the AFDC. The lease amount varies from approximately Rs 150,000.00 (in 1978) to approximately Rs 500,000.00 (in 2002). But in spite of this, due to mismanagement during the tenures of certain managing committees of the SBFCS, the society remains a "defaulter" in not being able to clear the revenue share that it is supposed to pay to the government. Nevertheless, during the tenures of some other managing committees, the society made a big profit and cleared the revenue share that it was supposed to pay to the government. The SBFCS raises revenue by leasing out "bandhs" to different groups and individuals. Fishing restrictions are imposed only in "bandhs" at DSL of the beel after the lease becomes effective. At FSL of the beel, any fisher can fish anywhere in the beel. Sometimes a law-and-order problem occurs when fishing in Sone Beel (particularly in the "bandhs" at DSL) becomes a bone of contention between rival groups.

13.9.12.5 In Chatla Haor

Concomitant to Sone Beel, Chatla Haor, in Assam, is a much smaller seasonal floodplain wetland having a water spread area of approximately 1600 ha at FSL. It does not have any DSL and is almost completely dry during the winter. There are only a few insignificant fishing centers (locally called "kheo" or "jheng") on the NW side of the haor where fishing is to be completed and all fishes are to be harvested in autumn. Fishes of this wetland are also widely affected in the event of an EUS outbreak. Incidentally, "sagar" are typical wetlands in Tripura and fishes in the "sagar" are also known to be affected by EUS. On the other hand, a "dil" is a wetland in the hilly province of Mizoram. Fishes in the "dil" have not been much affected by EUS.

13.9.12.6 In the Anuas

Notwithstanding the above, the river-formed oxbow wetlands, locally called "anua," are significant fish habitats harboring a rich diversity of fishes in them. EUS also causes devastating effects on fishes of the anuas.

The "anuas," though not enormous in dimension compared with Sone Beel and Chatla Haor, have high potential for fish production mainly because of their great depth (comparable to that of River Barak), having been part of the course of River Barak once upon a time. Fishers belonging to different communities, notably the Maimals, Patnis, and Meiteis live in the villages around the anuas. They operate mainly the Lar barshi, Patan jal, and Jhaki jal for fishing in the anuas. Socioeconomically, they are poor and most live hand to mouth. However, the anuas are generally leased out by the government to fish traders who are rich and employ the poor fishers as laborers in fishing without giving much benefit to them.

13.9.12.7 Fish Marketing and Transportation

Transportation of fish may be a means of spreading EUS. Diseased fishes may appear in the market and may lead to more spread of the disease.

13.9.13 EUS, Ornamental Fishes, and the Economy of the Nation

Colorful and peculiar fishes are generally called the "ornamental fishes." These fishes, kept in aquarium, are used for decorating the house. Of late, they have occupied a significant place in the commercial trade, particularly in earning foreign exchange. In India, Ichthyospecies like *Botia dario, Botia rostrata, Colisa fasciata, Trichogaster labiosa*, and various species of *Nemacheilus* and *Schistura* are considered colorful ornamental fishes. However, *C. fasciata* has been affected by EUS on a quite large-scale. Further, attempts have been made to study in vitro sexual dimorphism of certain ornamental fish species in NE India.

13.9.14 Farming Systems and Rural Livelihoods

A big portion of the rural population in Asia earns its livelihood through farming systems. As an example, landholdings in Lao PDR appear to be of comparatively similar type, due mainly to large movements of people during the period of war followed by possible intrafamilial distribution of land packages by the people. It is quite unlikely to find farmers with either very large landholdings or complete landlessness. In fact, the quality of the land and the type of crops cultivated make the difference in the landholdings among the inhabitants.

Concomitant to the above, rice happens to be a staple food of the people of Lao PDR. They cultivate both wet rice and hill rice, generally as annual crops. Further, flat or terraced rice paddy is typical in most valleys. Moreover, corn and cassava are also cultivated on the hillsides, in addition to rice as part of their lifestyle of shifting cultivation. As such, the cycle between cultivation and fallowing is steadily decreasing. This possibly leads to soil erosion and soil deterioration. This, consequently, may result in siltation of the water bodies and thus may hamper AQC. As such, stabilization of shifting cultivation may be considered a possible solution, but this may need further confirmation.

In addition to the above, there is usually a limited area of irrigated paddy, which is principally confined to the lowland provinces around the bank of the Mekong River. This allows for production of a second rice crop. Further, some arrangement has been installed in the flood-prone provinces in order to permit dry-season farming as a single annual crop.

13.9.15 Epizootic Ulcerative Syndrome and Society

In Lao PDR, rural families are usually quite big in size, with 6.7−8.3 persons/household on average. The age structure also reflects that the national averages usually portray c 50% of the household members of <15 years of age, while c 10% are >51 years.

13.9.16 Ethnic Status

Lao PDR is the home to >40 ethnic groups with different types of livelihoods and traditions. These groups could be broadly categorized into three types, viz., (1) lowland, (2) middle, and (3) high land inhabitants, according to their tendency to inhabit different parts of valleys, slopes, and mountains of Lao. All three groups are believed to be involved in AQC to some extent. Further, the team LAO/97/007 worked with 17 different ethnic groups in Lao PDR.

13.9.17 Issues Related to Gender

Lowland Lao is matriarchal, matrilineal, and matrifocal with regard to land ownership through inheritance. However, it is not very clear among the other ethnic groups. There is division of labor, with the female folk mainly engaged in the marketing and economic affairs of the family. However, there are differences according to ethnicity. Nevertheless, participation of c 8% of the females in AQC-related program, LAO/97/007, had clearly demonstrated a general interest and involvement of women in AQC activities, and most had ponds close to their homes. On the contrary, it was not possible for the women to take part in fishery-related activities if the ponds were distantly located because, in addition to security concerns, women have to devote time to domestic activities as well (Murray et al., 1998).

13.9.18 Economic Issues

Farmers in Lao PDR are largely self-sufficient, although they are not very affluent. Farmers are mostly rice growers, although c 84% have fishponds with a good resource base. The hill farmers are less likely to be engaged in AQC because of obvious reasons of unsuitability of terrain. Nevertheless, there are examples of ponds made from dammed streams by the hill farmers. These may, however, collapse due to runoff water during the rainy season.

13.9.19 Dietary Matters

During the rainy season, Lao PDR has extensive water resources in the form of rivers, streams, wetlands, paddy fields, etc. The foodstuff grown during the rainy season provides food support to the people during the long dry season when food becomes less abundant. Incidentally, consumption of rice for five provinces has been estimated to be 189−458 kg/capita/year. Farmers in provinces like Sekong, Sayaboury, etc., had small landholdings in contrast to farmers in provinces like Xieng, Khouang, etc., where farmers possessed large landholdings with good quality soils. Incidentally, fish consumption ranged between 10 and 19 kg/capita/year.

13.9.20 Pond AQC

Most of the ponds in Lao PDR are manually constructed by excavation, the damming of small streams, or the conversion of terraced paddy fields. Peculiarly, machine-dug ponds are formed as a side effect of earth cutting for road construction, and such ponds are not very suitable for AQC.

13.9.21 Fish Health Issues Reported

There are said to be few predisposing factors in Lao PDR that might be causing losses in intensive AQC, as the subsistence fish culture in seasonally stocked ponds is done usually with low nutrient inputs and low stocking densities. However, *c* 28% of farmers had encountered fish mortality in their ponds. Nevertheless, some of the health-related issues, as reported from Lao PDR, are briefly discussed below:

13.9.21.1 Epizootic Ulcerative Syndrome

In Lao PDR, a substantive number of farmers reported ulcerated fishes, similar to those with EUS, generally during November and February and also mostly from Xieng Khouang Province.

13.9.21.2 Stress-Related Disease

Excess fertilization of the shallow ponds had been reported to cause red spots on the scales due to stressed condition. The situation is aggravated in the cooler upland areas, where fishes eat less vis-a-vis high feed input by farmers, thus leading to fouling of the water bodies and causing health hazards to fishes.

13.9.21.3 Trematode Infection

Examination of the gills of some sick fishes revealed infestation with metacercariae of trematodes.

13.9.22 Health Management Issues

Lao PDR is a small nation in SE Asia. AQC happens to be a lucrative avocation among the people of Lao PDR. Fishes in its water bodies are often attacked by diseases, notably EUS. As such, the following points may be discussed with regard to health management issues:

1. Specific disease conditions such as bacterial, fungal, or viral infections, or parasites such as helminths or *Lernaea*, are often difficult to treat mainly because of a lack of strong infrastructure and cost intensiveness.
2. Exclusion of wild fish is often recommended mainly because they compete with the stocked fish. This may also reduce the chances of disease inputs.
3. Notwithstanding the above, a positive health management feature of rain-fed ponds in Lao PDR is that they dry completely for some period during the year. This could therefore limit the transmission of the diseases.

13.9.23 Landless Farmers and Their Seasonal Occupations

About 80% of the population in Bangladesh is poor and landless, their livelihood uncertain, and living mostly hand to mouth. As such, they have to be always on the move in the lookout for improved living conditions. They often may migrate to the urban areas for so-called economic opportunities. Some of such villagers may settle in the "Char" areas, which may be found to be uninhabited, and earn their livelihood mainly through fishing. Further, they operate traps in the rivers during the fish-breeding season (February—July) for catching juvenile carps (notably *Catla catla* and *Cirrhinus mrigala*), which are sold usually to the pond-owners on the mainland. In addition, the landless people also fish in the canals and open waters during the rainy season.

13.9.24 Communities Who Are Dependent on Fisheries

It is a common idea that fishery-dependent people dwell in boats, floating houses or on the banks of the rivers and shore of the wetlands. They are dependent exclusively on fishing or fish-related avocations, which may include, among others, fabrication and construction of fishing gears (fish-catching devices) and crafts (mainly boats); growing and processing of fishes; fish trading; etc. However, the people belonging to these communities are usually socio-economically poor and downtrodden.

13.9.25 Services Becoming More Focused

In view of the escalating costs, there have been attempts to adopt a more businesslike approach in some countries, like the outsourcing of works to private laboratories rather than doing the same works in the Govt. Laboratories. The state of Victoria in Australia nearly halved the cost of routine serological tests. Similarly, in New Zealand, some extension services that had earlier been provided by the government free of cost to farmers have of late been privatized and now operate on a tariff basis.

13.9.26 Government Involvement Reduced

In view of global economic crisis, there has been an increasing trend in the reduction of government subsidies in different countries, notably Australia, New Zealand, Zimbabwe, India, and so on.

13.9.27 How Important Is the Small-Scale Livestock and AQC Sector?

As reported, the percentage of the population involved in agriculture ranges from 50% to 90% in the Asian region (FAO, 1986). Approximately 10—20% of the GDP is derived from livestock in these countries. Further, small-scale producers raise 60—90% of the livestock.

13.9.28 Primary Health Care

Primary health care may be defined as that kind of care that could be considered practical, community-based, scientifically sound, socially acceptable, and last, but not the least, appropriate to the needs of the small-scale farmers. In others words, "primary health care" generally, refers to an appropriate approach for small-scale AQC, which reduces the risk of disease outbreaks and concomitant loss of production. Primary health care also stresses that the emphasis should be on "people" and not narrowed down to pathogens and technology only.

13.9.29 Case Study One

There have been case studies related to issues in carp hatcheries and nurseries in Bangladesh, with special reference to health management with emphasis on primary aquatic animal health care in rural, small-scale AQC development.

There have been various studies related to socioeconomics and EUS. A number of studies had been conducted in Bangladesh in this regard. Hasan and Ahmed (2002) performed a case study carried out in 180 hatcheries and nurseries in the NE and SW regions of Bangladesh over a 30-day period during August–September 1999. Study of various aspects of management in small-scale carp hatcheries and nurseries, with particular reference to their health management, was the principal objective of the survey. Three IMCs (Rohu, Catla, and Mrigal) and three exotic carps (common, silver, and grass) were the Ichthyospecies cultured in most hatcheries and nurseries in Bangladesh. In this context, the average production of spawn in the hatcheries was 844 kg/ha, while in the nurseries, it was dependent on the size of the fry. Nevertheless, the average production of fry in nurseries was found to be 1.722, 1.339, and 0.837 million/ha respectively for early fry, fry, and fingerlings. Moreover, the average survival of spawn, fry, and fingerlings in hatcheries and nurseries had been reported to be reasonably high, varying between 74% and 82%. The study had further portrayed that the major source of spawn for the nurseries was hatcheries, while the hatchery brood stock had been reported to be collected mostly from farmers' grow-out ponds. In general, the hatcheries had been reported to be more profitable than the nurseries. Further, profitability of the nursery operations appeared to be fluctuating mainly due to high variability in market price of fry and fingerlings. Furthermore, the hatcheries and nurseries had been reported to provide full-time employment to farmers and labors. It is significant to note here that the average contribution of AQC to the household income of hatchery and nursery owners had been reported to vary between 79.3% (nursery owners) and 95.1% (hatchery owners).

13.9.30 Case Study Two

Funge-Smith and Dubeau (2002) reported about the aquatic animal health management issues in rural AQC development in Lao PDR as a distinct case study. The salient feature of the study is detailed below:

This communication had tried to highlight the role of small-scale AQC in subsistence farming systems in rural Lao PDR. In fact, small AQC is a popular component of subsistence farming systems in Lao PDR. Moreover, rice cultivation happens to be a principal activity during the rainy season. Concomitantly, collection of aquatic products from rice fields is also quite common. Results from a production-cum-consumption survey among subsistence farmers of rural Lao PDR revealed that many (c 84%) were engaged in fish culture. However, consumption of fish and other aquatic products was estimated to be 13–48 kg per capita annually, thereby accounting for c 22–55% of the total consumption of animal products. Further, fish production and livestock farming are the principal sources of income generation, and the value of fish production/household annually was US$ 81.0 on average. In addition, the overall income of the family had ranged between US$ 372 and US$ 594 per household/year.

13.10 CONCLUSION

In conclusion, the potential of fish diseases to cause the incurment of colossal financial losses has been brought to the notice of various governments (ADB/NACA, 1991). The governments in turn had sanctioned a substantial amount of money for managing EUS. In this regard, it could thus be opined that a positive impact of EUS was the hike in fund allocation meant for tackling of EUS fish disease in fishes. This increased fund could often enable many of the workers in the field of EUS research to enhance and upgrade their infrastructural facilities for conducting more in-depth studies with EUS fish disease. The EUS pandemic has demonstrated to national authorities the ability of fish disease to cause major financial losses, and a positive aftereffect of EUS outbreaks has been the increased funding allocated to fish disease research and diagnostic facilities in Asia by governments and international organizations.

Not many surveys have been conducted to accurately assess the impact of EUS on fish populations and associated fisher communities. However, a review of previous information about various aspects of EUS is incomplete to some extent, in view of the advocated fact that HP-based diagnosis of EUS had not been applied in most studies on EUS before 1994. However, information from the Bangladesh Food Action Plan 17 (FAP 17) Project, which possibly spanned from October 1992 to March 1994, portrayed that c 26% of the 34,451 FW fishes examined had lesions of some kind. Further, a recent cross-sectional survey in Bangladesh displayed that c 80% of the 471 fishes with lesions sampled from 84 sites were diagnosed as EUS-positive. This points out that studies on EUS in Bangladesh, which examined fishes with lesions, had not been grossly overestimating the prevalence of EUS disease.

REFERENCES

ADB/NACA, 1991. Fish Health Management System in Asia-Pacific. Report on a Regional Study and Workshop on Fish Disease and Fish Health Management. ADB Agriculture Department Report Series No.1. Network of Aquaculture Centres in Asia-Pacific, Bangkok.

Anon, 1991a. Vietnam Status Report. ADB/NACA Fish Health Management in Asia-Pacific. Report on a Regional Study on Workshop on Fish Disease and Fish Health Management. ADB Agriculture Department Report Series No.1. Network of Aquaculture Centres in Asia-Pacific, Bangkok, Thailand, pp. 273–280.

Anon, 1991b. Philippines status Report. ADB/NACA Fish Health Management in Asia-Pacific. Report on a Regional Study on Workshop on Fish Disease and Fish Health Management. ADB Agriculture Department Report Series No.1. Network of Aquaculture Centres in Asia-Pacific, Bangkok, Thailand, pp. 213–228.

Anon, 1991c. Sri Lanka status Report. ADB/NACA Fish Health Management in Asia-Pacific. Report on a Regional Study on Workshop on Fish Disease and Fish Health Management. ADB Agriculture Department Report Series No.1. Network of Aquaculture Centres in Asia-Pacific, Bangkok, Thailand, pp. 235–248.

Anon, 1991d. Nepal status Report. ADB/NACA Fish Health Management in Asia-Pacific. Report on a Regional Study on Workshop on Fish Disease and Fish Health Management. ADB Agriculture Department Report Series No.1. Network of Aquaculture Centres in Asia-Pacific, Bangkok, Thailand, pp. 89–92.

AAHRI, ACIAR, IoA and NACA, 1997. Epizootic Ulcerative Syndrome (EUS) of Fishes in Pakistan. A Report of the Findings of an ACIAR/DFID-funded Mission to Pakistan. 9–19 March 1997.

Barua, G., 1990. Bangladesh report. In: Regional Research Programme on Relationships between Epizootic Ulcerative Syndrome in Fish and the Environment, 13–26 August 1990. NACA, Bangkok, pp. 2–4.

Bhaumik, U., Pandit, P.K., Chatterjee, J.G., 1990. Impact of epizootic ulcerative syndrome on the society. In: The National Workshop on Ulcerative Disease Syndrome in Fish. 6–7 March 1990. Calcutta, India.

Coxton, F.E., Cowden, D.J., 1950. Applied General Statistics. Sir Isaac Pitman and Sons, London, p. 944.

Dey, S.C., 1981. Studies on the Hydrobiological Conditions of Some Commercially Important Lakes (Beels) of Kamrup District of Assam and Their Bearing on Fish Production, p. 177, 18 Figs., 8 pls., Final Technical Report. North-Eastern Council, Govt. of India, Shillong.

Dey, S.C., Kar, D., 1989. Fishermen of Lake Sone in Assam: their socio-economic status. Sci. Cult. 55, 395–398.

FAO, 1962. Rapport au Government de la Re' publique du Niger sur la situation et' evolution de la pêche au Niger: Base' sur le travail de M. Jaques M.A. Daget. Rapp. FAO/PEAT 1525, 27.

FAO, 1986. Report of the Expert Consultation on Ulcerative Fish Diseases in the Asia-Pacific Region (TCP/RAS/4508). Bangkok, 5–9 August 1986. FAO Regional Office for Asia and the Pacific, Bangkok.

Funge-Smith, S., Dubeau, P., 2002. Aquatic animal health management issues in rural AQC development in lao PDR, pp. 41–54. In: Arthur, J.R., Phillips, M.J., Subasinghe, R.P., Reantaso, M.B., MacRae, I.H. (Eds.), Primary Aquatic Animal Health Care in Rural, Small-scale, AQC Development. FAO Fisheries Tech, p. 54. Pap No. 406.

Hasan, M.R., Ahmed, G.U., 2002. Issues in carp hatcheries and nurseries in Bangladesh, with special reference to health management, pp. 147–164. In: Arthur, J.R., Phillips, M.J., Subasinghe, R.P., Reantaso, M.B., MacRae, I.H. (Eds.), Primary Aquatic Animal Health Care in Rural, Small-Scale Aquaculture Development. FAO Fish. Tech. Pap. No. 406.

ICSF, 1992. Enigma of EUS. Consultation on Epizootic Ulcerative Syndrome Vis-a-Vis the Environment and the People. 25–26 May 1992. Trivandrum, Kerala. International Collective in Support of Fishworkers, Madras, India, 40pp.

Iqbal, M.M., Tajima, K., Ezura, Y., 1998. Phenotypic identification of motile *Aeromonas* isolated from fishes with epizootic ulcerative syndrome in Southeast Asian countries. Bull. Fac. Fish. Hokkaido Univ. 49, 131–141.

Kar, D., 1990. Limnology and fisheries of Lake sone in the Cachar district of Assam (India). Matsya 15–16, 209–213.

Kar, D., 2000. Socio-economic development of the fisherwomen through aquaculture with emphasis on integrated farming in the villages around Chatla Haor wetland in Silchar, Assam. In: Seminar Presented on DBT-Sponsored Awarness Workshop on Biotechnology-Based Programmes for Women and Rural Development, NEHU, Shillong, 1, pp. 13–14.

Kar, D., 2003a. An Account of the Fish Biodiversity in South Assam, Mizoram and Tripura along with a Brief Account of Epizootic Ulcerative Fish Disease Syndrome in Freshwater Fishes. UGC-Sponsored Invited Lecture in Dept. of Environmental Engg., Guru Jambeswar University, Hissar, Haryana.

Kar, D., 2003b. Peoples' perspective on fish conservation in the water bodies of South Assam, Mizoram and Tripura, pp. 325–328. In: Mahanta, P.C., Tyagi, L.K. (Eds.), Participatory Approach for Fish Biodiversity Conservation in North-East India. National Bureau of Fish Genetic Resources (ICAR), Lucknow. pp. v + 412.

Kar, D., 2007. Fundamentals of Limnology and Aquaculture Biotechnology. pp. xiv + 609. Daya Publishing House, New Delhi.

Kar, D., 2013. Wetlands and Lakes of the World. pp. xxx + 687. Springer, London. Print ISBN: 978-81-322-1022-1; e-Book ISBN: 978-81-322-1923-8.

Kar, D., Dey, S.C., 1988a. A critical account of the recent fish disease in the Barak valley of Assam. In: Proc. Regional Symp. on Recent Outbreak of Fish Diseases in North-East India, vol. 1, p. 8.

Kar, D., Dey, S.C., 1988b. Impact of recent fish epidemics on the fishing communities of Cachar district of Assam. In: Proc. Regional Symp. on Recent Outbreak of Fish Diseases in North-East India, vol. 1, p. 8.

Kar, D., Dey, S.C., 1990a. Fish disease syndrome: a preliminary study from Assam. Bangladesh J. Zool. 18, 115–118.

Kar, D., Dey, S.C., 1990b. A preliminary study of diseased fishes from Cachar district of Assam. Matsya 15–16, 155–161.

Kar, D., Dey, S.C., 1990c. Epizootic ulcerative syndrome in fishes of Assam. J. Assam. Sci. Soc. 32 (2), 29–31.

Kar, D., Dey, S.C., Kar, S., Bhattacharjee, N., Roy, A., 1993. Virus-like particles in epizootic ulcerative syndrome of fish. In: Proc. International Symp. on Virus-Cell Interaction: Cellular and Molecular Responses, vol. 1, p. 34.

Kar, D., Roy, A., Dey, S.C., Menon, A.G.K., Kar, S., 1995a. Epizootic ulcerative syndrome in fishes of India. World Congr. In Vitro Biol. In Vitro 31 (3), 7.

Kar, D., Kar, S., Roy, A., Dey, S.C., 1995b. Viral disease syndrome in fishes of North-East India. In: Proc. International Symp. Of International Centre for Genetic Engg. and Biotechnology (ICGEB) and the University of California at Irvine, vol. 1, p. 14.

Lilley, J.H., Phillips, M.J., Tonguthai, K., 1992. A Review of Epizootic Ulcerative Syndrome (EUS) in Asia. Aquatic Animal Health Research Institute and Network of Aquaculture Centres in Asia-Pacific, Bangkok, p. 73.

Macintosh, D.J., 1986. Environmental background to the ulcerative disease conditionin south-east Asia. In: Roberts, R.J., Macintosh, D.J., Tonguthai, K., Boonyaratpalin, S., Tayaputch, N., Phillips, M.J., Millar, S.D. (Eds.), Field and Laboratory Investigations into Ulcerative Fish Diseases in the Asia-Pacific Region. Technical Report of FAO Project TCP/RAS/4508. Bangkok, pp. 61–113.

Mazid, M.A., 1995. Bangladesh, pp. 61–82. In: Report on a Regional Study and Workshop on the Environmental Assessment and Management of Aquaculture Development (TCA/RAS/2253). Food and Agriculture Organization of the United Nations/Network of Aquaculture Centres in Asia-Pacific, Bangkok, Thailand.

Murray, U., Sayasane, K., Funge-Smith, S., 1998. In: Jordans, E. (Ed.), Gender and AQC in Lao PDR: A Synthesis of a Socio-Economic and Gender Analysis of the UNDP/FAO AQC Development Project, LAO/97/007. FAO, Rome, p. 40.

Prasad, P.S., Sinha, J.P., 1990. Status paper on the occurrence of ulcerative disease syndrome in fishes of Bihar. In: The National Workshop on Ulcerative Disease Syndrome in Fish, 6–7 March 1990. Calcutta, India.

Priyadi, A., 1990. Indonesia report, pp. 15–17. In: Regional Research Programme on Relationships between Epizootic Ulcerative Syndrome in Fish and the Environment, 13–26 August 1990. NACA, Bangkok.

Tonguthai, K., 1985. A Preliminary Account of Ulcerative Fish Diseases in the Indo-Pacific Region (A Comprehensive Study Based on Thai Experiences). National Inland Fisheries Institute, Bangkok, Thailand, p. 39.

Welcomme, R.L., 1979. Fisheries Ecology of the Floodplain Rivers. Longmans, London, pp. viii + 317.

Chapter 14

Conclusions and Recommendations

14.1 INTRODUCTION

Epizootic ulcerative syndrome (EUS) has been sweeping through the Asia—Pacific region since 1972 (Roberts et al., 1986). Red spot disease (RSD) among the mullets of estuaries and freshwater in Australia has been recorded since 1972 as well (Rodgers and Burke, 1981). Indonesia suffered from large-scale fish kill during 1980—1981 (Anonymous, 1981). Outbreaks of similar nature occurred among the gudgeons and the mullets in Papua New Guinea during 1981—1982, and an outbreak similar to but less virulent than RSD in Australia had been reported from the southern rivers of Papua New Guinea during 1972—1974 (Coates et al., 1989). Ulcerative disease among FW fishes in Malaysia had been recorded during 1980. A major fish epizootic had been occurring in Thailand since 1981 and spreading to Lao PDR and Myanmar (Tonguthai, 1985). Subsequently, EUS with hemorrhagic ulcerations had been occurring among FW fishes of Bangladesh since March—April 1988 (Ahmed and Rab, 1995). From Bangladesh, EUS is believed to have entered neighboring India during July—August 1988 (Kar and Dey, 1988, 1990a,b,c; Kar, 2007, 2010, 2013; Kumar et al., 1991).

Concomitant to the above, the Indian subcontinent has been the region hit most hard by EUS. Diseases in freshwater fishes are a great threat to achieve optimum production and become a limiting factor to economic success in aquaculture. Open-water capture fish had been suffering from various types of diseases such as EUS, septicemia, tail and fin rot disease, gill rot disease, viral disease, bacterial disease, and fungal disease (Chowdhury, 1993). Fungi, which cause fungal diseases, are present in salt or freshwater. In most cases, fungi serve a valuable ecological function by processing dead organic debris. However, fungi could become a problem, if fish are stressed by poor nutrition, pressure of population and overexploitation. Fungal infections of freshwater fish are common and distributed worldwide. Fungal diseases are easily recognized by relatively superficial, colony of fluffy growth on the skin and gill of fishes. Miyazaki and Egusa (1973) observed an invasive component in histological section of ulcerative disease affected as MG. Similarly, McKenzie and Hall (1976) are said to have identified mycotic granulomas in histological sections of the early outbreak of EUS in Australasia and later in SE Asia. *Achlya, Saprolegnia*, and *Aphanomyces* were commonly identified from the lesion

surface of affected fish (Roberts et al., 1993). In addition to these, bacterial studies (Kar et al., 1999; Kar, 2007) and virological studies (Kar et al., 1993, 1995a,b; Kar, 2007, 2013) had revealed the occurrence of different types of bacteria and virus(es) in the EUS-affected fishes. Incidentally, pathogenic attacks are associated with immune suppression.

Notwithstanding the above, fisheries have been a lucrative sector in the world today, in view of burgeoning population pressure with increasing demands for finfish and shellfish. In fact, pisciculture in different countries seems to be progressing toward semi-intensive AQC. Conversely, shrimp culture has been moving toward an improved traditional system. In fact, uncalculated, unresisted, and indiscriminate use of fertilizers and feeds had been causing severe stress on fishes due to changes in the water quality parameters in the aquatic domain, thereby rendering fishes susceptible to pathogens in foul water. At the same time, not many rich databases seem to be available on the disease and health problems of finfish and shellfish in the region, although mortality and economic losses are the inevitable consequences of disease problems. Mass mortality of carp juveniles due to protozoan and metazoan parasites is often reported. Nevertheless, EUS has been a virulent, enigmatic, and *hitherto* unknown disease among fishes, and has been spreading through the FW fish population semiglobally since around 1972. Concomitantly, white spot disease, believed to be caused by white spot syndrome virus and reported as a systematic epidermal and mesodermal baculovirus, is believed to have caused the loss of c 44.4% of shrimp production alone during the mid-1990s, in Asia generally and the Indian subcontinent in particular. It is believed that the shrimp culture industry has suffered a setback since then. About 350,000 people associated with shrimp culture have been thrown out of employment. Consequently, it had appeared that diseases among the finfishes and shellfishes had severe economic impacts on the lives of people in low-income groups.

EUS was a new occurrence at the time of its initiation during the late 1980s in this region, and is still a *hitherto* unknown, enigmatic, and virulent fish disease on the Indian subcontinent. As such, fish consumers, particularly at the initial period of outbreak, had been apprehending a health disaster in the event of consumption of EUS-affected fish, a fear that later proved to be unfounded (Kar, 2007, 2013;

Epizootic Ulcerative Fish Disease Syndrome. http://dx.doi.org/10.1016/B978-0-12-802504-8.00014-6

Kar and Dey, 1988, 1990a,b,c; Kar et al., 1990, 1993, 1995a,b). Nevertheless, there was an alarming decline in fish prices and sales during the initial period of the EUS outbreak. This gave a severe blow to the economy of fish farmers and fishers at large. Moreover, this situation might have had also quite detrimental effects on the nutritional status of fish consumers because fish serves as an item of staple food and a potential source of animal protein in their diets. As reported, c 250 million families in SE Asia cultivate paddy (rice) as their principal crop, which is often, also, accompanied with fish culture; that is, paddy-cum-fish culture. Incidentally, much of the fish harvested from this paddy-cum-fish culture go into the diets of growers as potential nutritional supplements (Macintosh, 1986). As an example, the economic loss due to the ongoing EUS outbreak was estimated as 118.3 million taka in Bangladesh (US$ 4.3 million; 1 US$ = 35 taka at that time) during 1988—1989. The disease occurred with less severity and magnitude during 1989—1990, when the economic loss was estimated at 88.2 million taka (US$ 2.2 million). Incidentally, the price of fishes dropped to 25—40% of the predisease level during the first outbreak (Barua, 1990).

14.2 DISEASE AND ITS ETIOLOGY

"Etiology"—that is, what actually causes a disease—may also be considered a kind of challenge. In fact, the late nineteenth century saw the development of a deterministic view of causality—that is, agent "X" causing disease "Y." There was an implication of specificity for both cause and effect. Concomitantly, the formulation of Henle-Koch's postulates strengthened this view and rendered assistance in establishing the link between microbes and disease with the following salient features:

1. The causative agent should be present in every case of the disease when isolated in pure culture.
2. The same agent should not be present in other diseases.
3. The agent should be able to induce the disease in experiments, once it is isolated.
4. The agent should be recovered from the experimentally induced disease.

Concomitant to the above, there was advancement in the scientific understanding of disease processes throughout the twentieth century. Consequently, Henle-Koch's postulates were believed to be too restrictive when thinking about causality for a number of diseases in humans, for a number of reasons, some of which are briefly stated below:

1. Multiple etiologic factors, for example, the combination of many factors resulting in heart disease
2. Multiple effects of single factor, for example, the relationship of smoking to both cancer and heart disease
3. Carrier states, for example, *hepatitis* B

4. Quantitative causal factors, for example, the amount and period of smoking
5. Nonagent factors, for example, age and sex

The above-mentioned account paved the way for a broader definition of "cause" in order to embrace this broadened understanding. This could also facilitate epidemiological studies that might uncover less direct interventions to mitigate disease. Such a view widened the concept of cause. A "cause" now may include an event, condition, or characteristic that plays an essential role in producing an occurrence of the disease in question.

14.2.1 Etiology of EUS

The pathogenic *Aphanomyces* has been reported under various names such as *Aphanomyces piscicida* (Hatai, 1980), *Aphanomyces invaderis* (Willoughby et al., 1995), and EUS-related *Aphanomyces* sp. (Lumanlan-Mayo et al., 1997). The *Index of Fungi* (1997) renamed *A. invaderis* to *A. invadans*, and in 1998 indicated that *A. invadans* is the only valid name.

14.2.2 An Alternative View that EUS is a Polymicrobial Infection

The concept of polymicrobial disease is well accepted in human and veterinary medical sciences. However, EUS in fishes could also be typified under this category based on available evidence and suggestions. In particular, EUS could be regarded as a polymicrobial disease, supposed to be precipitated by environmental stresses. The syndrome, according to some, is the result of a combination of an early, primarily viral, etiology followed by a subsequent secondary mycotic invasion that is terminated by secondary or tertiary bacterial involvement. On the other hand, a cross section of other workers (Kar, 2007, 2013; Kar et al., 2003, 2007) considered EUS to be an outcome of complex interaction between viruses as the primary etiology. These were in turn associated with secondary bacterial and/or fungal and/or parasitic infections, and under the overall influence of environmental stresses in which temperature and total alkalinity of water appear to act as "stress factors."

Notwithstanding the above, according to some authors (Brogden and Guthmiller, 2002), polymicrobial diseases may be considered the outcome of clinical and pathological manifestations being induced by a number of microbes. They are often not easily identifiable and treatable. Further, polymicrobial infections may be the outcome of viral, bacterial, fungal, parasitic, etc., etiology coupled with disease-induced immunosuppression. Some significant underlying mechanisms in pathogenesis are as follows:

1. The host may be predisposed to polymicrobial diseases due to physical, physiological or metabolic disorders.

2. Colonization may occur by opportunistic biota due to changes on the body surface of another organism.
3. Other microbes may colonize a host, being made possible by pro-inflammatory cytokines from different microbes, which could not only trigger the severity of the disease, but also reactivate latent infections.
4. Damage to tissues may occur due to sharing of determinants among organisms.
5. Opportunistic pathogens may be able to colonize the host body due to alterations in the immune system of the host by other microbes.

14.3 DEFINITION OF EPIZOOTIC ULCERATIVE SYNDROME

Notwithstanding the above, designation of EUS as a polymicrobial disease might get a setback mainly due to works and views to the contrary. That is, there are studies, which emphasized that EUS is caused by single putative pathogens; for example, fungi (Roberts et al., 1993, Willoughby and Roberts, 1994) and virus(es) (Kanchanakhan et al., 1999; Lio-Po et al., 2000; Kar, 2007). In fact, exposed ulcers may always be colonized by opportunistic pathogens, like bacteria, etc. Further, different types of tissue damage may be due to liquefaction of tissue by highly proteolytic bacteria (Karunasagar and Karunasagar, 1994; Karunasagar et al., 1995).

Concomitant to the above, **case definition** is an important aspect of a disease etiology. Nevertheless, it may not require identification of the primary etiology of the disease as a prerequisite. The primary etiology of EUS is yet to be confirmed accurately. But Koch's postulates need to be satisfied under all circumstances. As such, there have been attempts to find an alternative view and case definition of EUS.

14.3.1 Framing of a Case Definition

A case definition can be pinpointed with no confusion. As such, the main problems in framing and establishing a case definition lie in ascertaining (1) when an organism is actually going to be considered to have disease in question, and (2) what actually "causes" that disease. Nevertheless, a "disease" in a very wide sense could be defined as "any condition that impacts an animal and concomitantly may be deleterious to animal or human health." Furthermore, with regard to infectious ailments, there could be an additional issue of ascertaining, at what stages in the entire process from initial infection to eventual outcome (recovery, disability, or death), the animal could be designated as "diseased." For instance, could we label a fish to be "EUS-affected" only when the gross lesions are visible to the naked eye. Or a fish could be designated as "EUS-affected" the moment pathological changes could be detected in HP sections subjected to an immunoassay.

14.3.2 A Case Definition to Be Established

A case definition is an agreed set of rules that allows investigators to uniformly conclude that a particular individual has or does not have a specific disease as defined. The "as defined" part is important here. Moreover, it may be convenient to frame a set of rules that could be able to distinguish both doubtful and confirmed cases. In light of the foregoing account, a "case definition" could thus be considered a set of standard criteria for deciding, whether an individual of interest has a particular disease. The test unit could be an individual of a group of animals such as a pond fish, etc. However, a useful case definition for aquatic animal diseases had been mentioned. In fact, an optimal case definition depends on criteria that could be applied to any potential case in the population that acts as the source. Few cases could display a complete range of disease criteria. However, there could always be some noncases in which some criteria—for example, clinical signs—may be similar to those of the specific disease being studied.

Notwithstanding the above, a definition based on field observations like history, clinical signs, gross pathology, etc. could be useful for a "suspect" case, whereas a "confirmed case" could preferably be based on laboratory findings, in addition to field observations that are equally important components for confirmation of a result. In addition, when a hitherto unknown vis-à-vis potentially virulent syndrome is being studied, it is usually suggested to use a very broad definition initially (having higher sensitivity but lower specificity) in order to minimize the risk of missing any cases.

In view of the above, a **tentative case definition for EUS** could be "a fish with necrotizing, granulomatous dermatitis and/or myositis and/or granulomas in internal organs with *A. invadans* (=*A. piscicida*) found within the lesion. On the other hand, an approximate case definition of RSD could be "a fish with one or more surface lesions, each of which could be described as a 'red spot.'"

14.3.3 Alternative Case Definitions for EUS

ODA (1994) defined EUS as "a seasonal epizootic condition of freshwater and estuarine warm water fish of complex infectious etiology characterized by the presence of invasive *Aphanomyces* infection and necrotizing ulcerative lesions typically leading to a granulomatous response." In view of the expert discussions and based on the large body of scientific information, two groups of case definitions, applicable in different situations, are proposed below. It may be noted that the lists are not exclusive, and that each proposed case definition has its own sensitivity and specificity.

14.3.3.1 Case Definitions for Screening Programs, Surveys, etc.

A fish with focal to locally extensive cutaneous ulceration
A fish with focal to locally extensive cutaneous erythema or ulceration

14.3.3.2 Case Definitions for Definitive Diagnosis

The principal criteria are:

1. A fish with necrotizing granulomatous dermatitis and myositis associated with hyphae of *A. invadans*
2. A fish with necrotizing, granulomatous dermatitis and/or myositis and/or granulomas in internal organs with *A. invadans* (=*A. piscicida*) found within the lesion
3. A fish with necrotizing, granulomatous dermatitis, myositis, and/or granulomatous response in internal organs, associated with the presence of *A. invadans* (=*A. piscicida*) hyphae
4. A seasonal epizootic affecting fresh and brackish-water fish species involving a specific fungal pathogen *A. invadans* (=*A. piscicida*) characterized by necrotizing surface ulcerative lesions and typical mycotic granulomatous response

An epizootic fungal infection, where formation of ulcer is secondary, aseptic fungus and granulomas are always observed in the lesion, and the pathogen is a fungus of the genus *Aphanomyces*; confirmatory diagnosis is by polymerase chain reaction (PCR) test.

There have been common names for the disease. Two new common names have been proposed. These are:

(i) Epizootic granulomatous aphanomycosis (EGA) or ulcerative aphanomycosis (UA). There could be synonymy in *Aphanomyces* sp. causing the disease. In view of the above, and since synonymy is a matter of opinion without formal nomenclatural requirements, we propose that in other than purely taxonomic contexts, the *Aphanomyces* causing EUS, MG, RSD, and UM should be initially referred to as *A. invadans* (=*A. piscicida*).

14.3.4 EUS Defined Further (As It Stands)

EUS was defined at a DFID Regional Seminar in Bangkok in 1994 as "a seasonal epizootic condition of freshwater and estuarine warm water fish of complex infectious etiology characterized by the presence of invasive *Aphanomyces* infection and necrotizing ulcerative lesions typically leading to a granulomatous response" (Roberts et al., 1994). However, research since that time suggest that a complex etiology is not necessarily involved in all cases. Reference to a specific fungal pathogen (*A. invadans*) could also now be included in the case definition for EUS. With these developments, EUS could be considered as

characterized beyond the level of a syndrome. However, the name "epizootic ulcerative syndrome" is well known among fish health workers and also among workers who work on EUS. Further, EUS is an OIE-listed disease, and notification to the World Animal Health Organisation (or OIE) is required in the event of an outbreak.

14.3.5 A Direct Question: What Is Epizootic Ulcerative Syndrome?

EUS, or "red-spot" as it is known colloquially, is an ulcerative syndrome of fish that affects a range of native species in the Northern Territory. The disease also occurs in New South Wales, Queensland, and Western Australia, as well as in many Asian countries. It begins as a small area of reddening over a single scale, which subsequently spreads to involve a number of adjacent scales; this is the characteristic "red spot." As the condition progresses, the "red-spot" expands and deepens, giving a deep ulcer, which sometimes extends into the abdominal cavity. Some fish, especially barramundi (*Lates calcarifer*) may develop unilateral or bilateral cloudiness of the cornea; these changes in the eye may or may not be accompanied by lesions in the skin. Some cases of EUS heal spontaneously, but many affected fish, especially juveniles, die.

Further, EUS is an infection caused by oomycete fungi known as *A. invadans* or *A. piscicida*. *Aphanomyces* is a member of a group of organisms formerly commonly known as water molds; they are currently recognized as belonging to the group of heterokonts or stramenopiles (OIE, 2006). EUS is an epizootic condition affecting wild and farmed freshwater and estuarine finfish since it was first reported in 1971. EUS is also known, as discussed in Chapter 1 of this book, as RSD, MG, and UM, and in 2005 it was suggested that EUS be renamed as epizootic granulomatous aphanomycosis (EGA) (Baldock et al., 2005). A single row of primary zoospores formed within a zoosporangium is then released through the sporangium to encyst at the apical tip to form achlyoid clusters. The main free-swimming stage of *Aphanomyces* spp. is the secondary zoospore that is discharged from the encysted primary zoospores.

14.3.6 Definition of Suspect Case of EUS

A suspect case of EUS disease is defined as the presence of typical clinical signs, a single or multiple red spots or ulcers on the body, in a population of susceptible fish at water temperatures between 18 and 25 °C.

14.3.7 Definition of Confirmed Case

A confirmed case of EUS is defined as a suspect case that has produced typical mycotic granulomas in affected tissues or organs OR that has been identified as positive by the PCR or fluorescence in situ hybridization detection techniques OR that *A. invadans* has been isolated and confirmed by bioassay, PCR, or sequence analysis.

14.4 CONCLUSION

It may be appreciated that the small-scale aquaculturists are marginally oppressed, economically poor, and socially downtrodden people and need to be cared for. Disease among fishes gives a serious blow to them, as they have to face the brunt of the impact. Hence, the following recommendations are put forward considering the above points:

1. Level of awareness about the disease problems is to be raised among small and marginal fish farmers.
2. Any news about mass mortality of fishes is to reach the nearby fish health center at the earliest.
3. Fish health center/farm managers are to be vigilant, and they may help small aquafarmers in mitigating the disease problems.
4. A quick line of communication may be established in the form of a network among small-scale farmers, big farmers, departments of fisheries, and fisheries research institutes in order to manage disease problems effectively.

There is the possibility of more disease occurrences rather than less in AQC systems as they expand. However, the underlying endemic diseases may be managed at the farm level. But rapidly sweeping infectious diseases are to be managed at different levels. The producers, the consumers, the stakeholders, and nearly the whole nation could be affected by the impact of virulent diseases such as the hitherto unknown, enigmatic, and virulent EUS.

Concomitant to the above, solutions are supposed to lie at all levels of production, marketing, and policy framing. Treatment with permitted drugs, proper hygienic measures, adequate quarantine steps, appropriate policies, etc., could go a long way in proper healthy management of the farms.

It may be a common notion that there could be a number of similarities between a small holder of livestock and a big AQC system. In both systems, similar outcomes may result after the application of the same principle of health management. Thus, it may appear appropriate that the lessons learnt from the livestock sector could be applied to AQC sector and vice versa. However, improvement in the situation of disease prevention and control may happen only when there is a clear understanding of the etiology of the various disease syndromes and their epidemiology. In this regard, it could be stated that effective measures could be adopted only when the main reasons could be found out through various kinds of studies including epidemiological surveys. Programs, adopted at the village level, may gradually be extended to national level. These

disease-control strategies could form strong database for present and future works on fish health management. However, such databases, for those countries, should invariably be based on competent diagnosis, well-designed disease reporting systems and specific field surveys. Indeed, collection of high-quality representative basic data from the farms and villages forms a prerequisite need for building a sound database.

In addition to the above, EUS has caused significant losses to both wild and farmed FW fishes. EUS has been expanding its ambit of spread beyond the Indian subcontinent to other parts of the world, at the same time recurring where it had appeared earlier. The devastating social and economic impacts coupled with difficulties in diagnosing and controlling the epidemic had been emphasizing the urgent need for developing and implementing improved infrastructure and skilled workers to combat this hitherto unknown, virulent, and enigmatic EUS fish disease. Concomitantly, the lessons learnt from dealing with EUS could be widely applied to other kinds of fish disease problems, which could be threatening the development of aquaculture in the region. Many of the issues had been deliberated upon at different National and International Fora, notably the ADB-NACA collaborative study *Fish Diseases and Fish Health Management in the Asia—Pacific Region* (ADB/NACA, 1991). The recommendations are expected to go a long way in formulating tangible actions required at the regional, national, and international levels, including capacity building through skill development. These could create the work force to tackle the present difficulties and future problems related to aquatic diseases.

Concomitantly, South Africa has many well-managed conservation areas. Many lie in the drainage pathways of major river catchments. With increasing urbanization and industrialization, many of the major catchments in South Africa have been degraded through anthropogenic activity and pollution (Ashton, 2007). Aquatic diseases such as EUS have the potential to spread along such river courses, and conservation authorities must be aware of the risk of such diseases entering national parks. The subsequent outbreaks in the Okavango Delta and in the Western Cape Province highlighted the urgent need to implement effective surveillance. In countries further north in Africa, fish from local fisheries form a major source of protein for the population, with fishing as an important income generator. As natural stocks become increasingly overharvested, the expansion of EUS into Africa threatens potential development of the fish farms.

Notwithstanding the above, it appeared that parasites, pathogens, etc., are often associated with the outbreak of EUS (Tonguthai, 1985). In fact, a large number of *Epistylis* sp., had caused red spots in the snakeheads. Further, apart from *Epistylis* sp., various other parasitic forms had been associated with the EUS-affected fishes, but none could be established as the primary causative factor. Nevertheless, the attachment of *Epistylis* sp. and *Trichodina* sp. associated with damage to the epidermis may have been responsible for pathogenesis in some specimens. Several forms of bacteria and virus have been isolated from EUS-affected fishes, but none have as yet been confirmed as the primary etiology with a cut-to-the-bottom guarantee—although the author of this present treatise opines that EUS is primarily a viral and secondarily a bacterial/fungal/parasitic disease (Kar and Dey, 1990a,b,c; Kar, 2007, 2013). It was reported by workers from Pakistan that the water-quality parameters did not provide any significant information suggesting a role in the outbreak of EUS, although they further opined that a role in EUS could not be totally ruled out. Thus in a general sense, EUS had caused a significant threat to the newly developing aquaculture industry in many countries including India, Bangladesh, Pakistan, Thailand, and so on.

Some important concluding remarks are briefly given below:

1. Randomization and pathology-based case diagnosis may be more suitable for studies related to EUS disease prevalence and its monitoring.
2. There had been a somewhat decreasing trend of prevalence and severity, but there has been continued widespread occurrence of EUS in the Indian subcontinent and its environs.
3. EUS has been continuing to affect new areas.
4. On the Asian continent in general, and the Indian subcontinent in particular, EUS sometimes has more significant impact on extensive low-input systems and wild fisheries than on controlled intensive fish culture.
5. Rural traders and consumers seemed to be more affected by EUS than the urban/suburban communities in South and SE Asia were.
6. Prevention of EUS disease by managing the potential "risk factors" seems to be cheaper and more effective than its treatment.
7. EUS seems to have some effects on the diversity of species in certain areas, but its long-term effects are unknown.

The major issues from the above account are briefly concluded below:

1. Both hatcheries and nurseries appear to be profitable enterprises despite sometimes incurring losses due to disease.
2. Both hatcheries and nurseries are potential institutions for employment generation.
3. Hatchery and nursery operators are usually better aware of their problems and limitations than their counterparts in other kinds of avocations are. As such, the former generally have better and more access to government-provided extension services.

4. Operation of nurseries appears to be a little uncertain (compared with hatcheries), in view of the fluctuating market prices of their products.
5. In view of less scientific background, the hatchery owners/operators often use the same brood stock for raising spawn. This could lead to deterioration of larval quality.
6. Inbreeding is a common practice in many hatcheries. However, hatchery owners do not generally exchange brood stock among themselves in order to maintain genetic diversity.
7. The use of chemicals and drugs indiscriminately may lead to environmental hazards in the long run. Moreover, indiscriminate use of antibiotics may lead to antibiotic resistance.

In addition to the above, investigations so far conducted by certain schools on EUS-infected fishes have shown that the spread of the disease is due to *A. invadans*, a fungus that is capable of producing a proteolytic enzyme that helps it to penetrate the fish tissue to cause the shallow to deep ulcers. However, the fungus itself was not able to initiate the infection unless the integrity of the epidermis is hampered due to either biological or abiological agents. The report of a pathogenicity study conducted on Atlantic menhaden had indicated that the fungus could cause the infection in experimental fish, which is not subjected to any stress conditions. However, the EM observations revealed that the germinating spores could enter the epidermal layer of the fish, which has lost its integrity.

Cohabitation studies may usually be done to ascertain the communicable nature of a disease, notably the EUS in fishes. The striped snakehead fish (*Channa striata*) was found to transmit the EUS to healthy snakehead fish in two separate investigations, where virus isolates were recovered in both cases. The ranavirus isolated from ulcerated largemouth bass also produced skin and muscle ulcerations in a pathogenicity study conducted in the same fish without the presence of an invading fungus. Continued investigations may lead to recovery of more viral agents from EUS-infected fishes and throw more light on the involvement of viruses in the initiation of the disease. Such initiation is believed to involve either lowering the immune status of fish and consequently hampering the integrity of the integument, or inducing dermal ulcerations independently. The presence of persistent viral infections in the cell lines developed from EUS-susceptible tropical fishes also points to the possible role of viruses in the induction of EUS. Although retrovirus isolated from SSN1 cells differs from human and other vertebrate retroviruses, what role such viral persistence could play in the development of EUS is not clear. The incidence of EUS has been on the decline in the last few years, and the role of viruses has not been keenly pursued. Further investigations are therefore necessary with a panel of additional susceptible fish cell lines for isolating viruses, if any, from the infected fishes to delineate the role of viruses in EUS.

14.5 RECOMMENDATIONS

There had been a number of recommendations from different corners (Anon, 1997) related to EUS fish disease research. A gist of some of the principal recommendations is given below:

1. EUS remains a hitherto unknown, virulent, and enigmatic disease among freshwater fishes semiglobally.
2. Continuous comprehensive research works are to be continued for determining the accurate etiology of EUS, as well as to develop exact and effective prophylactic and therapeutic measures.
3. As an ongoing process, detailed studies are to be conducted on:
   a. the pathogens involved;
   b. fish populations at affected and unaffected sites;
   c. HP studies of affected fishes;
   d. preventive and curative measures;
   e. climatic variations at the affected sites; and
   f. prioritized environmental parameters.
4. Pollution in water bodies is to be prevented. That pollution is a threat to the environment has been appreciated in almost all corners. Further, there have been suggestions from different quarters to conduct detailed study of the effect of pollution on fisheries, aquaculture, and human health.
5. There should be a mechanism to monitor the epidemiological spread of EUS and to establish a mechanism to disseminate information about EUS. There should be a central coordinating office to render this service, including providing an early warning system about EUS outbreaks. International organizations such as the FAO, NACA, and OIE could kindly think of such measures.
6. Regional research programs studying EUS must be continued.
7. There is possibly a need to develop adequate infrastructure and a skilled work force to effectively combat the problems arising out of EUS outbreaks. This could help not only in tackling the EUS problem, but also in managing any possible fish disease problem.

The broad recommendations outlined above have been deliberated upon widely in a number of national and international seminars, symposia, workshops, etc., including the various workshops organized by FAO, NACA, and other international organizations, involving various countries and organizations around the globe. These could be considered a distilled product of the deliberations.

14.5.1 Suggestions and Recommendations for Future Work

In view of the importance and continuing extension of the area of outbreak of this serious enigmatic EUS disease, it has been recommended in several meetings that the etiology of EUS should be a high priority for future works, particularly the confirmation of human-induced impacts, if any, pertaining to the root cause of origin of this hitherto unknown, virulent, and enigmatic fish epidemic.

14.5.2 Recommendations for Treatment

Lime is commonly in use to control EUS. However, there is a need to understand the currently inadequate control methods in order to improve upon them.

14.5.3 Recommendations for Control of EUS

As reported by some schools, EUS in fishes is believed to be caused by the fungus *A. invadans,* which does not seem to cause any disease in humans who consume fish. However, fish is to be cooked with proper precautionary measures.

However, HP studies of EUS-affected fish tissues, collected from around the ulcer, are to be done for confirmation of EUS. Nevertheless, the following measures could be adopted, to reduce not only the risk of EUS outbreaks, but also fish disease in general.

14.5.4 Measures for Preventing EUS Outbreak

1. Ichthyospecies, which are resistant to EUS, should be the only stock allowed in regions already affected by EUS.
2. The stocking ponds are to be drained, dried, and limed before stocking with fish.
3. Disease-free hatchery-reared fry (3″−4″ in size) pretreated with a 1% salt bath for 1 h are preferable for stocking.
4. Tube well water or treated incoming water is to be used.
5. Wild fishes are to be excluded from the culture ponds.
6. Farm equipment and shoes of visitors are to be disinfected with hypochlorite or iodophor disinfectants.
7. Good water quality with high DO is to be maintained.
8. Stress to fish is to be minimized. This could include low stocking density, healthy food (nutrition), and no excess handling or netting of the stocked fishes.
9. Special care should be taken during periods of low water temperature (winter) with regard to monitoring of potential "risk factors" such as water quality and parasitic infestations.

10. Farmers are to report immediately to fishery experts in the event of a suspected EUS outbreak, in order to take steps for reducing losses.

14.5.5 Further Recommendations

That a study be conducted to establish a baseline incidence for illness in commercial fishers in the Gladstone area and possibly other areas of Queensland. This is essential if any outbreak of disease is to be identified in the future.

That appropriate health and epidemiological statistics be routinely collected for the fish industry.

That appropriate best practice and guidelines for fishing and fish handling be developed in collaboration with the commercial fishing industry.

The Panel noted that identifying the causes of the diseases and prevalence of parasites on fish in Gladstone Harbor is a complex and difficult task. This task is further complicated by the extreme flood events of the 2010–2011 summers and the historical and ongoing industrial development of the Harbor, which have changed the local environment.

Determining conclusively whether any environmental changes have anything to do with the reported fish health problems. It is a formidable and perhaps impossible undertaking, given the available data for fish and human diseases that were collected from fish species in the Gladstone area. This includes establishing baselines and trends during "normal" periods, and appropriate areas outside the Harbor to act as a form of control for comparative analysis, that is, the use of more sophisticated study designs. The Panel suggested that ongoing pathological studies are a priority to support epidemiological studies and as a component of more in-depth investigations, including the development of a case definition for the observed skin discoloration or "reddening."

A panel reviewed the data for the parasite (*Neobenedenia* sp.), which was affecting the eyes and skin, particularly in barramundi. The parasite had been reported previously in Australia, had wide host specificity, and was known to cause mass mortalities in aquaculture cages in many countries. Reports of high prevalences in wild fish are unusual, but may simply be a reporting issue. While the presence of *Neobenedenia* on barramundi explains many of the lesions reported, the reasons for the high prevalence and abundance of the parasite were unclear but outbreaks are known to occur including in Hinchinbrook Channel, Queensland but not to the extent seen in this instance.

The fish health issue should be the ongoing focus of Queensland government studies.

As a priority, a conceptual model should be completed, if possible, including the cause–effect relationships to help guide studies and eliminate potential causal factors. The development of the concept maps is the first step in developing the conceptual model.

There is an immediate need to develop a case definition for the observed skin discoloration ("reddening") in fish.

The ongoing monitoring of the prevalence of the parasites, lesions, and skin discoloration and the associated pathology investigations should continue as a priority and be guided by the conceptual model.

Consideration should be given to experimental work with diseased fish and fish with *Neobenedenia* to better understand the parasite's taxonomy, biology, and pathogenesis, and studies on wild fish with lesions held in captivity and exposed to water of different quality.

14.5.6 Some Recommendations for the Prevention of Zoonotic Disease in Aquaculture Facilities

Guidelines established for the prevention of aquaculture-associated human disease are to be based upon fish health considerations, employee health considerations, and biosecurity.

14.5.6.1 Fish Health Considerations

For any fish-rearing or processing facility, it is reasonable to assume that the potential for zoonotic bacterial disease increases in proportion to the number of actively infected fish within the facility at a given time. The maintenance of disease-free stock is therefore a paramount concern for both the profit margin and the safety of fish handlers. One key to keeping fish healthy is to start with healthy fish. In this regard, procedures, such as segregated quarantine, prophylactic antiparasitic therapy, and the sacrifice of specimens for medical evaluation and bacterial culture of internal organs, were advocated as routine practices for all newly imported lots, even when fish are obtained from proven providers. Recognizing that most bacterial fish pathogens are opportunistic invaders, the mitigation of physiologic stressors is a universally acknowledged method of decreasing fish disease. Limiting stress usually involves committed attention to factors such as water quality, overcrowding, and excessive or inappropriate handling. A final animal health consideration is the generation of antibiotic resistance by bacteria. In fish disease outbreaks caused by resistant bacteria, effective treatment is often delayed, thus exposing workers to infected fish for extended amounts of time. Additionally, zoonotic infections acquired from repeatedly treated fish are more likely to be medically intractable. In order to decrease the occurrence of antibiotic resistance in aquaculture facilities, antibiotic therapy could be based strictly upon bacterial isolation and antibiotic sensitivity results.

14.5.6.2 Employee Health Considerations

The relevant literature consistently suggests that the incidence and severity of fish-associated zoonotic infections are dependent, in part, upon the immune status of the human hosts. Therefore, it is suggested that persons who handle large numbers of farmed fish on a regular basis receive an initial baseline health screen (including a tuberculin test), followed by scheduled periodic medical examinations. General safety and hygiene training of aquaculture personnel could include specific information relative to the management and reporting of fish-associated injuries. Additionally, it is imperative that persons with preexisting viral, neoplastic, metabolic, or other immune-compromising conditions be made aware of their increased risk for zoonotic infections.

14.5.6.3 Biosecurity

Infected stock is not the only potential source of pathogenic fish-associated bacteria. Biosecurity measures to decrease the introduction of pathogens into fish-holding facilities include: (1) the use of a sanitary and protected water supply; (2) physical barriers to prevent contamination by biological vectors such as birds, rodents, and human visitors; and (3) the education of employees in the principles of aquaculture hygiene.

14.6 SUMMARY

EUS is a seasonal epizootic condition of wild and farmed, fresh- and brackish-water fish of complex infectious etiology. It is characterized by the presence of invasive *Aphanomyces* infection and necrotizing ulcerative lesions typically producing a granulomatous response. The disease is now endemic in SE and South Asia, and has recently extended to West Asia. EUS is indistinguishable from RSD in eastern Australia (Anon, 2012) and MG in Japan.

EUS has been reported to occur in over 100 freshwater fish species and to a lesser extent in brackish-water fish. It is more prevalent in snakeheads and barbs, while Chinese carps are rarely affected, and tilapias seem to be resistant. The primary causative agent of EUS, according to certain schools of thoughts, appears to be one or more members of the fungal genus *Aphanomyces*. Isolates from EUS-infected fish in SE and South Asia have recently been described as *A. invaderis*, while in Japan, *A. piscicida* was described as the primary cause of MG. Studies had shown that these two species are indistinguishable culturally and pathogenically, and could probably prove to be conspecific. Additionally, the opportunistic bacteria *Aeromonas hydrophila* and *Aeromonas sobria* and one or more of a number of viruses may also be involved in the pathogenesis of EUS.

Diagnosis is based on clinical signs and histological evidence of the typical aggressive invasiveness of the nonseptate fungal hyphae, within the context of high mortality. Isolation of the fungus allows its characteristic growth profile to be used as an aid to identification.

Control of EUS in natural waters is probably impossible. In outbreaks occurring in small, closed water bodies, liming of water and improvement of water quality, together with removal of infected fish, is often effective in reducing mortality.

As reported:

1. Farm visits should be conducted to ascertain the extent of the spread of EUS in private- as well as public-sector aquaculture farms.
2. Preliminary studies had revealed that population of *Channa marulius* had specifically been affected by EUS to a quite large extent. Additional EUS-affected fishes include *Colisa* sp. as well as others.
3. Some important and feasible measures have been suggested by the Pakistan Mission in order to reduce the impact of EUS in Pakistan.
4. The etiology and other parameters of EUS are to be studied on a long-term basis.

14.6.1 Further Aspects

1. Most of the hatcheries and nurseries in Bangladesh go for three species of IMCs (Rohu, Catla, and Mrigal) and three species of exotic carps (common carp, silver carp, and grass carp). Other culturable species are orange-fin *labeo*, Java barb, Kuria *labeo,* etc.
2. Hatcheries act as sustained sources of spawn for nurseries. However, hatcheries generally obtain brood stock from farmer's grow-out ponds.
3. In Bangladesh, the average survival rate of spawn, fry, and fingerlings in hatcheries and nurseries appears to vary between 74% and 82%, which could be categorized as a "high" survival rate.
4. Generally, hatcheries appear to be more profitable than nurseries. However, market prices of fry and fingerlings often show wide ranges of variability. As a result, nursery operations often experience greater fluctuations and uncertainty in terms of profitability.
5. Hatcheries and nurseries have the potential for employment generation of farmers. The contribution of hatchery owners and nursery proprietors to the household incomes of people connected with AQC in the field are 79.3% and 95.1% respectively.
6. In general, hatcheries are owned by rich farmers, and nurseries are owned by small- to medium-scale farmers.
7. Hatchery and nursery operations are often familial avocations with the involvement of collaborative labor by family members in which the workload is shared among husband, wife, and children. The system also generates employment for hired labor, which usually works 18 hours per day.

8. Fish, being a rich source of animal protein and a potential source of avocation, attracts people from different walks of life, including extension officers of the government who, as reported, visit fish farms more frequently in comparison with their counterparts at other institutions.
9. The major challenges facing hatcheries and nurseries are disease, drought, and devastating floods. Interestingly, however, diseases have been reported to be less prevalent in hatcheries than in nurseries.
10. Categorization of diseases reveals that the major diseases reported by nurseries are white spot, tail and fin rot, EUS, sudden spawn mortality, gill rot, dropsy, and malnutrition. Conversely, the main diseases reported by hatcheries are sudden spawn mortality and argulosis.
11. As reported, the prevalence of diseases in Jessore appears to be less than it is in Mymensingh. This regional difference could be because farmers of Jessore have more experience in farm management than their counterparts in Mymensingh.
12. As reported, the economic loss due to diseases is about 7.6% of profit. Gill rot is responsible for the highest economic losses to affected farms, followed by spawn mortality, fish lice, EUS, and malnutrition.

REFERENCES

Ahmed, M., Rab, M.A., 1995. Factors affecting outbreaks of epizootic ulcerative syndrome in farmed and wild fish in Bangladesh. J. Fish Dis. 18, 263—271.

ADB/NACA, 1991. Fish health management system in Asia-Pacific. In: Report on a Regional Study and Workshop on Fish Disease and Fish Health Management. ADB Agriculture Department Report Series No. 1. Network of Aquaculture Centres in Asia-Pacific, Bangkok.

Anon, 1997. In: Report of the International Emergency Disease Investigation Task Force on a Serious FInfish Disease in Southern Africa, 18—26 May 2007. Food and Agriculture Organization of the United Nations, Rome. Viewed in 2011, from: http://www.fao.org/docrep/012/i0778e00.htm.

Anon, 2012. Gladstone Fish Health Scientific Advisory Panel, pp. 1—47.

Anonymous, 1981. Report of informal consultation on standardisation of *Bacillus thuringiensis* H-14. Mimeographed Documents. TDR/VBC/BT H-14, 811, Geneva, 23—26, p. 15.

Ashton, P.J., 2007. Riverine biodiversity and conservation in South Africa: current situation and future prospects. Aquat. Conserv. Mar. Freshwater Ecosyst. 17, 441—445.

Baldock, F.C., Blazer, V., Callinan, R., Hatai, K., Karunasagar, I., Mohan, C.V., Bondad-Reantaso, M.G., 2005. Outcomes of a short expert consultation on epizootic ulcerative syndrome (EUS): reexamination of causal factors, case definition and nomenclature. In: Diseases in Asian Aquaculture V. Fish Health Section, Asian Fisheries Society, Manila, pp. 555—585.

Barua, G., 1990. Bangladesh report. In: Regional Research Programme on Relationships between Epizooic Ulcerative Syndrome in Fish and the Environment. 13—26 August 1990. NACA, Bangkok, pp. 89—92.

Brogden, K.A., Guthmiller, J.M. (Eds.), 2002. Polymicrobial Diseases. ASM Press, Washington. D.C., 446 pp.

Chowdhury, M.B.R., 1993. Research priorities for microbial fish disease and its control in Bangladesh for fish health. In: Tollervey, A. (Ed.), Disease Prevention and Pathology, pp. 8—11.

Coates, D., Nunn, M.J., Uwate, K.R., 1989. Epizootic ulcerative disease of freshwater fish in Papua New Guinea. Sci. New Guinea 15, 1—11.

Hatai, K., 1980. Studies on pathogenic agents of saprolegniasis in fresh water fishes. Spec. Rep. Nagasaki. Pref. Inst. Fish. 8, 1—95.

Kanchanakhan, S., Saduakdee, U., Areerat, S., 1999. Virus isolation from epizootic ulcerative syndrome diseased fishes. Asian Fish. Sci. 12, 327—335.

Karunasagar, I., Sugumar, G., Karunasagar, I., 1995. Virulence characters of *Aeromonas* spp. isolated from fish affected by epizootic ulcerative syndrome. In: Shariff, M., Arthur, J.R., Subasinghe, T.P. (Eds.), Diseases in Asian Aquaculture II. Fish Health Section, Asian Fisheries Society, Manila, pp. 307—314.

Karunasagar, I., Karunasagar, I., 1994. Bacteriological studies on ulcerative syndrome in India. In: Roberts, R.J., Campbell, B., MacRae, I.H. (Eds.), Proceedings of the ODA Regional Seminar on Epizootic Ulcerative Syndrome. Bangkok, Thailand, pp. 158—170.

Kar, D., 2010. Biodiversity Conservation Prioritisation. Swastik Publishers, Delhi, 170 pp.

Kar, D., Dey, S., 1988. Preliminary electron microscopic studies on diseased fish tissues from Barak valley of Assam. Proc. Annu. Conf. Electron Microsc. Soc. India 18, 88.

Kar, D., Dey, S.C., 1990a. Fish disease syndrome: a preliminary study from Assam. Bangladesh J. Zool. 18, 115—118.

Kar, D., Dey, S.C., 1990b. A preliminary study of diseased fishes from Cachar district of Assam. Matsya 15—16, 155—161.

Kar, D., Dey, S.C., 1990c. Epizootic ulcerative syndrome in fishes of Assam. J. Assam. Sci. Soc. 32 (2), 29—31.

Kar, D., Dey, S.C., Michael, R.G., Kar, S., Changkija, S., 1990. Studies on fish epidemics from Assam, India. J. Indian Fisheries Assoc. 20, 73—75.

Kar, D., Dey, S.C., Kar, S., Bhattacharjee, N., Roy, A., 1993. Virus-like particles in epizootic ulcerative syndrome of fish. In: Proc. International Symp. On Virus-Cell Interaction: Cellular and Molecular Responses, vol. 1, p. 34.

Kar, D., Roy, A., Dey, S.C., Menon, A.G.K., Kar, S., 1995a. Epizootic ulcerative syndrome in fishes of India. World Congr. In Vitro Biol. In Vitro 31 (3), 7.

Kar, D., Kar, S., Roy, A., Dey, S.C., 1995b. Viral disease syndrome in fishes of North-East India. In: Proc. International Symp. of International Centre for Genetic Engg. and Biotechnology (ICGEB) and the Univ. of California at Irvine, vol. 1, p. 14.

Kar, D., Rahaman, H., Barnman, N.N., Kar, S., Dey, S.C., Ramachandra, T.V., 1999. Bacterial pathogens associated with epizootic ulcerative syndrome in freshwater fishes of India. Environ. Ecol. 17 (4), 1025—1027.

Kar, D., Baishya, A.K., Mandal, M., Arjun, J., Laskar, B.A., 2003. A preliminary account of the vegetations in Dargakona locality of Silchar in Assam. Environ. Ecol. 21 (2), 254—258.

Kar, D., Mazumdar, J., Barbhuiya, M.A., 2007. Isolation of mycotic flora from fishes affected by epizootic ulcerative syndrome in Assam, India. Asian J. Microbiol. Biotechnol. 9 (1), 37—39.

Kar, D., 2007. Fundamentals of Limnology and Aquaculture Biotechnology. Daya Publishing House, New Delhi, pp. vi + 609.

Kar, D., 2013. Wetlands and Lakes of the World. pp. xxx + 687. Springer, London. Print ISBN: 978-81-322-1022-1; e-Book ISBN: 978-81-322-1923-8.

Kumar, D., Day, R.K., Sinha, A., 1991. In: Sinha, V.R.P., Srivastava, H.C. (Eds.), Aquaculture Productivity. Oxford and IBH Publ. Co, New Delhi, p. 345.

Lio-Po, G.D., Traxler, G.S., Albright, L.J., Leaño, E.M., 2000. Characterization of a virus obtained from snakeheads *Ophicephalus striatus* with epizootic ulcerative syndrome (EUS) in the Philippines. Dis. Aquat. Org. 43, 191–198.

Lumanlan-Mayo, S.C., Callinan, R.B., Paclibare, J.O., Catap, E.S., Fraser, G.C., 1997. Epizootic ulcerative syndrome (EUS) in rice-fish culture systems: an overview of field experiments 1993–1995. In: Flegel, T.W., MacRae, I.H. (Eds.), Diseases in Asian Aquaculture III. Fish Health Section, Asian Fisheries Society, Manila, The Philippines, pp. 129–138.

Macintosh, D.J., 1986. Environmental background to the ulcerative disease condition in South-east Asia. In: Roberts, R.J., Macintosh, D.J., Tonguthai, K., Boonyaratpalin, S., Tayaputch, N., Phillips, M.J., Millar, S.D. (Eds.), Field and Laboratory Investigations into Ulcerative Fish Diseases in the Asia-Pacific Region. Technical Report of FAO Project TCP/RAS/4508, Bangkok, pp. 61–113.

McKenzie, R.A., Hall, W.T.K., 1976. Dermal ulceration of mullet (*Mugil cephalus*). Aust. Vet. J. 52, 230–231.

Miyazaki, T., Egusa, S., 1973. Histopathological studies of edwardsiellosis of the Japanese eel (*Anguilla japonica*). Suppurative interstitial nephritis form. Fish Pathol. 11, 33–43.

ODA, 1994. Regional Seminar on Epizootic Ulcerative Syndrome, 25–27 January, 1994. Aquatic Animal Health Research Institute, Bangkok.

OIE, 2006. Epizootic Haematopoietic Necrosis. Manual of Diagnostic Tests for Aquatic Animals 2006, 85–103.

Rodgers, L.J., Burke, J.B., 1981. Seasonal variation in the prevalence of red spot disease in estuarine fish with particular reference to the sea mullet, *Mugil cephalus* L. J. Fish Dis. 4, 297–307.

Roberts, R.J., Macintosh, D.J., Tonguthai, K., Boonyratpalin, S., Tayaputch, N., Phillips, M.J., Millar, S.D., 1986. Field and Laboratory Investigations into Ulcerative Fish Diseases in Asia-Pacific Region. Technical Report, FAO Project TCP/RAS/4508, pp. 214.

Roberts, R.J., Campbell, B., MacRae, I.H., 1993. In: Proceedings of the Regional Seminar on Epizootic Ulcerative Syndrome. 25–27 January 1994. The Aquatic Animal Health Research Institute, Bangkok.

Roberts, R.J., Frerichs, G.N., Tonguthai, K., Chinabut, S., 1994. Epizootic ulcerative syndrome of farmed and wild fishes. In: Muir, J.F., Roberts, R.J. (Eds.), Recent Advances in Aquaculture, vol. 5. Blackwell Science Ltd, Oxford, England, pp. 207–239.

Tonguthai, K., 1985. A Preliminary Account of Ulcerative Fish Diseases in the Indo-Pacific Region (A Comprehensive Study Based on Thai Experiences). National Inland Fisheries Institute, Bangkok, Thailand, pp. 39.

Willoughby, L.G., Roberts, R.J., 1994. Improved methodology for isolation of the *Aphanomyces* fungal pathogen of epizootic ulcerative syndrome (EUS) in Asian fish. J. Fish. Dis. 17, 541–543.

Willoughby, L.G., Roberts, R.J., Chinabut, S., 1995. *Aphanomyces invaderis* sp. nov., the fungal pathogen of freshwater tropical fish affected by epizootic ulcerative syndrome. J. Fish Dis. 18, 273–275.

Index

Printed in the United States
By Bookmasters